Sustainable Environmental Waste Management Strategies

Series Editors

Alok Prasad Das, Department of Life Sciences, Rama Devi Women's University, Bhubaneswar, Odisha, India

Eric van Hullebusch, Équipe ACE, Institut de Physique du Globe de Paris, Paris Cedex 05, France

Sustainable waste management is an absolute imperative in a world that is already dealing with the effects of climate change. This book series seeks to publish an extensive collection of scientific books targeting researchers, scientists, professionals, academics, students, and anyone else involved in the wide, diverse field of sustainable waste management. The series accepts monographs, edited volumes, textbooks, and conference proceedings. Topics covered include but are not limited to environmental pollution and remediation, sources of environmental waste, environmental and human health, sustainable waste management, circular economy, nature-inspired environmental solutions, pollution prevention, and energy-efficient technologies for waste management. The series puts a particular emphasis on developing countries/ the majority world but accepts proposals globally as long as the topic is in scope.

Alok Prasad Das
Editor

Bioconversion of Environmental Waste to Value Added Products

A Sustainable Approach

Springer

Editor
Alok Prasad Das
Department of Life Sciences
Rama Devi Women's University
Bhubaneswar, Odisha, India

ISSN 3005-1622 ISSN 3005-1630 (electronic)
Sustainable Environmental Waste Management Strategies
ISBN 978-3-032-14214-6 ISBN 978-3-032-14215-3 (eBook)
https://doi.org/10.1007/978-3-032-14215-3

This Springer imprint is published by the registered company Springer Nature Switzerland AG
The registered company address is: Gewerbestrasse 11, 6330 Cham, Switzerland

Contents

Climate-Resilient Bioconversion Through Insects: Turning Waste into Wealth

Jasti Sri Vishnu Murthy, Bindu Gudi Ramakrishna, Alagesan Keerthana, Mani Chellappan, and M. T. Ranjith

Abstract As climate change intensifies, global challenges concerning waste management, innovation, and resilience in solutions are indispensable for sustainability. This chapter presents the emerging potential of insect-based bioconversion as a climate-resilient strategy for waste valorization. Especially, viz. black soldier flies (BSFs), yellow mealworms, and houseflies are known for their ability to productively convert organic waste into wealthy bio-products. The various waste streams processed by our insect-based bioconversion systems include agricultural residues, food waste, and organic by-products. High-value outputs produced in the process include high-protein animal feed, biofuels, and soil-enhancing fertilizers. Economic analysis would show about 30–45% potential returns within the first two years of operations, with a considerable carbon footprint reduction compared to conventional methods of waste management. A few other circular economy benefits would be derived from the study, including relieving landfills of pressure, mitigating greenhouse gas emissions, and providing sustainable sources for protein in animal feed. This chapter demonstrates that insect-based bioconversion offers a promising, climate-resilient solution to waste management challenges while generating valuable products for agriculture and industry. Our chapter underlines the necessity for supportive policy frameworks and scaled implementation to fully realize the potential of this innovative approach in sustainable waste management.

Keywords Climate · Insect · Bioconversion · Waste · Sustainable management

J. S. V. Murthy (✉) · B. G. Ramakrishna · M. Chellappan · M. T. Ranjith
Department of Entomology, College of Agriculture, Kerala Agricultural University, Vellanikkara, Kerala, India
e-mail: srivishnu.entomon@gmail.com

A. Keerthana
Department of Entomology, Tamil Nadu Agricultural University, Coimbatore, Tamil Nadu, India

A. P. Das (ed.), *Bioconversion of Environmental Waste to Value Added Products*, Sustainable Environmental Waste Management Strategies,
https://doi.org/10.1007/978-3-032-14215-3_1

1 Introduction

Addressing the growing need for human food, animal feed, fuels, and effective waste management has become a significant global issue. The global population is expected to rise from 8 to 9.7 B (2050) and 10.4 billion by 2080, with nearly 2 billion people expected to increase over the following decades. According to the FAO, to meet human food requirements in 2050, agricultural production is expected to rise by 49%, and egg production by 64%. Additionally, poultry, pork, and food fish consumption are expected to increase by 173, 105, and 17.6%, respectively. The increase in agriculture and food production intensifies deforestation, climate change, urban expansion, reduction of arable land, and water scarcity (Dey et al., 2023; Mishra et al., 2023). Furthermore, increasing global agricultural production, rapid industrialization and urbanization, and population growth have collectively resulted in enormous waste generation (Priyadarsini & Das, 2024, 2025). In 2020, the generation of municipal solid waste was 2.13 billion MT, and it is projected to increase to 2.68 billion MT by 2030, 3.23 billion MT by 2040, and 3.78 billion MT by 2050. Additionally, global expenditures are expected to reach around $375.5 billion by 2025.

Around the world, a large quantity of waste is produced and disposed of incorrectly because of insufficient facilities and infrastructure, particularly in middle- and low-income nations. The management of organic waste has become a major global issue alongside climate change because of its effects on human health, causing diseases and environmental problems like pollution. In addition, organic waste significantly contributes to carbon dioxide emissions, with a global total of 3.3 billion tons of CO_2 equivalent produced each year. The current waste management methods include anaerobic digestion, composting, open dumping, landfilling, incineration, and converting waste into animal feed (Surendra et al., 2020). There are various drawbacks linked to present waste management practices (Siddiqui et al., 2022). For instance, in many developing countries, landfills and illegal open dumps are the preferred methods for disposing of or treating organic waste. This practice leads to various environmental, social, and economic issues (Siddiqui et al., 2022). Additionally, landfilling alone contributes to 8% of greenhouse gas emissions. The anaerobic and composite methods are widely used for organic waste treatment but produce unpleasant odors, such as volatile organic compounds, terpenes, alcohol, amines, and sulfur compounds. While combustion is effective, it necessitates advanced techniques and expensive equipment to manage gas emission by-products (Wei et al., 2013). Conventional waste disposal methods often require a significant amount of time for decomposition. Therefore, there is a pressing need for a fast, climate-resilient, environmentally sustainable, economically viable, and technical method is required for organic waste management; Mishra et al., 2019). Hence, insect-based conversions have recently been viable methods for organic waste management.

The use of insects for bioconversion is gaining popularity for treating biowaste because it is economically viable and environmentally beneficial. *Hermetia illucens* can process various types of waste, including up to 70% of materials such as animal remains, food scraps, and both fruit and vegetable waste, along with human waste.

Nonetheless, it is important to evaluate energy use to understand how insect-based bioconversion stacks up against conventional waste management methods. Energy is crucial for the breeding, processing, and transport of insects. However, this approach results in valuable by-products like animal feed, fertilizers, and biodiesel. In contrast, traditional waste management methods, such as landfilling and incineration, also require energy for transport and processing but do not produce useful by-products. This chapter highlights the bioconversion of various wastes using insects, including the types of insects employed for waste treatment and the beneficial by-products and economic opportunities arising from the bioconversion process.

2 Suitable Waste Categories for Insect-Based Bioconversion

Organic substrates are essential in the bioconversion process, serving as important sources of organic materials that offer a variety of benefits (Cucina et al., 2021). The European Food Safety Authority (EFSA) categorizes seven types of organic substrates that are appropriate for bioconversion:

1. approved feed for food-producing animals,
2. outdated or faulty food products initially meant for human consumption,
3. by-products from slaughterhouses that are safe but not used in human food,
4. food waste from restaurants, catering operations, and households,
5. animal manure and intestinal contents,
6. agricultural residue, garden, and forest waste,
7. plastic waste (Murthy et al., 2024),
8. human excrement and sewage sludge (Gałęcki et al., 2023).

Both high-quality and low-quality organic materials—including waste, by-products, and complex feeds—can be utilized for the production of insect biomass. The success of bioconversion is significantly impacted by the quality and composition of the organic substrate, particularly the amounts of fats (ether extract), carbohydrates, and fibers, which play a crucial role in determining the growth rate and biomass yield of the insects involved (Pinotti & Ottoboni, 2021).

3 Insects Used in Bioconversion of Waste

Currently, only a limited number of insect species are utilized for the bioconversion of organic waste. Insect farming on organic waste has emerged as a promising waste management solution, offering various by-products. Insects can efficiently convert organic waste from agriculture, household, food, and industrial sectors into valuable sources (Van Huis, 2020). Therefore, biomass could become a protein supplement for pets, fish, or chickens (Cappellozza et al., 2019; Hawkey et al., 2021). There is a potential for bioconversion by different insect species to transform low-value organic substrates into nutritious diets that will replace conventional, expensive feeds like fishmeal (Fowles, 2021). Here, we have briefly emphasized the insect species involved in the bioconversion of organic materials (Table 1).

Table 1 Insects utilized in bioconversion of waste

S. No.	Insect species	Targeted waste	End products	References
1	Black Soldier Fly, *Hermetia illucens* Linnaeus	Food waste, cattle manure, chicken manure, pig manure, agricultural residues, municipal solid waste, and plastic residues	Protein-rich larvae (for feed), biofertilizer (frass), biodiesel (larval fat)	Beesigamukama et al. (2022), Sheppard et al. (1994), Li et al. (2011), Diener et al. (2011)
2	Housefly, *Musca domestica* Linnaeus	Chicken manure, pig manure, food waste, solid urban waste	Protein-rich maggots (for animal feed), biofertilizer	
3	Mealworm, *Tenebrio molitor* Linnaeus	Pure wheat bran, food waste, agricultural residues	Edible larvae (protein source), biofertilizer	Li et al. (2015)
4	Superworm, *Zophobas morio* Fabricius	Fibrous plant waste, food waste, cattle manure, horse manure, polyethene	Protein biomass, degraded plastic by-products	Harsányi et al. (2020)
5	Flesh fly, *Sarcophaga carnaria* Linnaeus	Fish and poultry waste	Protein-rich larvae (for feed), biofertilizer	Yehuda et al. (2011)
6	Blowfly, *Lucilia sericata* Wiedemann	Fish and poultry waste; slaughter waste	Protein-rich larvae (for feed), biofertilizer	Yehuda et al. (2011)
7	House cricket, *Acheta domesticus* Linnaeus	Brewery's waste, food waste, and by-products from wine production	Edible protein source (for human and animal feed)	Jucker et al. (2022)
8	Mediterranean flour moth, *Ephestia kuehniella* Zeller	Agri-waste, broken cereal grains, eggshell debris, infertile eggs, dead embryos, and egg fluids	Protein-rich larvae (for feed)	
9	Termite, *Macrotermes nigeriensis*	Lignocellulosic biomass	Protein-rich larvae (for feed), biodiesel	Kalu-Uka et al. (2021)

3.1 *Black Soldier Fly (BSF),* Hermetia illucens *Linnaeus (Diptera: Stratiomyidae)*

BSF is one of the most commonly used insect species for organic waste management. *H. illucens* is a cosmopolitan distribution, especially in tropical regions. The BSF larvae feed on decomposing organic material, making them a natural source of animal proteins since adults are non-feeders and biters and do not carry diseases. BSF larvae can process various waste materials, including food, fruit and vegetable waste, poultry feed, animal, and human manure. Through these conversions, the larvae can be used for animal feed, the extraction of bioactive compounds, and the production of biofuels. Additionally, BSF larvae generate a significant amount of frass, which serves as a biofertilizer. BSF larvae feed on organic matter and efficiently convert it into substantial biomass rich in protein and fat (Ravi et al., 2020). The nutritional composition, including crude protein and fatty acid profiles, varies depending on the feeding substrates (Riekkinen et al., 2022; Siddiqui et al., 2022). When BSF larvae feed on different substrates, the produced frass fertilizer acts as a bio-stimulant when applied to plants. It contains humic substances and peptides that enhance biotic and abiotic stress resistance, improving plant vigor, growth, and photosynthetic activity. BSF larvae effectively transform agricultural by-products, such as cow manure and fermented plant materials, into valuable products, positioning them as an ideal option for large-scale waste conversion (Surendra et al., 2020). Moreover, BSF-based systems foster employment opportunities and promote sustainable manufacturing practices.

3.2 *Housefly,* Musca domestica *Linnaeus (Insecta: Diptera: Muscidae)*

M. domestica is a globally distributed species accompanying humans and animals. *M. domestica* is considered a nuisance pest because it transmits various diseases and infections (Čičková et al., 2015). The housefly maggots decompose waste naturally through bioconversion. Maggots destroy meal scraps, turning them into treasured assets such as proteins, oil, and fertilizers (Niu et al., 2017). According to the latest research, maggots decreased meal waste by 79–84% and reduced chicken manure by nearly 60% within a short period (Cheng et al., 2021; Hasan & Leong, 2018). Biodiesel is extracted from maggot oil, which has an excessive conversion efficiency (Shabani et al., 2020). Maggots can be used for fish feed and other purposes, indicating the potential of the larvae in sustainable waste management systems (Koné et al., 2017). Furthermore, those larvae can cause a vast range of substrates from garbage to animal feed, diminishing amounts while yielding wastes useful for agriculture or zootechnical applications. Future studies look to optimize maximum maggot production and nutrient recycling conditions for environmental and financial advantage.

3.3 Yellow Mealworms, Tenebrio molitor *Linnaeus (Coleoptera: Tenebrionidae)*

T. Molitor is a storage pest often feeding on various substrates (Oonincx et al., 2015). The European Food Safety Authority authorized yellow mealworms to be the first approved for human consumption (Rumbos & Athanassiou, 2023). These insects turn municipal solid and agricultural crop waste into valuable materials (Sangiorgio et al., 2022; Shojaaddini, 2024). Diets such as rapeseed, chicken, and wheat bran have induced superior feed conversion efficiency (Bordiean et al., 2022). Diets like rapeseed, chicken, and wheat bran have induced superior feed conversion efficiency (Bordiean et al., 2022). Although producing mealworms remains costly compared to traditional protein sources, they can transform organic waste into nutrient-dense biomass. Their versatility in converting agro-industrial by-products into precious feed highlights their eventuality for sustainable protein products.

3.4 Superworm Zophobas morio *Fabricius (Coleoptera: Tenebrionidae)*

Superworms are great at turning waste into helpful living matter. These creatures consume both plant and animal waste. They can live on low-quality single-type foods and still make important amino acids. This leads to their impressive feed conversion rate (Siddiqui et al. 2022). Their rapid growth and high metabolic activity during tissue growth improve their bioconversion capacity. In addition, the symbiotic relationship of nitrogen-fixing bacteria with superworms facilitates the recycling of nutrients and supports additional waste decomposition (Rumbos & Athanassiou, 2021). The type of waste used as a feed dramatically affects the growth rate and biomass production, which outlines their capacity for sustainable waste management and protein production.

3.5 House Cricket, Acheta domesticus *Linnaeus (Orthoptera: Gryllidae)*

Crickets are nuisance pests due to their chirping noise and materialistic damage. Nonetheless, it is essential to recognize their potential contributions to sustainable circular feed and food systems by offering protein for animal and human consumption. These are the most extensively farmed insects. During their rapid growth phase, crickets prefer a protein-rich diet comprising garden and vegetable waste and horse and cattle manure at ambient conditions (Oonincx et al., 2015). Many house crickets are successfully fed by products of agro-industries. Their nutritional composition and metabolic rates largely depend on the environmental conditions.

The rearing substrates always influence the nutrient quality of crickets, and poor-quality substrates can negatively affect their growth. For example, less nutrient feed exhibits high fats and lower protein. Hence, house crickets possess the ability to physiologically adjust to various environments, making them ideal candidates for farming. Therefore, it is proposed that house crickets, as a source of edible insects, hold considerable promise for cultivation alongside other organisms, representing a crucial advancement for the circular economy.

3.6 Flesh Fly **Sarcophaga dux** *Meigen (Diptera: Sarcophagidae)*

It is an effective decomposer, which converts manures into nutrient-rich residues with minimum environmental effect (Al-Khalaf et al., 2024). Substrates such as fish, pork, pork liver, and blood are generally used to induce oviposition. This species suggests excellent capability for manure recycling, lowering pollution even, and providing sustainable aid for agricultural packages (Hasan & Leong, 2018).

*3.7 Blow Flies (***Chrysomya putoria** *Meigen,* **Lucilia sericata***) (Diptera: Calliphoridae)*

Blow flies possess the ability to reduce organic waste while generating protein suitable for biofuel and animal feed. For instance, *C. Putoria* effectively decomposes numerous wastes, and *L. sericata* larvae effectively reduce fish and poultry waste by 5.5 and 2.5 times (Parry et al., 2017; Yehuda et al., 2011). These larvae are rich in protein, fats, and carbohydrates. Their capacity to transform herbal waste into nutrient-dense biomass makes them a promising choice for sustainable waste management. The growth rate and body composition of blowflies can be influenced by the garbage type they develop on, as well as the species of fly and the density of their population.

4 From Waste to Wealth: The Circular Economy of Insect Bioconversion

Insect bioconversion offers a sustainable solution for managing organic waste while generating valuable by-products. Some insects efficiently transform organic waste into nutrient-rich frass (organic fertilizer) and high-protein insect biomass, which can be used as animal feed. This process not only reduces landfill waste and greenhouse gas emissions but also promotes a circular economy by turning waste into

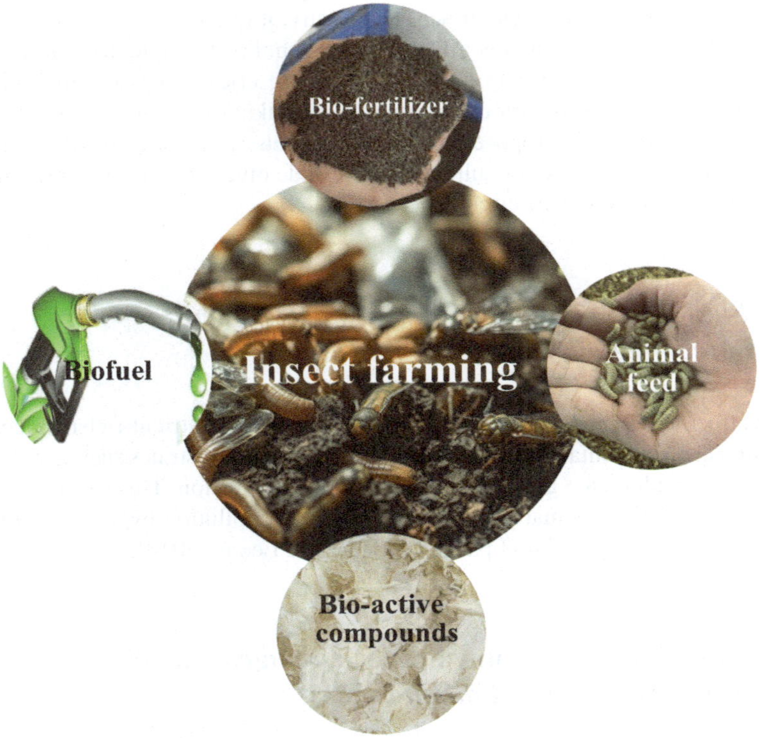

Fig. 1 Insect farming generates diverse forms of by-products

valuable resources. As global demand for sustainable waste management and alternative protein sources rises, insect bioconversion is emerging as an innovative and eco-friendly approach to resource recovery. Some insects which convert into valuable by-products are represented in Fig. 1 and also listed in Tables 1 and 2.

4.1 Insect Frass Fertilizer

The global population continues to rise, which places significant strain on natural resources like soil and water due to various dietary trends (Schmitt & de Vries, 2020). To meet the growing demand for food sustainably, there is a need for fertilizers that are high in nitrogen (N) and phosphorus (P) for agricultural production. However, chemical fertilizers fall short of meeting this demand and also lead to environmental issues such as soil contamination, eutrophication, coastal marine problems, and harmful algal blooms (Schmitt & de Vries, 2020). This highlights the need for sustainable agricultural practices to protect environmental health and maintain

Table 2 Insect bioconversion of different organic waste into wealth

Species	Type of organic waste	Organic waste	Bioconversion output	References
Black soldier fly (*Hermetia illucens*)	Agricultural residues (crop waste)	Restaurant waste (70%): Rice straw (30%)	Biofuel	Zheng et al. (2012a, 2012b)
		Rice straw	Biomass, fertilizer	Manurung et al. (2016)
		Corn stover	Biofuel	Wang et al. (2017)
		Corncob	Biomass	Li et al. (2015)
		Sorghum	Biomass	Tinder et al. (2017)
		Cowpea	Biomass	Tinder et al. (2017)
	Food Processing Waste	Cassava peel	Biomass	Supriyatna et al. (2016)
		Coffee pulp, husk	Biomass	Lardé et al. (1990); Suantika et al. (2017)
		Reject material from pears, banana, and cucumber (5:3:2)	Biomass	Mutafela (2015)
		Brewers' distiller grain	Biomass, fertilizer	Anyega et al. (2021)
		Vegetable trimmings, spent coffee grounds, and tea leaves	Biomass	Cheng (2016)
		Vegetables, peels of yam, cassava, plantain		Bonso (2013)
Yellow mealworm (*Tenebrio molitor*)	Agricultural Residues (Crop Waste)	Wheat straw, bruised cabbage leaves	Biomass	Fowles and Nansen (2020)
		Corn stover	Biofuel	Wang et al. (2017)
Housefly (*Musca domestica*)	Agricultural Residues: Household & Municipal Food Waste	Whole plant corn silage, sawdust (30%): Restaurant waste (70%)	Biomass, biofuel, fertilizer	Niu et al. (2017)

(continued)

Table 2 (continued)

Species	Type of organic waste	Organic waste	Bioconversion output	References
Codling moth (*Cydia pomonella*)	–	Starch and cheese wastewater sludge	Biomass	Brar et al. (2008)
Cambodian field crickets (*Teleogryllus testaceus*)	–	Cassava plant tops, spent grain, mung bean sprout waste, field weeds	Biomass	Miech et al. (2016)

ecosystem integrity (Hu et al., 2020). As a result, insect-based fertilizers, along with biofertilizers and vermicomposting, have emerged as promising solutions to enhance plant growth and soil health. Insect bioconversions have attracted interest because they can transform organic waste into fertilizers that possess beneficial physical and chemical properties, which improve soil fertility and encourage sustainable agriculture (Chavez & Uchanski, 2021; Fischer & Romano, 2021). A notable product from large-scale insect farming is frass, the excreta of insects, which is commonly viewed as an organic fertilizer and a potential livestock feed source. For example, Beesigamukama et al. (2021) demonstrated the use of BSF for the recycling of agro-industrial waste into nutrient-rich fertilizer. Similarly, Mostafaie et al. (2025) researched the use of BSF larvae to convert olive pomace, known for its phytotoxic characteristics, into valuable by-products like insect frass, proteins, and fats. Their results showed that the frass produced during this process did not negatively affect soil invertebrates and had minimal impact on rye grass and broccoli when used as fertilizer, indicating its potential as an environmentally friendly agricultural amendment. In a study by Song et al. (2021), BSF larvae were utilized to transform urban solid waste into a growth medium for vegetables. An experiment involving pak choi tested BSFL frass at three stages: fresh, composted with forced aeration, and naturally composted. The results indicated that composted frass enhanced plant growth, with the 10% naturally composted frass achieving the highest biomass. Germination tests and FTIR analysis revealed elevated levels of phytotoxins in fresh frass, thereby highlighting the need for effective composting methods to mitigate phytotoxicity. From the insects listed in Table 1, only frass from *H. illucens* and *T. molitor* harvested from waste has been tested as fertilizers, demonstrating improvements in soil fertility, crop yield, and nutritional quality. The various waste bioconversion methods using BSF resulted in frass with differing NPK compositions (Table 3) and different insect species in frass production (Table 4).

Table 3 NPK composition of different frass derived from black soldier fly (*H. illucens*)

Waste type	N (%)	P (%)	K (%)	C/N ratio	References
Gainesville diet	3.8	5.2	4.1	8.0	
Distiller's grains	3.4	0.8	1.1	-	
Brewery spent grain	3.6	0.5	0.3	10.7	Anyega et al. (2021)
Okara and wheat bran	3.1	0.7	0.9	9.6	Song et al. (2021)
Okara and wheat bran	2.2	0.5	0.8	9.6	Song et al. (2021)
Household waste	3.3	3.4	2.4	9.0	
Wheat bran	3.4	2.4	4.0	9.6	
Brewery spent grain	2.1	1.2	0.2	7.7	
Fresh okra	2.3	0.8	1.9	7.3	
Chicken manure	2.3	1.1	1.8	7.3	
Pig manure	2.9	2.0	1.4	7.8	
Chicken feed	2.4	1.8	1.9	7.5	
Grass cuttings	2.9	1.3	1.3	7.2	
Fruits and vegetables	1.8	1.3	1.5	6.8	
Cow manure	1.9	1.0	0.2	11.5	
Vegetables	2.7	1.3	1.5	7.6	

4.1.1 Frass Fertilizer: A Sustainable Fertilizer for Nutrient Enrichment and Stress Management

Nutrient limitations hinder plant growth due to their low availability, which is often a result of unsustainable agricultural practices, biological factors such as denitrification, microbial competition, and soil erosion (Goulding et al., 1998). Insect frass, a byproduct of waste, plays a vital role in supplying nutrients and supporting soil health. Its function as a nitrogen source in fertilizers has been extensively researched. One of the key constituents of insect frass is chitin, a compound that is known to promote plant growth. Amendments with insect frass and chitin significantly influence the soil microbiome, which is essential for enhancing plant growth and development. For example, the Eucalyptus defoliating beetle (*Paropsis atomaria*) and *Doratifera quadriguttata* can generate 160 and 270 kg/ha of frass, respectively, contributing around 2–4 kg/ha of nitrogen to the soil (Gherlenda et al., 2016). This increase in nitrogen supports plant growth and elevates nitrogen levels in plant tissues. Furthermore, insect frass not only aids in plant development but also enhances soil health. For instance, frass from the Malacosoma americanum has been shown to improve total carbon and nitrogen levels as well as ammonium content in the soil, while promoting microbial activity in red oak ecosystems (Frost & Hunter, 2004). Additionally, mealworm frass has been observed to boost chard growth by raising leaf chlorophyll levels, stem length and width, and the weight of the plant's above-ground portions (Poveda et al., 2019). Insect frass contains bioactive compounds that assist in plant growth and development. For example, frass from saproxylic cerambycid larvae like

Table 4 Waste conversion by insect species for frass fertilizer production

Insect species	Waste type	C/N ratio	N (kg/ha)	P (kg/ha)	K (kg/ha)	Germination rate (%)	Reference
Hermetia illucens	Mixture of Irish potato peels + brewery spent grain	13.2	147.5	63.2	203.7	93.3	Beesigamukama et al. (2022)
	Brewery spent grains	10.7	180.5	25	145	86	Anyega et al. (2021)
Tenebrio molitor	Wheat bran + chayote (*Sechium edule*)	20.0	124.3	69.8	104.8	56.7	Beesigamukama et al. (2022)
Musca domestica	Food waste	-	148.5,	11.5	15	–	Cheng et al. (2021)
Schistocerca gregaria	Wheat & barley seedlings + wheat bran	19.5	103.3	36.5	175.3	23.3	Beesigamukama et al. (2022)
Gryllus bimaculatus	Wheat bran + soybean + sweet potato vines + weeds	18.6	114.5	76.7	103.7	76.7	Beesigamukama et al. (2022)
Scapsipedus icipe	Wheat bran + soybean + sweet potato vines + weeds	17.4	120.7	74.2	122.7	63.3	Beesigamukama et al. (2022)
Pachnoda sinuata	Fresh cattle dung from a dairy farm	14.6	106.8	52.2	151.2	96.7	Beesigamukama et al. (2022)
Oryctes rhinoceros	Fresh cattle dung from a dairy farm	14.6	82.8	21.5	82	93.3	Beesigamukama et al. (2022)
Bombyx mori	Leaves of mulberry tree (*Morus spp.*)	19.1	117.2	8.5	86.5	60	Beesigamukama et al. (2022)
Gonimbrasia krucki	Brazilian pepper tree leaves (*Schinus terebinthifolia Raddi*)	22.9	107	9.8	70.3	86.7	Beesigamukama et al. (2022)

Chlorophorus annularis adds sugars, alkaloids, and phenols to the soil. These beneficial substances can be enhanced by microorganisms such as Trichoderma hamatum, which has been shown to improve seed germination and seedling growth in lettuce (Khan et al., 2016). This underscores the multiple benefits of insect frass in agriculture and highlights its potential as a sustainable soil amendment that enhances plant nutrition and growth. Insect frass is rich in microorganisms and biomolecules that play a crucial role in activating plant systemic resistance through the salicylic acid and jasmonic acid/ethylene pathways. These microorganisms support plant growth, boost resilience to abiotic stresses, and lessen the negative impact of biotic stresses (de Souza-Vandenberghe et al., 2017). The plant defense response is initiated when cellular receptors detect molecular patterns related to microorganisms and herbivores, such as chitin (Poveda, 2019). Mealworm frass has been noted to improve plant tolerance to abiotic stresses such as drought, flooding, and salinity. The microorganisms present in insect frass aid in nitrogen fixation, phosphorus and potassium solubilization, and the production of Auxins, siderophores, and ACC deaminase. For example, the frass from *Spodoptera frugiperda* triggers the expression of pathogenesis-related (pr) defense genes in corn against the pathogenic fungus *Cochliobolus heterostrophus* (Ray et al., 2015). Similarly, cowpea frass from *H. illucens* stimulates plant defense responses, reducing the incidence of Fusarium wilt disease due to the presence of chitin (Quilliam et al., 2020). Additionally, the bacterium *P. ananatis*, found in the frass of *S. frugiperda*, enhances the expression of the maize gene coding for the herbivore-induced proteinase inhibitor (mpi), which results in a decrease in insect attacks on tomato plants. The different insect species frass has different mechanisms which are mentioned in Table 5.

Table 5 Insect frass fertilizer and its mechanisms

S. No.	Name of the species	Crop	Mechanism involved	References
1	*Hermetia illucens*	Lettuce Cowpea Sugar beet	Activation of plant defensive responses	Lardé et al. (1990), Suantika et al. (2017), Quilliam et al. (2020)
2	*Musca domestica*	–		Niu et al. (2017)
3	*Paropsis atomaria*	Eucalyptus	Nitrogen supply to the plants	Gherlenda et al. (2016)
4	*Tenebrio molitor*	Beans, Barley	Plant growth and tolerance to abiotic stress	Poveda et al. (2019), Dulaurent et al. (2020), Houben et al. (2020)
5	*Zophobas morio*	Dragon fruit	Increases plant growth	Koh et al. (2020)
6	*Anoplophora glabripennis* *A. swainsoni*	*Salix babylonica*	Parasitoid attraction for volatile organic compounds	Wei et al. (2013)
7	*Agrotis ipsilon*	Potato	Presence of phenols and flavonoids	Ahmed et al. (2013)

4.2 Insects as Sustainable Nutrition

With the steady rise in the global population, achieving food security and proper nutrition has become increasingly challenging. Conventional livestock farming, which includes meat sources such as pork, beef, and fish, faces many production issues, largely due to high resource demands, environmental damage, and greenhouse gas emissions (Tullo et al., 2019). Thus, finding alternative protein sources is urgent. Edible insects are generally viewed as a valuable source of protein, fats, essential nutrients, vitamins, and minerals. They can be raised on low-cost feeds and have a low carbon footprint, making them a viable solution. Over 2000 insect species are recognized for human consumption. Among these, Coleoptera has the highest diversity with 468 species, followed by Hymenoptera with 351 and Orthoptera with 267. Edible insects can be consumed at various developmental stages, including eggs, larvae, pupae, and adults. Commonly consumed insect species include weaver ant eggs, different cricket varieties, cicadas, bamboo worms, and wasps (Kafle et al., 2024).

The cold-blooded characteristic of insects allows for effective feed conversion into protein. Insects account for 5–10% of animal protein sources and provide additional fats, calories, vitamins, and minerals for human diets (Hernandez Toxqui et al., 2021). Insects commonly utilized for transforming organic waste into protein-rich food include BSF larvae the most extensively used species (Wang & Shelomi, 2017), along with *M. domestica*, *T. molitor*, *Z. morio*, *Locusta migratoria*, and *S. gregaria*. These insects are recognized as alternative sources of protein and lipids (Mudalungu et al., 2021; Wang & Shelomi, 2017). A table detailing the edible insects consumed in various regions of India is presented as Table 6.

4.2.1 Nutrient Content in Insects

Insects are rich in nutrients and offer vital components like proteins, fats, vitamins, minerals, and sugars that are crucial for human growth and development (Dobermann et al., 2017). The nutritional makeup of insects can be influenced by several factors, including their diet, developmental stage, gender, species, environmental conditions, and the analytical techniques employed for assessment (van Huis & Oonincx, 2017). A comprehensive overview of the nutrient profiles of commonly eaten edible insects is presented in Table 7.

4.2.2 Protein

Proteins are essential for life and crucial for human health. They are integral to immune functions, including the production of antibodies, and to various biochemical processes involving enzymes. Composed of 20 different amino acids, proteins are primarily acquired through diet. In particular, insect proteins are noteworthy because

Table 6 Edible insect species

S. No.	Name of the insect	Life stage	Region	References
1	*Hermetia illucens*	Larvae	Globally	Thrastardottir et al. (2021)
2	*Tenebrio molitor*	Larvae, pupae	Europe, Asia, North America	Thrastardottir et al. (2021)
3	*Gryllus bimaculatus*	Nymphs, adults	Nagaland	Chowdhury et al. (2015)
4	*Oxya hyla hyla*	Nymphs, adults	Nagaland	Chowdhury et al. (2015)
5	*Bombyx mori*	Pupae	Assam, Nagaland	Chowdhury et al. (2015)
6	*Lethocerus indicus*	Adults	Assam, Manipur	Chowdhury et al. (2015)
7	*Oecophylla smaragdina*	Eggs, larvae adults	Assam, Odisha, Jharkhand, Tripura, Manipur	Chowdhury et al. (2015)
8	*Odontotermes obesus*	Winged adults	Karnataka, Tamil Nadu, Kerala, Meghalaya, Assam	Chowdhury et al. (2015)
9	*Apis cerana indica*	Larvae, pupae	Kerala and the Northeast regions	Chowdhury et al. (2015)
10	*Gryllotalpa africana*	Larvae	Assam	Gahukar (2018)
11	Melanopus sp.	Adults	Assam	Gahukar (2018)

they offer all the essential amino acids, thus serving as a complete and nutritionally advantageous protein source. Insects typically contain between 35 and 60% protein by dry weight and 10–20% by fresh weight, which surpasses the protein content found in cereals and legumes (Bukkens, 1997). Specifically, members of the Orthoptera order, such as crickets, locusts, and grasshoppers, have a particularly high-protein content (Rumpold & Schlüter, 2013). This positions insects as a sustainable and valuable option for human dietary protein (Liceaga et al., 2022).

4.2.3 Vitamins and Minerals

Edible insects represent a great source of vital vitamins and minerals, which could be beneficial as a dietary supplement for human micronutrient consumption (Kinyuru et al., 2015). Different insect species are abundant in significant vitamins, such as vitamins A, D2, D3, C, E, K, thiamin, riboflavin, pantothenic acid, niacin, pyridoxine, folic acid, D-biotin, and vitamin B12 (Finke, 2013; Mlček et al., 2014; Rumpold &

Table 7 Nutrient profile of different edible insect species

Name of the insect	Protein mg/100 g	Fat mg/100 g	Zinc mg/100 g	Iron mg/100 g	Thiamine mg/100 g	Riboflavin mg/100 g	References
Black soldier fly	175	140	125	90	7.7	16.2	Finke (2013)
Housefly	197	19			11.3	77.2	Finke (2013)
Cricket	205	68	19.9	9.1	0.4	34.1	Finke (2002)
Super worm	197	177	3.8	7.7	0.6	7.5	Finke (2002)
Mealworm	187	134	5.3	11.8	2.4	8.1	Finke (2002)
Giant mealworm	184	168			1.2	16.1	Finke (2002)
Waxworm	141	249	5.0	8.6	2.3	7.3	Finke (2002)
Silkworm	93	14	15.9	16.8	3.3	9.4	Finke (2002)

Schlüter, 2013). The concentration of vitamin B12 differs among species, with mealworms providing 1.08 μg/100 g, Gryllus assimilis at 2.88 μg/100 g, Locusta migratoria at 0.84 μg/100 g, and Shelfordella lateralis exhibiting the highest amount at 13.2 μg/100 g dry weight. Insects belonging to the Orthoptera and Coleoptera orders are particularly high in folic acid (Finke, 2013). Additionally, insects offer minerals like iron, zinc, calcium, copper, phosphorus, magnesium, and manganese, which can be found in species such as grasshoppers, crickets, termites, and mealworms (Mlček et al., 2014). Studies suggest that consuming insects could greatly assist in fulfilling daily mineral requirements, especially iron (Latunde-Dada et al., 2016). Crickets and termites are favored choices due to their elevated levels of iron and zinc. These aspects underline the potential of edible insects as a significant source of vitamins and minerals for human dietary needs.

4.2.4 Insects as Feed

Insect-based ingredients are gaining popularity in aquafeed due to their beneficial effects on immune function and gut health (Huyben et al., 2019). Insects that facilitate the recycling of organic waste can also be utilized as feed for pets and livestock, especially for fish and chickens (Cappellozza et al., 2019). The larvae of the black soldier fly and housefly are commonly incorporated into livestock diets. *H. illucens* is particularly notable for its high content of Omega-3 fatty acids, making it a valuable component of fish feed (Wang & Shelomi, 2017). *T. molitor* and *Z. morio* are regarded as significant sources of protein and fat for poultry feed (Benzertiha et al., 2020;

Kieronczyk et al., 2018). Additionally, the oil extracted from *T. molitor* serves as a feasible substitute for soybean oil, which is often used in broiler chicken diets. Furthermore, *M. domestica* maggots have been included in feed for broiler chickens (Sanchez-Muros et al., 2016).

4.3 Biofuels (Biodiesel Production)

First- and second-generation biofuels, which are produced from food crops and lignocellulosic biomass (primarily bioethanol), have been developed. However, these biofuels come with notable disadvantages, such as greenhouse gas emissions, increased food prices, high water usage, and complicated production methods (Searchinger et al., 2008). Thus, it is essential to investigate alternative biofuel sources and technologies. A promising, cost-effective, and environmentally friendly approach is biotransformation (Parodi et al., 2018). The use of insects for breaking down organic waste and accumulating fat-based biomass is gaining popularity due to their economic feasibility, widespread availability, high-fat yield, and low environmental footprint (Surendra et al., 2016). Certain insect species like *H. illucens*, *M. domestica*, and *T. molitor* have proven to be effective in converting organic waste into fat-rich biomass, making them strong candidates for biodiesel production (Li et al., 2011).

The swift growth of the global population has intensified the demand for traditional energy resources and has had a significant impact on the environment (Liddle, 2013; O'Neill et al., 2012). By 2040, global fuel consumption is expected to rise by 121 million barrels per day, which will lead to the depletion of fossil fuels, higher oil prices, energy security challenges, and negative environmental effects (Shell, 2025). Consequently, first- and second-generation biofuels, derived from food crops and lignocellulosic biomass (mainly bioethanol), have been introduced. However, these biofuels come with various drawbacks, including greenhouse gas emissions, increased food costs, substantial water usage, and complex production processes (Searchinger et al., 2008). Therefore, it is crucial to explore alternative sources and advancements in biofuel technology. One promising and sustainable method is biotransformation (Parodi et al., 2018). The utilization of insects to convert organic waste into high-fat biomass is emerging as a noteworthy solution due to its cost-effectiveness, accessibility, high-fat yield, and minimal environmental impact (Surendra et al., 2016). Specific insect species such as *H. illucens*, *M. domestica*, and *T. molitor* have been identified as efficient converters of organic waste into fat-rich biomass, making them valuable candidates for biodiesel production (Li et al., 2011).

Insects like the yellow mealworm and the black soldier fly are particularly effective at breaking down organic materials and converting waste into nutrient-dense larval biomass (Oonincx et al., 2015; Purschke et al., 2017). The larvae of yellow mealworms possess a high-fat content, ranging from 23 to 47%, making them a potential resource for biodiesel production. Research by Zheng et al. (2012a, 2012b)

Table 8 Insects used in biodiesel production

Insects used in biodiesel production	Biodiesel content	Substrate (1 kg)	References
Hermetia illucens	35.6, 57.8, and 91.4	Cattle, pig, and chicken manure	Manzano-Agugliaro, et al. (2012)
Tenebrio molitor	8.50	Corn stover	Wang et al. (2017)
Musca domestica	13.5	Chicken manure	Li et al. (2011)
Acheta domesticus	20.1	Wheat bran	

demonstrated the feasibility of producing eco-friendly biodiesel from yellow mealworms fed on decayed vegetables. Likewise, black soldier fly larvae have a fat content between 20 and 40% (Lalander et al., 2019; Mohan et al., 2023) and have been effectively utilized in biodiesel production using various waste materials, including animal manure, rice straw, restaurant refuse, and corncobs (Bulak et al., 2018; Cappellozza et al., 2019). Given their capability to utilize low-quality biomass from both plant and animal origins, fats derived from insects are increasingly being recognized as a valuable feedstock for biodiesel development (Beesigamukama et al., 2020; Bonelli et al., 2020; Schmitt & de Vries, 2020).

4.3.1 Production of Biodiesel from Insects (Table 8)

The fat found in insect larvae could serve as a promising resource for biodiesel production. The black soldier fly (BSF) is commonly found in both tropical and temperate regions and has a rapid growth cycle, reaching maturity in just three weeks from the egg stage. During their larval phase, BSF larvae consume food waste, converting the carbohydrates and lipids from this waste into their fat reserves, thanks to their accelerated metabolism (Čičková et al., 2015; Wang & Shelomi, 2017). Research indicates that the lipid content in BSF larvae ranges from 10 to 40% on a dry basis (Čičková et al., 2015; Wang & Shelomi, 2017). A diagram illustrating the biodiesel production process from black soldier fly larvae can be seen in Fig. 2. Additionally, a generalized protocol for biodiesel production has been developed and is outlined below.

4.3.2 Collection, Mass Production, and Processing of Insects

The chosen insect larvae were sourced from a reputable supplier to guarantee quality and consistency for large-scale production. They were cultivated on various organic materials, including manure from cattle, chickens, and pigs, as well as food waste. The larvae were kept in optimal temperature and humidity conditions for 10 days to promote their growth and development. After reaching maturity, the larvae were harvested and washed with distilled water to remove any contaminants. Following

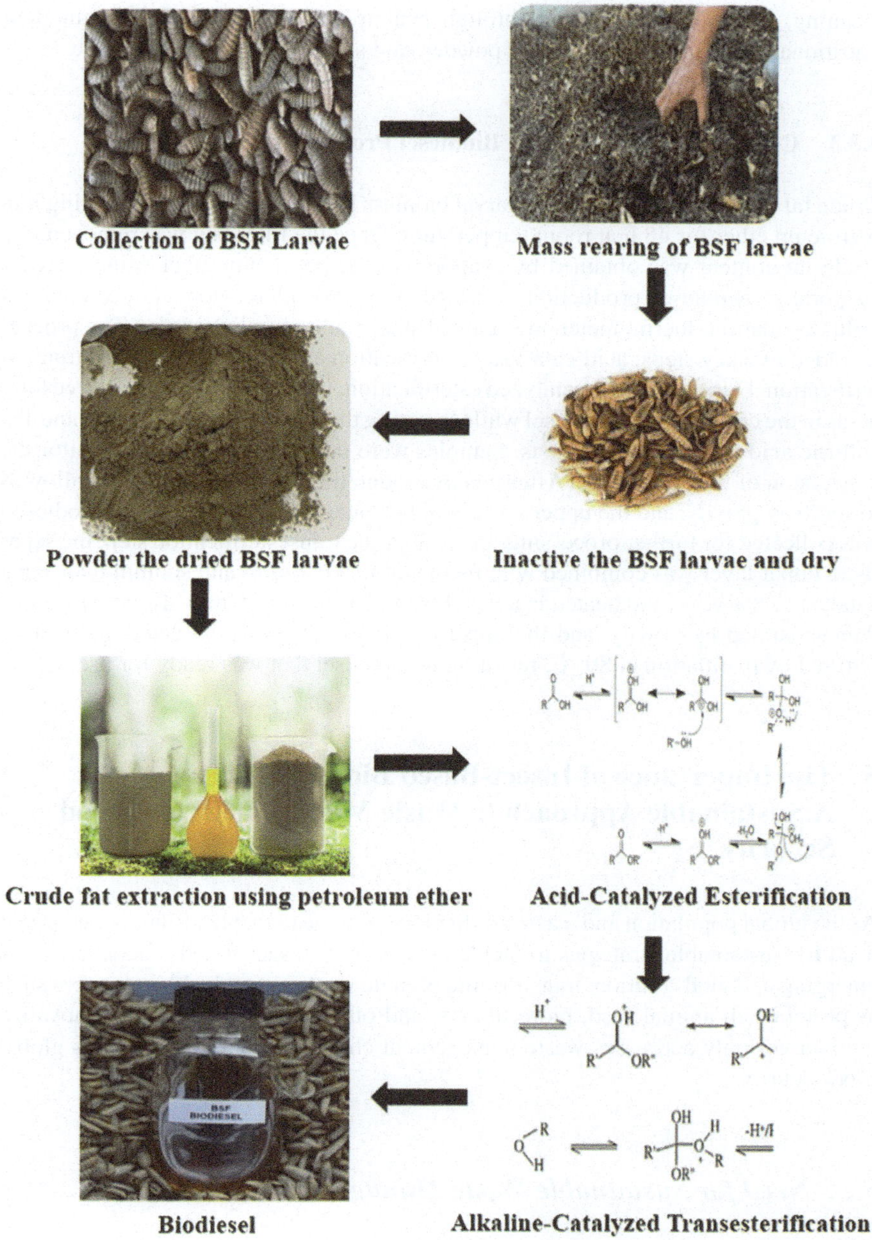

Fig. 2 Steps in biodiesel production from insects

cleaning, they were inactivated through heat treatment, dried to maintain their nutritional value, ground into a fine powder, and stored at 4 °C for future use.

4.3.3 Crude Fat Extraction and Biodiesel Production

Crude fat was extracted by placing larval biomass in a filter bag and submerging it in petroleum ether for 48 h at room temperature. Once the filter bag was removed, the crude fat content was obtained by evaporating the petroleum ether using a rotary evaporator. Biodiesel production occurred in a 100 ml reactor equipped with a reflux condenser, thermometer, mechanical stirrer, and sampling outlet. The process included two key steps: acid-catalyzed esterification and alkaline-catalyzed transesterification. During the acid-catalyzed esterification, the goal was to convert free fatty acids in the crude fat into biodiesel while lowering the acidity, using methanol and 1% sulfuric acid (H_2SO_4) as catalysts. Samples were taken periodically to monitor the conversion of free fatty acids. After this treatment, the reaction mixture was allowed to settle by gravity, and the upper layer, which contained the crude fat and biodiesel, was collected for further processing. In the alkaline transesterification step, the separated upper layer was combined with methanol in a 6:1 ratio and sodium hydroxide catalyst (0.8% w/w) and heated in a water bath at 65 °C for 30 min. The mixture was then separated by gravity, and the upper biodiesel layer was extracted and further purified by distillation at 80 °C, resulting in biodiesel that was ready for use.

5 The Importance of Insect-Based Bioconversion: A Sustainable Approach to Waste Management and Food Security

As the global population and waste production continue to increase, there is an urgent need for sustainable strategies to tackle these issues. Insect-based bioconversion is emerging as a vital solution, transforming organic waste into valuable resources such as protein-rich animal feed, biofertilizers, and other bio-products. This innovative method not only addresses waste management challenges but also supports global food security.

5.1 Need for Sustainable Waste Management

1. **Efficient Waste Conversion**: Insects have the capability to convert organic waste into protein efficiently, requiring less feed compared to traditional livestock and growing at a rapid rate. This makes them a scalable option for managing waste effectively.

2. **Environmental Advantages**: Raising insects produces lower greenhouse gas emissions than typical waste disposal methods. Moreover, it helps reduce landfill contributions and the associated methane emissions from decomposing organic materials.
3. **Circular Economy**: Insect bioconversion contributes to a circular economy by turning waste into valuable products, lowering waste management expenses, and generating new economic prospects.

5.2 *Need for Food Security*

1. **Source of Protein**: Insects represent a high-quality protein alternative for animal feed, which decreases reliance on traditional livestock farming and eases the burden on global food resources.
2. **Sustainability**: Insect farming demands significantly fewer land, water, and energy resources than conventional agricultural practices, making it a more sustainable approach to protein production.
3. **Resilience in Food Systems**: Incorporating insect-based products into food systems boosts resilience while minimizing the environmental footprint of agriculture, aiding in climate change mitigation efforts.

6 Future Directions

Despite the numerous advantages of insect-based bioconversion, several obstacles remain. The cultural stigma associated with insect consumption is a key barrier, highlighting the need for education and awareness initiatives to improve acceptance. Furthermore, advancements in insect farming technology are necessary to enhance efficiency, scalability, and automation. Collaboration among researchers, policy-makers, and industry players is essential for creating regulatory frameworks that ensure food safety and build public trust in insect-derived products. Future research should concentrate on optimizing rearing practices, investigating new substrates for insect feed, and expanding market possibilities for insect-based offerings.

7 Conclusion

Insect-based bioconversion represents a sustainable and economical method for managing organic waste while boosting food security. By integrating insects into waste management and food production systems, we can lessen environmental impacts, foster a circular economy, and provide alternative protein sources for generations to come. Addressing current challenges through technological progress, policy backing, and consumer acceptance will be vital in realizing the full potential of this revolutionary approach.

References

Ahmed, A. A. I., Hashem, M. Y., Mohamed, S. M., & Khalil, S. S. (2013). Protection of potato crop against *Phthorimaea operculella* (Zeller) infestation using frass extract of two noctuid insect pests under laboratory and storage simulation conditions. *Archives of Phytopathology and Plant Protection, 46*(18), 2409–2419. https://doi.org/10.1080/03235408.2013.795356

Al-Khalaf, A., Al Obthani, R. O., Al Rasheedi, H. N., & Rana Saeed Alghamdi, R. S. A. (2024). Review of the promising effectiveness of insects in waste management. *Biodiversity Research Journal, 1*(1), 16–34.

Anyega, A. O., Korir, N. K., Beesigamukama, D., Changeh, G. J., Nkoba, K., Subramanian, S., Van Loon, J. J., Dicke, M., & Tanga, C. M. (2021). Black soldier fly-composted organic fertilizer enhances growth, yield, and nutrient quality of three key vegetable crops in Sub-Saharan Africa. *Frontiers in Plant Science, 12*, 680312. https://doi.org/10.3389/fpls.2021.680312

Beesigamukama, D., Mochoge, B., Korir, N. K., Fiaboe, K. K. M., & Nakimbugwe, D. (2020). Exploring black soldier fly frass as a novel fertilizer for improved growth, yield, and nitrogen use efficiency of maize under field conditions. *Frontiers in Plant Science, 11*, 574592. https://doi.org/10.3389/fpls.2020.574592

Beesigamukama, D., Mochoge, B., Korir, N. K., Fiaboe, K. K., Nakimbugwe, D., Khamis, F. M., Subramanian, S., Wangu, M. M., Dubois, T., Ekesi, S., & Tanga, C. M. (2021). Low-cost technology for recycling agro-industrial waste into nutrient-rich organic fertilizer using black soldier fly. *Waste Management, 119*, 183–194. https://doi.org/10.1016/j.wasman.2020.09.043

Beesigamukama, D., Subramanian, S., & Tanga, C. M. (2022). Nutrient quality and maturity status of frass fertilizer from nine edible insects. *Scientific Reports, 12*(1), 7182. https://doi.org/10.1038/s41598-022-11336-z

Behie, S. W., & Bidochka, M. J. (2013). Insects as a nitrogen source for plants. *Insects, 4*(3), 413–424. https://doi.org/10.3390/insects4030413

Benzertiha, A., Kierończyk, B., Rawski, M., Mikołajczak, Z., Urbański, A., Nogowski, L., & Józefiak, D. (2020). Insect fat in animal nutrition—A review. *Annals of Animal Science, 20*(4), 1217–1240.

Bonelli, M., Bruno, D., Brilli, M., Gianfranceschi, N., & Tian, L. (2020). Black soldier fly larvae adapt to different food substrates through morphological and functional responses of the midgut. *International Journal of Molecular Sciences, 21*(15), 4955. https://doi.org/10.3390/ijms21154955

Bonso, N. K. (2013). *Bioconversion of organic fraction of solid waste using the larvae of the black soldier fly* (*Hermentia illucens*) (Doctoral dissertation).

Bordiean, A., Krzyżaniak, M., & Stolarski, M. J. (2022). Bioconversion potential of agro-industrial byproducts by Tenebrio molitor—Long-term results. *Insects, 13*(9), 810. https://doi.org/10.3390/insects13090810

Brar, S. K., Verma, M., Tyagi, R. D., Valéro, J. R., & Surampalli, R. Y. (2008). Wastewater sludges as novel growth substrates for rearing codling moth larvae. *World Journal of Microbiology and Biotechnology, 24*, 2849–2857. https://doi.org/10.1007/s11274-008-9818-z

Bukkens, S. G. (1997). The nutritional value of edible insects. *Ecology of Food and Nutrition, 36*(2–4), 287–319.

Bulak, P., Polakowski, M., Nowak, M., Wrzosek, M., & Bieganowski, M. (2018). *Hermetia illucens* as a new and promising species for use in entomoremediation. *Science of the Total Environment, 633*, 912–919. https://doi.org/10.1016/j.scitotenv.2018.03.252

Cappellozza, S., Leonardi, M. G., Savoldelli, S., Carminati, D., & Rizzolo, A. (2019). A first attempt to produce proteins from insects by means of a circular economy. *Animals, 9*(5), 278. https://doi.org/10.3390/ani9050278

Chavez, M., & Uchanski, M. (2021). Insect left-over substrate as plant fertilizer. *Journal of Insects as Food and Feed, 7*(4), 683–694. https://doi.org/10.3920/JIFF2020.0063

Cheng, Y. K. (2016). *Black soldier fly bioconversion for food waste recycling: Improved residue separation method and increased larval survival rate* (Doctoral dissertation), The Hong Kong University of Science and Technology, Hong Kong, China. http://repository.ust.hk/ir/Record/1783.1-86623. Accessed on February 26, 2019.

Cheng, Z., Yu, L., Li, H., Xu, X., & Yang, Z. (2021). Use of housefly (*Musca domestica* L.) larvae to bioconversion food waste for animal nutrition and organic fertilizer. *Environmental Science and Pollution Research, 28*(35), 48921–48928. https://doi.org/10.1007/s11356-021-14118-8

Chowdhury, S., Sontakke, P. P., Bhattacharjee, J., Bhattacharjee, D., Debnath, A., Kumar, A., & Datta, J. (2015). An overview of Edible insects in Eastern Himalayas: Indigenous traditional food of Tribal's. *Ecology Environment and Conservation, 21*(4), 1941–1946.

Čičková, H., Newton, G. L., Lacy, R. C., & Kozánek, M. (2015). The use of fly larvae for organic waste treatment. *Waste Management, 35*, 68–80. https://doi.org/10.1016/j.wasman.2014.09.026

Cucina, M., De Nisi, P., Trombino, L., Tambone, F., & Adani, F. (2021). Degradation of bioplastics in organic waste by mesophilic anaerobic digestion, composting and soil incubation. *Waste Management, 134*, 67–77. https://doi.org/10.1016/j.wasman.2021.08.016

de Souza-Vandenberghe, L. P., Garcia, L. M. B., Rodrigues, C., Camara, M. C., de Melo Pereira, G. V., de Oliveira, J., & Soccol, C. R. (2017). Potential applications of plant probiotic microorganisms in agriculture and forestry. *AIMS Microbiology, 3*(4), 629–648. https://doi.org/10.3934/microbiol.2017.3.629

Dey, S., Das, A., Mallick, K., Sahu, A., & Das, A. P. (2023). Environmental petroleum waste: Pollution. *Toxicity, Sustainable Remediation, 159–175*. https://doi.org/10.1007/978-3-031-48220-5_7

Diener, S., Studt Solano, N. M., Roa Gutiérrez, F., Zurbrügg, C., & Tockner, K. (2011). Biological treatment of municipal organic waste using black soldier fly larvae. *Waste and Biomass Valorization, 2*, 357–363. https://doi.org/10.1007/s12649-011-9079-1

Dobermann, D., Swift, J. A., & Field, L. M. (2017). Opportunities and hurdles of edible insects for food and feed. *Nutrition Bulletin, 42*(4), 293–308. https://doi.org/10.1111/nbu.12291

Dulaurent, A. M., Daoulas, G., Faucon, M. P., & Houben, D. (2020). Earthworms (*Lumbricus terrestris* L.) mediate the fertilizing effect of frass. *Agronomy, 10*(6), 783. https://doi.org/10.3390/agronomy10060783

Finke, M. D. (2002). Complete nutrient composition of commercially raised invertebrates used as food for insectivores. *Zoo Biology, 21*(3), 269–285. https://doi.org/10.1002/zoo.10031

Finke, M. D. (2013). Complete nutrient content of four species of feeder insects. *Zoo Biology, 32*(1), 27–36. https://doi.org/10.1002/zoo.21012

Fischer, H., & Romano, N. (2021). Fruit, vegetable, and starch mixtures on the nutritional quality of black soldier fly (*Hermetia illucens*) larvae and resulting frass. *Journal of Insects as Food and Feed, 7*(3), 319–327. https://doi.org/10.3920/JIFF2020.0100

Fowles, T. M. (2021). *Using insects in conversion of agricultural by-products and applications of their value-added products*. Davis: University of California.

Fowles, T.M., & Nansen, C. (2020). Insect-based bioconversion: Value from food waste. In: Närvänen, E., Mesiranta, N., Mattila, M., & Heikkinen, A. (Eds.), *Food waste management*. Palgrave Macmillan. https://doi.org/10.1007/978-3-030-20561-4_12

Frost, C. J., & Hunter, M. D. (2004). Insect canopy herbivory and frass deposition affect soil nutrient dynamics and export in oak mesocosms. *Ecology, 85*(12), 3335–3347. https://doi.org/10.1890/04-0003

Gahukar, R. T. (2018). Entomophagy for nutritional security in India. *Current Science, 115*(6), 1078–1084.

Gałęcki, R., Bakuła, T., & Gołaszewski, J. (2023) Foodborne diseases in the edible insect industry in Europe—New challenges and old problems. *Foods, 12*(4), 770. Available at: https://doi.org/10.3390/foods12040770

Gherlenda, A. N., Crous, K. Y., Moore, B. D., Haigh, A. M., Johnson, S. N., & Riegler, M. (2016). Precipitation, not CO_2 enrichment, drives insect herbivore frass deposition and subsequent nutrient dynamics in a mature *Eucalyptus* woodland. *Plant and Soil, 399*(1–2), 29–39. https://doi.org/10.1007/s11104-015-2683-2

Goulding, K. W., Bailey, N. J., Bradbury, N. J., Hargreaves, P., Howe, M., Murphy, D. V., et al. (1998). Nitrogen deposition and its contribution to nitrogen cycling and associated soil processes. *New Phytologist, 139*(1), 49–68. https://doi.org/10.1046/j.1469-8137.1998.00182.x

Harsányi, E., Juhász, C., Kovács, E., Huzsvai, L., Pintér, R., Fekete, G., & Gyuricza, C. (2020). Evaluation of organic wastes as substrates for rearing *Zophobas morio*, *Tenebrio molitor*, and *Acheta domesticus* larvae as alternative feed supplements. *Insects, 11*(9), 604. https://doi.org/10.3390/insects11090604

Hasan, H. A., & Leong, K. P. (2018). Growth of *Musca domestica* (Diptera: Muscidae) and *Sarcophaga dux* (Diptera: Sarcophagidae) larvae in poultry and livestock manures: implication for animal waste management. *Journal of Asia-Pacific Entomology, 21*(3), 880–884. https://doi.org/10.1016/j.aspen.2018.07.001

Hawkey, K. J., Lopez-Viso, C., Brameld, J. M., Parr, T., & Salter, A. M. (2021). Insects: a potential source of protein and other nutrients for feed and food. *Annual Review of Animal Biosciences, 9*(1), 333–354. https://doi.org/10.1146/annurev-animal-021419-083930

Hernández Toxqui, A. G., Ramírez Ramírez, J., Pino Moreno, J. M., Talamantes Gómez, J. M., Angeles Campos, S. C., & Ramírez Orejel, J. C. (2021). Development of nutraceutical ice creams using flour yellow worm larvae (*Tenebrio molitor*), Chia (*Salvia hispanica*), and quinoa (Chenopodium quinoa)—frontiers. *Veterinary Science, 8*, 629180. https://doi.org/10.3389/fvets.2021.629180

Houben, D., Daoulas, G., Faucon, M. P., & Dulaurent, A. M. (2020). Potential use of mealworm frass as a fertilizer: Impact on crop growth and soil properties. *Scientific Reports, 10*, 1–9. https://doi.org/10.1038/s41598-020-61765-x

Hu, Y., Su, M., Wang, Y., Cui, S., Meng, F., Yue, W., Liu, Y., Xu, C., & Yang, Z. (2020). Food production in China requires intensified measures to be consistent with national and provincial environmental boundaries. *Nature Food, 1*(9), 572–582. https://doi.org/10.1038/s41538-020-00107-0

Huyben, D., Vidaković, A., Hallgren, S. W., & Langeland, M. (2019). High-throughput sequencing of gut microbiota in rainbow trout (*Oncorhynchus mykiss*) fed larval and pre-pupae stages of black soldier fly (*Hermetia illucens*). *Aquaculture, 500*, 485–491. https://doi.org/10.1016/j.aquaculture.2018.10.056

Jucker, C., Belluco, S., Oddon, S. B., Ricci, A., Bonizzi, L., Lupi, D., … & Gasco, L. (2022). Impact of some local organic by-products on *Acheta domesticus* growth and meal production. *Journal of Insects as Food and Feed, 8*(6), 631–640. https://doi.org/10.3920/JIFF2021.0121

Kafle, K., Subedi, S., & Poudel, A. (2024). A review on edible insects diversity, abundance, and conservation. *Journal of Entomology and Zoology Studies, 12*(2), 137–142.

Kalu-Uka, G. M., Kumar, S., Kalu-Uka, A. C., Vikram, S., Okorafor, O. O., Kigozi, M., & Onwualu, A. P. (2021). Prospects for biodiesel production from *Macrotermes nigeriensis*: Process optimization and characterization of biodiesel properties. *Biomass and Bioenergy, 146*, 105980. https://doi.org/10.1016/j.biombioe.2021.105980

Khan, D., Sahito, Z. A., Dawar, S., & Zaki, M. J. (2016). Frass of saproxyliccerambycid larvae from dead twigs of Acacia stenophylla A. Cunn. EX. Benth. and its effects on germination and seedling growth of *Lactuca sativa* L. var. grand rapids. *International Journal of Biology and Biotechnology, 13*, 461–470. https://ijbb.pakbs.org/issue/2016%20No.3/8.html

Kierończyk, B., Rawski, M., Józefiak, A., Mazurkiewicz, J., Świątkiewicz, S., Siwek, M., Bednarczyk, M., Szumacher-Strabel, M., Cieślak, A., Benzertiha, A., & Józefiak, D. (2018). Effects of replacing soybean oil with selected insect fats on broilers. *Animal Feed Science and Technology, 240*, 170–183. https://doi.org/10.1016/j.anifeedsci.2018.04.019

Kinyuru, J. N., Mogendi, J. B., Riwa, C. A., & Ndung'u, N. W. (2015). Edible insects—A novel source of essential nutrients for human diet: Learning from traditional knowledge. *Animal Frontiers, 5*(2), 14–19. https://doi.org/10.2527/af.2015-0016

Koh, D. W. S., Ang, B. Y. X., Yeo, J. Y., Xing, Z., & Gan, S. K. (2020). Plastic agriculture using worms: Augmenting polystyrene consumption and using frass for plant growth towards a zero-waste circular economy. bioRxiv. https://doi.org/10.1101/2020.05.29.123521

Koné, N., Sylla, M., Nacambo, S., & Kenis, M. (2017). Production of house fly larvae for animal feed through natural oviposition. *Journal of Insects as Food and Feed, 3*(3), 177–186. https://doi.org/10.3920/JIFF2016.0044

Lalander, C., Diener, S., Zurbrügg, C., & Vinnerås, B. (2019). Effects of feedstock on larval development and process efficiency in waste treatment with black soldier fly (*Hermetia illucens*). *Journal of Cleaner Production, 208*, 211–219. https://doi.org/10.1016/j.jclepro.2018.10.017

Lardé, G. (1990). Recycling of coffee pulp by *Hermetia illucens* (Diptera: Stratiomyidae) larvae. *Biological Wastes, 33*(4), 307–310. https://doi.org/10.1016/0269-7483(90)90134-E

Latunde-Dada, G. O., Yang, W., & Vera Aviles, M. (2016). In vitro iron availability from insects and sirloin beef. *Journal of Agricultural and Food Chemistry, 64*(44), 8420–8424. https://doi.org/10.1021/acs.jafc.6b03891

Li, Q., Zheng, L., Cai, H., Garza, E., Yu, Z., & Zhou, S. (2011). From organic waste to biodiesel: Black soldier fly, *Hermetia illucens*, makes it feasible. *Fuel, 90*(4), 1545–1548. https://doi.org/10.1016/j.fuel.2010.11.016

Li, W., Li, Q., Zheng, L., Wang, Y., Zhang, J., Yu, Z., & Zhang, Y. (2015). Potential biodiesel and biogas production from corncob by anaerobic fermentation and black soldier fly. *Bioresource Technology, 194*, 276–282. https://doi.org/10.1016/j.biortech.2015.06.112

Liceaga, A. M., Aguilar-Toalá, J. E., Vallejo-Cordoba, B., González-Córdova, A. F., & Hernández-Mendoza, A. (2022). Insects as an alternative protein source. *Annual Review of Food Science and Technology, 13*(1), 19–34. https://doi.org/10.1146/annurev-food-052521-010613

Liddle, B. (2013). Population, affluence, and environmental impact across development: Evidence from panel cointegration modeling. *Environmental Modelling & Software, 40*, 255–266. https://doi.org/10.1016/j.envsoft.2012.09.010

Manurung, R., Supriatna, A., Esyanthi, R. R., & Putra, R. E. (2016). Bioconversion of rice straw waste by black soldier fly larvae (*Hermetia illucens* L.): Optimal feed rate for biomass production. *Journal of Entomology and Zoology Studies,4*(4), 1036–1041.

Manzano-Agugliaro, F., Sanchez-Muros, M. J., Barroso, F. G., Martínez-Sánchez, A., Rojo, S., & Pérez-Bañón, C. (2012). Insects for biodiesel production. *Renewable and Sustainable Energy Reviews, 16*(6), 3744–3753. https://doi.org/10.1016/j.rser.2012.03.020

Michaelsen, K.F., Hoppe, C., Roos, N., Kaestel, P., Stougaard, M., Lauritzen, L., Mølgaard, C., Girma, T., & Friis, H. (2009). Choice of foods and ingredients for moderately malnourished children 6 months to 5 years of age. *Food and Nutrition Bulletin, 30*(3_suppl3), S343–S404. https://doi.org/10.1177/15648265090303s303

Miech, P., Berggren, Å., Lindberg, J. E., Chhay, T., Khieu, B., & Jansson, A. (2016). Growth and survival of reared Cambodian field crickets (*Teleogryllus testaceus*) fed weeds, agricultural and food industry by-products. *Journal of Insects as Food and Feed, 2*(4), 285–292. https://doi.org/10.3920/JIFF2016.0028

Mishra, S., Ghosh, S., van Hullebusch, E. D., Singh, S., & Das, A. P. (2023). A critical review on the recovery of base and critical elements from electronic waste-contaminated streams using microbial biotechnology. *Applied Biochemistry and Biotechnology, 2023*, 1–30. https://doi.org/10.1007/S12010-023-04440-X

Mishra, S., Rout, P. K., & Das, A. P. (2019). Solar photovoltaic panels as next generation waste: A review. *Biointerface Research in Applied Chemistry, 9*(6).

Mlček, J., Rop, O., Borkovcova, M., & Bednářová, M. (2014). A comprehensive look at the possibilities of edible insects as food in Europe—A review. *Polish Journal of Food and Nutrition Sciences*. https://doi.org/10.2478/v10222-012-0095-0

Mohan, K., Sathishkumar, P., Rajan, D. K., & Ganesan, A. R. (2023). Black soldier fly (*Hermetia illucens*) larvae as potential feedstock for biodiesel production: Recent advances and challenges. *Science of the Total Environment, 857*, 159166. https://doi.org/10.1016/j.scitotenv.2022.159166

Mostafaie, A., Silva, A. R. R., Pinto, J. N., Prodana, M., Lopes, I. G., Murta, D., Brooks, B. W., Loureiro, S., & Cardoso, D. N. (2025). Towards circularity for agro-waste: Minimal soil hazards of olive pomace bioconverted frass by insect larvae as an organic fertilizer. *Journal of Environmental Management, 375*, 124151. https://doi.org/10.1016/j.jenvman.2025.124151

Mudalungu, C. M., Tanga, C. M., Kelemu, S., & Torto, B. (2021). An overview of antimicrobial compounds from African edible insects and their associated microbiota. *Antibiotics, 10*(6), 621. https://doi.org/10.3390/antibiotics10060621

Murthy, J. S. V., Keerthana, A., Logeswaran, K., Das, A., Choudhury, S., Ramakrishna, B. G., Chowdhury, S., Aggarwal, H., Saravanan, S., Pal, A., & Dubey, V. K. (2024). Harnessing insects mediated plastic biodegradation: Current insight and future directions. *Journal of Environmental Management, 372*, 123038. https://doi.org/10.1016/j.jenvman.2024.123038

Mutafela, R. N. (2015). *High value organic waste treatment via black soldier fly bioconversion: Onsite pilot study*. Master's thesis, Royal Institute of Technology. Available at: https://www.diva-portal.org/smash/get/diva2:868277/FULLTEXT02.pdf. Accessed February 26, 2019.

Niu, Y., Zheng, D., Yao, B., Cai, Z., Zhao, Z., Wu, S., & Yang, D. (2017). A novel bioconversion for value-added products from food waste using *Musca domestica*. *Waste Management, 61*, 455–460. https://doi.org/10.1016/j.wasman.2016.10.054

O'Neill, B. C., Liddle, B., Jiang, L., Smith, K. R., Pachauri, S., Dalton, M., & Fuchs, R. (2012). Demographic change and carbon dioxide emissions. *The Lancet, 380*(9837), 157–164. https://doi.org/10.1016/S0140-6736(12)60958-1

Oonincx, D. G. A. B., van Broekhoven, S., van Huis, A., & van Loon, J. J. A. (2015). Feed conversion, survival and development, and composition of four insect species on diets composed of food by-products. *PLoS ONE, 10*(12), e0144601. https://doi.org/10.1371/journal.pone.0144601

Parodi, A., Leip, A., De Boer, I. J. M., Slegers, P. M., Ziegler, F., Temme, E. H. M., & Van Zanten, H. H. E. (2018). The potential of future foods for sustainable and healthy diets. *Nature Sustainability, 1*, 782–789. https://doi.org/10.1038/s41893-018-0189-7

Parry, N. J., Pieterse, E., & Weldon, C. W. (2017). Longevity, fertility and fecundity of adult blow flies (Diptera: Calliphoridae) held at varying densities: Implications for use in bioconversion of waste. *Journal of Economic Entomology, 110*(6), 2388–2396. https://doi.org/10.1093/jee/tox251

Pinotti, L., & Ottoboni, M. (2021). Substrate as insect feed for bio-mass production. *Journal of Insects as Food and Feed, 7*, 585–596. https://doi.org/10.3920/JIFF2020.0110

Poveda, J., Jiménez-Gómez, A., Saati-Santamaría, Z., Usategui-Martín, R., Rivas, R., & García-Fraile, P. (2019). Mealworm frass as a potential biofertilizer and abiotic stress tolerance-inductor in plants. *Applied Soil Ecology, 142*, 110–122. https://doi.org/10.1016/j.apsoil.2019.04.016

Priyadarsini, S., & Das, A. P. (2024). Analytical and structural characterization of waste lithium-ion batteries for their effective recycling strategy. *Environmental Chemistry and Ecotoxicology*. https://doi.org/10.1016/j.enceco.2024.12.004

Priyadarsini, S., & Das, A. P. (2025). Lithium bioleaching: A review on microbial-assisted sustainable technology for lithium bio-circularity. *Journal of Water Process Engineering, 69*, 106744. https://doi.org/10.1016/j.jwpe.2024.106744

Purschke, B., Scheibelberger, R., Axmann, S., Adler, A., & Jäger, H. (2017). Impact of substrate contamination with mycotoxins, heavy metals, and pesticides on growth performance and composition of black soldier fly larvae (*Hermetia illucens*) for use in the feed and food value chain. *Food Additives & Contaminants: Part A, 34*(8), 1410–1420. https://doi.org/10.1080/19440049.2017.1325928

Quilliam, R. S., Nuku-Adeku, C., Maquart, P., Little, D., Newton, R., & Murray, F. (2020). Integrating insect frass biofertilisers into sustainable periurban agro-food systems. *Journal of Insects as Food and Feed, 6*, 315–322. https://doi.org/10.3920/JIFF2019.0049

Ravi, H. K., Degrou, A., Costil, J., Trespeuch, C., Chemat, F., & Vian, M. A. (2020). Larvae mediated valorization of industrial, agriculture and food wastes: Biorefinery concept through bioconversion, processes, procedures, and products. *Processes, 8*(7), 857. https://doi.org/10.3390/pr8070857

Ray, S., Gaffor, I., Acevedo, F. E., Helms, A., Chuang, W. P., Tooker, J., Felton, G. W., & Luthe, D. S. (2015). Maize plants recognize herbivore-associated cues from caterpillar frass. *Journal of Chemical Ecology, 41*, 781–792. https://doi.org/10.1007/s10886-015-0619-1

Riekkinen, K., Väkeväinen, K., & Korhonen, J. (2022). The effect of substrate on the nutrient content and fatty acid composition of edible insects. *Insects, 13*(7), 590. https://doi.org/10.3390/insects13070590

Rumbos, C. I., & Athanassiou, C. G. (2021). The superworm, *Zophobas morio* (Coleoptera: Tenebrionidae): A 'sleeping giant' in nutrient sources. *Journal of Insect Science, 21*(2), 13. https://doi.org/10.1093/jisesa/ieab014

Rumbos, C. I., & Athanassiou, C. G. (2023). Mealworms for food and feed: Are they only yellow? *Journal of Insects as Food and Feed, 9*(11), 1409–1416. https://doi.org/10.1163/23524588-230912ED

Rumpold, B. A., & Schlüter, O. K. (2013). Nutritional composition and safety aspects of edible insects. *Molecular Nutrition & Food Research, 57*(5), 802–823. https://doi.org/10.1002/mnfr.201200735

Sánchez-Muros, M., De Haro, C., Sanz, A., Trenzado, C. E., Villareces, S., & Barroso, F. G. (2016). Nutritional evaluation of *Tenebrio molitor* meal as fishmeal substitute for tilapia (*Oreochromis niloticus*) diet. *Aquaculture Nutrition, 22*(5), 943–955. https://doi.org/10.1111/anu.12313

Sangiorgio, P., Verardi, A., Dimatteo, S., Spagnoletta, A., Moliterni, S., & Errico, S. (2022). Valorisation of agri-food waste and mealworms rearing residues for improving the sustainability of *Tenebrio molitor* industrial production. *Journal of Insects as Food and Feed, 8*(5), 509–524. https://doi.org/10.3920/JIFF2021.0101

Schlüter, O., Rumpold, B., Holzhauser, T., Roth, A., Vogel, R. F., Quasigroch, W., Vogel, S., Heinz, V., Jäger, H., Bandick, N., & Kulling, S. (2017). Safety aspects of the production of foods and food ingredients from insects. *Molecular Nutrition & Food Research, 61*(6), 1600520. https://doi.org/10.1002/mnfr.201600520

Schmitt, E., & de Vries, W. (2020). Potential benefits of using Hermetia illucens frass as a soil amendment on food production and for environmental impact reduction. *Current Opinion in Green and Sustainable Chemistry, 25*, 100335. https://doi.org/10.1016/j.cogsc.2020.100335

Searchinger, T., Heimlich, R., Houghton, R. A., Dong, F., Elobeid, A., Fabiosa, J.,... & Yu, T. H. (2008). Use of U.S. croplands for biofuels increases greenhouse gases through emissions from land-use change. *Science, 319*(5867), 1238–1240. https://doi.org/10.1126/science.1151861

Shabani, J. M., Babajide, O. O., & Oyekola, O. O. (2020). An investigation into the potential of maggot oil as a feedstock for biodiesel production. *Valorization of Biomass to Value-Added Commodities: Current Trends, Challenges, and Future Prospects, 285–301*. https://doi.org/10.1007/978-3-030-38032-8_14

Shell. (2025). Shell expects 60% rise in global LNG demand by 2040 as Asia leads growth. *Reuters*. Retrieved from https://www.reuters.com/business/energy/shell-expects-60-rise-global-lng-demand-by-2040-2025-02-25/

Sheppard, D. C., Newton, G. L., Thompson, S. A., & Savage, S. (1994). A value-added manure management system using the black soldier fly. *Bioresource Technology, 50*(3), 275–279. https://doi.org/10.1016/0960-8524(94)90102-3

Shojaaddini, M. (2024). Applicability of black soldier fly and yellow mealworm in municipal food waste bioconversion: Assessment of efficiency, nutritional proficiency, and safety. *Journal of Asia-Pacific Entomology, 27*(3), 102306. https://doi.org/10.1016/j.aspen.2024.102306

Siddiqui, S. A., Snoeck, E. R., Tello, A., Alles, M. C., Fernando, I., Saraswati, Y. R., & Nagdalian, A. A. (2022). Manipulation of the black soldier fly larvae (*Hermetia illucens*; Diptera: Stratiomyidae) fatty acid profile through the substrate. *Journal of Insects as Food and Feed, 8*(8), 837–855. https://doi.org/10.3920/JIFF2021.0162

Song, S., Ee, A. W. L., Tan, J. K. N., Cheong, J. C., Chiam, Z., Arora, S., Lam, W. N., & Tan, H. T. W. (2021). Upcycling food waste using black soldier fly larvae: Effects of further composting on frass quality, fertilising effect and its global warming potential. *Journal of Cleaner Production, 288*, 125664. https://doi.org/10.1016/j.jclepro.2020.125664

Suantika, G., Putra, R. E., Hutami, R., & Rosmiati, M. (2017). November. Application of compost produced by bioconversion of coffee husk by black soldier fly larvae (*Hermetia illucens*) as solid fertilizer to lettuce (*Lactuca sativa* Var. *Crispa*). In *Proceedings of the International Conference on Green Technology* (Vol. 8, No. 1, pp. 20–26).

Supriyatna, A., Manurung, R., Esyanti, R. R., & Putra, R. E. (2016). Growth of black soldier larvae fed on cassava peel wastes, An agriculture waste. *Journal of Entomology and Zoology Studies, 4*(6), 161–165.

Surendra, K. C., Olivier, R., Tomberlin, J. K., & Khanal, S. K. (2016). Bioconversion of organic wastes into biodiesel and animal feed via insect farming. *Renewable Energy, 98*, 197–202. https://doi.org/10.1016/j.renene.2016.03.022

Surendra, K. C., Tomberlin, J. K., van Huis, A., Cammack, J. A., Heckmann, L. H. L., & Khanal, S. K. (2020). Rethinking organic wastes bioconversion: Evaluating the potential of the black soldier fly (*Hermetia illucens* (L.)) (Diptera: Stratiomyidae) (BSF). *Waste Management, 117*, 58–80. https://doi.org/10.1016/j.wasman.2020.07.050

Thrastardottir, R., Olafsdottir, H. T., & Thorarinsdottir, R. I. (2021). Yellow mealworm and black soldier fly larvae for feed and food production in Europe, with emphasis on Iceland. *Foods, 10*(11), 2744.

Tinder, A. C., Puckett, R. T., Turner, N. D., Cammack, J. A., & Tomberlin, J. K. (2017). Bioconversion of sorghum and cowpea by black soldier fly (*Hermetia illucens* (L.)) larvae for alternative protein production. *Journal of Insects as Food and Feed, 3*(2), 121–130. https://doi.org/10.3920/JIFF2016.0048

Tullo, E., Finzi, A., & Guarino, M. (2019). Environmental impact of livestock farming and precision livestock farming as a mitigation strategy. *Science of the Total Environment, 650*, 2751–2760. https://doi.org/10.1016/j.scitotenv.2018.10.018

Van Huis, A. (2020). Insects as food and feed, a new emerging agricultural sector: A review. *Journal of Insects as Food and Feed, 6*(1), 27–44. https://doi.org/10.3920/JIFF2019.0017

Van Huis, A., & Oonincx, D. G. (2017). The environmental sustainability of insects as food and feed. A review. *Agronomy for Sustainable Development, 37*, 1–14. https://doi.org/10.1007/s13593-017-0452-8

Van Huis, A. (2013). Potential of insects as food and feed in assuring food security. *Annual Review of Entomology, 58*(1), 563–583. https://doi.org/10.1146/annurev-ento-120811-153704

Wang, H., Rehman, K. U., Liu, X., Yang, Q., Zheng, L., Li, W., Cai, M., Li, Q., Zhang, J., & Yu, Z. (2017). Insect biorefinery: A green approach for conversion of crop residues into biodiesel and protein. *Biotechnology for Biofuels, 10*, 1–13. https://doi.org/10.1186/s13068-017-0986-7

Wang, Y. S., & Shelomi, M. (2017). Review of black soldier fly (*Hermetia illucens*) as animal feed and human food. *Foods, 6*(10), 91. https://doi.org/10.3390/foods6100091

Wei, J. R., Lu, X. P., & Jiang, L. (2013). Monoterpenes from larval frass of two Cerambycids as chemical cues for a parasitoid, *Dastarcus helophoroides. Journal of Insect Science, 13*, 59–12. https://doi.org/10.1673/031.013.5901

Yehuda, B., Marchaim, U., Glatman, L., Drabkin, V., Chizov-Ginzburg, A., Mumcuoglu, K. Y., & Gelman, A. (2011). Bioconversion of poultry and fish waste by *Lucilia sericata* and *Sarcophaga carnaria* larvae. *Asian Journal of Water, Environment and Pollution, 8*(4), 69–75. https://doi.org/10.3233/AJW-2011-8_4_09

Zheng, L., Hou, Y., Li, W., Yang, S., Li, Q., & Yu, Z. (2012a). Biodiesel pro duction from rice straw and restaurant waste employing black soldier fly assisted by microbes. *Energy, 47*(1), 225–229. https://doi.org/10.1016/j.energy.2012.09.006

Zheng, L., Li, Q., Zhang, J., & Yu, Z. (2012b). Double the biodiesel yield: Rearing black soldier fly larvae, *Hermetia illucens*, on solid residual fraction of restaurant waste after grease extraction for biodiesel production. *Renewable Energy, 41*, 75–79. https://doi.org/10.1016/j.renene.2011.10.004

Zheng, L., Hou, Y., Li, W., Yang, S., Li, Q., & Yu, Z. (2013). Exploring the potential of grease from yellow mealworm beetle (*Tenebrio molitor*) as a novel biodiesel feedstock. *Applied Energy, 101*, 618–621. https://doi.org/10.1016/j.apenergy.2012.06.067

Transforming Waste to Wealth: The Role of Vermicomposting in Sustainable Resource Management

Ipsita Priyadarsini Samal, Sameer Jena, B. Jyotirmayee, and Gyanranjan Mahalik

Abstract Vermicomposting is a promising biotechnological tool that is a prime example of transforming waste into wealth. Using earthworms' innate ability to collaborate with microorganisms, this environmentally friendly method turns organic waste into valuable resources. This review examines vermicomposting as a sustainable method of managing organic waste and recovering resources. Its dual potential to solve waste issues and produce valuable resources is reviewed. Vermicomposting, a bio-oxidative process powered by microorganisms and earthworms, is an exceptionally environmentally friendly method of turning organic waste into soil amendments and fertilizer-rich nutrients. The chapter examines the basic biochemical processes of vermicomposting, emphasizing the vital interactions between microorganisms and earthworms that promote adequate nutrient enrichment and organic matter decomposition. We assess a variety of organic waste streams, such as industrial bio-wastes, municipal solid waste, and agricultural residues, looking at their appropriateness and conversion into products with added value like vermicompost, vermiwash, and earthworm biomass. The research highlights explicitly the economic feasibility of vermicomposting activities at various scales, ranging from residential to commercial, showcasing its capacity to generate wealth while upholding the ideals of the circular economy. In addition to offering substantial economic opportunities through producing high-quality organic fertilizers, our discussion shows that vermicomposting provides an environmentally friendly waste management solution. This helps to support sustainable agriculture, conserve resources, and determine future directions for improving vermicomposting procedures and growing their use in sustainable waste management.

Keywords Vermicompost · Microorganisms · Decomposition · Organic · Sustainable

---

I. P. Samal · S. Jena · B. Jyotirmayee · G. Mahalik (✉)
Department of Botany, School of Applied Sciences, Centurion University of Technology and Management, Bhubaneswar, Odisha, India
e-mail: gyanranjan.mahalik@cutm.ac.in

A. P. Das (ed.), *Bioconversion of Environmental Waste to Value Added Products*, Sustainable Environmental Waste Management Strategies,
https://doi.org/10.1007/978-3-032-14215-3_2

1 Introduction

The international community must sustainably address waste management, considering the growing environmental issues of our time (Mishra et al., 2023; Mohanty et al., 2018). The World Bank predicts that by 2050, the garbage generated annually will have increased by 70% from 2016 to 3.4 billion metric tons, underscoring the pressing need for creative and sustainable waste management strategies. In this regard, vermicomposting shows promise as a biotechnology that solves several sustainability issues while converting organic waste into useful products (Mishra & Das, 2021; Mishra et al., 2022). Utilizing earthworms to bio-convert organic waste into nutrient-rich compost is known as vermicomposting, and it is a combination of conventional wisdom and contemporary scientific knowledge (Behera et al., 2022; Hajam et al., 2023). This method uses a complex interplay of mechanical and biochemical processes to break down organic materials by utilizing the natural eating behavior of earthworms, namely Eisenia fetida and Eisenia andrei. Vermicomposting, in contrast to conventional composting techniques, produces better final products with greater agricultural value while hastening the stabilization of organic waste. Vermicomposting is far more important than treating only garbage. It provides a powerful illustration of the circular economy's tenets in the context of sustainable resource management. The method turns what would otherwise be garbage into two useful resources: earthworm biomass, which can be used in pharmaceutical and animal feed applications, and vermicompost, a high-quality organic fertilizer full of nutrients plants can absorb and helpful microbes. Several Sustainable Development Goals of the UN are addressed by this conversion of trash into wealth, including SDG 2 (zero hunger), SDG 13 (climate action), and SDG 12 (responsible consumption and production). Vermicomposting's economic effects are especially significant in both developed and underdeveloped countries. In developed nations, it provides an affordable way to manage organic waste while generating valuable agricultural inputs (Shaikh & Shaikh, 2025). Vermicomposting offers prospects for sustainable entrepreneurship, job creation, and increased agricultural output in impoverished nations. This industry has significant economic potential, as the global market for organic fertilizers, including vermicompost, is expected to reach USD 15.9 billion by 2025. Several successful projects worldwide show that vermicomposting system implementation varies greatly between scales and situations. Examples from Europe demonstrate the potential for technological innovation in vermicomposting. Vermeer Technologies' commercial facility in the Netherlands is an example of how IoT sensors and automated monitoring systems are integrated. Their facility maintains ideal conditions with automated climate control and feeding systems while processing 50 tons of agricultural waste daily (Goswami et al., 2024). Compared to conventional ways, this high-tech approach has enhanced processing efficiency by 40% and guarantees consistent product quality that satisfies stringent EU organic farming criteria. One creative example is from Singapore, where organic waste from tall buildings is processed by a vertical vermicomposting system created by Sky Farms. Twenty residential complexes have adopted this space-efficient solution, which processes two tons of food waste daily while taking up tiny ground areas. The system's effectiveness in an urban setting shows how vermicomposting may be adjusted to various environmental and spatial limitations. These varied case studies demonstrate the adaptability and efficiency of vermicomposting at various scales and settings. Applications in the actual world, ranging from instructional programs to municipal trash management, have shown the technology's versatility and many advantages. In addition to demonstrating the technology's contribution to advancing sustainable resource management worldwide, these examples offer insightful information on effective implementation techniques and possible obstacles. Vermicomposting also concurrently tackles several important environmental issues. It enhances soil health and carbon sequestration capability, lessens dependency on chemical fertilizers, and lowers greenhouse gas emissions linked to the decomposition of organic waste in landfills (Singh & Sinha, 2022). By improving the soil's structure and the final product's ability to retain water, the technique also helps save water. This study analyzes the technical, financial, and environmental aspects of vermicomposting's complex role in sustainable resource management. We hope to show how vermicomposting may support a more sustainable and circular economy by thoroughly analyzing present practices, obstacles, and opportunities. The research synthesizes findings from various geographic settings and operational scales to offer insights into developing vermicomposting systems and integrating them into larger waste management and agricultural frameworks. This study explores how vermicomposting turns trash into useful resources, advancing our knowledge of sustainable waste management strategies and their possible effects on food security, economic growth, and environmental preservation. Policymakers, farmers, waste management specialists, and business owners looking to start or expand vermicomposting enterprises may find the findings especially pertinent.

2 Literature Review

A. **Historical Development of Vermicomposting**

i. **Ancient Origins and Traditional Practices**

The association between earthworms and soil fertility was first recorded in ancient civilizations, where the history of vermicomposting begins. Hieroglyphic depictions show that ancient Egyptian farmers understood the significance of earthworms in Nile Valley agriculture. Later, Aristotle acknowledged earthworms' contribution to soil improvement and called them "the intestines of the earth" in his writings. Earthworm activity was explicitly linked to the agricultural output of Chinese farm writings from the "Han Dynasty" (Sharma & Garg, 2019).

ii. **Traditional Agricultural Knowledge**

Indigenous agricultural knowledge across continents has long acknowledged earthworms' critical role in soil health, and Indigenous farming communities worldwide have acquired a comprehensive understanding of earthworm-based soil enrichment. Understanding the significance of earthworm populations for productive crop production, Native American tribes took great care to preserve abundant soil beds in these organisms. Similarly, traditional Indian farming systems recognized the importance of earthworms to soil fertility and purposefully included techniques that encouraged earthworm activity. By acknowledging and prioritizing regions with high earthworm populations as premium agricultural land, farming groups in Africa displayed advanced ecological awareness and passed this information down through the generations. These methods demonstrate a profound comprehension of soil ecology before contemporary scientific verification.

iii. **Contemporary Evolution (1970–Present)**

Notable advancements in science and technology have marked the current period of vermicomposting:

1970s–1980s

Mary Appelhof's development of continuous flow systems, which transformed large-scale worm composting operations, is mainly responsible for the current commercialization of vermicomposting (Singh & Singh, 2017). Her inventions caused a shift from conventional small-scale methods to industrial-scale organic waste processing by opening the first commercial vermicomposting operations. Scientists found that certain earthworm species, particularly Eisenia fetida and Eisenia andrei, were excellent candidates for waste processing through thorough investigation and testing because of their quick reproduction, high waste consumption rates, and versatility about different organic compounds.

1990s

A significant environmental development is the rise of vermicomposting as a common waste management technique. Because vermicomposting can lessen the load on landfills and provide valuable organic fertilizer, municipalities worldwide are increasingly including it in their waste management plans. As a result of this expansion, advanced industrial-scale processing technologies that can handle tons of organic waste every day were developed. In tandem with this growth, industry stakeholders and regulatory agencies created uniform vermicompost quality standards, guaranteeing consistent product quality and boosting consumer trust in commercial vermicompost goods.

2000s–2010s

Significant technological advancements in vermicomposting have been made in recent years, as seen by installing advanced automated monitoring systems that continuously measure vital factors, including pH, temperature, and moisture. Higher processing capacity in smaller footprints is now possible because of the development of vertical vermicomposting systems, which is especially advantageous in urban environments (Sharma & Garg, 2018). Additionally, creative fusion with anaerobic digestion technology has produced hybrid systems that optimize waste processing effectiveness, fusing the advantages of both techniques to produce high-quality vermicompost and biogas.

2010s–Present

With the advent of IoT-based monitoring and control systems, vermicomposting has entered a new era of digital transformation that enables remote management of composting operations and real-time data collection. Specialized equipment and materials, including automated harvesting systems and designed bedding materials, have been developed with this technological advancement. Vermicomposting has been incorporated into circular economy models and urban farming projects to create sustainable loops where organic waste is converted into valuable resources, promoting community involvement in sustainable practices and supporting local food production and waste reduction objectives (Smith, 2017). This historical progression shows how vermicomposting has changed from a traditional agricultural method to a contemporary biotechnology, underscoring the growing complexity of knowledge and application. New applications and technologies are being created in response to current environmental issues, and the sector is still evolving.

B. **Types of Waste Suitable for Vermicomposting**

A key component of effective vermicomposting operations is the selection of suitable waste materials; different kinds of organic waste work well for this process. Crop residues, straws, harvest leftovers, and agricultural processing byproducts are all considered forms of agricultural waste. When properly pre-composted and combined with nitrogen-rich materials to reach ideal carbon-to-nitrogen ratios, these materials rich in carbon and other nutrients make an excellent basis for vermicomposting. Household and kitchen organic waste is another crucial vermicomposting category (Garg et al., 2006). Paper goods, coffee grounds, tea bags, eggshells, and leftover fruit and vegetables are all included in this. These materials are beneficial because they are usually abundant in metropolitan areas and contain a lot of nitrogen. However, since they might cause odor issues and draw bugs, avoiding meat, dairy, and fatty foods is imperative. When suitably moistened and combined with nitrogen-rich materials, paper waste, including shredded cardboard, newspaper, and office paper, can be digested through vermicomposting. Fruit pulp, vegetable processing waste, brewery waste, and sugar industry leftovers are examples of industrial organic waste from food processing businesses that make excellent vermicomposting feed (Ali et al., 2015). These materials frequently include significant moisture and nutrients, making them suited for worm-eating. However, appropriate pre-treatment and mixing could be required to control moisture content and avoid too much acidity. After proper pre-composting to prevent heat damage to the worms, animal manure makes an excellent substrate for vermicomposting, primarily from cattle, horses, and rabbits. Because of its high nitrogen concentration, poultry manure must be aged correctly before usage. Vermicomposting efficiently manages garden and landscape waste, such as plant trimmings, leaves, and grass clippings. When blended correctly, these materials offer a healthy balance of carbon and nitrogen and are easily accessible in most environments. After appropriate segregation and pre-treatment, municipal solid waste, especially the organic fraction, can also be vermicomposted. This covers food waste from establishments and restaurants and garbage from the garden and market. To guarantee appropriate waste segregation and eliminate pollutants, thorough screening is required (Maharjan et al., 2022).

Vermicomposting biosolids and sewage sludge from wastewater treatment facilities is possible under controlled circumstances, but it necessitates close supervision and adherence to regional laws. Although these materials are high in nutrients, they may also contain diseases or heavy metals that require careful handling (Ghosh & Das, 2018). Vermicomposting can also be used to handle specialized industrial organic wastes, such as distillery effluents, paper mill sludge, and textile industry waste, provided that they are appropriately treated and combined with other appropriate materials. Proper trash selection, preparation, and mixing are essential for vermicomposting success because they provide the best circumstances for worm activity and reproduction while avoiding elements that could injure the worms or cause issues with operation (Singh et al., 2011). This thorough analysis shows the variety of organic wastes that can be vermicomposted, emphasizing every waste type's particular needs and factors to guarantee the best possible processing and high-quality final products.

3 Methodology

A. Analysis of Existing Vermicomposting Systems

i. System Design Comparison

Vermicomposting systems have substantially developed over time, with each design providing unique benefits for various operational scales and situations. Continuous flow systems, which are distinguished by their creative utilization of vertical space and earthworm migration patterns, are among the most effective vermicomposting techniques (Enebe & Erasmus, 2023). By using earthworms' inclination to gravitate higher when fed new material, these devices produce a self-sorting mechanism that makes harvesting much more manageable. Because space is at a premium in urban areas, the vertical design makes the most use of available space. The capacity of these systems to operate continuously adds to their effectiveness by enabling routine harvesting without upsetting the earthworm population or stopping the composting process. Systems for stacking bins provide exceptional versatility and flexibility for vermicomposting operations (Fornes et al., 2012). Based on waste volume and output demands, operators can readily expand or reduce their operations thanks to its modular architecture. Multiple processing levels are created while keeping a comparatively compact footprint because of these systems' stackable design. These systems are straightforward to maintain because individual bins may be isolated for repairs, monitoring, or troubleshooting without impacting the operation as a whole. The ability to scale is awe-inspiring because more units can be easily added to current configurations, which makes these systems perfect for businesses that expect to grow or experience seasonal fluctuations in waste production. Windrow systems are the foundation of large-scale vermicomposting operations and are especially well-suited for managing waste in municipalities and agriculture. These systems have unmatched capability for digesting massive volumes of organic waste, although they require significant land. From garbage application to vermicompost harvesting, the long, parallel rows make it easier to automate various tasks, significantly lowering the need for labor (Zhou et al., 2022). The open design makes machinery access simple, facilitating effective turning, harvesting, and monitoring activities. Although large-scale applications have economies of scale, environmental variables like moisture control and temperature regulation must be carefully considered throughout the large surface area. Through innovative design advancements that integrate contemporary technology and automation, the vermicomposting industry is progressing. Using conveyor belts or automated dispensers to distribute organic waste across the composting beds evenly ensures ideal feeding patterns and lowers labor costs. Automated feeding systems are a significant advancement. Intelligent monitoring technologies have wholly transformed process control by using sensors to continuously monitor vital factors like temperature, moisture content, and pH levels. When parameters depart from ideal ranges, these systems can notify operators or automatically modify circumstances. In addition to increasing efficiency, these developments make vermicomposting more approachable for operators with little prior expertise. The benefits of combining many configurations in hybrid systems result from integrating these diverse design approaches (Li, 2011). These innovative vermicomposting system designs often integrate modular components for scalability, automated monitoring for precise process control, and vertical configuration to maximize space efficiency. Continued advancement in vermicomposting system design is essential to expand the adoption of this sustainable waste management approach across differnt settings from urban households to large commercial facilities (Fig. 1).

ii. **Operational Parameters**

Maintaining the ideal moisture content in the system is essential to the effectiveness of vermicomposting operations. The optimal moisture range of 60–80% gives earthworms the conditions to digest organic matter effectively while maintaining sufficient oxygen availability (Kuznetsov et al., 2019). Techniques for measuring moisture have advanced from straightforward squeeze tests to complex moisture probes that offer data in real-time. Operators use management techniques to maintain ideal moisture levels, including manual misting for smaller units and overhead

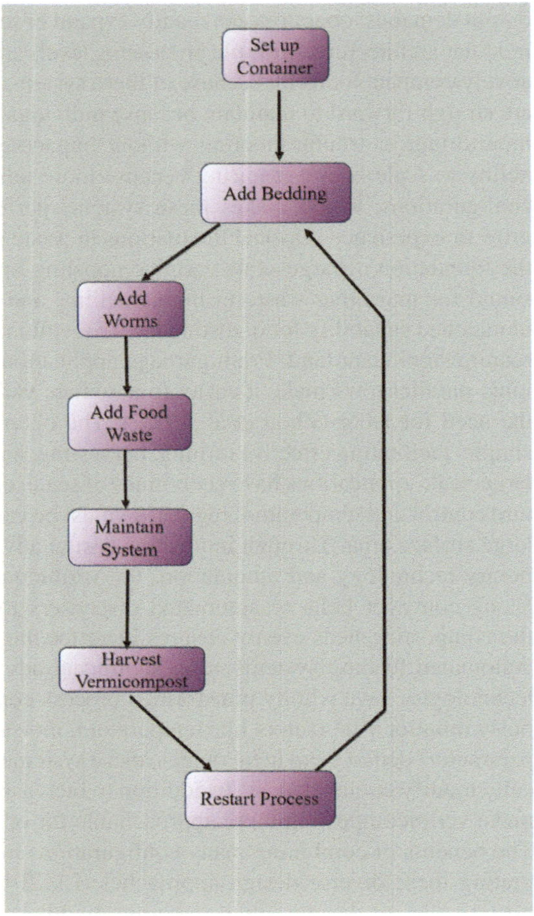

Fig. 1 System design for vermicomposting

sprinkler systems for more significant operations. While moisture-retaining materials like aged compost or coco coir assist in stabilizing humidity levels, waterlogging can be avoided by placing moisture barriers and appropriate drainage systems. Gradual rehydration methods are crucial during extreme drought to keep the earthworm population from experiencing shock. One of the main challenges in vermicomposting operations is controlling temperature, which calls for close consideration of microbial activity and seasonal fluctuations (Wu et al., 2023). Earthworms thrive in temperatures between 15 and 25 °C (59 and 77 °F). Hence, year-round active management techniques are required. Insulation methods like insulated coverings, straw bales, or greenhouse structures help prevent freezing temperatures throughout the winter. Summer management aims to avoid overheating by using ventilation systems and shade structures and carefully planning when to feed. Significant differences exist in the efficiency of various insulation materials; some contemporary synthetic materials provide better temperature stability than conventional alternatives (Srivastava et al., 2020). Zone-based approaches are frequently used in temperature management tactics, whereby various parts of significant operations maintain varying temperature profiles to maximize processing efficiency. Finding and controlling the proper feeding rates significantly impacts vermicomposting success. It is necessary to precisely calibrate organic waste input rates according to environmental factors, processing capability, and earthworm population density. When various organic waste products are carefully mixed, the ideal carbon-to-nitrogen ratio, which usually falls between 25:1 and 30:1, is reached. Batch systems may need weekly or bi-weekly additions, while continuous flow systems can typically handle daily feeding. Feeding frequency varies depending on the operational size and system architecture. Operators must keep a close eye on waste consumption rates and modify feeding schedules to avoid underfeeding, which could lower earthworm populations and processing efficiency, and overfeeding, which could result in anaerobic conditions (Ghasemi et al., 2020). Aeration is essential to sustain the aerobic conditions necessary for earthworm life and effective organic matter decomposition. Several ventilation techniques have been devised to guarantee a sufficient oxygen supply throughout the vermicomposting material. While active ventilation systems may use mechanical turning mechanisms or perforated pipes to drive air circulation, passive ventilation systems let natural airflow through carefully positioned perforations or mesh surfaces. Significant differences exist in the effects of various ventilation strategies on decomposition rates; generally, well-aerated systems produce more consistent product quality and faster processing times. Operators can optimize processing efficiency by adjusting ventilation methods with routine oxygen level monitoring, especially in deeper layers of the composting material (Maharjan et al., 2022). Specific sophisticated systems have automated ventilation controls that modify airflow in response to oxygen and temperature readings, guaranteeing ideal conditions with the least operator involvement.

iii. **Earthworm Species Analysis**

The many earthworm species employed in vermicomposting exhibit unique traits and capacities. *Red wigglers*, or *Eisenia fetida*, are notable for their remarkable reproduction rates. Under ideal circumstances, they can produce two to three cocoons each week and digest organic waste equivalent to their body weight per day (Sharma, 2003). These worms thrive in temperatures between 15 and 25 °C, demonstrating an exceptional ability to withstand environmental changes. While *Eisenia andrei* share many traits, under some circumstances, they frequently grow more quickly. Although they may be well suited to local temperatures, native earthworm species usually exhibit inferior processing efficiency when contrasted with specialized composting worms (Table 1). Combining several species in one system through hybrid techniques has produced encouraging results, especially in large-scale operations where various species might occupy diverse ecological niches, potentially boosting the resilience and efficiency of the system as a whole (Garg et al., 2006).

Table 1 Comparison of different earthworm species used in vermicomposting

Species	Optimal temperature (°C)	Processing capacity (kg/worm/day)	Reproduction rate (cocoons/worm/year)	Preferred waste type
Eisenia fetida	20–25	0.3–0.5	300–400	Animal manure, Kitchen waste
Lumbricus rubellus	15–20	0.2–0.4	250–300	Garden waste, Leaf litter
Eudrilus eugeniae	25–30	0.4–0.6	400–500	Agricultural waste
Perionyx excavatus	23–28	0.3–0.5	350–450	Kitchen waste, Garden waste

iv. Maintenance Requirements

Managing vermicomposting effectively necessitates a planned schedule of operating duties. Daily operations include critical monitoring processes such as leachate collection evaluation, feeding schedule adherence, and temperature and moisture checks. Every week, the bedding materials are turned, any processed vermicompost is removed, and any indications of disease or pest presence are examined (Chanu et al., 2018). The monthly assessments include comprehensive system health evaluations, earthworm population surveys, and vermicompost quality metrics analysis. Emergency procedures are designed to handle unforeseen problems, such as pest invasions, which necessitate prompt quarantine and treatment measures, or severe weather events, which call for the quick installation of protective measures like extra insulation or emergency irrigation systems to preserve the best possible conditions for earthworm survival (Enebe & Erasmus, 2023).

B. Case Studies from Different Geographical Regions

i. Tropical Regions

Environmental issues and how to handle them must be carefully considered for outdoor vermicomposting operations to be successful. Strategic placement of shade and cooling measures, such as installing shade cloths and sprinkler systems, helps regulate the temperature and preserve ideal conditions throughout hot weather. Rain management integrates carefully planned drainage networks with graded surfaces and collection systems to avoid waterlogging while maintaining the appropriate moisture levels. Organic pest management techniques, such as neem oil sprays and physical barriers, combat local pest problems, such as ants and rodents (Ghosh et al., 2018). To ensure consistent, high-quality vermicompost production, success is measured by regular testing for nutrient content, particle size uniformity, and maturity levels. Key performance indicators, such as organic waste processing speeds and final product quality metrics, are also regularly monitored.

ii. **Temperate Zones**

Seasonal dynamics and community involvement must be carefully considered for vermicomposting systems to be implemented successfully. To preserve ideal processing conditions throughout the cold months, winter protection techniques include temperature-controlled surroundings, strategically placed bedding depths, and insulated covers. While outdoor systems have more capacity but need more environmental care, indoor systems enable reliable year-round operations but have space constraints and ventilation issues. The availability of resources varies seasonally; food waste is comparatively stable throughout the year, whereas garden trash is plentiful in the autumn (Bhadauria et al., 2014). Community integration is essential, and its success depends on local support and involvement. This is accomplished through educational initiatives and demonstration projects highlighting the advantages of vermicomposting and its function in environmentally friendly waste management.

iii. **Arid Areas**

Effective irrigation techniques, such as drip systems, moisture sensors, and scheduled watering, are necessary for agricultural water conservation to reduce wastage. Developing environmentally friendly shelter designs that maximize insulation and natural ventilation is the primary goal of environmental protection. To preserve productivity in severe weather, modified farming techniques for extreme conditions entail modifying conventional farming practices through vertical farming, greenhouse agriculture, and climate-controlled habitats. Initial expenditures for specialized equipment, improved infrastructure, and continuing system maintenance are expensive. These adjustments frequently save money in the long run by reducing water usage, increasing crop yields, and strengthening resilience to environmental problems, even with a high initial cost (Kumari et al., 2022).

C. **Economic Viability Assessment**

i. **Initial Investment Analysis**

Infrastructure costs include purchasing necessary storage units and processing facilities and essential construction expenditures such as building materials, structural elements, and facility installation. The acquisition of crucial instruments such as temperature control technology, irrigation systems, monitoring devices, and automated machinery for effective operations is the main emphasis of equipment requirements. To guarantee ideal farming conditions, site preparation entails major land alteration activities like leveling, installing drainage systems, improving soil quality, and building access roads (Bhakuni et al., 2025). To maintain high operating standards and guarantee the correct application of contemporary agricultural methods, training costs cover thorough personnel preparation through specialized programs, safety certificates, equipment operation training, and continuous professional development.

ii. **Operational Cost Breakdown**

The labor needed for agricultural operations varies greatly depending on their size; small farms usually require 20–30 person-hours each week, while more significant enterprises may need more than 100 h for harvesting and general administration tasks. Utility expenditures include power for pumping, climate control, machine operation, and considerable water usage, which averages 2000–3000 gallons per acre monthly. Regular equipment maintenance, urgent repairs, and the recurring replacement of machinery and irrigation components are all included in maintenance expenses (Satish et al., 2024). The cost of input materials varies according to market conditions and supplier distance, and they represent continuous expenditures for seed procurement, fertilizers, pest control products, and transportation logistics.

iii. **Revenue Stream Analysis**

The vermicompost market's primary product sales show steady demand trends, shifting prices according to seasonal availability, quality, and regional considerations. Secondary goods diversify revenue streams by producing worm tea, offering breeding stock options, and providing specialized consulting services for new businesses. Home gardeners, organic agricultural enterprises, landscaping companies, and commercial greenhouse facilities are the probable consumer groups identified by the market development study. The logistics of transportation across different distances, the best packaging options to preserve product integrity, and storage needs all along the supply chain are all considered when choosing a distribution route (Das et al., 2025). The market has significant growth potential in sectors that prioritize sustainable agriculture methods and are environmentally conscientious.

iv. **Financial Metrics**

Scale, infrastructure quality, and market positioning affect the return on investment in vermicomposting businesses; careful planning is required to identify the best payback times. More extensive facilities enjoy economies of scale, whereas smaller enterprises can command premium rates through specialized market sectors. These disparities in profit margins are seen throughout operating scales. By taking predicted income sources, fixed expenditures, and operating expenses into account, break-even analysis assists in determining the least feasible operation size (Nisar et al., 2025). In addition to identifying potential obstacles like production variability, market price swings, and competition, risk assessment offers solutions like long-term customer contracts, diversified product lines, and emergency reserve maintenance.

D. **Environmental Impact Evaluation**

i. **Waste Reduction Metrics**

Vermicomposting is a waste processing method that effectively converts organic wastes into helpful soil nutrients while demonstrating notable volume reduction capabilities. Through the systematic processing and conversion of organic waste into usable products, diversion rates demonstrate a significant reduction in landfill waste. Environmental factors, worm population density, and waste makeup affect processing speed; effective systems use regulated temperature and moisture levels to provide the best possible decomposition (ul Baligah et al., 2025). To ensure consistent, high-quality output that satisfies horticultural and agricultural standards while preserving market competitiveness, quality assessment of the finished product looks at essential features such as nutrient content, microbial activity, particle size distribution, and overall stability.

ii. **Emissions Analysis**

According to carbon footprint studies, vermicomposting has significant environmental benefits over conventional waste management techniques, mainly because it requires less mechanical processing and allows for spontaneous breakdown. Redirecting organic waste from landfills, where anaerobic decomposition usually results in hazardous greenhouse gas emissions, has the potential to reduce methane emissions. Because vermicomposting facilities can function efficiently at local scales, reducing fuel consumption and related emissions and resulting in energy savings from fewer transportation needs; because vermicomposting systems are enclosed and provide natural filtering, air quality monitoring reveals better conditions than traditional waste handling, with less odor creation and particle matter discharge (Suhani et al., 2025).

iii. **Water Impact Studies**

Through intentional moisture management and optimized consumption patterns, vermicomposting's water usage efficiency exhibits impressive resource saving. Analysis of leachate composition shows little effect on the environment, and when collected and handled appropriately, the liquid residue frequently turns into a valuable fertilizer. Innovative techniques for collecting and reusing moisture, such as condensation collection and filtered runoff recirculation, are incorporated into water recycling systems. According to groundwater impact assessments, there is little chance of contamination provided appropriate containment mechanisms are implemented and routine monitoring guarantees that water quality criteria are maintained (Hajam et al., 2023). Organically filtering water via several organic layers supports sustainable water management techniques while lowering possible environmental effects.

iv. **Soil Enhancement Studies**

Several important factors are the focus of the thorough soil health assessment. To determine soil fertility, nutrient content analysis looks at vital micronutrients and NPK (nitrogen, phosphorus, and potassium) concentrations. Improvements in physical characteristics, such as improved aggregation, porosity, and water retention capacity, are revealed by soil structure examination (Fig. 2). Increased biological processes are shown by microbial activity assessment, with larger populations of beneficial species aiding in the breakdown of organic matter and the cycling of nutrients. With recorded increases in soil organic carbon, higher nutrient availability, and increased structural stability, long-term monitoring of these variables shows consistent soil improvement over time, improving overall soil health and agricultural output (Jjagwe et al., 2019).

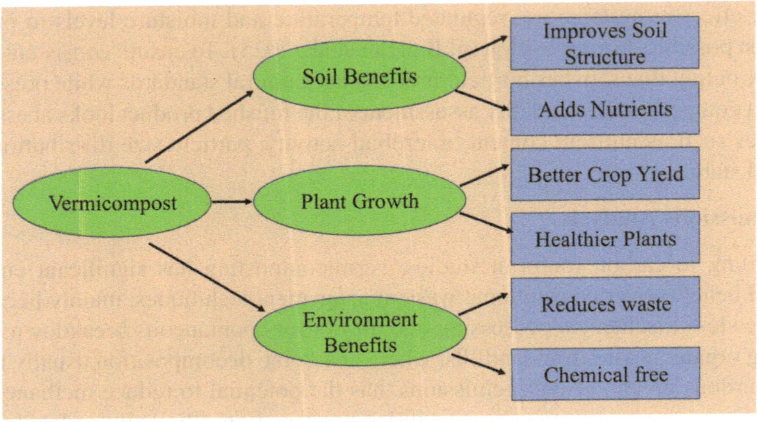

Fig. 2 Environmental impact evaluation of vermicomposting

v. **Biodiversity Assessment**

The ecological evaluation shows significant advancements in plant interactions and soil biodiversity. Studies of earthworm populations reveal sustained colony development and successful breeding patterns, which point to better soil conditions. According to the documentation of beneficial organisms, the diversity of microorganisms, including helpful bacteria, fungi, and other soil-dwelling species, has significantly increased. Pollinators, beneficial insects, and native plant species are positively impacted, and the effects on the local ecosystem go beyond the surrounding soil environment (Panda et al., 2022). Plant growth analysis in various crop types reveals improved vigor, nutrient uptake, and root development, leading to higher yields and higher-quality crops in different agricultural settings.

4 Benefits and Applications of Vermicompost

A. Advantages and Agricultural Benefits

Vermicomposting has many significant agricultural advantages that revolutionize farming methods and increase crop yields. By producing a nutrient-rich organic fertilizer, the procedure significantly increases soil fertility and structure, improving agricultural results for a variety of crop kinds. Earthworms transform organic matter into high-quality vermicompost, which includes key trace elements like zinc, copper, and manganese, and a balanced amount of macro- and micronutrients like potassium, phosphorus, and nitrogen necessary for plant growth (Dey Chowdhury et al., 2023). Given that vermicompost improves soil porosity, aeration, and water retention capacity, the increase in soil structure is significant. The stable aggregates formed by the worm castings provide the perfect conditions for plant growth and root development. Agricultural systems can use water more efficiently because of improved drainage, decreased erosion, and increased soil moisture retention. Additionally, enhanced soil structure encourages increased root penetration and growth, making it possible for plants to obtain nutrients throughout their life cycle more efficiently (Ducasse et al., 2022). Vermicomposting has more biological advantages than just supplying nutrients. By introducing advantageous microbes into the soil, the procedure produces a thriving soil ecosystem that promotes the health of plants (Fig. 3). These microbes create plant growth hormones that encourage the development of crops, improve nutrient cycling, and inhibit plant diseases. These beneficial organisms also enhance the soil's ability to suppress disease, which lessens the need for chemical fungicides and pesticides. Applying vermicompost regularly yields outstanding effects in terms of crop output. Crop yields usually rise by 20–40%, and farmers also see gains in their produce's size, nutritional value, and quality. Because vermicompost releases nutrients gradually, it provides a consistent supply of nutrients throughout the growing season, minimizing nutrient loss and the need for frequent fertilizer treatments (Singh et al., 2011). For many crops, this prolonged nutrient availability prolongs the harvest season and encourages more consistent crop growth. Vermicompost has significant financial benefits for agriculture. Farmers may increase their profit margins by drastically reducing their reliance on chemical fertilizers and saving input expenses. Because of the improved soil health and water retention qualities, less irrigation is needed, which lowers operating costs even more. Furthermore, when farmers can obtain organic certification, the enhanced crop quality frequently fetches higher prices in the marketplace. Vermicompost is an essential tool for sustainable farming methods because of its long-term soil improvement benefits, which provide a stable basis for ongoing agricultural productivity (Yasmin et al., 2022).

B. **Environmental and Economic Benefits**

Vermicomposting solves several important ecological issues while providing significant environmental advantages. Fundamentally, it offers an effective way to manage organic waste, thereby lessening the load on landfills and lowering dangerous methane emissions. The procedure successfully diverts tons of waste from conventional disposal methods, which organically transforms organic waste into beneficial soil supplements. Organic materials from homes, businesses, and agriculture are also reduced in garbage, resulting in a sustainable cycle of resource usage. Additionally, vermicomposting is essential for water conservation as it enhances soil structure and water retention capacity. Vermicompost-treated soils can hold onto 30–40% more water, drastically lowering the need for irrigation and lessening issues with erosion and water runoff (Singh et al., 2020). Vermicomposting has the potential to reduce climate change significantly. The procedure actively stores carbon in the soil by creating stable organic matter while lowering greenhouse gas emissions compared to traditional composting techniques. Vermicomposting reduces the

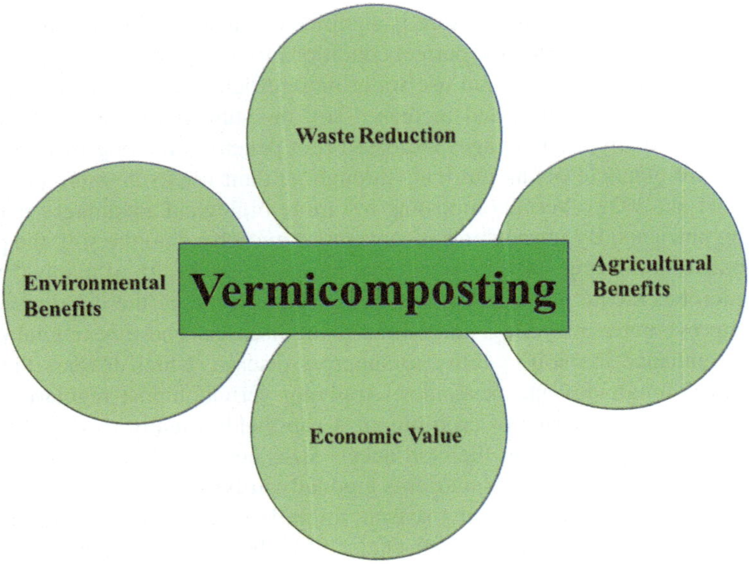

Fig. 3 Benefits and applications of vermicomposting

carbon footprint of agricultural activities by reducing the need for chemical fertilizers. Furthermore, it fosters biodiversity by establishing rich soil ecosystems that support beneficial insect populations and microbial variety, aiding in improved plant development and natural pest control. Vermicomposting operations' consulting and training components provide further revenue potential, resulting in a multifaceted business model. The cost-cutting advantages are just as striking. Vermicomposting systems usually result in a 50–100% decrease in chemical fertilizer costs and a 20–30% reduction in irrigation costs for agricultural businesses. These savings translate into lower pest control costs and fewer soil amendment requirements. Along with longer harvest seasons and better product quality, crop yields have increased by an average of 20–40%, indicating significant increases in agricultural production. Farmers can frequently command higher prices due to increased productivity, especially when obtaining organic certification (Monaghan, 2024). Vermicomposting's business features are incredibly alluring because it requires a relatively minimal initial investment. Due to its rapid return on investment period, typically 6–12 months, it is a viable business option. Multiple revenue streams and operational flexibility result in adaptable company structures that meet the market's needs. Additionally, vermicomposting supports local economic development by creating many job possibilities, ranging from direct operational occupations to professional positions in training and consultancy. Infrastructure savings for waste management agencies and municipalities are part of the broader economic benefit. Vermicomposting minimizes public infrastructure expenses by lowering the need for waste transportation and landfill growth. Opportunities in the vermicomposting industry are being further expanded by the expanding market for organic products and rising environmental consciousness, while new waste management laws are generating new demand for these services (Al-Rumaihi et al., 2020). Vermicomposting is becoming a more alluring option for sustainable waste management and agricultural development because of its many advantages for the environment and the economy.

5 Challenges and Solutions

Several important obstacles to vermicomposting deployment call for considerable thought and strategic planning. Keeping the ideal climatic conditions for earthworm survival and reproduction is one of the fundamental concerns. Because worms need particular temperature ranges (15–25 °C) to function at their best, temperature control is particularly challenging. Worm populations can be significantly impacted by extreme weather, which calls for new infrastructure for temperature control, raising operating costs and complexity. This problem is exacerbated in areas with severe winters or hot summers, necessitating the use of advanced environmental management technologies (Belliturk, 2018). Another significant obstacle in vermicomposting systems is managing moisture. It takes ongoing observation and modification to maintain the optimal moisture content (60–80%), especially in outdoor systems subjected to rain or dry weather. While too little moisture might stop decomposition and compel worms to move or die, too much moisture can cause anaerobic conditions and worm death. Regular maintenance is necessary to maintain this delicate equilibrium, frequently requiring purchasing irrigation systems and moisture monitoring equipment. There are unique difficulties associated with the feeding schedule, especially in large-scale operations. Careful consideration is needed to manage feed particle size, maintain appropriate carbon-to-nitrogen ratios, and achieve the ideal mix of organic components. While underfeeding restricts worm growth and production ability, overfeeding can result in material accumulation and perhaps hazardous situations. Furthermore, it might be challenging to source organic waste materials of regular quality and quantity throughout the year, particularly when garbage creation varies seasonally (Vukovic et al., 2021). Another major issue is labor intensity because vermicomposting must be monitored regularly. Because most tasks, including feeding, controlling moisture, controlling temperature, and harvesting, are done by hand, labor costs account for a sizable amount of operating costs. Workforce management is made more difficult by the requirement for knowledgeable personnel knowledgeable about worm biology and composting principles, especially in areas lacking vermicomposting experience. Issues with marketing and distribution can also impact the performance of vermicomposting projects. Even if more people know the advantages of organic farming, vermicompost products still need much work to establish their market (Shaikh & Shaikh, 2025). Market penetration may be challenging due to price competition with alternative soil amendments and conventional fertilizers, particularly in areas where chemical fertilizers are subsidized. Furthermore, the large size of vermicompost products makes storage and transportation difficult, impacting the logistics and distribution expenses. Managing pests, controlling diseases, and preserving ideal worm population densities are examples of technical difficulties. If left unchecked, common pests like ants, mites, and centipedes can majorly affect worm populations. Even though they are uncommon, disease outbreaks have the power to eradicate worm populations. While preserving the ideal population density is essential for effective processing, it also necessitates close observation and management. Harvesting may be labor-intensive and time-consuming without the right machinery, especially when removing worms from the completed compost. Infrastructure needs pose logistical and budgetary difficulties, particularly for large-scale activities. Significant investment is needed for processing areas, raw material and final product storage facilities, proper drainage systems, and weather protection. Zoning laws and space restrictions may restrict the size and location of vermicomposting businesses in urban regions, while transportation expenses and infrastructure accessibility may be issues in rural places. Process optimization is still a major technological issue, especially when preserving ideal conditions for earthworm activity (Nghituwamhata et al., 2023). To guarantee optimal performance in the vermicomposting process, accurate management of temperature regulation, moisture control, and feed material preparation is required. Quality control poses another major difficulty since the finished vermicompost product must adhere to stringent agricultural regulations for nutrient content, pathogen levels, and heavy metal concentrations (Ghosh et al., 2016). When moving from small-scale operations to commercial facilities, scale-up problems usually arise because it takes complex infrastructure and monitoring systems to maintain consistent conditions across higher volumes of organic waste. Economic issues have a big impact on vermicomposting technology uptake. Establishing a commercial-scale vermicomposting operation can involve a significant initial expenditure, which includes purchasing earthworms, building infrastructure, and purchasing specialized processing and monitoring equipment. Since success hinges on educating prospective buyers about the greater advantages of vermicompost over conventional composting solutions, market development needs careful consideration. Competition from chemical fertilizers is a constant problem since these traditional goods frequently maintain pricing advantages despite their negative environmental effects thanks to well-established production processes and government subsidies. Issues with implementation also arise in the social and policy spheres. For vermicomposting operations to be effective, personnel must get thorough training that covers everything from worm biology to process management and quality assessment (Oyege & Balaji Bhaskar, 2023). Regional cultural acceptance varies greatly, with some groups expressing opposition to employing worm-processed materials in food production or cooperating with worm-based systems. Since the success of vermicomposting projects frequently relies on supportive regulatory frameworks that acknowledge and incentivize organic waste processing processes, policy support is still essential.

6 Future Perspectives

Vermicomposting has a bright future thanks to growing markets and technical advancements. Automation and monitoring systems are becoming more complex, and technological advancements are happening quickly. These innovations include automated feeding systems, sensor-based moisture control, and real-time temperature and pH monitoring. Through rigorous selection and breeding operations, new earthworm species and hybrid systems are being created, offering enhanced waste processing effectiveness and climate flexibility (Zhou et al., 2022). Beyond standard vermicompost, product diversity is still growing and now includes liquid extracts, specialized formulations for various crops, and value-added products like worm biomass for animal feed. Opportunities in the market are increasing dramatically, especially in the organic farming industry, where there is a growing need for fertilizers devoid of chemicals. As communities look for sustainable ways to manage organic waste while creating high-quality soil amendments for local food production, urban agriculture offers a growing potential. As global markets acknowledge the exceptional quality of vermicompost products, especially in areas with robust organic farming sectors, export potential is growing. The main goal of policy proposals is to establish a supportive environment for the growth of vermicomposting. Financial incentives for setting up vermicomposting facilities, technical assistance programs, and research money for innovation are all examples of government support mechanisms (Meghvansi et al., 2015). The regulatory framework must be developed to ensure environmental protection and set quality requirements for vermicompost products. Small-scale adoption by municipalities and small-scale implementation by farmers and individuals should be encouraged via incentive programs.

7 Conclusion

A significant opportunity in sustainable resource management is converting waste into wealth through vermicomposting, which answers problems related to trash management and soil fertility. In addition to highlighting the difficulty of expanding this technology, the issues found in the technical, financial, and implementation domains also indicate certain areas that need targeted attention and resource allocation. There are significant ramifications for sustainable resource management, including less reliance on chemical fertilizers, better soil health, and lower landfill usage. Optimizing vermicomposting procedures for various waste streams, creating more effective harvesting and processing technologies, and looking into novel uses for vermicompost products should be the main goals of future research and development. It will take ongoing cooperation between researchers, practitioners, policymakers, and communities to successfully increase vermicomposting adoption. Vermicomposting is becoming increasingly important in sustainable resource management to create a more circular and regenerative economy as the world's waste management problems worsen and worries about soil degradation increase.

References

Ali, U., Sajid, N., Khalid, A., Riaz, L., Rabbani, M. M., Syed, J. H., & Malik, R. N. (2015). A review on vermicomposting of organic wastes. *Environmental Progress & Sustainable Energy, 34*(4), 1050–1062. https://doi.org/10.1002/ep.12100

Al-Rumaihi, A., McKay, G., Mackey, H. R., & Al-Ansari, T. (2020). Environmental impact assessment of food waste management using two composting techniques. *Sustainability, 12*(4), 1595. https://doi.org/10.3390/su12041595

Behera, S., Jyotirmayee, B., Mandal, U., Mishra, A., Mohanty, P., & Mahalik, G. (2022). Effect of organic fertilizer on growth, yield and quality of *Pisum sativum* L.: A review. *Ecology, Environment and Conservation, 28*, S233–S241.

Bellitürk, K. (2018). Vermicomposting in Turkey: Challenges and opportunities in future. *Eurasian Journal of Forest Science, 6*(4), 32–41. https://doi.org/10.31195/ejejfs.476504

Bhadauria, T., Kumar, P., Maikhuri, R., & Saxena, K. G. (2014). Effect of application of vermicompost and conventional compost derived from different residues on pea crop production and soil faunal diversity in agricultural system in Garhwal Himalayas India. *Natural Science, 6*(6), 433–446. https://doi.org/10.4236/ns.2014.66042

Bhakuni, M. S., Das, U., Bhakuni, P., Ramesh Kumar, S., & Das, A. (2025). An integrated MCDM-based mathematical modeling approach for sustainable practices of vermicomposting in educational institutions. *Process Integration and Optimization for Sustainability, 1–18*. https://doi.org/10.1007/s41660-025-00487-x

Chanu, L. J., Hazarika, S., Choudhury, B. U., Ramesh, T., Balusamy, A., Moirangthem, P., … & Sinha, P. K. (2018). A guide to vermicomposting-production process and socio economic aspects. *Extension Bulletin, 81*, 30.

Das, K., Dansena, S. K., Pottipati, S., Maturi, K. C., Dhal, G. C., & Kalamdhad, A. S. (2025). Sustainable remediation of abandoned coal mines using vermicompost: A case study in Ledo coal mine, India. *Environmental Geochemistry and Health, 47*(2), 51. https://doi.org/10.1007/s10653-024-02355-y

Dey Chowdhury, S., Suhaib, K. H., Bhunia, P., & Surampalli, R. Y. (2023). A critical review on the vermicomposting of organic wastes as a strategy in circular bioeconomy: Mechanism, performance, and future perspectives. *Environmental Technology, 1–38*. https://doi.org/10.1080/09593330.2023.2215458

Ducasse, V., Capowiez, Y., & Peigné, J. (2022). Vermicomposting of municipal solid waste as a possible lever for the development of sustainable agriculture. A review. *Agronomy for Sustainable Development, 42*(5), 89. https://doi.org/10.1007/s13593-022-00819-y

Enebe, M. C., & Erasmus, M. (2023). Vermicomposting technology—A perspective on vermicompost production technologies, limitations and prospects. *Journal of Environmental Management, 345*, 118585. https://doi.org/10.1016/j.jenvman.2023.118585

Fornes, F., Mendoza-Hernández, D., García-de-la-Fuente, R., Abad, M., & Belda, R. M. (2012). Composting versus vermicomposting: A comparative study of organic matter evolution through straight and combined processes. *Bioresource Technology, 118*, 296–305. https://doi.org/10.1016/j.biortech.2012.05.028

Garg, P., Gupta, A., & Satya, S. (2006). Vermicomposting of different types of waste using *Eisenia foetida*: A comparative study. *Bioresource Technology, 97*(3), 391–395. https://doi.org/10.1016/j.biortech.2005.03.009

Ghasemi, S., Mirzaie, M., Hasan-Zadeh, A., Ashrafnejad, M., Hashemian, S. J., & Shahnemati, S. R. (2020). Design, operation, performance evaluation and mathematical optimization of a vermifiltration pilot plan for domestic wastewater treatment. *Journal of Environmental Chemical Engineering, 8*(1), 103587. https://doi.org/10.1016/j.jece.2019.103587

Ghosh, A., Debnath, B., Ghosh, S. K., Das, B., & Sarkar, J. P. (2018). Sustainability analysis of organic fraction of municipal solid waste conversion techniques for efficient resource recovery in India through case studies. *Journal of Material Cycles and Waste Management, 20*, 1969–1985. https://doi.org/10.1007/s10163-018-0721-x

Ghosh, S., & Das, A. P. (2018). Metagenomic insights into the microbial diversity in manganese-contaminated mine tailings and their role in biogeochemical cycling of manganese. *Scientific Reports, 8*(1), 1–12. https://doi.org/10.1038/s41598-018-26311-w

Ghosh, S., Mohanty, S., Akcil, A., Sukla, L. B., & Das, A. P. (2016). A greener approach for resource recycling: Manganese bioleaching. *Chemosphere, 154*, 628–639. https://doi.org/10.1016/J.CHEMOSPHERE.2016.04.028

Goswami, J., Deka, M. K., Devi, R., Sarma, P. P., & Brahma, A. (2024). A review on optimizing organic waste management and income generation through vermicomposting and AI-powered vermicomposting: Insights from Guwahati, Assam. *International Journal, 13*(6). https://doi.org/10.30534/ijatcse/2024/011362024

Hajam, Y. A., Kumar, R., & Kumar, A. (2023). Environmental waste management strategies and vermi transformation for sustainable development. *Environmental Challenges, 13*, 100747. https://doi.org/10.1016/j.envc.2023.100747

Jjagwe, J., Komakech, A. J., Karungi, J., Amann, A., Wanyama, J., & Lederer, J. (2019). Assessment of a cattle manure vermicomposting system using material flow analysis: A case study from Uganda. *Sustainability, 11*(19), 5173. https://doi.org/10.3390/su11195173

Kumari, S., Manyapu, V., & Kumar, R. (2022). Recent advances in composting and vermicomposting techniques in the cold region: Resource recovery, challenges, and way forward. *Advanced Organic Waste Management, 131–154*. https://doi.org/10.1016/B978-0-323-85792-5.00005-8

Kuznetsov, E. V., Khadzhidi, A. E., Poltorak, Y. A., & Kuznetsova, M. (2019). Operator model to control process of obtaining vermicompost. *EurAsian Journal of BioSciences, 13*(1), 315–321.

Li, X. (2011). Rural organic waste treatment system design and analysis-based on vermicomposting technology.

Maharjan, K. K., Noppradit, P., & Techato, K. (2022). Suitability of vermicomposting for different varieties of organic waste: A systematic literature review (2012–2021). *Organic Agriculture, 12*(4), 581–602. https://doi.org/10.1007/s13165-022-00413-2

Meghvansi, M. K., Khan, M. H., Rajeev Gupta, R. G., Chaudhary, K. K., Sazada Siddiqui, S. S., & Vijay Veer, V. V. (2015). *Vermibiotechnology: relevance, challenges and future prospects for India.*

Mishra, S., & Das, A. P. (2021). Current treatment technologies for removal of microplastic and microfiber pollutants from wastewater. *Wastewater Treatment: Cutting-Edge Molecular Tools, Techniques and Applied Aspects, 237–251*. https://doi.org/10.1016/B978-0-12-821881-5.00011-8

Mishra, S., Ghosh, S., van Hullebusch, E. D., Singh, S., & Das, A. P. (2023). A critical review on the recovery of base and critical elements from electronic waste-contaminated streams using microbial biotechnology. *Applied Biochemistry and Biotechnology, 2023*, 1–30. https://doi.org/10.1007/S12010-023-04440-X

Mishra, S., Singh, R. P., Rout, P. K., & Das, A. P. (2022). Membrane bioreactor (MBR) as an advanced wastewater treatment technology for removal of synthetic microplastics. *Development in Wastewater Treatment Research and Processes, 45–60*. https://doi.org/10.1016/B978-0-323-85583-9.00022-3

Mohanty, S., Ghosh, S., Bal, B., & Das, A. P. (2018). A review of biotechnology processes applied for manganese recovery from wastes. *Reviews in Environmental Science and Bio/Technology, 17*(4), 791–811. https://doi.org/10.1007/S11157-018-9482-1

Monaghan, J. L. (2024). *Characterization of the environmental impact and economic analysis of a vermifiltration system used to treat liquid dairy cattle manure* (Master's thesis, Washington State University).

Nghituwamhata, S. N., Horn, L. N., & Ashipala, S. N. (2023). Vermicompost as a possible solution to soil fertility problems and enrichment in the semiarid zones of Namibia. In *Vermicomposting for sustainable food systems in Africa* (pp. 203–212). Springer Nature Singapore. https://doi.org/10.1007/978-981-19-8080-0_12

Nisar, A., Nazeer, S., Faridoon, S., Khan, A. M., Chaudhry, Z., Nazir, T.,... & Ali, H. (2025). Comparative assessment of vermicompost and chemical fertilizers on maize yield and nutritional quality. *Scholars Journal of Agriculture and Veterinary Sciences, 1,* 1–25. https://doi.org/10.36347/sjavs.2025.v12i01.001

Oyege, I., & Balaji Bhaskar, M. S. (2023). Effects of vermicompost on soil and plant health and promoting sustainable agriculture. *Soil Systems, 7*(4), 101. https://doi.org/10.3390/soilsystems7040101

Panda, A. K., Mishra, R., Dutta, J., Wani, Z. A., Pant, S., Siddiqui, S., … & Bisht, S. S. (2022). Impact of vermicomposting on greenhouse gas emission: A short review. *Sustainability, 14*(18), 11306. https://doi.org/10.3390/su141811306

Satish, P., Hussain, S. A., Ranka, T., & Shruti, K. Economic viability of vermicompost unit at college farm, PJTSAU Hyderabad. https://doi.org/10.33545/2618060X.2025.v8.i2a.2501

Shaikh, K. M., & Shaikh, R. U. (2025). Vermicompost: role in sustainable agriculture. *The Infinite Path: Research Across Disciplines, 125.*

Sharma, K., & Garg, V. K. (2018). Vermicomposting: A green technology for organic waste management. *Waste to Wealth, 199–235.* https://doi.org/10.1007/978-981-10-7431-8_10

Sharma, K., & Garg, V. K. (2019). Vermicomposting of waste: a zero-waste approach for waste management. In *Sustainable resource recovery and zero waste approaches* (pp. 133–164). Elsevier. https://doi.org/10.1016/B978-0-444-64200-4.00010-4

Sharma, S. (2003). Municipal solid waste management through vermicomposting employing exotic and local species of earthworms. *Bioresource Technology, 90*(2), 169–173. https://doi.org/10.1016/S0960-8524(03)00123-8

Singh, A., & Singh, G. S. (2017). Vermicomposting: A sustainable tool for environmental equilibria. *Environmental Quality Management, 27*(1), 23–40. https://doi.org/10.1002/tqem.21509

Singh, A., Karmegam, N., Singh, G. S., Bhadauria, T., Chang, S. W., Awasthi, M. K., … & Ravindran, B. (2020). Earthworms and vermicompost: An eco-friendly approach for repaying nature's debt. *Environmental Geochemistry and Health, 42*, 1617–1642. https://doi.org/10.1007/s10653-019-00510-4

Singh, R. P., Singh, P., Araujo, A. S., Ibrahim, M. H., & Sulaiman, O. (2011). Management of urban solid waste: Vermicomposting a sustainable option. *Resources, Conservation and Recycling, 55*(7), 719–729. https://doi.org/10.1016/j.resconrec.2011.02.005

Singh, S., & Sinha, R. K. (2022). Vermicomposting of organic wastes by earthworms: Making wealth from waste by converting 'garbage into gold' for farmers. In *Advanced organic waste management* (pp. 93–120). Elsevier. https://doi.org/10.1016/B978-0-323-85792-5.00004-6

Smith, A. (2017). *Creating wealth from waste: Resource use efficiency in climate-smart agriculture.* https://hdl.handle.net/10568/83007

Srivastava, V., Goel, G., Thakur, V. K., Singh, R. P., de Araujo, A. S. F., & Singh, P. (2020). Analysis and advanced characterization of municipal solid waste vermicompost maturity for a green environment. *Journal of Environmental Management, 255*, 109914. https://doi.org/10.1016/j.jenvman.2019.109914

Suhani, I., de Araujo, A. S. F., Kothari, R., Vaish, B., & Singh, R. P. (2025). Investigating the effect of compost and vermicompost on wheat plants and soil health under natural salt affected field conditions. *International Journal of Environmental Research, 19*(2), 42. https://doi.org/10.1007/s41742-024-00710-6

ul Baligah, H., Chesti, M. H., Ahmed Baba, Z., Mir, S., Wani, F. J., Bhat, J. A., & Khan, I. M. (2025). Vermicomposting technology as a dynamic strategy to mitigate environmental crisis: A bibliometric study of last three decades. *Environmental Technology, 46*(1), 72–86. https://doi.org/10.1080/09593330.2024.2339191

Vuković, A., Velki, M., Ečimović, S., Vuković, R., Štolfa Čamagajevac, I., & Lončarić, Z. (2021). Vermicomposting—Facts, benefits and knowledge gaps. *Agronomy, 11*(10), 1952. https://doi.org/10.3390/agronomy11101952

Wu, H., Li, A., Gao, S., Xing, Z., & Zhao, P. (2023). The performance, mechanism and greenhouse gas emission potential of nitrogen removal technology for low carbon source wastewater. *Science of the Total Environment, 903*, 166491. https://doi.org/10.1016/j.scitotenv.2023.166491

Yasmin, N., Jamuda, M., Panda, A. K., Samal, K., & Nayak, J. K. (2022). Emission of greenhouse gases (GHGs) during composting and vermicomposting: Measurement, mitigation, and perspectives. *Energy Nexus, 7*, 100092. https://doi.org/10.1016/j.nexus.2022.100092

Zhou, Y., Xiao, R., Klammsteiner, T., Kong, X., Yan, B., Mihai, F. C., … & Awasthi, M. K. (2022). Recent trends and advances in composting and vermicomposting technologies: A review. *Bioresource Technology, 360*, 127591. https://doi.org/10.1016/j.biortech.2022.127591

A State-of-the-Art Microbial Biotechnology Approach for Biofuel Production from Plastic and Microplastic Waste

Kamalesh Sen

Abstract The growing environmental threat posed by plastic waste (PWs) and microplastics (MPs) has spurred the need for innovative and sustainable waste management solutions. Biofuel production from PWs through microbial technology presents a promising approach to mitigate PW pollution while generating renewable energy. This review explores the current advancements in the bioconversion of PW and MPs waste into biofuels, focusing on microbial degradation and conversion mechanisms. Various microbial strains, including bacteria, fungi, and engineered microorganisms, have demonstrated their capacity to degrade PWs such as polyethylene (PE), polypropylene (PP), and polystyrene (PS) into smaller compounds, which can be further processed into biofuels. Key technologies, such as anaerobic digestion, fermentation, and enzymatic biodegradation, are reviewed for their efficiency in converting PW into bioenergy. The use of microbial consortia and genetically modified microbes enhances the biodegradation process, making it more efficient and feasible at industrial scales. Additionally, this review discusses the challenges associated with PW depolymerization, the toxicity of MPs, and the scalability of microbial technologies for large-scale biofuel production. Emphasis is placed on recent innovations in metabolic engineering to optimize microbial pathways for PW degradation and fuel production. The integration of microbial technology with waste management systems offers a circular economy approach, turning PW into valuable resources. This article provides a comprehensive overview of the potential and limitations of microbial technologies for biofuel production from PW, aiming to contribute to a sustainable future in waste-to-energy conversion.

Keywords PW · MPs · Microbial technology · Biofuel production · Biodegradation · Circular economy

---

K. Sen (✉)
Department of Environmental Science, Syamsundar College, Burdwan, West Bengal, India
e-mail: ksen.envs@gmail.com

A. P. Das (ed.), *Bioconversion of Environmental Waste to Value Added Products*, Sustainable Environmental Waste Management Strategies,
https://doi.org/10.1007/978-3-032-14215-3_3

1 Introduction

PW pollution has become a global environmental crisis, with millions of tons of PW accumulating in terrestrial and aquatic ecosystems (Abbasi et al., 2018; Abbasi & Turner, 2022). The issue is made worse by MPs pollution, which endangers both human health and wildlife (Ghosh et al., 2023; Mishra et al., 2021, 2022; Sahu & Das, 2024). Microbial biotechnology offers a novel way to deal with PW through biological decomposition and the creation of biofuel, even while conventional recycling techniques are inadequate (Ahuja et al., 2024; Cunha et al., 2020; Prus & Wilk, 2024). The goal of this study is to present a thorough examination of microbial approaches to PW breakdown and how they may be used in the production of biofuel (Devi et al., 2021; Kaykhaii et al., 2023).

The global environmental disaster of PW pollution has resulted in serious ecological and health issues (Mirsalami & Mirsalami, 2024). Because of their durability and adaptability, PW materials are widely used, which has led to a significant buildup of PW trash in both terrestrial and marine settings (Abinandan et al., 2023). PWs' enduring presence, especially MPs, worsens their environmental impact and poses major risks to human health and biodiversity (Abinandan et al., 2023). The growing production of single-use PWs and the inadequate handling of PW trash highlight how urgent it is to create sustainable waste-to-energy plans (Ahuja et al., 2024; Costa et al., 2024; Prus & Wilk, 2024; Ye et al., 2022).

There are inherent drawbacks to conventional PW management techniques such as landfilling, incineration, and mechanical recycling (Abomohra & Hanelt, 2022). While incineration emits toxic chemicals and destructive greenhouse gases, landfilling adds to pollution of soil and water (Ye et al., 2022). Despite being successful for some polymers, contamination problems, material deterioration, and the restricted recyclability of complex PW composites frequently limit mechanical recycling (Nath, 2024; Prus & Wilk, 2024). Alternative strategies that are both economically and environmentally feasible must be investigated in light of these difficulties (Mohanty et al., 2024).

A novel and promising approach to the breakdown of PW trash and its subsequent transformation into biofuels is provided by microbial biotechnology (Jain et al., 2023; Rosas-Vega et al., 2025; Zhou et al., 2020). PW may be used as a valuable feedstock for the production of bioenergy because microbes can break down complex PW polymers into smaller, more energy-rich molecules (Gao et al., 2024; Jain et al., 2023). The ability of a number of bacterial and fungal strains to break down major PW polymers, including PE, PP, PS, polyethylene terephthalate (PET), and polyvinyl chloride (PVC), has been reported (Allemann et al., 2024; Maurya et al., 2020). In order to create biofuels like ethanol, butanol, biodiesel, and biogas, these microbes use enzymatic pathways to break down PWs into simpler organic molecules (Beschkov, 2017; Prus & Wilk, 2024). These molecules may then be further processed by fermentation and anaerobic digestion (Beschkov, 2017).

Recent advancements in synthetic biology and metabolic engineering have further enhanced the efficiency of microbial PW degradation (Prus & Wilk, 2024). Engineered microbial strains with optimized metabolic pathways can improve the breakdown of recalcitrant PWs and enhance biofuel yields (Wang & Zhou, 2024). For instance, genetic modifications in bacteria such as Ideonella sakaiensis and Pseudomonas putida have demonstrated improved PET degradation and biofuel synthesis (Costa et al., 2024). Similarly, fungal species such as Aspergillus and Fusarium have shown promising potential for breaking down PW into bioenergy precursors.

The use of microbial consortia, where multiple microbial species work synergistically to degrade PWs, has emerged as an effective strategy to enhance biodegradation efficiency (Abomohra & Hanelt, 2022). These consortia leverage complementary metabolic capabilities, allowing for the complete breakdown of PW polymers and maximizing biofuel production. Studies have indicated that the combined activity of aerobic and anaerobic microbes can accelerate PW degradation and improve energy conversion efficiency (Ebrahimzade et al., 2021).

Anaerobic digestion, enzymatic hydrolysis, and microbial fermentation are important methods in the microbiological conversion of polymers into biofuels (Wang & Zhou, 2024). The microbial breakdown of organic intermediates into methane-rich biogas, a sustainable energy source, is facilitated by anaerobic digestion (Bala et al., 2024; Das et al., 2024; Sahu et al., 2024). Certain enzymes, such as cutinases and PETases, are used in enzymatic hydrolysis to catalyze the breakdown of PWs into the building blocks of biofuel. These intermediates are subsequently processed by microbial fermentation to produce liquid biofuels, which provide an alternative to energy sources based on fossil fuels (Carruthers & Lee, 2022).

Anaerobic digestion, enzymatic hydrolysis, and microbial fermentation are important methods in the microbiological conversion of polymers into biofuels (Bing et al., 2021; Menzel et al., 2020). The microbial breakdown of organic intermediates into methane-rich biogas, a sustainable energy source, is facilitated by anaerobic digestion (Menzel et al., 2020; Kabeyi & Olanrewaju, 2022). Certain enzymes, such as cutinases and PETases, are used in enzymatic hydrolysis to catalyze the breakdown of PWs into the building blocks of biofuel (Maurya et al., 2020; Rosas-Vega et al., 2025). These intermediates are subsequently processed by microbial fermentation to produce liquid biofuels, which provide an alternative to energy sources based on fossil fuels (Dhiman et al., 2024).

Future studies may concentrate on improving bioprocessing conditions, integrating microbial technology with current waste management infrastructures, and increasing microbial efficiency through genetic alterations (Carruthers & Lee, 2022; Wang & Zhou, 2024). Converting laboratory-scale results into industrial applications requires the creation of scalable and reasonably priced bioreactor systems (Benner et al., 2022). Furthermore, to promote the acceptance of microbial biotechnology as a workable solution for the management of PW and the creation of renewable energy, multidisciplinary cooperation between microbiologists, engineers, and legislators is crucial (Gao et al., 2024). Furthermore, microbial biotechnology offers a cutting-edge solution to the dual problems of PW pollution and sustainable energy generation. PW trash may be converted into useful biofuels by utilizing microorganisms' metabolic capabilities, which supports the circular economy framework (Costa et al., 2024). To fully utilize this technique, more developments in enzyme engineering, synthetic biology, and biofuel optimization are required (Wackett, 2021). The creation of microbial biofuel from PW trash can open the door to a cleaner and more sustainable future through creative research and well-timed regulatory changes (Mansouri et al., 2010).

The goal of this work is to present a thorough analysis of the most recent microbial biotechnology techniques for producing biofuel from PW and MPs waste (Gilani et al., 2023). Assessing the effectiveness of microbial strains in breaking down PW, investigating the metabolic pathways and enzymatic processes involved in PW biodegradation, analyzing technological developments in the production of microbial biofuel, and identifying obstacles and potential research areas are some of the main goals (Nath, 2024). Additionally, this study aims to emphasize how microbial bioconversion may be integrated with circular economy models to support renewable energy production and sustainable waste management (Abdel-Shafy & Mansour, 2018). This project is to support the continuous development of novel biotechnological methods for reducing PW pollution and improving the production of biofuel by combining contemporary breakthroughs in science.

2 Microbial Degradation of PW and MPs Waste

The permanence of PW and MPs trash makes them a major environmental hazard (Bing et al., 2021). Through the breakdown of these compounds into non-toxic metabolites, microbial degradation provides a sustainable alternative (Hessler et al., 2024). Enzymatic activity, genetic modifications that promote PW biodegradation, and specialized microbial consortia are all part of the process (Chow et al., 2023).

2.1 Microbial Consortia and Their Role in PW Degradation

Microbial consortia play a crucial role in the biodegradation of PW by harnessing the combined metabolic potential of different microorganisms (Mishra et al.,). Unlike single microbial strains, consortia exhibit synergistic interactions that enhance PW degradation efficiency (Table 1) (Costa et al., 2024; Wackett, 2021). They achieve this through cooperative enzyme production, metabolic exchange, and adaptation to environmental conditions (Cai et al., 2023). Microbial consortia consist of diverse bacterial and fungal species that complement each other's metabolic activities. Some microbes initiate the breakdown of PWs by secreting extracellular enzymes, while others utilize the degraded byproducts as carbon and energy sources (Beschkov, 2017; Temporiti et al., 2022). For instance, *Ideonella sakaiensis* produces PETase, which hydrolyzes PET into simpler compounds. Other bacteria, such as *Pseudomonas putida*, further metabolize these intermediates, preventing the accumulation of toxic byproducts (Satta et al., 2024). Similarly, fungal species like *Aspergillus niger* and *Penicillium simplicissimum* secrete oxidative enzymes such as laccases and peroxidases that facilitate PW depolymerisation (Daly et al., 2021). Microbial consortia improve PW degradation by addressing environmental challenges such as oxygen availability, pH fluctuations, and nutrient scarcity (Jagatee et al., 2024; Mishra et al., 2020). Aerobic and anaerobic microbes coexist within these consortia, allowing for degradation under different oxygen conditions (Shi et al., 2021). Certain microbes produce biosurfactants that increase PW surface hydrophilicity, enhancing microbial attachment and enzyme penetration (Samak et al., 2020). Furthermore, the metabolic cooperation between species ensures complete degradation of PW-derived intermediates, reducing the risk of MPs accumulation (Daly et al., 2021). Harnessing microbial consortia in bioreactors, composting systems, or contaminated sites can accelerate PW management (Hessler et al., 2024). Researchers are exploring genetic modifications to enhance enzyme production and microbial interactions, making microbial consortia a promising tool for sustainable PW reduction (Bing et al., 2021; Costa et al., 2024). The microbial breakdown of PPE and LLDPE film in soil and broth microcosms is also examined in this work. Changes in plastic structure over the course of 90 days indicate microbial activity and possible biodegradation routes (Fig. 1).

2.2 *Mechanisms of Biodegradation: Enzymatic Breakdown and Metabolic Pathways*

Microbial degradation of PW is an eco-friendly approach to mitigating PW pollution (Li et al., 2024). Microbial consortia, composed of bacteria and fungi, play a crucial role by working synergistically to break down complex PW polymers (Cai et al., 2023; Chi et al., 2021). These consortia enhance degradation efficiency through metabolic cooperation, enzyme secretion, and adaptation to environmental conditions (Thakur et al., 2023). For instance, *Pseudomonas putida, Ideonella sakaiensis, Aspergillus niger*, and *Penicillium* species produce PW-degrading enzymes such as PETase, laccases, and cutinases (Gaylarde et al., 2023; Mohanan et al., 2020). This collective action overcomes limitations like oxygen availability and nutrient deficiency, accelerating PW breakdown in various ecosystems (Barcoto & Rodrigues, 2022). Advanced biotechnological approaches, such as genetic modifications and enzyme immobilization, further enhance microbial consortia efficiency for large-scale applications. Figure 2 shows how microbial hydrocarbon degradation occurs. Aliphatic hydrocarbons undergo β-oxidation to produce Acetyl CoA for the TCA cycle, whereas aromatic hydrocarbons undergo dioxygenation to produce catechol and enter central metabolism (Edwards et al., 2022).

Table 1 Various microbial consortia, microbial species, degraded PW types, and their enzymatic mechanisms

Microbial consortium	Microbial species	PW types degraded	Enzymatic mechanism	Additional functions	References
Bacterial-fungal consortium	*Pseudomonas putida, Aspergillus niger*	PE, Polyurethane (PU)	Esterases, hydrolases, and laccases for polymer breakdown	Fungi aid in extracellular degradation, bacteria assist in monomer assimilation	(Lokesh et al., 2023)
Bacterial consortium	*Ideonella sakaiensis, Pseudomonas* species	PET	PETase hydrolyzes PET to mono(2-hydroxyethyl) terephthalate (MHET), further degraded by MHETase	Mutual metabolic pathway sharing for efficient degradation	(Carr et al., 2020; Pinto et al., 2021)
Fungal consortium	*Penicillium, Aspergillus*	Polystyrene, Polyvinyl chloride (PVC)	Laccases and peroxidases oxidize polymer chains	Fungal species produce extracellular enzymes for efficient PW breakdown	(Temporiti et al., 2022)
Mixed bacterial consortium	*Rhodococcus ruber, Sphingomonas*	Polyvinyl chloride (PVC), PP	Oxygenases, esterases, and cutinases cleave polymeric bonds	Microbes degrade various monomers and enhance degradation through metabolic collaboration	(Gluth et al., 2022)
Arthrobacter and pseudomonas consortium	*Arthrobacter, Pseudomonas citronellolis*	PE, PP	Esterases and hydrolases break down long-chain hydrocarbons	Microbial strains effectively degrade both PW polymers and plasticizers	(Edwards et al., 2022; Tiso et al., 2022)

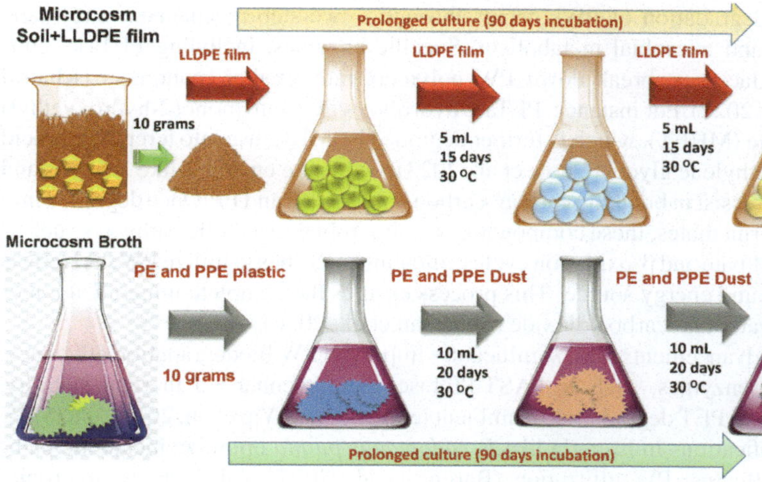

Fig. 1 Biodegradation of LLDPE film, PE, and PPE plastic in soil and broth microcosm systems under prolonged incubation (90 days). The experimental setup includes incubation with microbial cultures at 30 °C, with periodic sampling intervals to assess degradation progress

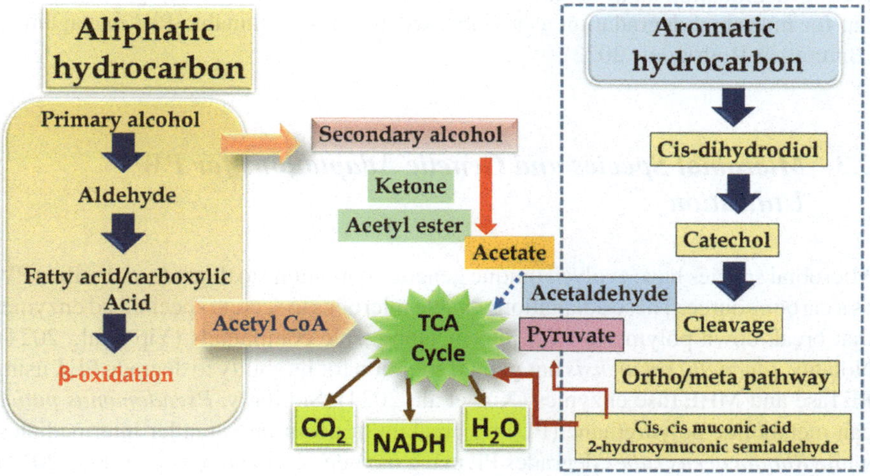

Fig. 2 Metabolic pathways for the degradation of aliphatic and aromatic hydrocarbons. Aliphatic hydrocarbons undergo β-oxidation to form Acetyl CoA, while aromatic hydrocarbons are degraded via catechol intermediates through the ortho/meta pathway, ultimately entering the TCA cycle

The biodegradation of PWs occurs through a two-step mechanism: enzymatic hydrolysis and microbial metabolism. Specific enzymes, including PETase, cutinases, and laccases, break down PW polymers into smaller monomers (Khairul Anuar et al., 2022). For instance, PETase hydrolyzes PET into mono(2-hydroxyethyl) terephthalate (MHET), which is further degraded by MHETase into terephthalic acid (TPA) and ethylene glycol (Chow et al., 2023). Oxidative enzymes like laccases and peroxidases assist in breaking carbon–carbon bonds in PE and PP. Once degraded into smaller intermediates, these compounds enter microbial metabolic pathways such as the β-ketoadipate and β-oxidation cycles, allowing microbes to utilize PW derivatives as a carbon and energy source. This process ensures the complete mineralization of PWs into water and carbon dioxide (Sudarsan et al., 2016).

Recent advancements have significantly improved PW biodegradation efficiency. Engineered enzymes, such as FAST-PETase, exhibit enhanced thermal stability, allowing rapid PET degradation at ambient temperatures (Yip et al., 2024). CRISPR-based modifications in bacteria like *Pseudomonas putida* optimize metabolic pathways for effective PW utilization (Barone et al., 2024). Additionally, microbial consortia engineering has improved synergy among PW-degrading microorganisms, leading to more efficient breakdown processes. The application of enzyme-polymer binding modifications has further enhanced enzyme affinity for PW surfaces, accelerating degradation (Yip et al., 2024). These breakthroughs highlight the potential for microbial degradation as a viable solution for sustainable PW management (Barcoto & Rodrigues, 2022).

2.3 Microbial Species and Genetic Adaptations for PW Utilization

Microbial species have evolved unique genetic adaptations to degrade and utilize PW as a carbon source. These adaptations enable microbes to secrete specialized enzymes that break down polymeric PWs into metabolizable compounds (Yip et al., 2024). Notably, *Ideonella sakaiensis* has gained attention for its ability to degrade PET using PETase and MHETase enzymes (Xue et al., 2024). Similarly, *Pseudomonas putida* can metabolize polyurethane (PU) by breaking it down into smaller intermediates, while *Rhodococcus ruber* degrades PE using oxygenase enzymes (Peng et al., 2014). These microbes possess specific genes that allow them to produce PW-degrading enzymes, facilitating the assimilation of PW-derived monomers into their metabolic pathways (Ma et al., 2024).

The genetic basis of PW biodegradation involves various enzymatic pathways and regulatory mechanisms (Yip et al., 2024). For example, the ISF6_4831 and ISF6_0224 genes in *Ideonella sakaiensis* encode PETase and MHETase, which hydrolyze PET into terephthalic acid (TPA) and ethylene glycol (EG). In *Pseudomonas putida*, the phaC and fadE genes play a crucial role in the degradation of plasticizers and polyurethane-derived intermediates (Loll-Krippleber et al., 2022; Tanaka et al.,

2024). Additionally, oxidative enzymes like laccases and alkane monooxygenases, encoded by genes in *Rhodococcus ruber* and *Bacillus subtilis*, enable the degradation of PE and PP (Tanaka et al., 2024). These microbes exhibit adaptive genetic mechanisms such as plasmid-mediated gene transfer and horizontal gene acquisition, which improve their ability to break down synthetic PWs in diverse environments (Khalid & Elsherif, 2023).

(i) The phaC gene encodes PHA synthase, an enzyme responsible for the biosynthesis of polyhydroxyalkanoates (PHAs), biodegradable biopolymers that bacteria produce as carbon storage reserves (Yip et al., 2024). During PW degradation, such as polyurethane (PU) breakdown, microbes convert the resulting carbon sources into hydroxyalkanoates (HAs), which are precursors for PHA synthesis. The process begins with the degradation of PU into intermediates like adipic acid and fatty acids (Yip et al., 2024). These intermediates are then converted into acetyl-CoA and hydroxyacyl-CoA by bacterial pathways. The PHA synthase (encoded by phaC) then polymerizes these compounds into PHA granules, which the bacteria can store for energy or use in their growth. This ability allows microbes to convert PW into biodegradable PWs, promoting sustainability (Mitra et al., 2020).

(ii) The fadE gene encodes acyl-CoA dehydrogenase, a crucial enzyme in the β-oxidation pathway, enabling microbes to break down plasticizers and long-chain hydrocarbons from PW polymers like polyurethane (Ru et al., 2020). Plasticizers often consist of phthalates or long-chain hydrocarbons, which microbes oxidize before further degradation. The FadE enzyme catalyzes the initial step in this process, converting acyl-CoA into trans-2-enoyl-CoA by introducing a double bond. This molecule then enters the β-oxidation cycle, where it is further broken down to provide energy and metabolites that microbes can use for growth (Kato et al., 2024). This process helps reduce the accumulation of plasticizers in the environment, enabling their safe degradation and assimilation (Baldera-Moreno et al., 2022; Ye et al., 2022).

(iii) Microbial strains' capacity to break down PW has been greatly increased by genetic engineering (Yip et al., 2024). The conversion of PW into biodegradable PHAs, which can be utilized as substitutes for synthetic PWs, is increased, for instance, when *Pseudomonas putida* is modified to overexpress the phaC gene (Wen et al., 2024). Similarly, bacteria that overexpress fadE are better able to degrade polyolefins and plasticizers, which speeds up the degradation process as a whole (Wen et al., 2024; Zhang et al., 2022). These changes not only increase the microorganisms' capacity for PW biodegradation, but they also present viable ways to produce biodegradable polymers on an industrial scale (Zhang et al., 2022). Furthermore, by degrading PW polymers and plasticizers, these genetically modified microorganisms can be employed in bioremediation to purify areas contaminated by PW. This genetic engineering strategy has potential for protecting the environment and managing PW trash sustainably.

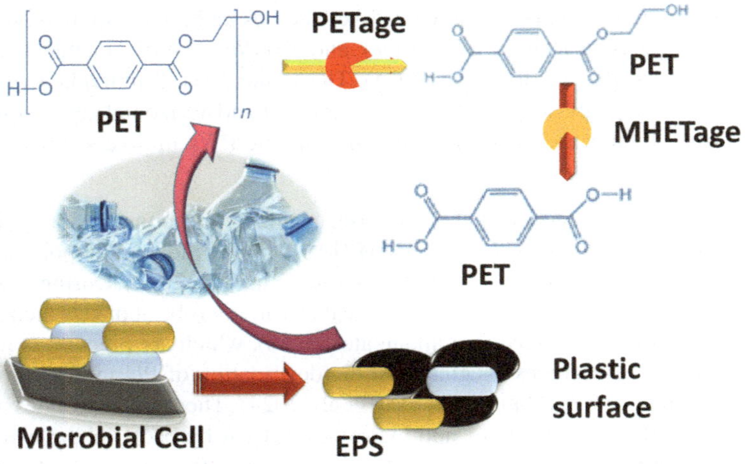

Fig. 3 Microbial degradation of PET through enzymatic activity. PETase and MHETase catalyze the breakdown of PET into smaller monomers, while microbial cells attach to the plastic surface via extracellular polymeric substances (EPS) to enhance biodegradation

Advancements in genetic engineering have further enhanced microbial PW degradation. Researchers have successfully modified *Pseudomonas putida* using CRISPR-Cas9 to optimize its metabolic pathways for more efficient PW utilization (Sun et al., 2020; Zhou et al., 2020). Synthetic biology approaches have also led to the engineering of multi-enzyme systems that combine PETase, MHETase, and laccases to accelerate PW breakdown (Sun et al., 2020; Zhang et al., 2022). Additionally, gene-editing tools have been employed to enhance enzyme stability and efficiency, allowing microbes to function under extreme environmental conditions. Instead, PETase and MHETase depolymerize PET enzymatically, promoting the biodegradation of plastic. One of the most important factors in starting and maintaining breakdown is microbial adherence to the plastic surface through EPS (Fig. 3). These advancements open the door for extensive microbial bioremediation techniques, which provide a long-term remedy for PW contamination (Khairul Anuar et al., 2022; Xue et al., 2024).

3 Bioconversion of PW-Derived Intermediates into Biofuels

The bioconversion of PW-derived intermediates into biofuels is an emerging strategy aimed at addressing PW while producing renewable energy (Table 2) (Meena et al., 2021). Through this process, microbes and engineered pathways can transform PW products into valuable biofuels such as ethanol, butanol, biodiesel, and biogas (Nath, 2024). The breakdown of PWs like PE, PP, and PET by microbial degradation or enzymatic hydrolysis yields intermediate compounds (Mohanan et al., 2020) such as fatty acids, hydrocarbons, and alcohols, which can then be converted into biofuels through various metabolic processes (He et al., 2003). This bioconversion provides a dual advantage by mitigating PW pollution while contributing to sustainable energy production and associated mechanisms (Table 2).

3.1 Metabolic Engineering for Enhanced Bioconversion

Metabolic engineering plays a pivotal role in enhancing the efficiency of PW-derived biofuel production (Lu et al., 2022). By modifying the genetic pathways of microorganisms, researchers can improve their ability to degrade PWs and convert the intermediates into biofuels (Devi et al., 2021). Specific genes involved in hydrocarbon degradation and fermentation pathways are upregulated to optimize biofuel yields (Joshi & Mishra, 2022). For instance, *Escherichia coli* and *Saccharomyces cerevisiae* have been genetically engineered to convert PW-derived fatty acids into ethanol or butanol more efficiently (Atsumi & Liao, 2008). Additionally, pathways that normally process plant-derived sugars are adapted to utilize hydrocarbons from PW degradation (Joshi & Mishra, 2022). This involves the introduction of novel enzymatic pathways, including those for oxidation, hydrolysis, and acetyl-CoA utilization, enabling microbes to assimilate PW intermediates and produce biofuels on a larger scale (Joshi & Mishra, 2022).

3.2 Fermentation and Bioreactor Technologies

Fermentation and bioreactor technologies are essential for scaling up the bioconversion of PW into biofuels (Devi et al., 2021). Fermentation is the primary process used by microorganisms to convert substrates into biofuels such as ethanol and butanol (Kazmi et al., 2025). This process can be optimized by utilizing engineered microbes in bioreactors, where factors such as temperature, pH, and nutrient availability are controlled to maximize the yield of biofuels (Wierckx et al., 2019). In bioreactors, microbes ferment PW-derived intermediates like fatty acids or alcohols, producing biofuels that can be easily harvested and purified (Wan Osman et al., 2024). Moreover, continuous fermentation systems can be integrated with biodegradation processes to allow for the continuous breakdown of PWs while simultaneously producing biofuels (Devi et al., 2021). This integration of PW degradation and biofuel production in industrial-scale bioreactors presents a promising solution for both waste management and renewable energy production. The metabolic processes that enable the bioconversion of intermediates obtained from PW into biofuels are described in Table 3. Through different metabolic routes, several microbial species help convert PW into ethanol, butanol, biodiesel, and biogas (Kazmi et al., 2025).

Table 2 Summarizes various PW types, their degradation intermediates, associated microbial species, metabolic pathways involved, and the corresponding biofuel produced

PW type	Degradation intermediates	Microbial species	Metabolic pathway	Biofuel produced	References
PE	Long-chain hydrocarbons, fatty acids	*Pseudomonas putida*, *Rhodococcus ruber*	β-oxidation, fatty acid metabolism	Biodiesel	(Alvarez et al., 2021)
PET	Terephthalic acid (TPA), ethylene glycol (EG)	*Ideonella sakaiensis*, *Pseudomonas aeruginosa*	TPA and EG metabolism via TPA dioxygenase and glycolysis	Ethanol, Butanol	(Kalathil et al., 2022)
PP	Alkanes, alkenes, carboxylated hydrocarbons	*Sphingomonas*, *Acinetobacter*	Hydrocarbon degradation via alkane hydroxylase	Biodiesel, Biogas	(Alaidaroos, 2023)
PS	Styrene monomers, benzoic acid	*Rhodococcus*, *Pseudomonas*	Styrene oxidation, aromatic ring cleavage	Bioethanol, Biogas	(Luo et al., 2023)
Polyurethane (PU)	Adipic acid, amines, alcohol derivatives	*Bacillus subtilis*, *Pseudomonas chlororaphis*	Urethane bond hydrolysis, β-oxidation	Ethanol, Butanol	(Brzeska & Piotrowska-Kirschling, 2021; Bech et al., 2024)
Polyvinyl Chloride (PVC)	Chlorinated hydrocarbons, acetaldehyde	*Clostridium*, *Dehalococcoides*	Reductive dechlorination, anaerobic fermentation	Biogas (Methane, Hydrogen)	(Bhaskar et al., 2006; He et al., 2003)

Table 3 Biochemical reactions involved in the bioconversion of PW-derived intermediates into biofuels

Biofuel type	PW types used	Microbial species
Ethanol	PET, Polyurethane (PU), PS	$PET + H_2O \xrightarrow{\text{PETase}} MHET$ $MHET + H_2O \xrightarrow{MHETase} TPA + EG2.$ $\text{Fermentation}: C_2H_6O_2 \xrightarrow{\text{Microbial Fermentation}} 2CH_3CH_2OH + CO_2$
Butanol	Polyurethane (PU), PP, PE	$\text{PU Hydrolysis}: PU + H_2O \xrightarrow{\text{Esterase, Urease}} \text{Adipic acid} + \text{Amines2}$ $\text{ABE Fermentation}: C_6H_{10}O_4 \xrightarrow{\text{Clostridium}} \text{Butanol} + CO_2 + H_2$
Biodiesel	PE, PP	$\text{Fatty Acid Formation}: C_nH_{2n+2} + O_2 \xrightarrow{\text{Oxygenase}} \text{Fatty Acids} + \text{Alcohols}$ $\text{Transesterification}: \text{Fatty Acid} + \text{Methanol} \xrightarrow{\text{Lipase}} \text{Biodiesel} + \text{Glycerol}$
Biogas (Methane, Hydrogen)	Polyvinyl chloride (PVC), PE, PS	$\text{Fermentation}: \text{Organic Waste} \xrightarrow{\text{Fermentation}} \text{Acetic acid} + H_2 + CO_2$ $\text{Fermentation}: CH_3COOH \xrightarrow{\text{Methanogens}} CH_4 + CO_2$ $\text{Methanogenesis}: CO_2 + 4H_2 \xrightarrow{\text{Methanogenesis}} CH_4 + 2H_2O$

3.3 Types of Biofuels Produced from PW (Ethanol, Butanol, Biodiesel, Biogas)

PW can be converted into several types of biofuels through microbial degradation and bioconversion processes (Neupane, 2023). Ethanol and butanol are alcohol-based biofuels commonly produced from PW-derived hydrocarbons, with genetically engineered microbes like *E. coli* and *Clostridium acetobutylicum* playing a central role in their production (Mouterde & Allais 2020). Biodiesel can be derived from the breakdown of plasticizers and long-chain hydrocarbons found in PWs like polyurethane, using microbial transesterification processes. Microbes such as *Rhodococcus* and *Pseudomonas* are known for their ability to break down fatty acid residues, which are then converted into biodiesel (Neupane, 2023). Biogas, primarily composed of methane, is another biofuel produced by anaerobic digestion of PW intermediates by specialized bacteria. The conversion of PWs to biogas helps reduce landfill accumulation and offers a clean energy source (Mouterde & Allais 2020). Table 4 highlights the production of various biofuels from PW through microbial degradation pathways (Kalathil et al., 2022; Tiso et al., 2022). Different microbial species metabolize PW into intermediates, which are further processed via glycolysis, fermentation, β-oxidation, and methanogenesis to generate ethanol, butanol, biodiesel, and biogas (Kazmi et al., 2025).

4 Synthetic Biology and Genetic Engineering in Microbial Biofuel Production

Synthetic biology and genetic engineering play a transformative role in enhancing microbial biofuel production by creating optimized microorganisms capable of more efficient degradation of PW and conversion into biofuels (Saini et al., 2020; Joshi & Mishra, 2022). By leveraging techniques like CRISPR/Cas9, omics technologies, and systems biology approaches, researchers can engineer microbial strains to meet industrial-scale biofuel demands while simultaneously tackling PW pollution (Rosas-Vega et al., 2025; Saini et al., 2020). These advancements enable the development of microorganisms with novel pathways and improved metabolic efficiency, allowing for higher biofuel yields from PW-derived intermediates (Nath, 2024).

4.1 CRISPR and Gene Editing for Enhanced Degradation Pathways

CRISPR/Cas9 technology has revolutionized genetic engineering by enabling precise edits in microbial genomes to enhance PW degradation pathways (Zhou et al., 2020). By utilizing CRISPR to knockout or overexpress genes, researchers can

Table 4 Summary of various biofuels produced from PW

Biofuel type	PW types used	Microbial species	Degradation intermediates	Metabolic pathway	References
Ethanol	PET, Polyurethane (PU), PS	*Ideonella sakaiensis*, *Pseudomonas aeruginosa*, *Bacillus subtilis*	Terephthalic acid (TPA), ethylene glycol, adipic acid	Glycolysis, fermentation	(Kalathil et al., 2022; Tiso et al., 2022)
Butanol	Polyurethane (PU), PP, PE	*Clostridium acetobutylicum*, *Pseudomonas putida*	Adipic acid, long-chain hydrocarbons, fatty acids	Acetone-butanol-ethanol (ABE) fermentation	(Jones, 2024; Khamaiseh et al., 2014)
Biodiesel	PE, PP	*Rhodococcus ruber*, *Pseudomonas putida*	Fatty acids, alkanes, esters	β-oxidation, Transesterification	(Mouterde & Allais, 2020; Gluth et al., 2022)
Biogas (Methane, Hydrogen)	Polyvinyl chloride (PVC), PE, PS	*Clostridium*, *Methanobacterium*, *Dehalococcoides*	Acetaldehyde, organic acids, methane precursors	Anaerobic fermentation, Methanogenesis	(Sawada et al., 2023)

optimize microbes to degrade PW more efficiently (Otoupal & Chatterjee, 2018). For example, CRISPR can be used to introduce specific PW-degrading enzymes into microbial strains, allowing them to break down complex PWs like polyethylene or polyurethane (Mayorga-Ramos et al., 2023). Additionally, CRISPR can modify microbial metabolic pathways to improve the conversion of PW intermediates into biofuels such as ethanol or butanol (Verdezoto-Prado et al., 2025). This targeted genetic editing not only improves the microbes' ability to utilize PW but also enhances their tolerance to harsh environmental conditions often associated with bioreactors, leading to higher yields and faster degradation rates (Otoupal & Chatterjee 2018). Moreover, CRISPR facilitates the development of multi-functional consortia, where different engineered strains work synergistically to break down diverse PW types and optimize biofuel production. In CRISPR-mediated gene editing, Cas9 introduces site-specific cuts while being guided by gRNA (Mayorga-Ramos et al., 2023; Sun et al., 2020). However it can be observed that the CRISPR–Cas9 system, highlighting the components of the guide RNA (gRNA; tracrRNA + crRNA, ~100 nt), the 20-nt target-binding region, and the Cas9 endonuclease (1308 a.a). Target recognition occurs via RNA–DNA hybridization adjacent to the protospacer adjacent motif (PAM; 3 nt). The CRISPR-associated protein Cas9 introduces a double-strand break in the DNA. The cell subsequently repairs this break either through non-homologous end joining (NHEJ) or homology-directed repair (HDR). The red circled region represents NHEJ, which directly rejoins the broken DNA ends, whereas HDR uses a homologous DNA template to accurately repair the break by inserting the desired DNA segment between the two ends. The diagram provides information about CRISPR-associated programs for editing DNA and currently being developed precision editors using prime editing technology (Fig. 4).

4.2 Omics Technologies and Systems Biology Approaches

Omics technologies such as genomics, proteomics, and metabolomics are integral to understanding and optimizing the pathways involved in microbial PW degradation and biofuel production (Otoupal & Chatterjee, 2018). By studying the genomes of microorganisms, scientists can identify key genes and pathways responsible for PW degradation, which can then be targeted for modification (Sawada et al., 2023; Tanaka et al., 2024). Proteomics allows for the analysis of protein expression levels, identifying enzymes that are crucial for the breakdown of PW polymers and the conversion of intermediates into biofuels (Majumder et al., 2024). Metabolomics, nevertheless, provides insights into the metabolic profiles of engineered microbes, helping identify bottlenecks in the biofuel production process and allowing for more efficient pathway engineering (Alvarez et al., 2021; Nath, 2024). Systems biology approaches integrate data from all omics layers to construct a comprehensive understanding of the metabolic network within the microbe, facilitating the design of synthetic metabolic pathways (Majumder et al., 2024; Okoro et al., 2022). By combining omics data

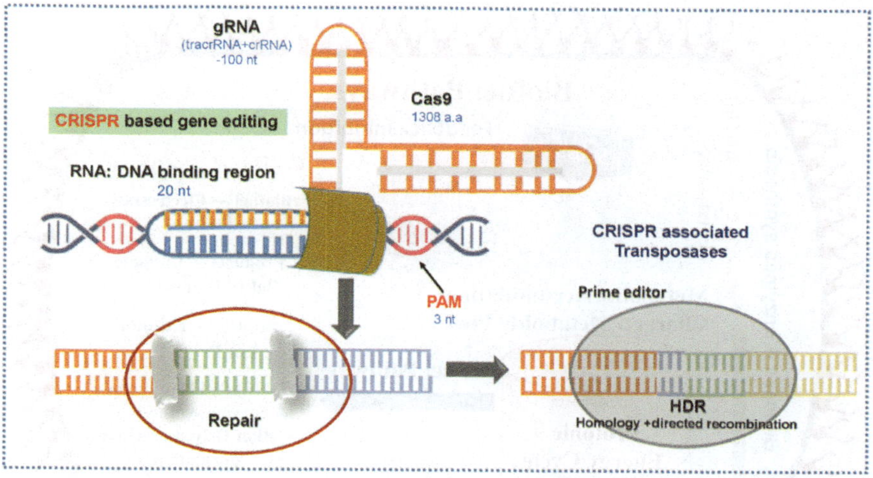

Fig. 4 Mechanism of CRISPR-based gene editing using Cas9 and CRISPR-associated transposases. The guide RNA (gRNA) directs Cas9 to the target DNA sequence, where a double-strand break is induced at the PAM site, followed by repair via homology-directed recombination (HDR) or prime editing

with computational modeling, researchers can predict how engineered microbes will behave under various conditions and fine-tune their biofuel production capabilities.

4.3 *Engineering Microbes for High-Yield Biofuel Production*

To achieve high-yield biofuel production from PW, microbial engineering is essential to optimizing key factors such as substrate utilization, enzyme activity, and metabolic flux (Atsumi & Liao, 2008; Lu et al., 2022; Nath, 2024). Genetic modifications are made to enhance microbes' ability to utilize PW-derived intermediates like fatty acids, aromatic compounds, and alcohols. For instance, genetic enhancements in microbes like *Escherichia coli* or Saccharomyces cerevisiae can improve their efficiency in converting PW-derived hydrocarbons into bioethanol or butanol (Lu et al., 2022; Shi et al., 2021). Additionally, advanced metabolic engineering approaches can involve re-routing carbon flux toward biofuel production pathways while minimizing byproduct formation. Optimizing cofactor availability and reducing energy consumption during the fermentation process are also crucial to improving biofuel yields (Meena et al., 2021). With the help of synthetic biology, microbes can be engineered to tolerate high concentrations of toxic intermediates that often accumulate during PW degradation, thus improving the overall efficiency of PW-to-biofuel conversion (Devi et al., 2021). With an emphasis on charge-dependent side reactions and ionic energy conversion, this figure illustrates how metabolic and electrostatic pathways interact during the production of biofuel (Lu et al., 2022; Meena et al.,

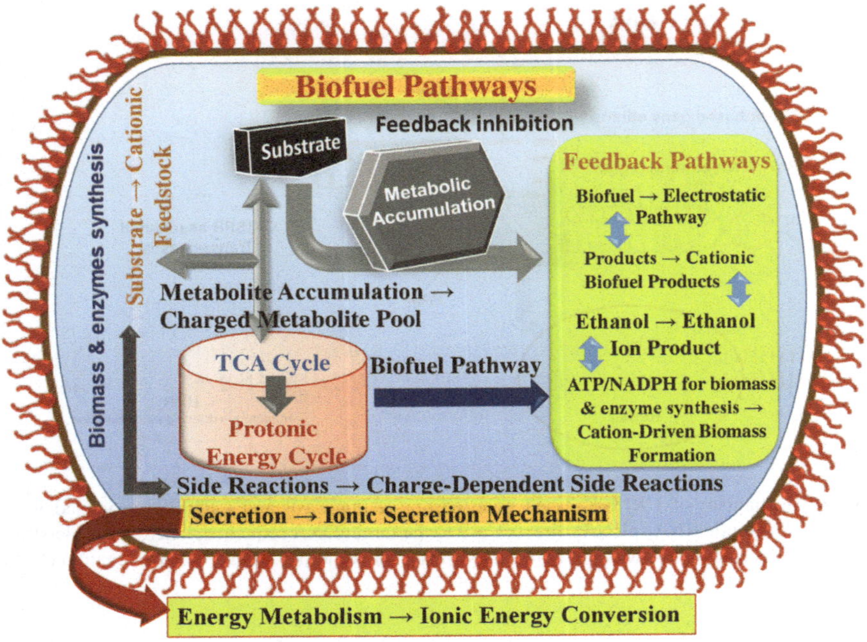

Fig. 5 Schematic representation of biofuel pathways highlighting metabolic accumulation, energy metabolism, and ionic secretion mechanisms. The protonic energy cycle within the TCA cycle drives biofuel production, while feedback pathways regulate metabolite accumulation and biomass synthesis

2021). Biomass generation and enzyme synthesis are impacted by ionic processes that control the release of biofuel (Fig. 5). These advancements hold the potential to make microbial biofuel production a scalable and sustainable solution to both PW management and renewable energy generation (Joshi & Mishra, 2022; Neupane, 2023).

5 Challenges and Future Perspectives

While microbial biodegradation and biofuel production from PW hold significant promise, there are several challenges that must be addressed to ensure their successful implementation at scale (Wong et al., 2015). These challenges range from the technical difficulties in optimizing microbial processes to broader concerns about economic feasibility, environmental impact, and regulatory frameworks. Overcoming these hurdles will require continued research, technological innovation, and coordinated policy efforts (Bai et al., 2019; Wong et al., 2015).

5.1 Scaling up Microbial Biodegradation and Biofuel Synthesis

Scaling up microbial biodegradation and biofuel synthesis is one of the major challenges for the industrial application of this technology (Mouterde & Allais, 2020). The laboratory-scale processes of PW biodegradation often do not translate well to industrial-scale operations due to the complexity of microbial consortia, the heterogeneity of PW, and the need for specific growth conditions (Atsumi & Liao, 2008; Meena et al., 2021). For instance, maintaining optimal environmental conditions (e.g., temperature, pH, oxygen levels) in large bioreactors can be difficult, and scaling up degradation pathways without compromising microbial activity and biofuel production rates presents additional challenges (Wong et al., 2015). Furthermore, large quantities of PW need to be effectively broken down and converted into biofuels, which requires high microbial efficiency and long-term stability (Bai et al., 2019). Process optimization, including bioreactor design, substrate pretreatment, and improved microbial strains, is essential for achieving high yields in industrial settings. Developing automated monitoring systems and real-time control mechanisms to track microbial growth and biodegradation is also key to improving scalability (Mouterde & Allais, 2020).

5.2 Economic and Environmental Feasibility of Microbial Biofuel Production

The economic and environmental feasibility of microbial biofuel production from PW depends on several factors (Bai et al., 2019). From an economic standpoint, the cost of PW collection, microbial cultivation, and bioreactor maintenance must be outweighed by the revenue generated from biofuel production (Anthony et al., 2019). Currently, the cost of biofuels produced through microbial processes is often higher than that of conventional fuels or biofuels derived from food crops (Mouterde & Allais, 2020). However, as technologies improve, costs may come down through optimization of microbial strains, efficiency improvements in fermentation processes, and economies of scale (Beluhan et al., 2023). Environmental concerns, such as energy consumption during PW degradation and fermentation, must also be carefully evaluated to ensure the overall sustainability of the process (Li et al., 2022; Neupane, 2023). The use of renewable energy in the process could improve the environmental footprint, making microbial biofuel production a more viable option. Additionally, ensuring that PW biodegradation does not release harmful byproducts into the environment is critical for making this technology environmentally sound (Wong et al., 2015).

5.3 Regulatory Frameworks and Policy Implications

As microbial biofuel production from PW is a relatively new and evolving field, there is a need for comprehensive regulatory frameworks to guide research, commercialization, and environmental impact (Bai et al., 2019; Peng et al., 2022). Policies must address several issues, such as biotechnology safety, environmental protection, and intellectual property associated with genetically engineered microorganisms (Li et al., 2022; Neupane, 2023; Verdezoto-Prado et al., 2025). Regulatory authorities must also develop standards for biofuel quality and environmental safety, ensuring that microbial processes meet the required performance metrics (Li et al., 2022; Saini et al., 2020). Additionally, governments will need to incentivize the research and development of PW-to-biofuel technologies through grants, tax breaks, and research funding. Collaborations between academia, industry, and government agencies will be crucial for establishing policies that promote innovation while protecting public health and the environment. International cooperation will be necessary to ensure that global standards are met, especially as PW pollution is a transboundary issue.

5.4 Integrating Microbial Biotechnology with Circular Economy Models

Integrating microbial biotechnology into circular economy models offers a promising pathway for sustainable PW management and biofuel production (Meena et al., 2021). In a circular economy, waste products are seen as valuable resources rather than pollutants, and microbial biotechnologies can play a central role in converting PW into useful products, such as biofuels and biodegradable materials. This integration requires developing efficient waste collection and sorting systems to supply PW to bioreactors, where microbes can break it down and convert it into biofuels. Additionally, the biorefinery model, which integrates microbial PW degradation with other forms of biomass processing (e.g., agricultural waste or algae), could help create a diverse portfolio of biofuels and chemicals. A circular economy model also promotes the idea of zero waste, where the output of one process serves as the input for another, reducing the environmental burden. Government incentives for sustainable innovation, investments in green technologies, and a strong focus on waste-to-energy technologies will be essential for successfully integrating microbial biotechnology with circular economy models, contributing to both PW reduction and sustainable biofuel production.

6 Conclusions

One possible approach to solving the worldwide PW pollution problem and promoting sustainable energy generation is the microbial biodegradation of PW and its subsequent conversion into biofuels. It is possible to optimize bacteria to break down complex PW polymers and transform the resultant intermediates into useful biofuels like ethanol, butanol, and biodiesel by using synthetic biology, genetic engineering, and bioconversion technologies. This strategy offers a solution to the environmental problem caused by PW trash in addition to a renewable energy source. But there are a lot of obstacles to overcome, especially when it comes to expanding the use of microbial processes beyond lab studies to industrial settings. For the production of microbial biofuel to compete with conventional energy sources, two important issues must be resolved: economic viability and environmental sustainability. Furthermore, in order to guarantee the appropriate and safe use of biofuels and genetically modified microorganisms in a variety of applications, regulatory frameworks must be developed. Looking ahead, the combination of circular economy models with microbial biotechnology presents an intriguing prospect for a more sustainable future in which PW is viewed as a useful resource rather than a problem. Microbial biofuel production might be crucial in lowering PW pollution, producing clean energy, and promoting the ideas of a circular economy if microbial efficiency is maintained, bioreactor systems are optimized, and international collaboration is encouraged. This area has a bright future, but in order to reach its full potential, more research, funding, and supportive legislation are needed.

References

Abbasi, S., & Turner, A. (2022). Sources, concentrations, distributions, fluxes and fate of microplastics in a hypersaline lake: Maharloo, south-west Iran. *Science of the Total Environment, 823*, Article 153721. https://doi.org/10.1016/j.scitotenv.2022.153721

Abbasi, S., Soltani, N., Keshavarzi, B., et al. (2018). Microplastics in different tissues of fish and prawn from the Musa Estuary, Persian Gulf. *Chemosphere, 205*, 80–87. https://doi.org/10.1016/j.chemosphere.2018.04.076

Abdel-Shafy, H. I., & Mansour, M. S. M. (2018). Solid waste issue: Sources, composition, disposal, recycling, and valorization. *Egyptian Journal of Petroleum, 27*, 1275–1290. https://doi.org/10.1016/j.ejpe.2018.07.003

Abinandan, S., Praveen, K., Venkateswarlu, K., & Megharaj, M. (2023). Microalgae–Microplastics interactions at environmentally relevant concentrations: Implications toward ecology, bioeconomy, and UN SDGs. *Water Research, 247*, Article 120778. https://doi.org/10.1016/j.watres.2023.120778

Abomohra, A., & Hanelt, D. (2022). Recent advances in micro-/nanoplastic (MNPs) removal by microalgae and possible integrated routes of energy recovery. *Microorganisms, 10*, 2400. https://doi.org/10.3390/microorganisms10122400

Ahuja, V., Singh, P. K., Mahata, C., et al. (2024). A review on microbes mediated resource recovery and bioplastic (polyhydroxyalkanoates) production from wastewater. *Microbial Cell Factories, 23*, 187. https://doi.org/10.1186/s12934-024-02430-0

Alaidaroos, B. A. (2023). Advancing eco-sustainable bioremediation for hydrocarbon contaminants: Challenges and solutions. *Processes, 11*, 3036. https://doi.org/10.3390/pr11103036

Allemann, M. N., Tessman, M., Reindel, J., et al. (2024). Rapid biodegradation of microplastics generated from bio-based thermoplastic polyurethane. *Scientific Reports, 14*, 6036. https://doi.org/10.1038/s41598-024-56492-6

Alvarez, H. M., Hernández, M. A., Lanfranconi, M. P., et al. (2021). Rhodococcus as biofactories for microbial oil production. *Molecules, 26*, 4871. https://doi.org/10.3390/molecules26164871

Anthony, W. E., Carr, R. R., DeLorenzo, D. M., et al. (2019). Development of Rhodococcus opacus as a chassis for lignin valorization and bioproduction of high-value compounds. *Biotechnology for Biofuels, 12*, 192. https://doi.org/10.1186/s13068-019-1535-3

Atsumi, S., & Liao, J. C. (2008). Metabolic engineering for advanced biofuels production from *Escherichia coli. Current Opinion Biotechnology, 19*, 414–419. https://doi.org/10.1016/j.copbio.2008.08.008

Bai, B., Jin, H., Fan, C., et al. (2019). Experimental investigation on liquefaction of plastic waste to oil in supercritical water. *Waste Management, 89*, 247–253. https://doi.org/10.1016/j.wasman.2019.04.017

Bala, K., Dogra, A., Lata, P., & Das, A. P. (2024). Discussing sources and biological-chemical remediation approaches for microplastic pollution. In *Sustainable microbial technology for synthetic and cellulosic microfiber bioremediation* (pp. 37–59). https://doi.org/10.1007/978-3-031-62600-5_3

Baldera-Moreno, Y., Pino, V., Farres, A., et al. (2022). Biotechnological aspects and mathematical modeling of the biodegradation of plastics under controlled conditions. *Polymers, 14*, 375. https://doi.org/10.3390/polym14030375

Barcoto, M. O., & Rodrigues, A. (2022). Lessons from insect fungiculture: From microbial ecology to plastics degradation. *Frontiers Microbiology, 13*, Article 812143. https://doi.org/10.3389/fmicb.2022.812143

Barone, G. D., Rodríguez-Seijo, A., Parati, M., et al. (2024). Harnessing photosynthetic microorganisms for enhanced bioremediation of microplastics: A comprehensive review. *Environmental Science and Ecotechnology, 20*, Article 100407. https://doi.org/10.1016/j.ese.2024.100407

Bech, T. B., Donslund, B. S., Kristensen, S. K., & Skrydstrup, T. (2024). Chemical separation of polyurethane via acidolysis—Combining acidolysis with hydrolysis for valorisation of aromatic amines. *Green Chemistry, 26*, 8395–8404. https://doi.org/10.1039/D4GC00819G

Beluhan, S., Mihajlovski, K., Šantek, B., & Ivančić Šantek, M. (2023). The production of bioethanol from lignocellulosic biomass: Pretreatment methods, fermentation, and downstream processing. *Energies, 16*, 7003. https://doi.org/10.3390/en16197003

Benner, P., Meier, L., Pfeffer, A., et al. (2022). Lab-scale photobioreactor systems: principles, applications, and scalability. *Bioprocess Biosystems Engineering, 45*, 791–813. https://doi.org/10.1007/s00449-022-02711-1

Beschkov, V. (2017). Biogas, biodiesel and bioethanol as multifunctional renewable fuels and raw materials. In *Frontiers in bioenergy and biofuels*. IntechOpen.

Bhaskar, T., Negoro, R., Muto, A., & Sakata, Y. (2006). Prevention of chlorinated hydrocarbons formation during pyrolysis of PVC or PVDC mixed plastics. *Green Chemistry, 8*, 697–700. https://doi.org/10.1039/B603037H

Bing, R. G., Sulis, D. B., Wang, J. P., et al. (2021). Thermophilic microbial deconstruction and conversion of natural and transgenic lignocellulose. *Environmental Microbiology Reports, 13*, 272–293. https://doi.org/10.1111/1758-2229.12943

Brzeska, J., & Piotrowska-Kirschling, A. (2021). A brief introduction to the polyurethanes according to the principles of green chemistry. *Processes, 9*, 1929. https://doi.org/10.3390/pr9111929

Cai, Z., Li, M., Zhu, Z., et al. (2023). Biological degradation of plastics and microplastics: A recent perspective on associated mechanisms and influencing factors. *Microorganisms, 11*, 1661. https://doi.org/10.3390/microorganisms11071661

Carr, C. M., Clarke, D. J., & Dobson, A. D. W. (2020). Microbial polyethylene terephthalate hydrolases: Current and future perspectives. *Frontiers Microbiology, 11*. https://doi.org/10.3389/fmicb.2020.571265

Carruthers, D. N., & Lee, T. S. (2022). Translating advances in microbial bioproduction to sustainable biotechnology. *Frontier Bioengineering Biotechnology, 10*. https://doi.org/10.3389/fbioe.2022.968437

Chi, J., Zhang, H., & Zhao, D. (2021). Impact of microplastic addition on degradation of dibutyl phthalate in offshore sediments. *Marine Pollution Bulletin, 162*, Article 111881. https://doi.org/10.1016/j.marpolbul.2020.111881

Chow, J., Perez-Garcia, P., Dierkes, R., & Streit, W. R. (2023). Microbial enzymes will offer limited solutions to the global plastic pollution crisis. *Microbial Biotechnology, 16*, 195–217. https://doi.org/10.1111/1751-7915.14135

Costa G dos, S., Martinez-Burgos, W. J., dos Reis, G. A. et al. (2024). Advances in biomass and microbial lipids production: Trends and prospects. *Processes, 12*, 2903. https://doi.org/10.3390/pr12122903

Cunha, C., Lopes, J., Paulo, J., et al. (2020). The effect of microplastics pollution in microalgal biomass production: A biochemical study. *Water Research, 186*, Article 116370. https://doi.org/10.1016/j.watres.2020.116370

Daly, P., Cai, F., Kubicek, C. P., et al. (2021). From lignocellulose to plastics: Knowledge transfer on the degradation approaches by fungi. *Biotechnology Advances, 50*, Article 107770. https://doi.org/10.1016/j.biotechadv.2021.107770

Das, A., Ray, S. K., Mohanty, M., Mohanty, J., Dey, S., & Das, A. P. (2024). Ecotoxicity of microplastic wastes and their sustainable management: A review. *Environmental Chemistry and Ecotoxicology, 6*, 144–152. https://doi.org/10.1016/J.ENCECO.2024.05.003

Devi, D. S., Kumar, R., & Rajak, U. (2021). Plastic waste as a biofuel feedstock-A conceptual study. *IOP Conference Series: Material Science Engineering, 1116*, Article 012029. https://doi.org/10.1088/1757-899X/1116/1/012029

Dhiman, S., Kaur, P., Narang, J., et al. (2024). Fungal bioprocessing for circular bioeconomy: Exploring lignocellulosic waste valorization. *Mycology, 15*, 538–563. https://doi.org/10.1080/21501203.2024.2316824

Ebrahimzade, I., Ebrahimi-Nik, M., Rohani, A., & Tedesco, S. (2021). Higher energy conversion efficiency in anaerobic degradation of bioplastic by response surface methodology. *Journal of Cleaner Production, 290*, Article 125840. https://doi.org/10.1016/j.jclepro.2021.125840

Edwards, S., León-Zayas, R., Ditter, R., et al. (2022). Microbial consortia and mixed plastic waste: Pangenomic analysis reveals potential for degradation of multiple plastic types via previously identified PET degrading bacteria. *International Journal Molecular Science, 23*, 5612. https://doi.org/10.3390/ijms23105612

Gao, W., Xu, M., Zhao, W., et al. (2024). Microbial degradation of (Micro)plastics: Mechanisms, enhancements, and future directions. *Fermentation, 10*, 441. https://doi.org/10.3390/fermentation10090441

Gaylarde, C. C., de Almeida, M. P., Neves, C. V., et al. (2023). The importance of biofilms on microplastic particles in their sinking behavior and the transfer of invasive organisms between ecosystems. *Micro, 3*, 320–337. https://doi.org/10.3390/micro3010022

Ghosh, S., Sinha, J. K., Ghosh, S., Vashisth, K., Han, S., & Bhaskar, R. (2023). Microplastics as an emerging threat to the global environment and human health. *Sustainability, 15*, 10821. https://doi.org/10.3390/SU151410821

Gilani, I. E., Sayadi, S., Zouari, N., & Al-Ghouti, M. A. (2023). Plastic waste impact and biotechnology: Exploring polymer degradation, microbial role, and sustainable development implications. *Bioresource Technology Reports, 24*, Article 101606. https://doi.org/10.1016/j.biteb.2023.101606

Gluth, A., Xu, Z., Fifield, L. S., & Yang, B. (2022). Advancing biological processing for valorization of plastic wastes. *Renewable and Sustainable Energy Reviews, 170*, Article 112966. https://doi.org/10.1016/j.rser.2022.112966

He, J., Ritalahti, K. M., Aiello, M. R., & Löffler, F. E. (2003). Complete detoxification of vinyl chloride by an anaerobic enrichment culture and identification of the reductively dechlorinating population as a Dehalococcoides species. *Applied Environmental Microbiology, 69*, 996–1003. https://doi.org/10.1128/AEM.69.2.996-1003.2003

Hessler, T., Harrison, S. T. L., Banfield, J. F., & Huddy, R. J. (2024). Harnessing fermentation may enhance the performance of biological sulfate-reducing bioreactors. *Environmental Science Technology, 58*, 2830–2846. https://doi.org/10.1021/acs.est.3c04187

Jagatee, S., Priyadarshini, S., Rath, C. C., & Das, A. P. (2024). Synthetic microfiber: An enduring environmental problem linked to sustainable development. *Environmental Science and Engineering, Part F2333*, 93–112. https://doi.org/10.1007/978-3-031-51792-1_6

Jain, R., Gaur, A., Suravajhala, R., et al. (2023). Microplastic pollution: Understanding microbial degradation and strategies for pollutant reduction. *Science of the Total Environment, 905*, Article 167098. https://doi.org/10.1016/j.scitotenv.2023.167098

Jones, D. T. (2024). The industrial fermentation process and clostridium species used to produce biobutanol. *Applied Microbiology, 4*, 894–917. https://doi.org/10.3390/applmicrobiol4020061

Joshi, S., & Mishra, S. (2022). Recent advances in biofuel production through metabolic engineering. *Bioresource Technology, 352*, Article 127037. https://doi.org/10.1016/j.biortech.2022.127037

Kabeyi, M. J. B., & Olanrewaju, O. A. (2022). Biogas production and applications in the sustainable energy transition. *Journal of Energy, 2022*, Article e8750221. https://doi.org/10.1155/2022/875 0221

Kalathil, S., Miller, M., & Reisner, E. (2022). Microbial fermentation of polyethylene terephthalate (PET) plastic waste for the production of chemicals or electricity. *Angewandte Chemie Interantional Edition English, 61*, Article e202211057. https://doi.org/10.1002/anie.202211057

Kato, R., Takenaka, Y., Ohno, Y., & Kihara, A. (2024). Catalytic mechanism of trans-2-enoyl-CoA reductases in the fatty acid elongation cycle and its cooperative action with fatty acid elongases. *Journal Biology Chemistry, 300*, Article 105656. https://doi.org/10.1016/j.jbc.2024.105656

Kaykhaii, M., Honarmandrad, Z., & Gębicki, J. (2023). Effect of microplastics pollution on hydrogen production from biomass: A comprehensive review. *Industrial Engineering Chemistry Research, 62*, 3835–3843. https://doi.org/10.1021/acs.iecr.2c04499

Kazmi, A., Sultana, T., Ali, A., et al. (2025). Innovations in bioethanol production: A comprehensive review of feedstock generations and technology advances. *Energy Strategy Reviews, 57*, Article 101634. https://doi.org/10.1016/j.esr.2024.101634

Khairul Anuar, N. F. S., Huyop, F., Ur-Rehman, G., et al. (2022). An overview into polyethylene terephthalate (PET) hydrolases and efforts in tailoring enzymes for improved plastic degradation. *International Journal Molecular Science, 23*, 12644. https://doi.org/10.3390/ijms232012644

Khalid, S. A., & Elsherif, W. M. (2023). Types of microorganisms for biodegradation. In G. A. M. Ali & A. S. H. Makhlouf (Eds.), *Handbook of biodegradable materials* (pp. 195–220). Cham: Springer International Publishing.

Khamaiseh, E. I., Abdul Hamid, A., Abdeshahian, P., et al. (2014). Enhanced butanol production by clostridium acetobutylicum NCIMB 13357 grown on date fruit as carbon source in P2 medium. *The Scientific World Journal, 2014*, Article 395754. https://doi.org/10.1155/2014/395754

Li, N., Liu, H., Cheng, Z., et al. (2022). Conversion of plastic waste into fuels: A critical review. *Journal of Hazardous Materials, 424*, Article 127460. https://doi.org/10.1016/j.jhazmat.2021.127460

Li, X., Li, G., Wang, J., et al. (2024). Elucidating polyethylene microplastic degradation mechanisms and metabolic pathways via iron-enhanced microbiota dynamics in marine sediments. *Journal of Hazardous Materials, 466*, Article 133655. https://doi.org/10.1016/j.jhazmat.2024.133655

Lokesh, P., Shobika, R., Omer, S., et al. (2023). Bioremediation of plastics by the help of microbial tool: A way for control of plastic pollution. *Sustainable Chemistry for the Environment, 3*, Article 100027. https://doi.org/10.1016/j.scenv.2023.100027

Loll-Krippleber, R., Sajtovich, V. A., Ferguson, M. W., et al. (2022). Development of a yeast whole-cell biocatalyst for MHET conversion into terephthalic acid and ethylene glycol. *Microbial Cell Factories, 21*, 280. https://doi.org/10.1186/s12934-022-02007-9

Lu, H., Yadav, V., Zhong, M., et al. (2022). Bioengineered microbial platforms for biomass-derived biofuel production—A review. *Chemosphere, 288*, Article 132528. https://doi.org/10.1016/j.chemosphere.2021.132528

Luo, X., Zhan, J., Mei, Q., & Zhang, S. (2023). Selective oxidative upgrade of waste polystyrene plastics by nitric acid to produce benzoic acid. *Green Chemistry, 25*, 6717–6727. https://doi.org/10.1039/D3GC00865G

Ma, M., Ma, S., Zeng, D., et al. (2024). Temperature-dependent microbial mechanism and accumulation of volatile fatty acids in primary sludge pretreated with peroxymonosulfate. *Bioresource Technology, 408*, Article 131201. https://doi.org/10.1016/j.biortech.2024.131201

Majumder, D., Dey, A., Ray, S., et al. (2024). Use of genomics and proteomics in studying lipase producing microorganisms and its application. *Food Chemistry: Molecular Sciences, 9*, Article 100218. https://doi.org/10.1016/j.fochms.2024.100218

Mansouri, J., Harrisson, S., & Chen, V. (2010). Strategies for controlling biofouling in membrane filtration systems: Challenges and opportunities. *Journal Material Chemistry, 20*, 4567–4586. https://doi.org/10.1039/B926440J

Maurya, A., Bhattacharya, A., & Khare, S. K. (2020). Enzymatic remediation of polyethylene terephthalate (PET)–Based polymers for effective management of plastic wastes: An overview. *Frontiers Bioengineering Biotechnology, 8*. https://doi.org/10.3389/fbioe.2020.602325

Mayorga-Ramos, A., Zúñiga-Miranda, J., Carrera-Pacheco, S. E., et al. (2023). CRISPR-cas-based antimicrobials: Design, challenges, and bacterial mechanisms of resistance. *ACS Infection Disease, 9*, 1283–1302. https://doi.org/10.1021/acsinfecdis.2c00649

Meena, M., Shubham, S., Paritosh, K., et al. (2021). Production of biofuels from biomass: Predicting the energy employing artificial intelligence modelling. *Bioresource Technology, 340*, Article 125642. https://doi.org/10.1016/j.biortech.2021.125642

Menzel, T., Neubauer, P., & Junne, S. (2020). Role of microbial hydrolysis in anaerobic digestion. *Energies, 13*, 5555. https://doi.org/10.3390/en13215555

Mirsalami, S. M., & Mirsalami, M. (2024). Achieving optimal output of microplastic petroleum waste and improving energy and environmental indicators by optimizing the pyrolysis process and modifying the fuel composition. *Fuel, 376*, Article 132699. https://doi.org/10.1016/j.fuel.2024.132699

Mishra, S., Singh, R. P., Rath, C. C., & Das, A. P. (2020). Synthetic microfibers: Source, transport and their remediation. *Journal of Water Process Engineering, 38*, Article 101612. https://doi.org/10.1016/J.JWPE.2020.101612

Mishra, S., Swain, S., Sahoo, M., Mishra, S., & Das, A. P. (2021). Microbial colonization and degradation of microplastics in aquatic ecosystem: A review. *Geomicrobiology Journal*. https://doi.org/10.1080/01490451.2021.1983670

Mishra, S., Dash, D., & Das, A. P. (2024). Aquatic microbial diversity on plastisphere: colonization and potential role in microplastic biodegradation. *Geomicrobiology Journal, 41*(4), 312–323. https://doi.org/10.1080/01490451.2023.2209750

Mishra, S., Dash, D., & Das, A. P. (2024). Isolation and molecular characterization of polyester-degrading bacterial strains from municipal wastewater and their application in microplastic degradation in contaminated wastewater. *Geomicrobiology Journal*. https://doi.org/10.1080/01490451.2024.2376101

Mishra, S., Singh, R. P., Rout, P. K., & Das, A. P. (2022). Membrane bioreactor (MBR) as an advanced wastewater treatment technology for removal of synthetic microplastics. In *Development in wastewater treatment research and processes* (pp. 45–60). https://doi.org/10.1016/B978-0-323-85583-9.00022-3

Mitra, R., Xu, T., Xiang, H., & Han, J. (2020). Current developments on polyhydroxyalkanoates synthesis by using halophiles as a promising cell factory. *Microbial Cell Factories, 19*, 86. https://doi.org/10.1186/s12934-020-01342-z

Mohanan, N., Montazer, Z., Sharma, P. K., & Levin, D. B. (2020) Microbial and enzymatic degradation of synthetic plastics. *Frontier Microbiology, 11*. https://doi.org/10.3389/fmicb.2020.580709

Mohanty, M., Tripathy, B., Koteswara Reddy, G., & Das, A. P. (2024). Greener technologies for establishing restraint over microplastic pollution. In *Sustainable microbial technology for synthetic and cellulosic microfiber bioremediation* (pp. 239–258). https://doi.org/10.1007/978-3-031-62600-5_13

Mouterde, L. M. M., & Allais, F. (2020). Chapter 5—Bio-based production of chemicals through metabolic engineering. In C. M. Galanakis (Ed.), *Biobased products and industries* (pp. 171–202). Elsevier.

Nath, S. (2024). Biotechnology and biofuels: Paving the way towards a sustainable and equitable energy for the future. *Discovery Energy, 4*, 8. https://doi.org/10.1007/s43937-024-00032-w

Neupane, D. (2023). Biofuels from renewable sources, a potential option for biodiesel production. *Bioengineering, 10*, 29. https://doi.org/10.3390/bioengineering10010029

Okoro, V., Azimov, U., & Munoz, J. (2022). Recent advances in production of bioenergy carrying molecules, microbial fuels, and fuel design—A review. *Fuel, 316*, Article 123330. https://doi.org/10.1016/j.fuel.2022.123330

Otoupal, P. B., & Chatterjee, A. (2018). CRISPR gene perturbations provide insights for improving bacterial biofuel tolerance. *Frontiers Bioengineering Biotechnology, 6*. https://doi.org/10.3389/fbioe.2018.00122

Peng, Y.-H., Shih, Y., Lai, Y.-C., et al. (2014). Degradation of polyurethane by bacterium isolated from soil and assessment of polyurethanolytic activity of a Pseudomonas putida strain. *Environment Science Pollution Research International, 21*, 9529–9537. https://doi.org/10.1007/s11356-014-2647-8

Peng, Y., Wang, Y., Ke, L., et al. (2022). A review on catalytic pyrolysis of plastic wastes to high-value products. *Energy Conversion and Management, 254*, Article 115243. https://doi.org/10.1016/j.enconman.2022.115243

Pinto, A. V., Ferreira, P., Neves, R. P. P., et al. (2021). Reaction mechanism of MHETase, a PET degrading enzyme. *ACS Catalysis, 11*, 10416–10428. https://doi.org/10.1021/acscatal.1c02444

Prus, Z., & Wilk, M.: Microplastics in sewage sludge: Worldwide presence in biosolids, environmental impact, identification methods and possible routes of degradation, including the hydrothermal carbonization process. *Energies, 17*, 4219 (2024). https://doi.org/10.3390/en17174219

Rosas-Vega, F. E., Pozzan, R., Martínez-Burgos, W. J., et al. (2025). Enzymes produced by the genus aspergillus integrated into the biofuels industry using sustainable raw materials. *Fermentation, 11*, 62. https://doi.org/10.3390/fermentation11020062

Ru, J., Huo, Y., & Yang, Y. (2020) Microbial degradation and valorization of plastic wastes. *Frontiers Microbiology, 11*. https://doi.org/10.3389/fmicb.2020.00442

Sahu, S., & Das, A. P. (2024). In *Microplastics degradation mechanism by aquatic plastiomes: A review* (pp. 213–226). https://doi.org/10.1007/978-3-031-50840-0_12

Sahu, A., Dey, S., Mallick, K., Priyadarsini, S., Behera, I. D., & Das, A. P. (2024). A review on advanced technology for sustainable management of synthetic microplastic waste. In *Sustainable management of urban plastic waste through circular economic approaches* (pp. 262–279). https://doi.org/10.1201/9781003465447-15

Saini, R., Hegde, K., Brar, S. K., & Vezina, P. (2020). Advanced biofuel production and road to commercialization: An insight into bioconversion potential of *Rhodosporidium* sp. *Biomass and Bioenergy, 132*, Article 105439. https://doi.org/10.1016/j.biombioe.2019.105439

Samak, N. A., Jia, Y., Sharshar, M. M., et al. (2020). Recent advances in biocatalysts engineering for polyethylene terephthalate plastic waste green recycling. *Environment International, 145*, Article 106144. https://doi.org/10.1016/j.envint.2020.106144

Satta, A., Zampieri, G., Loprete, G., et al. (2024). Metabolic and enzymatic engineering strategies for polyethylene terephthalate degradation and valorization. *Reviews Environmental Science Biotechnology, 23*, 351–383. https://doi.org/10.1007/s11157-024-09688-1

Sawada, H., Tomita, H., Okazaki, F., & Tamaru, Y. (2023). Comparison of different carbon sources on biomethane production with clostridium cellulovorans and methanogens. *Applied Microbiology, 3*, 493–503. https://doi.org/10.3390/applmicrobiol3020035

Shi, L., Liu, H., Gao, S., et al. (2021). Enhanced extracellular production of IsPETase in *Escherichia coli* via engineering of the pelB signal peptide. *Journal Agricultural Food Chemistry, 69*, 2245–2252. https://doi.org/10.1021/acs.jafc.0c07469

Sudarsan, S., Blank, L. M., Dietrich, A., et al. (2016). Dynamics of benzoate metabolism in Pseudomonas putida KT2440. *Metab Engineering Communications, 3*, 97–110. https://doi.org/10.1016/j.meteno.2016.03.005

Sun, J., Lu, L.-B., Liang, T.-X. et al. (2020). CRISPR-assisted multiplex base editing system in pseudomonas putida KT2440. *Frontier Bioengineering Biotechnology, 8*. https://doi.org/10.3389/fbioe.2020.00905

Tanaka, Y., Hiraga, K., & Inui, M. (2024). Regulation of the expression of MHETase and TPA degradation genes involved in the degradation of PET in Ideonella sakaiensis. *FEBS Journal, 291*, 4489–4500. https://doi.org/10.1111/febs.17240

Temporiti, M. E. E., Nicola, L., Nielsen, E., & Tosi, S. (2022). Fungal enzymes involved in plastics biodegradation. *Microorganisms, 10*, 1180. https://doi.org/10.3390/microorganisms10061180

Thakur, B., Singh, J., Singh, J., et al. (2023). Biodegradation of different types of microplastics: Molecular mechanism and degradation efficiency. *Science of the Total Environment, 877*, Article 162912. https://doi.org/10.1016/j.scitotenv.2023.162912

Tiso, T., Winter, B., Wei, R., et al. (2022). The metabolic potential of plastics as biotechnological carbon sources—Review and targets for the future. *Metabolic Engineering, 71*, 77–98. https://doi.org/10.1016/j.ymben.2021.12.006

Verdezoto-Prado, J., Chicaiza-Ortiz, C., Mejía-Pérez, A. B., et al. (2025). Advances in environmental biotechnology with CRISPR/Cas9: Bibliometric review and cutting-edge applications. *Discover Applied Sciences, 7*, 167. https://doi.org/10.1007/s42452-025-06609-x

Wackett, L. P. (2021). The future of microbial biotechnology. *Microbiology Biotechnology, 15*, 79–80. https://doi.org/10.1111/1751-7915.13920

Wan Osman, W. N. A., Rosli, M. H., Mazli, W. N. A., & Samsuri, S. (2024). Comparative review of biodiesel production and purification. *Carbon Capture Science and Technology, 13*, Article 100264. https://doi.org/10.1016/j.ccst.2024.100264

Wang, H., & Zhou, Q. (2024). Bioelectrochemical systems—A potentially effective technology for mitigating microplastic contamination in wastewater. *Journal of Cleaner Production, 450*, Article 141931. https://doi.org/10.1016/j.jclepro.2024.141931

Wen, Q., Chen, J., Li, J., et al. (2024). A single-plasmid-based, easily curable CRISPR/Cas9 system for rapid, iterative genome editing in Pseudomonas putida KT2440. *Microbiology Cell Fact, 23*, 349. https://doi.org/10.1186/s12934-024-02634-4

Wierckx, N., Narancic, T., Eberlein, C., et al. (2019). Plastic biodegradation: Challenges and opportunities. In R. J. Steffan (Ed.), *Consequences of microbial interactions with hydrocarbons, oils, and lipids: Biodegradation and bioremediation* (pp. 333–361). Cham: Springer International Publishing.

Wong, S. L., Ngadi, N., Abdullah, T. A. T., & Inuwa, I. M. (2015). Current state and future prospects of plastic waste as source of fuel: A review. *Renewable and Sustainable Energy Reviews, 50*, 1167–1180. https://doi.org/10.1016/j.rser.2015.04.063

Xue, K., Bai, Z., Fordour, E., et al. (2024). Bacterial surface display of PETase mutants and MHETase for an efficient dual-enzyme cascade catalysis. *Bioresource Technology, 408*, Article 131177. https://doi.org/10.1016/j.biortech.2024.131177

Ye, J., Chen, Y., Gao, C., et al. (2022). Sustainable conversion of microplastics to methane with ultrahigh selectivity by a biotic-abiotic hybrid photocatalytic system. *Angewandte Chemie, 134*, Article e202213244. https://doi.org/10.1002/ange.202213244

Yip, A., McArthur, O. D., Ho, K. C., et al. (2024). Degradation of polyethylene terephthalate (PET) plastics by wastewater bacteria engineered via conjugation. *Microbial Biotechnology, 17*, Article e70015. https://doi.org/10.1111/1751-7915.70015

Zhang, N., Ding, M., & Yuan, Y. (2022). Current advances in biodegradation of polyolefins. *Microorganisms, 10*, 1537. https://doi.org/10.3390/microorganisms10081537

Zhou, Y., Lin, L., Wang, H., et al. (2020). Development of a CRISPR/Cas9n-based tool for metabolic engineering of Pseudomonas putida for ferulic acid-to-polyhydroxyalkanoate bioconversion. *Communication Biology, 3*, 1–13. https://doi.org/10.1038/s42003-020-0824-5

Biofuel Production of Plastic Waste Through Microbial Technology

Neethu Asokan, Santosh Kumar Singh, Rajeshwari Ullagaddi, and Kritish De

Abstract With the escalating population the plastic waste generated globally was estimated to about 367 million tons in 2020 and expected to triple by 2060. India witnesses particularly alarming plastic pollution of 9.3 million tons annually accounting for the one-fifth global plastic waste contributor. There is hence an emergent need for waste management strategies. This study explores the conversion of plastic waste into valuable fuels specifically hydrogen gas and diesel through microbial technology, advanced pyrolysis, and catalytic depolymerization technologies. Pyrolysis effectively breaks down different plastic waste at high temperature to produce gas and oil. Diesel produced can be used as fuel for vehicles with existing diesel engines. While hydrogen gas produced will reduce green house emissions if integrated into energy systems like fuel cells or vehicles powered with hydrogen. To optimize plastic degradation microbial enzyme engineering is employed to break down plastics to value-added products like biofuel along with bioreactor optimization for ideal cultivation of microorganisms. With this research it is estimated to produce an approximate of 60:50 percent ratio of hydrogen gas to diesel oil per ton of plastic waste. The research highlights the potential for combining chemical engineering and life science to address plastic pollution. By implementing this we can significantly reduce the landfills while fostering energy production and sustainability. This research underscores to combat pollution and contribute to a greener future.

Keywords Plastic waste · Diesel · Hydrogen gas · Microbial enzymes · Optimization · Biofuel

N. Asokan (✉) · S. K. Singh · R. Ullagaddi · K. De
Department of Life Sciences, Sri Satya Sai University for Human Excellence, Kalaburagi, Karnataka, India
e-mail: neethu.asokan@sssuhe.ac.in

A. P. Das (ed.), *Bioconversion of Environmental Waste to Value Added Products*, Sustainable Environmental Waste Management Strategies,
https://doi.org/10.1007/978-3-032-14215-3_4

1 Introduction

Garbage is one of the most significant problems of today. It is generated by the ongoing accumulation of plastic debris in various areas. Although organic waste can be allowed to biodegrade, plastic waste is non-biodegradable and persists in the ecosystem for thousands of years, with quite severe environmental consequences. At present, as plastic is employed in several industries and economic sectors, particularly in packaging and construction, the generated volume of plastic is more than 460 million metric tons of waste annually and is rising logarithmically (Ghosh et al., 2023; Mishra et al., 2021; Sahu & Das, 2024). Several kinds of plastics, mostly single-use plastics, should be of worry concerning plastic pollution hazards; these include bottles, straws, and plastic bags (Mishra et al., 2022; Mohanty et al., 2024a, 2024b). A shocking 55% of these plastics are dumped in the landfill while 25 are burnt and therefore emit fumes into air and are also sources of greenhouse gases (Jones & Patel, 2021). Mismanagement of plastics result to pollution of the terrestrial and aquatic ecosystems which in turn goes ahead to pollute food chains, hence are dangerous to the health of humans (Bala et al., 2024; Das et al., 2024b; Ray et al., 2024). However, most of these places lack proper waste management technology to help contain the problem from worsening. There is a common lack of technical experience and assets in many countries to properly sort, collect, and recycle plastics (Mishra et al., 2024a; Sahu et al., 2024). This means a high capacity of plastics land up in landfills or is irresponsibly discharged in the natural habitats, areas where it is causing the loss of species and habit degradation. Plastic waste problem can only be solved where there is a composite approach to the enhancement of waste management, minimizing the use of plastics which have single use only and citizens' education. It's not a question of how we deal with the waste arising from the use of this material in all aspects of life but about finding ways to lessen the impact of such a material on the planet (Doe et al., 2020).

1.1 Global Plastic Waste Scenario

Globally, the plastic waste situation has become worrisome an area of which highlights a major conservation concern in the world today. Currently, the global plastic production stands at about 460 Mt. rising from 2 Mt. in 1950. It is expected that without intervention, production of plastics could rise to nearly triple to about 1.23 billion tons by 2060 (Landrigan et al., 2023). Plastic waste globally has expanded at a rate of 1.2 times over twenty years leading to 156 million tons in 2000 and reaching 353 million tons in 2019. This waste management situation makes for critical reading since only fifteen percent received collection services for recycling, yet nine percent succeeded in recycling. The plastic recovery efforts led to landfill deposition of fifty percent of the materials and twenty-five percent disposal at unsanitary locations or through thermal treatment. Environmental degradation mostly stems from improper waste handling, and it poses the greatest threat to land and aquatic habitats (Mishra & Das, 2021; Tripathy et al., 2024). The usage of plastic has intensified since the Covid-19 outbreak when medical tools and vacuum packing materials added to existing amounts of waste (Acharya et al., 2024). This caused an increased influx of plastic in the oceans and more frequently from facial masks and gloves. Historically, SSPs' wastes have triggered enormous global controversy not as ordinary wastes to be disposed of but as threats to biological diversity as well as human health. Future projections for the future the OECD expects that amount of the plastic waste will rise to around 1014 Mt. by 2060 (Landrigan et al., 2023). It is projected that short-life use items such as packaging and consumer items that would be most of the total plastics waste stream at the time identified as 25%, would constitute approximately 65% of the total waste stream. Thus, non-OECD nations will be responsible for 65% of total BAU plastic waste; among which, the emergent economies in Asia and Africa will grow the fastest. Even though the recycling infrastructure has evolved, the recycling rate is projected to increase only slightly from 9% (2019) to 17% (2060).

Conversely, landfilling will still constitute one of the dominant disposal methods at 50%, followed by incineration at 18% of total waste management strategies. Furthermore, unmanaged plastic waste will increase from 79 Mt. in 2019 to 153 Mt. in 2060 primarily because of the inadequate structure in fast-developing regions (Pradhan et al., 2024; Priyadarshini & Das, 2024a). Environmental impact of plastic pollution pose a major concern as it affects every aspect of the environment. Some of them are just as general such that they cause pollution while others release chemicals that harm natural settings. A paper observed that the present generation of plastics is more than that of the planet's animals together. Moreover, emissions from plastic are seen to grow rapidly and, in the absence of changes in the consumption and production patterns, could add 6.5 billion tons of CO_2 equivalent by 2050 (Landrigan et al., 2023).

1.2 Plastic Pollution in India: Current Trends and Challenges

Generating about 9.3 million tons of plastic waste yearly, India is one of the nations worldwide affected by plastic pollution. Around 20% of global emissions are linked to this figure, with Nigeria and Indonesia responsible for 3.5 million tons and 3.4 million tons, respectively (Times of India, 2024). High population growth, urbanization, and consumerism are said to be behind the over plastic waste handling. The rising consumption of single-use plastics and packaging materials and trends in plastic litter generation are to blame for an approximate per capita plastic waste of 11 kg/year in urban areas. Dealing with waste management is a challenging problem in India since incineration releases noxious chemicals including dioxin, which causes many fires and other health issues like lung disease. This leads to an extensive problem with waste disposal. Each year, plastic waste is released onto the land, air, or water approximately 3.5 million times more 16. India's inability to expand its waste facilities has led to a plastic pollution crisis, with the major factors being: inefficient waste management infrastructure and inadequate plant capacity. Research indicates that roughly 40% of municipal waste is disposed in open space landfills, while 77% of urban waste falls into these landfill sites (Zhang et al., 2024). It should be noted that around 60% of the plastic waste generated are recycled, and the majority of this recycling occurs in the black market, albeit not properly managed (Zhang et al., 2024). Open burning practices and landfill operations are notorious for their significant role in air pollution. Roughly 30% of all the plastic waste generated is unregulated, and a significant number of these sites are leaded with toxic chemicals into the surrounding environment. Regulations are hindered by the fact that some bisphenol A-containing products, such as single-use plastics have been banned by government since 2022. Because of their low cost and easy availability, these components still account for nearly half (49%) of the total plastic waste generated in India today newspaper. Sadly, there is a gap in reporting on the accurate and produced plastic waste as well as its management. While the official record shows a collection rate of 95%, studies and surveys indicate that the actual statistics are at an average of 81%, which highlights the system's flaws. Information regarding their environmental and health effects is important to share (NITI Aayog, 2021). The negative impact of plastic pollution on India's ecosystem is severe, as it obstructs the seas, water, and drainage systems, leading to frequent floods in urban areas. The impact of plastics on marine life has been found to be extremely harmful, with an estimated 80% of the marine litter in India being made up of these materials. According to recent research, approximately 83% of tap water samples in India contain microplastics. The transfer of microplastics into agricultural soils via contaminated irrigation water can have adverse effects on soil quality and food security (ACS Omega, 2023). Government of India have launched Swachh Bharat mission and also started the waste to wealth mission where research focussed on waste to value added products were encouraged along with plastic waste management amendment rules. The government shifted from open defecation to garbage free cities through Swachh Bharath Mission (Government of India, 2021; The Ministry of Housing and Urban Affairs, Government of India, 2021).

1.3 Environmental Health Hazards

Plastic combustion generates toxic emissions that pose health risks for local inhabitants. In the future, it is anticipated that microplastics in the environment could pose a risk to human health. India's plastic pollution problem necessitates comprehensive solutions, including the establishment of effective waste management systems (Paul et al., 2024; Singh et al., 2020). Students may find information that explains the replacement of informal systems and tactics with regulated systems to increase recycling rates and reduce environmental harm. Enhancing Regulatory Systems: The complete enforcement of current laws against single-use plastic is still essential. The topic may hold significance in the creation of affordable substitutes for these products. The implementation of public awareness campaigns can enhance community involvement in the fight against plastic pollution by educating individuals about the effects of plastic contamination and promoting adaptive behaviors. India should join international agreements because international cooperation stands as the answer to worldwide plastic pollution. When designing an anticipatory approach for plastic pollution in India it is essential to understand existing environmental and societal concerns regarding plastic pollution impacts.

1.4 The Urgency for Sustainable Waste Management

The need for a sustainable waste management solution demands immediate action because it affects both India and the global community. Higher population density from urbanization intensifies waste production levels, thus posing environmental and health risks to human beings (Dey et al., 2024; Mishra et al., 2024b). This document presents the current disposal of waste in India and worldwide while explaining the necessity for active participation with innovative eco-friendly waste management tactics (Mishra et al., 2021, 2022; Mishra & Das, 2023). This paper examines the waste generation emergency in India together with its resulting effects. Only about 23% of India's waste produced each year is effectively managed, whereas the rest is not. The growing population of cities is making the situation worse, and it is projected that urban areas will generate an estimated 125 million tons of waste annually by 2031 due to their rapid growth (NITI Aayog). Even though there has been some progress since the year of 2014, there are still significant issues that need to be addressed. In 2021, the proportion of waste treatment capacity being between 40 and 70% is estimated to be around 18%. A brief effort to promote clean India through SWM-led interventions was launched in 2014, the SBM program. Even so, the latest calculations indicate that waste processing activity is still below 40%, indicating significant variation between the legislation and its actual implementation. Globally, the problem of managing waste is just as severe. In the absence of remedial measures, global solid waste disposal will increase to 3.40-billion tons by 2050, as projected according to the World Bank. This will largely be due to the poor waste disposal infrastructure of L&M ICs. The global society faces significant challenges in managing and recycling waste, as well as handling hazardous wastes. Waste disposal practices that adhere to hygiene practices have a direct impact on the environment, including causing diseases and climate change (Fig. 1).

A circular economy is now seen as a practical solution to the world's waste issue. Waste reduction and efficient resource utilization are promoted by practices like recycling and controlled consumption (Priyadarsini & Das, 2024b, 2025). The implementation of circular economy strategies for waste management in India can transform challenges into opportunities and preserve the environment (Ellen MacArthur Foundation, 2013). According to the World Economic Forum, a circular economy transition could yield economic benefits of over $4.5 trillion by 2030. The degradation

Fig. 1 Waste to fuel

rate of plastics depends on the specific manner microbes react to them. PE and PET plastics alongside PS and PPV are the four main types of plastic waste which accumulate in the environment. The scientific community credits plastic breakdown to microbial species including *Ideonella scarleta* and *Rhodococcus ruber* together with Pseudomonas aeruginosa (Ru et al., 2020). These organisms are diverse. Researchers have studied the plastic degradation capabilities of fungi such as Aspergillus while also including Penicillium species in their examination. Plastics undergo reduction to break smaller organic molecules which fermenting microorganisms or thermochemical conversion processes like pyrolysis and gasification transform into biofuels. Multiple scientific research studies demonstrate the usefulness of microbial isolates in breaking down PE plastic wastes (Sen & Raut, 2016). Therefore, a sustainable strategy should incorporate circular economy principles and community involvement, as well as technological innovation. Policies aimed at aiding waste management should be implemented by governments, with additional funds allocated for infrastructure and technology. The responsibility of citizens as responsible consumers includes engaging in judicious consumption and managing local waste. Moving forward, it's time for India and the world to acknowledge that we can't separate our environment from development. If we adopt today's sustainable waste management practices, the planet will be cleaner and healthier in the future generations.

1.5 Microbial Technology for Plastic Degradation

Because pure plastics don't dissolve in water and are chemically pretty inactive, they're generally not very toxic. However, some plastic items can be toxic because of the extra ingredients mixed in (Das et al., 2024a; Mishra & Das, 2022). For instance, things like adipates and phthalates, which are plasticizers, are often added to plastics like polyvinyl chloride (PVC) to make them less brittle and more flexible. Tiny amounts of these chemicals can leak out of the product. Scientists think that chemicals seeping out of polystyrene food containers might mess with hormones and could even cause cancer. The final plastic product itself might be safe, but the building blocks (monomers) used to make the original polymers can be harmful. An experiment by Arkatkar in 2010 looked at whether biodegraded polythene is toxic to plants. They checked this by seeing how well seeds like groundnut, sunflower, safflower, sesame, and soybean sprouted after being exposed to it. The study found that fewer seeds sprouted in the treated group. But when they tested it on larvae (specifically, *Chironomous* species), they didn't see any signs of toxicity.

Biodegradation is a green process as well no hazardous byproducts are linked to it. Conversely, the oxalic acid biodegradation must take place. Prior to biodegradation, plastic waste that is not biocompatible can be considered a non-biodegradable material through natural organisms. Oxo-biodegradation is an abiotic. This reaction causes plastic to be biodegradable and non-biodegradable. The slow biodegradation of plastics is attributed to the structure of long polymers. Additives, including antioxidants, are present in chains (Zhang et al., 2024). Plastics undergo oxidative degradation where oxygen exposure accelerated by heat or UV radiation cause the breakage of polymer chains and form oxygen-containing functional groups like carboxylic acids, hydroxy acids, esters, aldehydes and alcohols. The oxo-biodegradation of plastic is the rate determining step as microbes act efficiently only once it is fragmented. Plastic waste undergo fragmentation via both photodegradation and thermal degradation facilitating subsequent microbial plastic breakdown.

Non-biodegradable plastics can undergo decomposition after oxo-biodegradation. Three distinct biodegradation roots, namely invertebrates, microorganisms, and plants have been involved in the process and treated with composite. For example, certain invertebrates. Degradation of plastics can occur in some organisms, such as mealworms (e.g., Tenebrio obscurus and others). Achroia californiens and lesser waxworms are also used in the production of poly (Shield, St. John's Facuum), while Zophobas atratus is another common species. (Grisella for high-density polyethylene). Microorganisms also mineralize. Biodeterioration of plastics causes the conversion of them into CO_2, H_2O, and biomass. Breaking down the polymer material into smaller particles, depolymerization occurs. The breakage of the polymer chain into oligomers, dimers, or other components occurs during the enzyme reaction. Microbial cells absorb small molecules via bioassimilation, which is facilitated by the existence of monomers. Create primary and secondary metabolites, along with mineralization. Release final products such as CO_2, CH_4, H_2O, and other metabolites (gases). Polyethylene is among the rare substances that microorganisms break down into their structure. Polyethylene terephthalate (PET), HDPE/LDPE, and other polymers are available. In the case of P. citronellolis and B, polyvinyl chloride is used as a solvent. Both polyurethane (PU) and polystyrene are used in industrial applications such as Aspergillus spp., while PU is utilized in the latter. Yang et al., 2014 used organism with the mixing of mature. After weighting the composite with plastics, it is treated at a temperature of 40 to 60 degrees, the microorganisms that exist are already present.

Mucor circinilloides together with Aspergillus flavus were isolated from municipal landfill areas. Organisms isolated from municipal landfill area demonstrated effective degradation of low density polythene when analyzed through CO_2 evolution tests. The researchers also used scanning electron microscopy and colonization studies (Pramila & Ramesh, 2011). The researchers tested low-density polythene degradation through a CO_2 evolution examination. The researchers evaluated the low-density polythene degradation through SEM (Scanning Electron Microscopy) technology and microbiological identification analyses, (Pramila & Ramesh, 2011). In vitro degradation of six strains of Aspergillus spp. together with two Fusarium spp. were used for LDPE degradation. The biodegradation of plastics has been identified in research through the investigation of two Fusarium spp. with the resulting evidence showing the cause of material destruction. The algal strain has active enzymes which it produces to degrade polyethylene as demonstrated by Kumar et al. (2017). The degradation process of polythene and plastic has shown different levels of completion. Pseudomonas spp. (37.09 and 28.42%), Streptomyces spp. (46.16 and 35.78%), and Aspergillus spp. (20.96 and 16.84%) in 6-month period in liquid (shaker) culture (Usha et al. 2011). Degradation of plastic cups and polythene bags: The study involved testing bacteria together with fungi to break down plastic materials during a one-month period. The bacterial strain known as Pseudomonas spp. showed ability to decompose 20.54% of polythene and 8.16% of plastics during this period. Polythene and 8.16% of plastics, while fungal species, Aspergillus glaucus brought polythene breakdown to 28.80% along with 7.26% plastic destruction (Shah et al. 2013). Summary of some studies on the evaluation of plastic biodegradation: The capacity of researchers evaluated the degradation potential of some fungal strains obtained from plastic-contaminated areas. The microorganisms underwent tests to determine their growth capabilities on polyurethane and urease, protease, esterase, and laccase activity. Rhodococcus ruber alongside with Brevibaccillus borstelensis exhibited this capability. Biological studies have shown these organisms possess the ability to break down the CH_2 backbone structure while using polyethylene as their carbon source, and polyethylene as its sole carbon source (Loredo-Trevino et al. 2011).

Microorganisms are utilized in biodegradation to break down synthetic plastics, which are not naturally able to decompose. The rise in plastic waste worldwide, which is estimated to be more than 360 million tons annually, has led to growing environmental concerns and health issues. The inadequacy and traditional disposal methods which include landfilling alongside incineration procedures present dangerous threats to the environment, thus biological alternatives emerge as solutions, being investigated. The degradation of plastics occurs in microorganisms by means of multiple biological reactions and specific enzymatic reactions. Microbes begin the breakdown process by attaching themselves to plastic surfaces. The bacteria create surface coatings known as biofilms that accelerate the degradation process. Hydrolysis marks the first stage which biocatalysts subsequently use to perform the decomposition. The process of hydrolysis combined with depolymerases produces smaller oligomers or monomers from polymer chains. Once absorbed, the microbial cells break down their gained monomers into basic substances including carbon dioxide and water and biomass during assimilation. Biodegradation speed of plastic depends on plastic classification combined with environmental conditions together with the particular microbial strains present. These environmental conditions together with the present microbial species determine the speed of plastic breakdown. Some bacterial strains demonstrate adequate potential for waste management through microbial techniques. Scientific studies demonstrate these species' capability for breaking down polystereine (PS) and polyethylene (PE) in addition to PEP plastics and PEP. Microbial technology, in combination with genetic engineering and polymerase chain reaction PCR technology joins environmental sustainability by supporting its efforts alongside other important contributors. The importance of various fields demonstrate equal importance for the microbial approach to degrade plastic. Microbial technology has the potential to be utilized in waste management, as some bacterial strains have been shown to degrade common plastic materials such as polyethylene (PE), polystereine (PS), and PEP (Thakur et al., 2022). Microbial technology, in combination with genetic engineering and polymerase chain reaction (PCR) technology is an important contributor to environmental sustainability (Thakur et al., 2022). The importance of microbial plastic degradation is shared among various fields. Recycling plastics by microbes through microbial recycling reduces pollution and restores the balance of the environment by producing harmless products like carbon dioxide and water. In sustainable waste management, this biological approach can be a valuable addition to existing methods that aim to reduce landfill discharge. Furthermore, microbial degradation offers an opportunity for biovalorization, where some microbes not only decompose plastic but also convert it into valuable chemicals, in line with the circular economy model. Finally, the biotechnological use of microbes in the degradation processes of plastic waste is a very promising step forward on the path of environmental biotechnology (Thakur et al., 2022). By focusing on the optimization of microbial capabilities and scalability, this technology provides a sustainable and eco-friendly answer to the global plastic waste crisis.

Polyethylene degradation produces different molecular mechanisms including chemical thermal photo and biodegradation. Microbes naturally perform transformation of different compounds including hydrocarbons during their secondary metabolic process. The microbial capacity includes receiving multiple compounds which stretch to hydrocarbons: (PAHs), pharmaceutical substances, and metals. The enzymes transform hydrolyzed polyethylene compounds into necessary functional groups that promote better hydrophilicity which breaks down main chain sequences to produce weak elastic products of low molecular weight. This material becomes less accessible for microbial use because weak mechanical properties and weight reduction combine to create barriers for microbial assimilation (Shah et al., 2013). Plastic degradation through particular enzymatic systems produces enzymatic breaking of polymer chains into oligomeric matter and final monomer fragment products. These materials go through enzymatic decomposition until they become either oligomeric chains or organic intermediate substances. During biodegradation the toxic substances including acids and alcohols and ketones develop as byproducts. The water-soluble cleaved products get absorbed by microbial cells through an intake process. The compounds serve as bacterial dietary for cell-based biological decomposition processes.

The end products from cellular aerobic metabolism included both carbon water and carbon dioxide, whereas anaerobic and aerobic metabolism generates carbon dioxide and water while anaerobic activity yields carbon dioxide and water combined with methane. The mechanical strain of polymers exposed to temperature, pressure, and moisture content brings about biological-friendly material separation through physical methods. This process becomes activated through external metabolism-triggered enzymes as well as additional metabolites according to (Yang et al. 2014). Plastics tend to break down through surface erosion because enzymes cannot easily penetrate polymeric materials, thus enzymes work only on the outside area of exposed plastic surfaces. The pro-oxidants use molecular oxygen to form free radicals in plastic degradation following their contact with polyethylene where the polyethylene becomes vulnerable to oxygen attacks within its matrix (Skariyachan et al., 2015). The determination of polymer biodegradation happens by examining CO_2 evolution rate and O_2 uptake and polymer properties alterations together with organism growth rates (Mohan, 2011).

1.6 Conversion Processes for Biofuel Production

Various techniques, including pyrolysis and catalytic depolymerization, have been developed to break down plastic polymers into useful liquid fuels. Plastics undergo thermal degradation, referred to as pyrolysis, due to the absence of oxygen and the subsequent breakdown of smaller hydrocarbon molecules when heated at high temperatures (300–800 °C). Depending on the reactor type, temperature, and duration of residence, bio-oil, syngas, or char are the main outputs from pyrolysis. What is an average product? Bio-oil that is gathered can be transformed into fuels that are similar to conventional diesel and gasoline. Conversely, the use of non-catalytic pyrolysis often produces an extensive array of hydrocarbons with impurities, necessitating further refinement. To improve fuel quality and process efficiency, catalysts like zeolites, silica-alumina, or metal oxides are utilized in catalytic depolymerization to break down plastic polymers at lower temperatures. This technique reduces unwanted byproducts and increases the yield of high-quality liquid fuel products by improving reaction selectivity. Catalysts enable processes to become more energy-efficient at the same time reducing operational costs. The extensive research within this field depends on optimization methods because catalyst types and reaction environmental factors significantly affect fuel composition outputs. The process of converting plastic materials into biofuel works by employing both pyrolysis and catalytic depolymerization as established methods that support energy requirements and plastic waste solutions.

The synthetic plastics exist in two distinct degradation pathway groups according to whether they include carbon-based structures (fuscation) or not. PE together with PS and PP along with PVC plastics all depend only on carbon atoms as their basic construction materials. PET and PU plastics possess heteroatoms together with the backbone components of their primary structures. Scientists have developed several biodegradable aliphatic polyesters such as Polyhydroxyalkanoate (PHA) and polylactic acid (PLA) which can function as alternative materials for specific petroplastics (Bhatia et al., 2019a, 2019b). The sustainable nature of polyhydroxyalkanoates (PHAs) exists because they show excellent biodegradability across multiple environments and they possess biocompatibility as well as diversity in chemical composition and they derive from renewable carbon sources and produce nontoxic harmless end products when degraded (Bhatia et al., 2019a, 2019b).

Plastics are suitable for depolymerization into monomers which can lead to recycling and through mineralization microbes create brand new bioproducts. The method shows great potential according to research by Grima et al. (2000). The microorganism excretes extracellular enzymes to break down plastic while they adhere to plastic surfaces before starting hydrolysis which produces short polymer intermediates that provide carbon source for generating CO_2 through microbial cell processes. Recent studies show that microorganisms have started showing indications of polymer degradation even though plastics are non-natural materials. Studies corroborate that laboratory research proves more than 90 types of microorganisms with both bacterial and fungal characteristics that show petroleum-based plastic deposition abilities (Jumaah, 2017). Bio-reactors involved in plastic decomposition count as living organisms. Microbes activate the enzymes that split polymer structures while the enzymes help the microbes break down these hydrolysis materials. Scientists do not have clear insight about the mechanisms of enzyme-substrate chemical reactions as well as the identification of enzymes that degrade petro-polymers such as PE and PP. Some advancements have been achieved to identify and differentiate between the enzymes that take part in petro-polymerization. The polymer chains experience hydrolysis/oxidation/hydroxylation until their breakdown into oligomers and monomers is achieved through enzymatic processes. Petro-polymers exist as two distinct groups which include hydrolyzable polymers such as PET and PUR and non-hydrolyzed polymers consisting of PET and its derivatives as well as PS, PP, and PVC. The classification system determines the enzymes which can degrade these materials.

Biofuel production can be achieved through pyrolysis, which is one of the most extensively researched processes and efficient ways to recycle plastic waste. A thermochemical process is used to break down long-chain plastic polymers into smaller hydrocarbon molecules, operating at temperatures between 300 and 800 °C without oxygen. Depending on the plastic type, reactor conditions, and catalyst usage, the process produces three main products: pyrolysis oil (50–80%), syngas (10–30%), or solid char (10–20%). The various forms of pyrolysis are as follows: The process of slow pyrolysis occurs at temperatures ranging from 300 to 500 °C, with extended residence times that produce more solid char. The use of rapid pyrolysis, which occurs at temperatures ranging from 400 to 600 °C and requiring brief periods of residence, can result in a bio-oil yield of up to 80%. To enhance the refinement, selectivity, and decrease impurities, catalysts such as zeolites, silica-alumina, or metal oxides are utilized in catalytic pyrolysis. The utilization of microwave radiation for uniform heating, a process known as microwave pyrolysis, reduces energy usage and improves the fuel composition. Pyrolysis oil (Plastic-Derived Oil, PDO): The main product of the process is pyrolysis oil, also called plastic-derived oil (PDO), which has properties that are similar to those of conventional fossil fuels. The range of PDO's calorific value is 35–45 MJ/kg, comparable to diesel (45 MJ/kg) and gasoline (42 MJ/kg). Hydro-treatment, fractional distillation, or catalytic upgrading must be used to meet the fuel quality standards of raw pyrolysis oil which often contains high levels of oxygen and sulfur as well as chlorine. Depending on the feedstock composition, type of reactor, residence time, heating rate, and catalyst selection, the efficiency of pyrolysis can be determined. Challenges involve energy consumption, plastic feedstock variability, and process scalability. Plastic pyrolysis for biofuel production is becoming more economically and environmentally viable due to emerging trends like AI-driven optimization, fluidized bed reactors, and microwave-assisted pyromyrolytic digestion.

The use of catalysts in catalytic depolymerization, an advanced thermochemical process, enables the conversion of plastic polymers into biofuels, leading to increased reaction efficiency. By using catalytic depolymerization instead of traditional pyrolysis, which requires high temperatures (300–800 °C) to break down plastics, the process decreases energy consumption, facilitates the selection of valuable hydrocarbons, and prevents unwanted byproducts such as char and wax. The conversion of plastic waste into liquid biofuels, syngas, and chemical feedstocks is a promising approach that can help combat plastic pollution and dependence on fossil fuels. Catalytic depolymerization comprises three primary phases: plastic polymers undergo thermal cleavage at high temperatures (250–500 °C) to form free radicals and oligomers. What are the consequences? The cracking and reforming of long-chain hydrocarbons are made possible by catalysts, which also reduce the selectivity for fuel-like compounds. By taking this step, the production of C5–C23 hydrocarbons is enhanced to the same level as that in gasoline and diesel fuels. Higher stability and energy density are achieved by increasing the efficiency of fuel quality through hydrogenation and upgrading, as well as other catalytic processes like hydrodeoxygenation (HDO), hydrocracking, and isomerization. This is a crucial factor for factors such as the type of catalyst used and their impact on product distribution, fuel properties, or process efficiency. Plastic depolymerization utilizes various catalysts, such as: microporous aluminosilicate catalysts with high surface area and acidity, known as zeolites (ZSM-5, HY, Beta), enhance cracking efficiency. ZSM-5 enhances the yield (30–50%) of aromatic hydrocarbons, which makes it a viable option for producing high-octane gasoline. The use of silica-alumina and metal oxides (Al_2O_2, TiO_2, CeO_2, and ZnO) catalysts can promote cracking reactions and decrease coke formation, which in turn boosts the yield of diesel-range hydrocarbons. Moreover, the use of alumina catalysts can assist in eliminating chlorine and other impurities that come from plastic-based fuels. The utilization of transition metal catalysts (Ni, Mo), Co, Pd, and Pt) enables the hydrocracking and dehydrogenation reactions, which increases the availability of hydrogen for syngas production and biohydrosynthesis. Many hydrogen-rich syngas are made using catalysts primarily of nickel, and the reduction of sulfur and nitrogen compounds by palladium or platinum is used to improve fuel stability. Through the use of zinc oxide and metal oxide catalysts, acidic cracking sites and hydrogenation functions based on metals are combined to produce higher fuel yield (up to 80%) and lower impurity content. They are particularly useful for processing mixed plastic waste, where multiple polymer types must be combed down at the same time. Several challenges must be overcome in order for it to be commercially successful, including the use of catalytic depolymerization. Key areas for improvement include: catalyst deactivation is characterized by the accumulation of carbon deposits and impurities, which can cause efficiency to decline over time. To maintain performance, the catalyst must be periodically regenerated through steam treatment, oxidation, or solvent washing. Various plastics such as PE, PP, and PS (PE, PET, PVC) decompose differently, leading to the production of different fuel compositions. More sophisticated sorting and pretreatment techniques, such as solvent-based separation and thermal pre-cracking are being developed to optimize feedstock quality. High biofuel yields have been demonstrated in laboratory-scale studies, but increasing reactor design improvements (fluidized bed, microwave-assisted, and plasma-aided reactors) are necessary for efficient catalyst utilization and consistent heating. The energy consumption can be reduced by opting for catalytic depolymerization over non-catalysis, but AI-driven process control, heat integration, and catalyst recycling can lead to higher energy efficiency. Plastic waste can be transformed into biofuels through catalytic depolymerization, which offers advantages such as lower energy consumption, better selectivity, and fewer impurities than traditional pyrolysis. The implementation of advanced catalysts, reaction conditions, and scalability issues make this technology highly beneficial for sustainable waste management and renewable energy production.

1.7 Role of Microorganisms in Plastic Breakdown

Plastics are broken down into smaller molecules by diverse enzymes, which are typically able to break polymeric materials. There are two main phases of degradation: primary degradation, where enzymes break large polymer chains into oligomers and monomeric groups, and secondary degradation in which microbes have broken these chains down to simpler organic compounds for biofuel production. The degradation rate of plastics depends on the specific manner microbes react to them. PE and PET plastics alongside PS and PPV are the four main types of plastic waste which accumulate in the environment. The scientific community credits plastic breakdown to microbial species including Ideonella scarleta and Rhodococcus ruber together with Pseudomonas aeruginosa (Ru et al., 2020). These organisms are diverse. Researchers have studied the plastic degradation capabilities of fungi such as Aspergillus while also including Penicillium species in their examination. Plastics undergo reduction to break smaller organic molecules which fermenting microorganisms or thermochemical conversion processes like pyrolysis and gasification transform into biofuels. Palleti et al., (2024) have comprehensively reviewed the possibility of bioconversion of plastic to biofuels and value added products. They have discussed on the methods and process, ongoing initiatives, microbial conversion of plastic to fuel and the challenges.

Multiple scientific research studies demonstrate the usefulness of microbial isolates in breaking down PE plastic wastes (Sen & Raut, 2016). Hydrolyzing PE-capable organisms have been isolated from soil, seawater, compost, and activated sludge (Montazer et al., 2020). Cassone et al. have reported that Galleria melonella bacteria can hydrolyze polyethylene (PE) in the gut of the greater wax worm, according to recent research. However, in addition to PET, pyrolysis and biotechnological transformations of PE into polyhydroxyalkanoate have been observed in various Pseudomonads (Montazer et al., 2019). Some forms of pretreatment led to the depolymerization (in bacteria) of PE, as shown by researchers at Agrobacteria spp. and Abrusci et al. (2013). Fungi also showed depolymerism during cell growth in yeast and fungal interactions with Aspergillus and Fusarium fungi respectively. Some others, including Pseudomonas putida IRN22, Acinetobacter pittiv, Micrococcus luteus INN20 (Montazer et al., 2019), and species of Phosphoronuma aerosols have been found to degrade untreated PE through microbial breakdown in various organisms. However, none of these organism types are known to do this activity as well as other bacteria.

Degraded plastic monomers, such as ethylene glycol and lactic acid, can be transformed into bioethanol or other alcohol-based biofuels through microbial fermentation. Bio-oil can be produced through pyrolysis, which involves heating degraded plastic waste in the absence of oxygen to further refine it into biodiesel and other biofuels. The gasification process converts plastic compounds into syngas that subsequently evolves into a mixture of hydrogen, carbon monoxide, and methane suitable for biofuel synthesis (Wei & Zimmermann, 2017). Biofuels produced from plastic waste destruction by bacteria represent a sustainable method to handle plastic waste during the rising energy demand for renewable solutions. Multiple plastic materials undergo decomposition into biofuel precursors under the metabolic activity of Ideonella sakaiensis, Pseudomonas aeruginosa, and Rhodococcus ruber microorganisms. Research progress in waste management along with bioengineering could establish microbial plastic degradation as an acceptable solution for future socioeconomic development (Wei & Zimmermann, 2017).

Research on PP biodegradation is limited with only two exceptions being E4 (Yoon et al., 2012) and bacterial strains from the genera Comamonas, Delftia, and Stenotrophomonaria (Peixoto et al., 2017). The genera Pseudomonas and Vibrio along with Astericin together with various bacteria degrade PP according to Cacciari et al., 1993. Pretreated PP is the primary material used in most studies. By using g-irradiation, UV-irrigation (Alariqi et al., 2006), or thermal treatment (Ramis yelling, 2004), the pretreatment techniques can be achieved and results indicate that these methods decrease hydrophobicity of the polymer, or they introduce groups like C=O or –OH, which are more susceptible to degradation. New groups (carbonyl and hydroxyl) are formed during the degradation process, leading to a decrease in viscosity (Alariqi et al., 2006). UV-pretreated PP can biodegrade to Bacillus flexus, as demonstrated by Arkatkar et al. (2010). Polypropylene has been able to biodegrade better through the use of polymer blends that contain carbohydrates, starch, or cellulose blend. These strains are known to form dense biofilm on the plastic film surface, resulting in a decrease in typical molecular weight of PVC film.

Polymer chain biodegradation is suggested by a decrease in mean molecular weight (Mn) and the presence of driest biofilm (Wilkes & Aristilde, 2017). Exocellular enzymes that are released in the culture medium cause decreases in molecular weight by hydrolyzing polymers at the ends of backbone chains and within them (Wilkes & Aristilde, 2017). The use of vinyl chloride monomers as the carbon source for growth in Pseudomonas putida strain AJ was reported by Verce et al. (2000).

1.8 Different Biofuels from Plastic

By converting the plastic waste into biofuels, we can tackle plastic pollution and meet the increasing demand for sustainable energy. However, there are few practical solutions. Due to the fact that more than 360 million tons of plastic are produced globally each year, improper disposal can result in significant environmental risks. Due to their high hydrocarbon content, plastics are excellent thermal and catalytic sources for biofuel production. There are multiple forms of biofuels that can be created from plastics, such as pyrolysis oil, syngas, solid char, biohydrogen (BPR), biodiesel, or biobutanol. A liquid biofuel, plastic-derived oil, or pyrolysis (PDO) is produced by heating plastics in an oxygen-free environment at temperatures ranging from 300 to 800 °C. Refined oil can be transformed into gasoline, diesel, and jet fuel equivalents, with yields that fluctuate between 50 and 80% depending on the plastic type and process conditions. Its high oxygen content and impurities necessitate catalytic upgrading for commercial purposes.

Syngas, a gas produced by gasification at temperatures of 800–1200 °C with controlled oxygen, is primarily composed of hydrogen (H_2), carbon monoxide (CO), carbon dioxide (CO_2), and methane (CH_4). Syngas has a range of 10–20 calories. It is a viable option for power generation and synthetic fuel production due to its low MJ/m^3 concentration. A high energy density clean fuel with low smog emissions (typically 142%) can be converted into biohydrogen. MJ/kg), through water-gas shift reactions. The gasification and pyrolysis process produces solid char, which typically accounts for 10–30% of the total yield. It has a high fixed carbon content (60–90%) and can be utilized as refractory fuel, activated carbon, or biochar for soil improvement. In addition, solid char may be used in electrode materials and for carbon sequestration. Despite their traditional origins in biomass, biodiesel and biobutanol can be produced using plastic-derived intermediates. By using plastic liquefaction to extract FAMEs, biodiesel can be produced, while the use of renewable biofuels is still possible due to their higher energy content (29.2%). (MJ/L) compared to ethanol (21.1). Through microbial fermentation, plastic-derived hydrocarbons can be transformed into MJ/L. Through the use of optimal catalyst selection and processing conditions, plastic waste can be transformed into a diverse range of biofuels, providing 80% of fossil fuel consumption through sustainable practices. Reactor design, process integration, and waste valorization are essential for the commercial success of plastic-to-biofuel technologies. Hence, progress is ongoing.

1.9 Optimization of Biofuel Production

To optimize processes, it is important to consider both the selection of appropriate catalysts and feedstocks as well as the fine-tuning of operational conditions. Bio-oil composition is influenced by temperature and heating time, but the efficiency of plastic decomposition is affected by residence duration. The fuel quality and process scalability of reactors are heavily influenced by their design, which can be shaped to meet specific needs. With the integration of artificial intelligence and machine learning into biofuel research, improved efficiency and cost-effectiveness are now attained through process optimization models that model and predict optimal process parameters. One of the primary considerations is that biofuel manufacturing from plastic faces economic and environmental success as the leading priority. Process optimization has significantly improved but three vital aspects like energy consumption as well as feedstock unpredictability alongside policy constraints still require resolution. LCA assessments evaluate the processes by showing how technological progress should match environmental sustainability metrics.

1.10 Plastic Waste Management Strategies

Plastic waste management strategies development stands as the essential solution for solving the ever-growing international plastic pollution crisis. These strategic frameworks unite prevention along with correction and creativity to achieve two main objectives including waste reduction and increased collection while advocating sustainable waste alternatives. Environmental authorities must implement biodegradable material and reuse programs first to prevent plastic waste buildup at its initial point (Hopewell et al., 2009). Plastic bans together with Extended Producer Responsibility (EPR) rules provide manufacturers with responsibility for post-consumer waste to ensure organized waste management approaches (OECD, 2016). Preventing environmental damage and building plastic recycability rates through mechanical and chemical recycling methods with various other approaches (Al-Salem et al., 2009). The public participation in waste disposal education as well as business sustainability initiatives and awareness programs drive responsible disposal practices which strengthen overall plastic waste management effectiveness (UNEP, 2018). Multiple comprehensive countermeasures built to address plastic pollution and protect environmental sustainability have the potential to create successful outcomes.

1.11 Traditional Waste Management Techniques: Limitations and Impacts

Landfills, burning, and mechanical recycling techniques are the basis of typical plastic waste management. These methods have several drawbacks however. Though among the more popular options, landfilling has devastating effects on the ecology. Slowly decomposing plastics causes long-term environmental damage by contaminating the land and filling great amounts of landfill area with plastic waste (Geyer et al., 2017). Landfills also worsen atmospheric change and pollution worries by generating noxious leachates and greenhouse gases. Another very common approach, incineration helps to cut waste volume even more but has major disadvantages. Plastic burning emits poisonous gases—including dioxins and furans—that present health risks and add to global warming and air pollution (Astrup et al., 2009). Furthermore, since it leads to the permanent loss of useful materials, incineration does not underpin a circular economy approach. Several difficulties exist for mechanical recycling, which is the gathering, sorting, and rethinking of plastic waste into fresh items. Contaminants and degradation during the recycling process lowers the quality of recycled plastics and therefore their reusability and economic feasibility. Moreover, difficult in sorting and processing is the mixed plastic waste streams, thereby making it tough to reach high recycling levels. Due their complicated compositions, many plastic types are not fit for mechanical recycling, which results in much garbage being dumped or burned instead of being properly recycled (Hopewell et al., 2009).

A core strategy for more sustainable plastic waste management is the Reduce, Reuse, and Recycle (3Rs) principle. Crucial in limiting trash production is less plastic use, which can be attained by encouraging choices including biodegradable items, reusable bags, and packaging-free goods. By motivating practices such as refilling containers and repurposing plastic goods, reusing plastic items lengthens their life span and reduces waste accumulation. Although recycling is absolutely essential in diverting plastic waste from landfills and incinerators, its efficiency depends on the advancement of sophisticated recycling technologies that convert waste plastics into high-quality reusable materials, therefore lowering reliance on virgin plastic production. Although the 3Rs strategy is essential for dealing with plastic waste, to effectively solve the increasing plastic pollution catastrophe it must be supplemented by creative waste management technologies and stricter legislation (Geyer et al., 2017; Hopewell et al., 2009).

1.12 Emerging Technologies in Plastic Waste Conversion

Modern technologies have introduced effective methods to deal with plastic pollution. A chemical recycling method enables excellent material retrieval by turning plastic items into monomer materials or fuel products. The manufacture of biodegradable as well as compostable plastics enables long-term waste accumulation reduction

(European Bioplastics, 2021). Three Waste-to-energy conversion technologies such as pyrolysis and gasification along with co-processing turn plastic waste into energy thus decreasing fossil fuel use (Astrup et al., 2009). The evolving technologies in plastic waste conversion are establishing sustainable alternatives for the traditional procedures of waste management. The objective of complex recycling methods is multifold because they seek to recover resources efficiently while minimizing pollution and improving recycling efficiency. Chemical recycling stands out as one of the most advanced plastic treatment approaches since it breaks down plastics into their original monomers or alternative useful materials with the applications of pyrolysis, hydrolysis, and solvolysis (Al-Salem et al., 2009). Through chemical recycling operations manufacturers create premium recycled materials that function equally to virgin plastics in production lines although mechanical recycling results in plastic material deterioration with each recycling cycle according to Hopewell et al. (2009). Through pyrolysis the transformation of plastic waste produces usable fuels such as diesel and petrol which decreases dependence on fossil fuels along with managing plastic accumulation (Astrup et al., 2009). Biodegradation represents a rising technology that uses microorganisms to convert plastic waste into non-hazardous byproducts which include both water and carbon dioxide. The rise of natural polymers including PLA and PHA in making sustainable biodegradable plastics has established these materials as alternatives to traditional plastics according to European Bioplastics (2021). A diverse set of environmental factors prevents effective decomposition of biodegradable plastics after their release into natural settings where microplastics persist (UNEP, 2018).

Scientific waste-to-energy techniques like gasification and plasma arc treatment have earned credibility for turning plastic waste into electricity and heat along with synthetic fuels (Al-Salem et al., 2009). Plastic gasification results in syngas composition that contains carbon monoxide and hydrogen which then produces electricity cleaner than ordinary incineration operations (Astrup et al., 2009). Plasma arc treatment breaks down plastics through extremely hot temperatures to produce essential industrial materials alongside reduced emission pollution (Geyer et al., 2017). A novel approach involves developing specific enzymes to break down plastic polymers effectively as part of enzymatic plastic degradation. The current rise of biotechnology enables the development of PETase and MHETase enzymes which boost polyethylene terephthalate (PET) decomposition rates within bottles and packaging structures (Hopewell et al., 2009). The enzymatic process helps sustain closed-loop recycling systems which present a probable solution for oceanic and terrestrial plastic accumulation. Governments with organizations need to use plastic waste management policies while raising public awareness and promoting circular economy standards. More than several nations worldwide have adopted plastic restrictions and guidelines aimed to cut down production waste while controlling environmental air and water contamination (UNEP, 2018). Under EPR policies manufacturers must accept full responsibility for plastic products from manufacturing until post-consumer waste retrieval and recycling (OECD, 2016). Companies receive financial rewards through subsidies to use eco-friendly waste management and packaging solutions (Hopewell et al., 2009). Public education about waste disposal along with community-led recycling activities serve as essential factors to develop improved waste management conduct among the public (Jambeck et al., 2015). The circular economy strategy works to decrease plastic waste through designs for long-lasting materials together with bike-loop recycling operations and business models featuring plastic buy-back programs (Ellen MacArthur Foundation, 2016). Multiple adopted measures enhance plastic waste management systems which help decrease environmental disruptions while promoting sustainable operations in the long term.

1.13 Environmental and Economic Impacts

Solid waste generation exceeds two billion tons yearly and recycling of municipal solid waste (MSW) reaches just 19% according to United Nations Environment Programme (2024). The bulk of waste produced by society consists of plastic while this substance has rapidly integrated itself into everyday life. Each year, more than 460 million metric tons of plastic are produced for various applications (IUCN, 2024). However, due to improper disposal approximately 20 million metric tons of plastic litter enter the environment each year, and this figure is anticipated to rise significantly by 2040 (IUCN, 2024). Plastic pollution has emerged as a severe global issue due to its long-lasting environmental impact, affecting ecosystems, infrastructure, and economies (Cottom et al., 2024). Recognizing the gravity of this issue, the United Nations recently ratified a draft resolution to establish an internationally legally binding instrument, known as the Plastics Treaty, aimed at eradicating plastic pollution. Scientists have already declared the current era as the "plastic age" and predict that global plastic production will double by 2050 (Mansuri et al., 2023). Phthalates and bisphenol A, commonly used in plastic production, are ubiquitous in everyday items, causing widespread harm to humans. These compounds pose significant health hazards, disrupting reproductive and hormone systems, increasing cancer risks, impairing development, and reducing fertility (United Nations Environment Programme, 2024). As plastic pollution escalates, urgent action is required to mitigate its detrimental effects and safeguard environmental and human health. Microbial biofuel production from plastic waste offers a promising approach to address the triple challenges of plastic landfill, greenhouse gas emission, and migration toward circular economy.

1.14 Reduction in Plastic Landfill Dependency

Landfills are designated sites where waste materials are disposed of and buried. These engineered systems are designed to minimize the environmental impact of waste decomposition. Landfills typically feature protective barriers to prevent leachate

(Gastelo et al., 2023), a toxic liquid produced by decomposing waste, from contaminating groundwater. Layers of soil or other cover materials are applied to manage odors, pests, and debris.

A plastic landfill is a designated site for the disposal of plastic waste, functioning as an engineered system designed to contain and manage waste materials. It is one of the most common methods of waste disposal globally, particularly for non-recyclable plastics (Saju et al., 2024). Structurally, a landfill is a carefully constructed facility that isolates waste from the surrounding environment to minimize contamination. This isolation is achieved through a combination of physical barriers, such as liners and covers, and controlled processes that manage the decomposition of organic materials. In the case of plastics, which are synthetic polymers, their resistance to biodegradation makes them unique compared to organic waste. Plastics can persist in landfills for decades or even centuries, undergoing slow physical and chemical changes influenced by environmental factors (Chamas et al., 2020; Wojnowska-Baryła et al., 2022).

The structure of a landfill is designed to manage waste in distinct layers. At the base, a liner system—typically made of compacted clay or synthetic materials like high-density polyethylene (HDPE)—prevents leachate from contaminating soil and groundwater (Gastelo et al., 2023). Above the liner, a drainage layer collects and channels leachate to treatment facilities. The waste layer follows, where plastics are mixed with other types of municipal solid waste (MSW). Over time, this layer is compacted to conserve space and reduce air pockets that could accelerate decomposition in organic materials (Siddiqua et al., 2022). A daily cover, often made of soil or alternative materials like tarps or foams, is applied to minimize odor, deter pests, and reduce the risk of fires (Njoku et al., 2019).

Despite their role in waste management, plastic landfills have several limitations. One major drawback is their finite capacity, as landfill space is limited and plastic waste continues to accumulate at an alarming rate. Additionally, plastics are non-biodegradable and can persist in landfills for hundreds of years, occupying valuable land resources without breaking down (Evode et al., 2021; Tammemagi, 1999). The leachate produced from plastic landfills can still pose a risk of contamination if not properly managed, potentially polluting nearby soil and water sources. Furthermore, the presence of microplastics in landfill waste can lead to long-term environmental concerns as these tiny particles may eventually seep into ecosystems (Singh et al., 2023). The cost of constructing and maintaining plastic landfills is also significant (Low et al., 2024), requiring continuous investment in monitoring, maintenance, and expansion. While landfill gas collection systems exist, plastic waste itself does not significantly contribute to biogas production, limiting the potential for energy recovery compared to organic waste landfills (Gao et al., 2024).

Plastic waste continues to be one of the most significant contributors to landfill overflow, and its improper disposal results in severe environmental hazards. Traditional disposal methods, such as landfilling and incineration, pose significant environmental risks, including soil and groundwater contamination, air pollution, and the release of greenhouse gases. In contrast, repurposing plastic into biofuel provides an efficient way to tackle the plastic crisis (Agarwal et al., 2022) by diverting waste from landfills. This approach not only helps manage the ever-growing volume of plastic waste but also decreases the need for extensive landfill sites, preserving land for more productive uses such as agriculture or urban development. Additionally, since landfills require constant maintenance, financial resources allocated for their expansion and upkeep could be redirected toward more sustainable initiatives such as waste-to-energy programs, further reducing plastic dependency on landfills.

The integration of biofuel production into existing waste management systems presents an opportunity for innovation in resource recovery. By implementing systems that prioritize the conversion of plastic waste into biofuels, communities can develop more sustainable practices that emphasize recycling and reusing materials rather than discarding those (Liu et al., 2021). This approach fosters a culture of sustainability where individuals and businesses are encouraged to think critically about their consumption patterns and waste generation. Education and awareness campaigns highlighting the benefits of converting plastic waste into biofuels can further enhance community engagement in recycling efforts (Sharma et al. 2024), ultimately leading to a more responsible approach to resource management.

1.15 Mitigation of Greenhouse Gas Emissions

The atmosphere contains greenhouse gases from natural and human-caused sources that both absorb and emit radiation which falls within the spectrum of infrared wavelength emissions from Earth and atmospheric components and clouds. The greenhouse effect develops due to this property. The Earth's atmosphere primarily consists of five main greenhouse gases including H_2O water vapor and CO_2 carbon dioxide, and N_2O nitrous oxide and methane CH_4 and O_3 ozone (UNEP, 2024). Different from natural greenhouse gases are human-produced chemicals consisting of halocarbons alongside other chlorine and bromine substances that reside in the atmosphere. Earth depends on greenhouse effect to maintain its suitable temperature but human-generated GHG productions trigger excessive increases within this vital natural process (Filonchyk et al., 2024). The accumulation of pollutants in Earth's atmosphere results in global warming which drives temperature increase effects together with intensified weather conditions and ice melting and rising sea levels. Atmospheric CO_2 rise causes disturbance to ocean carbon cycles thus leading to toxic ocean water conditions that harm marine species. Excessive GHG emissions have accelerated climate change which creates worldwide serious threats to both ecosystems and the socioeconomic systems (Nie et al., 2024).

Greenhouse gas released at plastic landfills stems from time-dependent degradation processes happening under different environmental conditions. Plastics made from fossil fuel-based polymers such as PE, PP, PET, PS, and PVC exist as chemically stable substances that do not decompose until long periods later (Kaur et al., 2024). The combination of photodegradation and thermal oxidation and both microbial action and leachate processes enables landfill environments to release greenhouse gases throughout extended degradation periods. UV radiation in sunlight activates photodegradation which breaks down polymer chains when oxygen is present thus forming carbonyl groups that finally convert into atmospheric carbon dioxide (CO_2) (Kaur et al., 2024). Thermal degradation processes found at landfills under high temperatures destroy plastic molecular structures into CO_2 along with generating VOCs according to Nzimande et al. (2024), Wang et al. (2024a, 2024b). Microbial decomposition functioning under anaerobic conditions deep inside landfills serves as a key process for creating methane (CH_4) which proved to be a strong greenhouse gas that emits heat at 28–36 times CO_2 heat levels during 100-year timeframes (Ali et al., 2024). The landfill ecological system creates nitrous oxide (N_2O) emission through the nitrification and denitrification processes of microorganisms which specifically act on nitrogen-containing materials present in the site (Na & Byun, 2024). The N_2O has a GWP approximately 265–298 times greater than CO_2 and contributes to both climate change and ozone layer depletion. Additionally, plastics in landfills degrade into microplastics and nanoplastics, which, while not directly emitting GHGs, can alter microbial communities in soil and water, potentially influencing CH_4 and CO_2 fluxes by modifying carbon cycling processes (Chang et al., 2024). Another indirect pathway of emissions occurs through leachate generation, where organic compounds dissolved from plastics into landfill leachate can undergo microbial fermentation, releasing further methane and carbon dioxide. Furthermore, when landfills reach capacity or are improperly managed, the spontaneous or intentional burning of plastics releases not only CO_2 but also toxic gases such as dioxins and furans, which have severe environmental and health implications (Baca et al., 2023). Landfill gas collection systems can be detrimental to climate change as they allow large amounts of methane to escape into the atmosphere.

Using specialized microorganisms capable of breaking down the synthetic polymers and converting them into useful biofuel by recycling plastic waste offers an innovative way to reduce greenhouse gas emissions from landfill. Among the bacteria and plant species are PETase, PSE-DEMOTEXYlose, Laccasedextero (Nazeer et al., 2024) enzymes that break down plastic polymers into smaller monomers and small oligomers in their own bodies (Sarsan et al., 2024). Bioethanol, biobutanol and the precursors of biodiesel can be produced from these degradation products by a biochemical pathway (-oxidation), glycolysis, or the TCA cycle. Through the use of metabolic engineering methods, synthetic pathways are introduced to increase microbial efficiency and yield for specific biofuels, (2024). During anaerobic digestion, microbial consortia (such as methanogens and acetogen) convert the plastic-derived intermediate materials into biohydrogen and biogas, which reduces emissions of meth that would otherwise result from uncontrolled plastic degradation in landfills. In addition, extremophiles such as thermophilic Clostides make plastic-to-biofuel conversion at high temperatures more efficient, reducing process energy requirements (Zhu et al., 2020). MECs and BESs utilize bacterial electron transfer mechanisms to improve plastic manufacturing and biofuel production, leading to energy efficiency improvements, 2016. By utilizing microbial bioreactors in waste treatment facilities, plastic degradation can be reduced, which is beneficial for preventable landfills and methane release Batarseh et al., (2010). Optimizing biofuel yields and sustainability is achieved by merging microbial processes with synthetic biology advancements, such as CRISPR-based metabolic pathway editing. To support the circular economy, it is important to use microbial biofuels as a means of producing renewable energy while also reducing landfill emissions through their valorization strategies.

1.16 Contribution to a Circular Economy

The circular economy is a system that seeks to reduce the consumption of energy, natural resources, and waste generation in order to close loop patterns through the flow of utilities and materials. This involves reducing the inputs of raw materials, decreasing waste output, and eventually reintegrating products into the system once they have been used up (Fan et al., 2019). Due to the inefficiencies of the traditional linear economy, a form of circular economy is associated with environmental degradation and negatively impacts ecosystems through resource depletion, pollution, and dismantling. Environmental degradation can be mitigated by the circular economy, which aims to close material loops and reduce environmental impact through resource efficiency (Fan et al., 2019). One of the key strategies it employs to prevent degradation is by limiting the extraction of raw materials, which is a significant contributor to deforestation, soil erosion, and biodiversity loss. Circular strategies, which aim to maximize resource recovery and promote secondary material use, aim at reducing land-use change while also preserving natural habitats. Design innovations in the circular economy led to improved durability, reparability, and recyclability, which minimize environmental degradation. (GACERE, 2024, Kandpal et al., 2024).

Extended producer responsibility (EPR) policies encourage manufacturers to minimize environmental impacts by taking into account end-of-life disposal in product design, resulting in less pollution from electronic waste, textiles, and packaging materials. By turning plastic waste into valuable resources, the circular economy can be further advanced through the development of microbial biofuel made from plastic (Bora, 2020). To reduce waste and create a sustainable source of energy, Tamoor et al. suggest that the utilization of specialized microorganisms to break down plastic polymers into valuable biofuels can be beneficial. By converting end-of-life plastics into biofuels through biological processes, this approach aligns with the circular economy principles, reducing the need for landfill and increasing resource efficiency.

Plastic waste management and recycling systems can be integrated with biofuel production to enhance resource efficiency and create a closed-loop approach. Rajavanshi et al., (2025) have emphasized recycling of waste especially plastic combining recycling with cost effective sustainable approaches like biofuel production to address plastic accumulation. The economic impact of microbial biofuel production can be minimized by diverting plastic waste from landfills, which can reduce the financial burden on municipalities and governments with waste disposal and remediation costs (Arpia et al., 2021). Also, the establishment of a demand for plastic-based biofuels produces financial benefits that promote waste collection and classification strategies, leading to increased investment in recycling infrastructure and technology.

This is advantageous. Jobs in biotechnology, bioengineering, and waste-to-energy industries can be created by the commercialization of biofuels made from plastic waste, which has been shown to promote economic growth and workforce development (Lange et al., 2021). The local economy can be boosted by the economic benefits of microbial biofuel production from plastic waste, which can result in more efficient and less expensive transportation and environmental impact reduction. This revenue, combined with infrastructure and community initiatives, creates a positive feedback loop that fosters long-term sustainability by reinforcing research and development into biofuels (Chowdhury et al., 2023; Loizia et al., 2021).

2 Conclusion

The use of microbial biofuel production is an innovative approach to managing plastic waste, as it involves breaking down plastic polymers into simpler compounds that can be transformed into biodiesel such as ethanol and butanol. By reducing the need for landfill, it also offers a sustainable energy source that reduces dependence on fossil fuels and minimizes environmental damage. One of the main benefits of this approach is that it reduces greenhouse gas emissions. It reduces the amount of methane released, a potent greenhouse gas, by diverting plastic waste from landfills and replaces fossil fuels with biofuels, which also decreases carbon dioxide emissions from transportation and industry. Microbial biofuel production is a significant factor in the mitigation of climate change, given its dual impact. Furthermore, it supports the circular economy by recycling plastic waste into valuable resources that reduces the need for virgin plastic production and encourages energy self-reliance. We can reduce pollution, conserve natural resources, and promote a more regenerative economy by including microbial biofuel production in our waste management strategies. It meets the requirements for sustainability and contributes to an environmentally friendly future. Microbial biofuel production is an effective means to manage plastic waste and energy requirements, providing environmental and economic benefits that support a circular economy. This approach has the potential to be both environmentally and economically beneficial.

References

ACS Omega. (2023). *Microplastics in agricultural soils*. ACS Publications. Retrieved from https://pubs.acs.org/doi/10.1021/acsomega.3c01214.

Abrusci, C., Pablos, J. L., Marín, I., Espí, E., Corrales, T., & Catalina, F. (2013). Comparative effect of metal stearates as pro-oxidant additives on bacterial biodegradation of thermal- and photo-degraded low density polyethylene mulching films. *International Biodeterioration and Biodegradation, 83*, 25–32. https://doi.org/10.1016/j.ibiod.2013.04.002

Acharya, B., Behera, A., Deshmukh, K., & Moharana, S. (2024). Plastic waste management during and post COVID-19 pandemic: Challenges and strategies. In *Plastic waste management: Methods and applications* (pp. 173–199).

Agarwal, P., Soni, R., Kaur, P., Madan, A., Mishra, R., Pandey, J., Singh, S., & Singh, G. (2022). Cyanobacteria as a promising alternative for sustainable environment: Synthesis of biofuel and biodegradable plastics. *Frontiers in Microbiology, 13*. https://doi.org/10.3389/fmicb.2022.939347.

Al-Salem, S. M., Lettieri, P., & Baeyens, J. (2009). Recycling and recovery routes of plastic solid waste. *Waste Management, 29*(10), 2625–2643.

Alariqi, S. A. S., Pradeep Kumar, A., Rao, B. S. M., & Singh, R. P. (2006). Biodegradation of γ-sterilized biomedical polyolefins under composting and fungal culture environments. *Polymer Degradation and Stability, 91*, 1105–1116. https://doi.org/10.1016/j.polymdegradstab.2005.07.004

Ali, S. S., Mustafa, A. M., Elsamahy, T., Khalil, M. A., Al-Zahrani, M., & Sun, J. (2024). Efficient methane production from anaerobic co-digestion of lignin-based feedstock and waste plastic: Digester performance and microbial community structures. *Fuel, 371*, 131926. https://doi.org/10.1016/j.fuel.2024.131926

Arkatkar, A., Juwarkar, A. A., Bhaduri, S., Uppara, P. V., & Doble, M. (2010). Growth of pseudomonas and bacillus biofilms on pretreated polypropylene surface. *International Biodeterioration and Biodegradation, 64*, 530–536. https://doi.org/10.1016/j.ibiod.2010.06.002

Arpia, A. A., Chen, W.-H., Ubando, A. T., Naqvi, S. R., & Culaba, A. B. (2021). Microplastic degradation as a sustainable concurrent approach for producing biofuel and obliterating hazardous environmental effects: A state-of-the-art review. *Journal of Hazardous Materials, 418*, 126381. https://doi.org/10.1016/j.jhazmat.2021.126381

Astrup, T., Tonini, D., Turconi, R., & Boldrin, A. (2009). Plastic waste as a fuel—CO_2-neutral or not? *Waste Management, 30*(1), 2–5.

Baca, D., Monroy, R., Castillo, M., Elkhazraji, A., Farooq, A., & Ahmad, R. (2023). Dioxins and plastic waste: A scientometric analysis and systematic literature review of the detection methods. *Environmental Advances, 13*, 100439. https://doi.org/10.1016/j.envadv.2023.100439

Bala, K., Dogra, A., Lata, P., & Das, A. P. (2024). Discussing sources and biological-chemical remediation approaches for microplastic pollution. In Sustainable Microbial Technology for Synthetic and Cellulosic Microfiber Bioremediation (pp. 37–59). https://doi.org/10.1007/978-3-031-62600-5_3.

Batarseh, E. S., Reinhart, D. R., & Berge, N. D. (2010). Sustainable disposal of municipal solid waste: Post bioreactor landfill polishing. *Waste Management, 30*, 2170–2176. https://doi.org/10.1016/j.wasman.2010.06.015

Bhatia, S. K., Gurav, R., Choi, T. R., Jung, H. R., Yang, S. Y., & Moon, Y. M. (2019a). Bioconversion of plant biomass hydrolysate into bioplastic (polyhydroxyalkanoates) using Ralstonia eutropha 5119. *Bioresource Technology, 271*, 306–315. https://doi.org/10.1016/j.biortech.2018.09.122

Bhatia, S. K., Gurav, R., Choi, T. R., Jung, H. R., Yang, S. Y., Song, H. S., et al. (2019b). Poly(3-hydroxybutyrate-co-3-hydroxyhexanoate) production from engineered Ralstonia eutropha using synthetic and anaerobically digested food waste derived volatile fatty acids. *International Journal of Biological Macromolecules, 133*, 1–10. https://doi.org/10.1016/j.ijbiomac.2019.04.083

Bora, D. K. (2020). Rise of the sustainable circular economy platform from waste plastics: A biotechnological perspective. *MRS Energy & Sustainability, 7*. https://doi.org/10.1557/mre.2020.28.

Cacciari, P., Quatrini, G., Zirletta, E., Mincione, V., Vinciguerra, P., & Lupattelli, P. (1993). Isotactic polypropylene biodegradation by a microbial community: Hysicochemical characterization of metabolites produced. *Applied and Environmental Microbiology, 59*, 3695–3700. https://doi.org/10.1128/AEM.59.11.3695-3700.1993

Chamas, A., Moon, H., Zheng, J., Qiu, Y., Tabassum, T., Jang, J. H., Abu-Omar, M., Scott, S. L., & Suh, S. (2020). Degradation rates of plastics in the environment. *ACS Sustainable Chem. Eng., 8*, 3494–3511. https://doi.org/10.1021/acssuschemeng.9b06635

Chang, S., Zhou, A., Hua, Z., Meng, H., Zhu, F., Li, S., & He, H. (2024). Microplastics alter soil carbon cycling: Effects on carbon storage, CO_2 and CH_4 emission and microbial community. *Cambridge Prisms Plastics, 2*. https://doi.org/10.1017/plc.2024.5

Chowdhury, H., Loganathan, B., Mofijur, M., & Rahman, S. M. A. (2023). Integration of anaerobic digestion with algal biofuel production: A sustainable approach to reduce nutrient demand and greenhouse gas emissions. *Sustainability, 15*(3), 1767.

Cottom, J. W., Cook, E., & Velis, C. A. (2024). A local-to-global emissions inventory of macroplastic pollution. *Nature, 633*, 101–108. https://doi.org/10.1038/s41586-024-07758-6

Das, A., Ray, S. K., Mohanty, M., Mohanty, J., Dey, S., & Das, A. P. (2024b). Ecotoxicity of microplastic wastes and their sustainable management: A review. *Environmental Chemistry and Ecotoxicology, 6*, 144–152. https://doi.org/10.1016/J.ENCECO.2024.05.003

Das, A., Dey, S., & Das, A. P. (2024a). Ecotoxicological impacts of synthetic microfiber pollutants and development of sustainable mitigation strategies. *Environmental Chemistry and Ecotoxicology*. https://doi.org/10.1016/J.ENCECO.2024.12.003

Dey, S., Bissoyi, A., Koteswara Reddy, G., & Das, A. P. (2024). Synthetic microfibers: An urban emerging pollutant demanding for sustainable environmental solutions. In *Sustainable microbial technology for synthetic and cellulosic microfiber bioremediation* (pp. 107–121). https://doi.org/10.1007/978-3-031-62600-5_7.

Doe, J., Smith, A., & Brown, K. (2020). Plastic pollution and its impact on biodiversity. *Environmental Research Journal, 45*(3), 112–130.

Ellen MacArthur Foundation. (2013). *Towards the circular economy vol. 1: Economic and business rationale for an accelerated transition.*

Ellen MacArthur Foundation. (2016). *The new plastics economy: Rethinking the future of plastics.*

European Bioplastics. (2021). *Biodegradable plastics: Opportunities and challenges.*

Evode, N., Qamar, S. A., Bilal, M., Barceló, D., & Iqbal, H. M. N. (2021). Plastic waste and its management strategies for environmental sustainability. *Case Studies in Chemical and Environmental Engineering, 4*, 100142. https://doi.org/10.1016/j.cscee.2021.100142

Fan, Y. V., Lee, C. T., Lim, J. S., Klemeš, J. J., & Le, P. T. K. (2019). Cross-disciplinary approaches towards smart, resilient and sustainable circular economy. *Journal of Cleaner Production, 232*, 1482–1491. https://doi.org/10.1016/j.jclepro.2019.05.266

Filonchyk, M., Peterson, M. P., Zhang, L., Hurynovich, V., & He, Y. (2024). Greenhouse gases emissions and global climate change: Examining the influence of CO_2, CH_4, and N_2O. *Science of the Total Environment, 935*, 173359. https://doi.org/10.1016/j.scitotenv.2024.173359

GACERE. (2024). *Circular design of plastic products—policy brief*. United Nations Industrial Development Organization.

Gao, Z., Qian, H., Cui, T., Ren, Z., & Wang, X. (2024). Comprehensive meta-analysis reveals the impact of non-biodegradable plastic pollution on methane production in anaerobic digestion. *Chemical Engineering Journal, 484*, 149703. https://doi.org/10.1016/j.cej.2024.149703

Gastelo, J., Li, D., Tian, K., Tanyu, B. F., & Erol Guler, F. (2023). Hydraulic conductivity of GCL overlap permeated with saline solutions. *Waste Management, 157*, 348–356. https://doi.org/10.1016/j.wasman.2022.12.028

Geyer, R., Jambeck, J. R., & Law, K. L. (2017). Production, use, and fate of all plastics ever made. *Science Advances, 3*(7), e1700782.

Ghosh, S., Sinha, J. K., Ghosh, S., Vashisth, K., Han, S., & Bhaskar, R. (2023). Microplastics as an emerging threat to the global environment and human health. *Sustainability, 15*(14), 10821. https://doi.org/10.3390/SU151410821.

Government of India. (2021). *Waste to wealth mission: Innovation and circular economy in waste management.*

Grima, S., Bellon-Maurel, V., Feuilloley, P., & Silvestre, F. (2000). Aerobic biodegradation of polymers in solid-state conditions: A review of environmental and physicochemical parameter settings in laboratory simulations. *Journal of Polymers and the Environment, 8*, 183–195. https://doi.org/10.1023/A:1015297727244

Hopewell, J., Dvorak, R., & Kosior, E. (2009). Plastics recycling: Challenges and opportunities. *Philosophical Transactions of the Royal Society B: Biological Sciences, 364*(1526), 2115–2126.

IUCN (2024). *IUCN issues briefs: Plastic pollution.* Gland, Switzerland: International Union for Conservation of Nature and Natural Resources. https://iucn.org/sites/default/files/2024-05/plastic-pollution-issues-brief-may-2024-update.pdf.

Jambeck, J. R., Geyer, R., Wilcox, C., Siegler, T. R., Perryman, M., Andrady, A., & Law, K. L. (2015). Plastic waste inputs from land into the ocean. *Science, 347*(6223), 768–771.

Jones, M., & Patel, R. (2021). The environmental and health effects of plastic incineration. *Journal of Environmental Science, 56*(2), 215–230.

Jumaah, O. S. (2017). Screening of plastic degrading bacteria from dumped soil area. *IOSR Journal of Environmental Science Toxicology Food Technology, 11*, 93–98. https://doi.org/10.9790/2402-1105029398

Kandpal, V., Jaswal, A., Santibanez Gonzalez, E. D. R., & Agarwal, N. (2024). Circular economy principles: Shifting towards sustainable prosperity. Circular economy and sustainability (pp. 125–165). Springer. https://doi.org/10.1007/978-3-031-52943-6_4.

Kaur, A., Bala, R., & Kansal, S. K. (2024). Recent advances in Photodegradation of various plastics. ACS Symposium Series, pp. 185–221. https://doi.org/10.1021/bk-2024-1489.ch009.

Kumar, R. V., Kanna, G. R., & Elumalai, S. (2017). Biodegradation of polyethylene by green photosynthetic microalgae. *Journal of Bioremediation & Biodegradation, 8*(381), 2.

Landrigan, P. J., Raps, H., Cropper, M., Bald, C., Brunner, M., Canonizado, E. M., & Dunlop, S. (2023). The Minderoo-Monaco commission on plastics and human health. *Annals of Global Health, 89*(1), 23.

Lange, L., Connor, K. O., Arason, S., Bundgård-Jørgensen, U., Canalis, A., Carrez, D., Gallagher, J., Gøtke, N., Huyghe, C., Jarry, B., Llorente, P., Marinova, M., Martins, L. O., Mengal, P., Paiano, P., Panoutsou, C., Rodrigues, L., Stengel, D. B., van der Meer, Y., & Vieira, H. (2021). Developing a sustainable and circular bio-based economy in EU: By partnering across sectors, upscaling and using new knowledge faster, and for the benefit of climate, environment & biodiversity, and people & business. *Frontiers in Bioengineering and Biotechnology, 8*. https://doi.org/10.3389/fbioe.2020.619066

Liu, Y., Cruz-Morales, P., Zargar, A., Belcher, M. S., Pang, B., Englund, E., Dan, Q., Yin, K., & Keasling, J. D. (2021). Biofuels for a sustainable future. *Cell, 184*, 1636–1647. https://doi.org/10.1016/j.cell.2021.01.052

Loizia, P., Voukkali, I., Kylili, A., Zorpas, A. A., & Christoforou, E. (2021). Economic and environmental assessment of plastic waste pyrolysis products and biofuels as substitutes for fossil-based fuels. *Frontiers in Energy Research, 9*, 676233.

Loredo-Treviño, A., García, G., Velasco-Téllez, A., Rodríguez-Herrera, R., & Aguilar, C. N. (2011). Polyurethane as substrate for fungal strains. *Advances in Bioscience and Biotechnology, 2*(2), 52–58.

Low, J. K., Hernandez, G., & Berry, T.-A. (2024). Plastic waste characterisation to maximise landfill diversion from a New Zealand residential construction site. *Frontiers in Sustainability, 5*. https://doi.org/10.3389/frsus.2024.1455480

Mansuri, A., Kansara, K., Raiyani, D., Mazmudar, D., & Kumar, A. (2023). New insight into long-term effects of phthalates microplastics in developing zebrafish: Evidence from genomic alteration and organ development. *Environmental Toxicology and Pharmacology, 99*, 104087. https://doi.org/10.1016/j.etap.2023.104087

Ministry of Housing and Urban Affairs, Government of India. (2021). *Swachh Bharat mission 2.0 guidelines.*

Mishra, S., & Das, A. P. (2022). Treatment of the wastewater polluted with synthetic microfiber released from washing machine. In B. B. Das, H. Hettiarachchi, P. K. Sahu, & S. Nanda (Eds.), *Lecture notes in civil engineering* (Vol. 207, pp. 109–117). Springer Science and Business Media Deutschland Gmb H. https://doi.org/10.1007/978-981-16-7509-6_9/COVER

Mishra, S., & Das, A. P. (2023). Microbial remediation of synthetic microfiber contaminated wastewater. In M. P. Shah (Ed.), *Microbial technologies in industrial wastewater treatment* (pp. 337–350). Springer. https://doi.org/10.1007/978-981-99-2435-6_17

Mishra, S., Dash, D., & Das, A. P. (2022). Detection, characterization and possible biofragmentation of synthetic microfibers released from domestic laundering wastewater as an emerging source of marine pollution. *Marine Pollution Bulletin, 185*, 114254. https://doi.org/10.1016/J.MARPOLBUL.2022.114254

Mishra, S., Dash, D., & Das, A. P. (2024a). Aquatic microbial diversity on Plastisphere: Colonization and potential role in microplastic biodegradation. *Geomicrobiology Journal, 41*(4), 312–323. https://doi.org/10.1080/01490451.2023.2209750

Mishra, S., Dash, D., & Das, A. P. (2024b). Microbial degradation of polyester microfibers using indigenously isolated bacterial strain Exiguobacterium Sp. *CLEAN–Soil, Air Water, 52*(12), e202300343. https://doi.org/10.1002/CLEN.202300343

Mishra, S., Rout, P. K., & Das, A. P. (2021). Emerging microfiber pollution and its remediation. In R. Prasad (Ed.), *Environmental and microbial biotechnology* (pp. 247–266). Springer. https://doi.org/10.1007/978-981-15-5499-5_9

Mishra, S., & Das, A. P. (2021). Current treatment technologies for Removal of microplastic and microfiber pollutants from wastewater. In *Wastewater treatment: Cutting-edge molecular tools, techniques and applied aspects* (pp. 237–251). https://doi.org/10.1016/B978-0-12-821881-5.00011-8.

Mishra, S., Singh, R. P., Rout, P. K., & Das, A. P. (2022). Membrane bioreactor (MBR) as an advanced wastewater treatment technology for removal of synthetic microplastics. In *Development in wastewater treatment research and processes* (pp. 45–60). https://doi.org/10.1016/B978-0-323-85583-9.00022-3.

Mishra, S., Swain, S., Sahoo, M., Mishra, S., & Das, A. P. (2021). Microbial colonization and degradation of microplastics in aquatic ecosystem: A review. *Geomicrobiology Journal*. https://doi.org/10.1080/01490451.2021.1983670

Mohan, K. (2011). Microbial deterioration and degradation of polymeric materials. *Journal of Biochemical Technology, 2*(4), 210–215.

Mohanty, J., Dey, S., Bala, K., & Das, A. P. (2024a). Recent advances in microplastic pollution for its sustainable management. In *Sustainable management of urban plastic waste through circular economic approaches* (pp. 27–48). https://doi.org/10.1201/9781003465447-3.

Mohanty, M., Tripathy, B., Koteswara Reddy, G., & Das, A. P. (2024b). Greener technologies for establishing restraint over microplastic pollution. In *Sustainable microbial technology for synthetic and cellulosic microfiber bioremediation* (pp. 239–258). https://doi.org/10.1007/978-3-031-62600-5_13.

Montazer, Z., Habibi Najafi, M. B., & Levin, D. B. (2019). Microbial degradation of low-density polyethylene and synthesis of polyhydroxyalkanoate polymers. *Canadian Journal of Microbiology, 65*, 1–11. https://doi.org/10.1139/cjm-2018-0335

Montazer, Z., Habibi Najafi, M. B., & Levin, D. B. (2020). Challenges with verifying microbial degradation of polyethylene. *Polymers, 12*, 123. https://doi.org/10.3390/polym12010123

NITI Aayog. (2021). *Plastic waste management in India*. NITI Aayog Reports.

Na, Y., & Byun, K. (2024). Estimation of potential nitrous oxide emissions from landfills in the United States: 2010–2020. *Sustainability, 16*, 4810. https://doi.org/10.3390/su16114810

Nazeer, A., Ahmad, A., Ahmad, F., & Ahmad, S. (2024). Microbial biodegradation of plastics and microplastics: recent development. In *Plastic pollution* (pp. 231–248). Springer. https://doi.org/10.1007/978-981-97-5528-8_14.

Nie, P., Zhao, K., Ma, D., Liu, H., Amin, S., & Yasin, I. (2024). Global climate change, mental health, and socio-economic stressors: Toward sustainable interventions across regions. *Sustainability, 16*, 8693. https://doi.org/10.3390/su16198693

Njoku, P. O., Edokpayi, J. N., & Odiyo, J. O. (2019). Health and environmental risks of residents living close to a landfill: A case study of Thohoyandou landfill, Limpopo Province, South Africa. *IJERPH, 16*, 2125. https://doi.org/10.3390/ijerph16122125

Nzimande, M. C., Mtibe, A., Tichapondwa, S., & John, M. J. (2024). A review of weathering studies in plastics and biocomposites—Effects on mechanical properties and emissions of volatile organic compounds (VOCs). *Polymers, 16*, 1103. https://doi.org/10.3390/polym16081103

OECD. (2016). *Extended producer responsibility: Updated guidance for efficient waste management.*

Palleti, S., Varshini, M., Bhargavi, L., & Tejasri, M. (2024). Bio-conversion of plastic waste into sustainable biofuels: A comprehensive review of microbial degradation approaches. *IOSRJESTFT, 18*, 29–34. https://doi.org/10.9790/2402-1810022934

Paul, A., Priyadarsini, S., Karakuş, S., Behera, J. K., Gnanasekaran, R., & Das, A. P. (2024). Fate and accumulation of synthetic microfibers in the environment. In *Sustainable microbial technology for synthetic and cellulosic microfiber bioremediation* (pp. 61–76). https://doi.org/10.1007/978-3-031-62600-5_4.

Peixoto, J., Silva, L. P., & Krüger, R. H. (2017). Brazilian Cerrado soil reveals an untapped microbial potential for unpretreated polyethylene biodegradation. *Journal of Hazardous Materials, 324*, 634–644. https://doi.org/10.1016/j.jhazmat.2016.11.037

Pradhan, G., Dey, S., Priyadarsini, S., Kumar, M. S., & Das, A. P. (2024). Characterization and quantification of microplastics pollutants in sediment samples from Daya River of Odisha state in India for their appropriate management. *Environmental Science and Engineering, Part, F2333*, 75–92. https://doi.org/10.1007/978-3-031-51792-1_5

Pramila, R., & Ramesh, K. V. (2011). Biodegradation of low density polyethylene (LDPE) by fungi isolated from marine water a SEM analysis. *African Journal of Microbiology Research, 5*(28), 5013–5018.

Priyadarshini, B., & Das, A. P. (2024a). The adverse impact of synthetic microplastics released from disposable face masks pollutants on aquatic microbiota, aquatic species and humans: A review. *Environmental Science and Engineering, Part, F2827*, 67–81. https://doi.org/10.1007/978-3-031-50840-0_5

Priyadarsini, S., & Das, A. P. (2025). Lithium bioleaching: A review on microbial-assisted sustainable technology for lithium bio-circularity. *Journal of Water Process Engineering, 69*, 106744. https://doi.org/10.1016/j.jwpe.2024.106744

Priyadarsini, S., & Das, A. P. (2024b). Analytical and structural characterization of waste lithium-ion batteries for their effective recycling strategy. Environmental Chemistry and Ecotoxicology. https://doi.org/10.1016/j.enceco.2024.12.004.

Rajavanshi, J., Sogani, M., Tziouvaras, G., Kumar, A., Syed, Z., Sonu, K.,... & Sen, H. (2025). An analytical review on revamping plastic waste management: exploring recycling, biodegradation, and the growing role of biobased plastics. Environmental Science and Pollution Research, 32(28), 17243-17261.

Ray, S. K., Dey, S., Das, A. P., & John, A. (2024). Microplastic waste management through greener technologies: A review. In *Sustainable management of urban plastic waste through circular economic approaches* (pp. 237–261). https://doi.org/10.1201/9781003465447-14.

Ru, J., Huo, Y., & Yang, Y. (2020). Microbial degradation and valorization of plastic wastes. *Frontiers in Microbiology, 11*, 442.

Sahu, A., Dey, S., Mallick, K., Priyadarsini, S., Behera, I. D., & Das, A. P. (2024). A review on advanced Technology for Sustainable Management of synthetic microplastic waste. In *Sustainable management of urban plastic waste through circular economic approaches* (pp. 262–279). https://doi.org/10.1201/9781003465447-15.

Sahu, S., & Das, A. P. (2024). Microplastics degradation mechanism by aquatic Plastiomes: A review. pp. 213–226. https://doi.org/10.1007/978-3-031-50840-0_12.

Saju, J. A., Bari, Q. H., Lorber, P., Rafizul, I. M., & Kraft, E. (2024). Integrated approach of waste analysis and life cycle assessment for the management of non-recyclable plastics in recycling shops. *Cleaner Environmental Systems, 15*, 100229. https://doi.org/10.1016/j.cesys.2024.100229

Sarsan, S., Kodaparthi, A., & Birru, S. (2024). Microbial enzymes in plastic degradation. In *Microbial essentialism* (pp. 207–242). Academic Press. https://doi.org/10.1016/B978-0-443-13932-1.00005-2.

Sen, S. K., & Raut, S. (2016). Microbial degradation of low density polyethylene (LDPE): A review. *Journal of Environmental Chemical Engineering, 3*, 462–473. https://doi.org/10.1016/j.jece.2015.01.003

Shah, Z., Hasan, F., Krumholz, L., Aktas, D. F., & Shah, A. A. (2013). Degradation of polyester polyurethane by newly isolated Pseudomonas aeruginosa strain MZA-85 and analysis of degradation products by GC–MS. *International Biodeterioration & Biodegradation, 77*, 114–122.

Sharma, R., Singla, L., & Singh, G. (2024). Challenges and sustainable solutions for the detection and bioremediation of microplastic pollution. In *Challenges and sustainable solutions in bioremediation* (pp. 125–142). CRC Press. https://doi.org/10.1201/9781003407317-8.

Siddiqua, A., Hahladakis, J. N., & Al-Attiya, W. A. K. A. (2022). An overview of the environmental pollution and health effects associated with waste landfilling and open dumping. *Environmental Science and Pollution Research, 29*, 58514–58536. https://doi.org/10.1007/s11356-022-21578-z

Singh, S., Malyan, S. K., Maithani, C., Kashyap, S., Tyagi, V. K., Singh, R., Malhotra, S., Sharma, M., Kumar, A., Panday, B. K., & Pandey, R. P. (2023). Microplastics in landfill leachate: Occurrence, health concerns, and removal strategies. *Journal of Environmental Management, 342*, 118220. https://doi.org/10.1016/j.jenvman.2023.118220

Singh, R. P., Mishra, S., & Das, A. P. (2020). Synthetic microfibers: Pollution toxicity and remediation. *Chemosphere, 257*, 127199. https://doi.org/10.1016/j.chemosphere.2020.127199

Skariyachan, S., Megha, M., Kini, M. N., Mukund, K. M., Rizvi, A., & Vasist, K. (2015). Selection and screening of microbial consortia for efficient and ecofriendly degradation of plastic garbage collected from urban and rural areas of Bangalore, India. *Environmental Monitoring and Assessment, 187*, 1–14.

Sunil Kumar, K., Babu, J. M., Venu, H., & Muthuraja, A. (2022). Waste plastic as a source of biofuel for stationary diesel engine: A critical review. *International Journal of Ambient Energy, 43*(1), 8577–8591.

Tammemagi, H. Y. (1999). *The waste crisis: Landfills, incinerators, and the search for a sustainable future.* Oxford university press.

Thakur, S., Mathur, S., Patel, S., & Paital, B. (2022). Microplastic accumulation and degradation in environment via biotechnological approaches. *Water, 14*(24), 4053.

Tripathy, B., Sahoo, P. P., Sundaray, H., & Das, A. P. (2024). In-vitro biodegradation of discarded marine microplastics across the eastern coast of the Bay of Bengal, India using Exiguobacterium sp. *Environmental Chemistry and Ecotoxicology, 6*, 236–247. https://doi.org/10.1016/J.ENCECO.2024.07.003

United Nations Environment Programme. (2024). *Global waste management outlook 2024: Beyond an age of waste—Turning rubbish into a resource.* Nairobi. https://wedocs.unep.org/20.500.11822/44939.

United Nations Environment Programme-UNEP. (2018). Single-use plastics: A roadmap for sustainability.

Usha, R., Sangeetha, T., & Palaniswamy, M. (2011). Screening of polyethylene degrading microorganisms from garbage soil. *Libyan Agriculture Research Center Journal International, 2*(4), 200–204.

Verce, M. F., Ulrich, R. L., & Freedman, D. L. (2000). Characterization of an isolate that uses vinyl chloride as a growth substrate under aerobic conditions. *Applied and Environmental Microbiology, 66*, 3535–3542. https://doi.org/10.1128/AEM.66.8.3535-3542.2000

Wang, S., Hadji-Thomas, A., Adekunle, A., & Raghavan, V. (2024a). The exploitation of bioelectrochemical system and microplastics removal: Possibilities and perspectives. *Science of the Total Environment, 930*, 172737. https://doi.org/10.1016/j.scitotenv.2024.172737

Wang, Y., Zhao, C., Lu, A., Dong, D., & Gong, W. (2024b). Unveiling the hidden impact: How biodegradable microplastics influence CO_2 and CH_4 emissions and volatile organic compounds

(VOCs) profiles in soil ecosystems. *Journal of Hazardous Materials, 471*, 134294. https://doi.org/10.1016/j.jhazmat.2024.134294

Wei, R., & Zimmermann, W. (2017). Microbial enzymes for the recycling of recalcitrant petroleum-based plastics: How far are we? *Microbial Biotechnology, 10*, 1308. https://doi.org/10.1111/1751-7915.12710

Wilkes, R. A., & Aristilde, L. (2017). Degradation and metabolism of synthetic plastics and associated products by pseudomonas sp.: Capabilities and challenges. *Journal of Applied Microbiology, 123*, 582–593. https://doi.org/10.1111/jam.13472

Wojnowska-Baryła, I., Bernat, K., & Zaborowska, M. (2022). Plastic waste degradation in landfill conditions: The problem with microplastics, and their direct and indirect environmental effects. *IJERPH, 19*, 13223. https://doi.org/10.3390/ijerph192013223

Yang, J., Yang, Y., Wu, W. M., Zhao, J., & Jiang, L. (2014). Evidence of polyethylene biodegradation by bacterial strains from the guts of plastic-eating waxworms. *Environmental Science & Technology, 48*, 13776–13784. https://doi.org/10.1021/es504038a

Yoon, M. G., Jeon, J. H., & Kim, M. N. (2012). Biodegradation of polyethylene by a soil bacterium and AlkB cloned recombinant cell. *Journal of Bioremediation & Biodegradation, 3*, 145.

Zhang, Z., Chen, Z., Zhang, J., Liu, Y., Chen, L., Yang, M., et al. (2024). Municipal solid waste management challenges in developing regions: A comprehensive review and future perspectives for Asia and Africa. *Science of the Total Environment, 172794.*

Zhu, D., Adebisi, W. A., Ahmad, F., Sethupathy, S., Danso, B., & Sun, J. (2020). Recent development of Extremophilic bacteria and their application in biorefinery. *Frontiers in Bioengineering and Biotechnology, 8*. https://doi.org/10.3389/fbioe.2020.00483

Fermentation Technology: Utilizing Microorganisms on Water Resources to Produce Enzymes or Organic Acids

Prasamita Samantaray, Sailaja Priyadarsini, and Alok Prasad Das

Abstract A new field of biotechnological research is the application of waste materials for the fermented microbial synthesis of organic acids, which is motivated by the demand for economical and environmentally friendly production techniques. Lactic acid, succinic acid, citric acid, and acetic acid are examples of organic acids varieties, those are used as crucial industrial intermediates in chemicals, food, pharmaceuticals, and biodegradable plastics. These organic acids have traditionally been produced using petrochemical or food-grade sugars, but the increasing need for sustainable alternatives has raised interest in using waste biomass and byproducts as the initial ingredients for fermentation processes. Microbial fermentation of waste materials, including food waste, lignocellulosic biomass, industrial byproducts, and agricultural residues, offers a low-cost feedstock for the synthesis of organic acid while also posing a threat to the environment. Under ideal fermentation circumstances, different types of microorganisms, yeast, bacteria including fungi, transform complicated waste materials into valuable organic acids. The waste resource's high protein, carbohydrate, and lignocellulose content necessitates the careful selection of microbial strains with particular enzymatic capabilities for effective fermentation and breakdown. This book chapter examines the latest developments in the microbial fermentation of waste materials to produce organic acid, emphasizing important microbes, metabolic processes, and fermentation methods. The potential advantages of this strategy, such as lower costs, less waste, and a more sustainable environment, highlight its potential as a crucial technology for upcoming bio-based industrial production.

Keywords Fermentation · Microorganism · Water · Enzymes · Organic acid

P. Samantaray · S. Priyadarsini · A. P. Das (✉)
Department of Life Sciences, Rama Devi Women's University, Bhubaneswar, Odisha, India
e-mail: alok1503@gmail.com

A. P. Das (ed.), *Bioconversion of Environmental Waste to Value Added Products*,
Sustainable Environmental Waste Management Strategies,
https://doi.org/10.1007/978-3-032-14215-3_5

1 Introduction

A traditional technique for extending food's shelf life and improving its nutritional value is fermentation. A wide range of foods made from plants and animals have been preserved using it. Fermentation-based milk products like cheese, curd were produced by European, Middle Eastern, and mostly in Indian cultures. Animal husbandry, on the other hand, was primarily limited to Eastern nations like Korea, Japan, and China. Social, economic, cultural, and religious factors also affected the variety of ingredients used to make fermented meals and alcoholic beverages. Fermentation is based on the controlled activity of microbes on a food substrate. Microorganisms that are frequently present in raw materials can cause fermentation on their own (. Traditional fermented foods made by spontaneous fermentation include kimchi, sourdough bread, and sauerkraut (Augustin et al., 2024). Fermentation is already a well-established process because of biotechnology advancements. It is possible to obtain a standardized food product by using starter cultures (Kumar et al., 2023). Strain engineering for food applications frequently uses filamentous fungi, yeasts such as *Saccharomyces cerevisiae*, *Komagataella phaffii*, and *Kluyveromyces* spp., or innocuous microbes like *Bacillus* spp. (Ghosh et al., 2023; Mohanty et al., 2016). Fermentation helps to produce certain food components, such as vitamins, enzymes, tastes, colorants, and antioxidants. Culinary waste is being used more and more as a substrate of fermentation in the production of various types of enzymes, pigments, antioxidants, bacteriocins, and other culinary ingredients (Chai et al., 2022; Gomes et al., 2018).

Eliminating food waste and creating innovative, high-value assets through the application of an underutilized or untapped resource are directly related (Biswal et al., 2017; Das & Swain ; Khatun et al., 2023). In exchange, scientific research has provided knowledge on the use of fermentation technology to create edible fermented insects to increase food security. Edible preparations that contain fermented insects, like pastes, powders, and dishes that are fermented with the potential to be expanded, are already on the market. *Locusta migratoria* and *Galleria mellonella* are two examples of these fermented sauces. Comparing these sauces to traditional fish sauces, they were quite satisfying (Mouritsen et al., 2017; Yang et al., 2022).

Fermentation creates new bioactive components originating from the food matrix, alters the profile of antioxidants, and increases the effectiveness of component extraction. Precision fermentation has recently gained attention as a key development in the food industry's impending fourth industrial revolution. Fermentation with accuracy is gaining popularity to modern food production due to its capacity to improve the efficiency of the creation of proteins, fats, and carbohydrates while lowering waste. The cultural and ecological relevance of fermentation, global technological developments in fermented food and beverage, health advantages, nutritional worth, as well as microbiological understanding into different fermented foods will all be covered in this overview. We will look at many fermented meals that come from edible insects, plants, animals, and microalgae. We will talk about the bacteria that ferment each type of product, consumer acceptance, provenance, and consumption.

Additionally covered are how the industrial revolution affected the manufacturing of fermented foods and how fermentation may be used to value food waste.

Polyols, alcohols, amino acids, and polysaccharides, including other organic molecules that comprise commodity chemicals, are exceptionally soluble in water, and the bioconversion reactions that yield them typically occur in hydrophilic environments (Ball, 2017). As a result, most industrial processes that use natural reaction systems and renewable resources to manufacture the corresponding biomolecules in vast quantities are conducted in water suspension (Liew et al., 2016). Nowadays, the manufacture of value-added compounds using fermentation substrates derived from renewable resources leads to fairly complicated aqueous systems, from which the products must be separated. Because of this, these technical methods necessitate the use of sophisticated techniques like downstream processing that raises manufacturing costs (Gu et al., 2014). Furthermore, extremely diluted systems are used to provide higher yields and get around mass transfer problems or inhibitory phenomena. Reducing or even eliminating the water is essential for subsequent processing, particularly for platform molecules that act as intermediates in several chemical industry value chains. Logistical considerations also influence the product concentration for efficient storage and transportation to end users.

2 Microorganisms for Enzyme and Organic Acid Production

The creation of microbial cell factories to synthesize chemicals and fuels from renewable biomass has been prompted by worldwide concerns about the environmental consequences of unsustainable chemical production that relies on fossil fuels (Lian et al., 2018). However, there are certain difficulties with the microbial synthesis of organic acids. Generally speaking, bacteria cannot withstand low pH levels. Gram-negative bacteria, including *Escherichia coli*, have outer as well as inner membranes that specifically encapsulate the periplasm (Hong et al., 2012). Protons can readily diffuse through non-specific transporters, including porins that are present in the outer membrane. Because the periplasm contains essential transporters and enzymes for the intake of food molecules, the acidification of the periplasm caused by the proton accumulation may break the periplasmic proteins down and impair their functioning, which would impede cell growth (Yin et al., 2015). This means that even though using bacteria to produce organic acids can achieve very high titres and yields. After that, organic acids are converted to salt forms. Consequently, an acidification phase employing a strong acid (such sulfuric acid) is necessary to get the undissociated forms of chemical compounds; however, this step produces gypsum, which raises the cost of downstream processing (Priyadarsini et al., 2025; Priyadarsini & Das, 2024, 2025).

Despite having smaller titers and outputs of organic acids than bacteria, yeasts can tolerate much lower pH levels as plasma membranes of yeasts are impervious to protons. In contrast, organic acids can freely permeate over the yeast cell membrane in considerable amounts when they are at low pH with their undissociated forms. Yeasts may potentially become poisonous if organic acid anions build up. Furthermore, strong microbial resistance to suppressors found in lignocellulosic hydrolysates is an additional crucial requirement when using lignocellulosic biomass as a feedstock (Jin & Cate, 2017; Yin et al., 2015). Microorganisms' limited ability to withstand acid stress and low pH presents difficulties for microbial organic acid synthesis. Achieving economical and efficient fermentation processes for the production of organic acid components requires strengthening the robustness of microbial cell factories. ALE, transcriptome analysis-guided engineering, and genome-scale engineering have all been successfully applied to improve the acid tolerance of various microorganisms (Guan et al., 2020). Microbes have also developed a variety of defense mechanisms, such as the control of energy metabolism, modifications to cell membranes, and pH homeostasis, to survive in severely acidic environments. Even though their strategies may be similar, some bacteria are nevertheless better than others at coping with and adjusting to acid stress. We may be able to learn more about the mechanism of tolerance of bacteria that exhibit exceptional resilience to acid stress and low pH thanks to developments in synthetic and systems biology. For instance, the unusual the yeast I. orientalis is well-known for its remarkable resistance to both low pH and high amounts of organic acids (Tran et al., 2019). It is now feasible to understand how this species copes with acid stress thanks to the emergence of numerous essential genetic tools. It may be possible to increase the generated organic acids by transferring these improved tolerance mechanisms to the factories of microbes that now have limited acid stress tolerance. Finally, genome-scale engineering tools like change are particularly helpful for determining for growth-associated traits involving resistance to high the levels of organic acidic solutions and extremely acidic environments, even though they have been employed solely to increase *S. cerevisiae's* acceptability to lignocellulosic inhibitors. Through growth enhancement in low pH or acidic settings, mutants with great endurance to these challenges can be enriched, and mutant libraries containing the entire genome can be created using genome-scale engineering approaches. Furthermore, these genome-scale engineering methods have a much shorter timeframe than ALE, and the plasmids of the enriched mutants may be sequenced to quickly determine the genetic alterations that result in increased acceptance (Cao et al., 2020; Lian et al., 2019) Change to Overview of various food waste substrates, associated microbial strains, and fermentation procedures employed for bioresource recovery (Table 1).

3 Water Resource Utilization in Fermentation

The numerous health advantages of deep ocean water (DOW), which is well-known for its high nutritional content, have been highlighted in numerous studies. These include preventing cardiovascular diseases and arteriosclerosis, curing dermatitis, and reducing obesity (Hwang et al., 2008; Lee et al., 2017, 2019; Wang et al., 2015).

Table 1 Valuation of food waste

Wasted food	Microbes	Procedure for fermentation	References
Peel of a potato	*Xanthomonas citri*	Fermentation in solid states	Vidhyalakshmi et al. (2012)
Waste from pomegranates, peanuts, and brandy distilleries	*Punica granatum, Aspergillus awamori, Rhizopus oryzae, Aspergillus niger, and Rhizopus oligosporus*	Solid-state fermentation	Sadh et al. (2018)
Waste from bakeries and olive pomace	*Xantophylomyces dendrorhous, Sporidiobolus salmonicolor, Monascus purpureus, Yamadazyma guilliermondii*, and *Yarrowia lipolytica*	Fermentation in solid states	Dursun and Dalgiç (2016)
Bran made from rice	*Rhizopus oryzae* and *Aspergillus oryzae*	Fermentation in solid states	Das et al. (2016)
Whey, or dairy waste	Aerius*Cryptococcus albidus* sp.	Fermentation in liquid	Németh et al. (2015)
Banana and pineapple waste	*Saccharomyces ceroviae*	Fermentation that is solid	Verotta et al. (2018)
Mango waste, orange peel, banana skin, and sweet potato waste	*Candida tropicalis, Lactobacillus acidophilus, Saccharomyces* sp., and *Saccharomyces cerevisiae*	Fermentation, both liquid and solid	Dulf et al. (2016)
Solid waste from pulp and paper	1526 *Rhizopus oryzae*	Fermentation in solid states	Das et al. (2016)
Apple pomace and apricot pomace	*Rhizopus oligosporus* (ATCC-22959), *Aspergillus awamori*, and *Aspergillus niger* (ATCC-6275); *Phanerocheate chrysosporium*	Fermentation in solid states	Sadh et al. (2018)

In addition to these health advantages, DOW can be utilized to create microbial fermentation that supports well-being and facilitates the production of multifunctional health foods. With DOW, manufacturing expenditures can be reduced, brewing efficiency can be increased, and the value of products and health benefits can be greatly increased. Additionally, mineral supplements enhanced mycelial development and cordycepin production in the fermentation of *Cordyceps militaris* when DOW was applied, increasing the effects on the advancement of hepatic fibrosis (Hung et al., 2017). Throughout the growth of *Epichloë festuca*, the organisms absorbed the minerals from DOW and converted them into organic minerals. In particular, organic magnesium, which is readily absorbed to raise magnesium levels in the brain, stimulated the brain's production of magnesium transport proteins, which in turn enabled more magnesium ions to enter neurons. Magnesium supplementation improved memory and learning capacities by lowering the expression of markers linked to Alzheimer's disease (Wu & Lee, 2021). This is becoming a useful tool in the food fermentation sector. In order to improve flavor and efficiency, DOW is frequently added to yeast fermentation in a variety of fermentation processes, including wine-making and bread fermentation. The supply and concentration of DOW, however, might have an impact on the fermentation outcomes, and the requirements and adaptabilities of various yeast strains may vary with respect to DOW. The use of DOW in yeast fermentation has not yet been the subject of any relevant research. The future growth of the food fermentation industry therefore depends on examining the effects of DOW on fermentation-related proteome expression, alcohol dehydrogenase activity, and yeast growth. We will utilize the abundant DOW resources in the Pacific Ocean off Taiwan's east coast. The DOW used in the trials will come from two sources: Taitung, which is 300 m deep, and Hualien, which is 600 m deep. By fermenting three different species of yeast with different DOW sources, we intend to compare alcohol dehydrogenase activity and development of yeast as well as if the yeast can absorb mineral components from the DOW.

Water is an essential component of the fermentation medium because it serves as an adhesive in which vitamins and nutrients, substrates, and essential trace elements are dissolved. Carbohydrate molecules, producers of nitrogen, vitamins, and minerals that microorganisms like fungal and bacterial species require for cellular development and energy utilization must dissolve in water. The fermentation process, which is often carried out in liquid media, converts substrates into useful metabolic products such as organic acids, ethanol, enzymes, or antibiotics. The wide spectrum of nutrient solubility in water ensures that microorganisms can get the elements they require to thrive. Water is also necessary to maintain the structural integrity of the microbial cells. Water is necessary for temperature regulation, waste removal, product recovery from addiction, and sustainability projects in addition to acting as a substrate for microbial development. The efficient use of water resources is essential for improving fermentation processes, increasing yields, and reducing environmental impacts. As industry keeps expanding up fermented operations and addresses growing environmental concerns, advancements in water management, including as reclamation and purification technology, will be crucial to guarantee that water is used efficiently and sustainably. Ultimately, the effective and environmentally friendly consumption of water in fermentation is critical to the future of bacteria-based biotechnology and its applications in sectors such as the cultivation of food, bioenergy, and pharmaceuticals.

The composition of the fermentation medium is directly influenced by the water concentration. The physical characteristics of the environment and substrate concentration, which are dictated by the water content to solute ratio, can have an effect on microbial activity. To prevent extremely concentrated circumstances that would limit microbial activity or substrate dilution that could obstruct microbial growth, the proper balance must be maintained. Water is also necessary to optimize fermentation conditions. In fermenting in solid-state (SSF), microorganisms flourish on solid, wet substrates like grains or agricultural waste, which frequently contain less water. Conversely, submerged fermentation employs a liquid medium–high heat, with water acting as the primary component of the culture broth. For both types of fermentation to promote microbial growth and enable the efficient elimination of byproducts, the water content must be carefully regulated.

Temperature control is crucial in fermentation processes because enzymatic reactions and microbial growth rates are highly sensitive to temperature variations. During fermentation, microorganisms generate heat, which, if not controlled, can cause overheating, stunt the growth of the germs, or even kill them. Both inside the fermentation vessel and in external cooling loops, water serves as a cooling agent in fermentation systems. By facilitating heat dissipation, water helps keep the fermenter's temperature steady. Because some fermentation processes, particularly those done on an industrial scale, generate a lot of heat, water is essential for preserving the system's optimal temperature range. By guaranteeing that the fermentation-related catalysts work efficiently and that microbial growth proceeds at the proper rate, this water-based temperature management optimizes product yield.

A range of byproducts, such as ethanol, which is acidic substances (like acetic acid), ions (like carbon dioxide), and additional waste products from metabolism, are produced by microorganisms during fermentation. Water is necessary for diluting these byproducts to keep them from accumulating to hazardous levels that could hinder further microbe development or fermentation efficiency. In order to maintain an atmosphere that is favorable for microbial development, water also helps move waste products out of the production vessel. In some cases, water is added to the fermentation broth to change its pH. Many fermentation processes result in the production of acidic or basic byproducts, which can cause pH shifts that have an adverse effect on microbial activity. Water can help dilute these byproducts by buffering pH changes and preserving an atmosphere that is conducive to microbial development and product formation.

Water is used to recover and cleanse the intended product once the process of fermentation is finished. For instance, the distillation process extracts ethanol from the process of fermenting broth using water. Water is also commonly used in scrubbing and filtration operations to remove pollutants, residual nutrients, and microbial biomass from the final product. Sometimes extraction processes call for the use of water as a solvent. For instance, when enzymes or antibiotics ferment, water is used to dissolve the result for further separation and purification. The water used in these operations needs to be pure to avoid contaminating the final product. As concerns about water scarcity and environmental sustainability gain urgency, the handling of the water supply in fermentation becomes increasingly important. In large-scale fermentation, substantial amounts of water are required for both the actual procedure of fermentation and the recuperation of the post-fermentation product. Many businesses utilize water recycling systems to reduce water consumption and their environmental impact. These systems allow water to be reused for extraction, cooling purposes, or even as a substrate for fermentation by filtering and treating wastewater to remove contaminants. Another strategy to increase water efficiency is to use closed-loop systems, which recycle water inside the structure and do away with the need for new water input. Additionally, developing environmentally friendly fermentation technologies and optimizing the consumption of water through process layout and management systems are two crucial steps toward sustainable in the fermentation industry.

The water's purity is very important for fermentation processes. Contaminated or unclean water can introduce undesirable substances, particles, or bacteria into the fermentation system, leading to pollution or a lower-quality product. The process of filtration sterilization because of reverse osmosis is therefore commonly used to remove impurities from fermentation water. Maintaining the appropriate bacteria population and avoiding spoiling or the development of an off flavor depend on clean, sterile water. When making high-value items like pharmaceuticals, food additives, or biofuels, water purity is particularly important in situations where even trace amounts of contamination can significantly affect the safety, effectiveness, and quality of the finished product.

4 Fermentation Engineering and Optimization

Several commercially significant chemicals utilized in the food, chemical, pharmaceutical, and energy production industries are synthesized using fermentation technology. Despite the fact that fermentation techniques have been used for many years, there is always a challenge because products must be produced economically and sustainably while yet satisfying consumer demands. The availability of fermented products that meet market demand is the most crucial element for every fermentation-based product. Numerous primary and secondary metabolites have been reported to be created by various microorganisms, albeit in minuscule quantities. Processes that have effectively employed a variety of high yielding techniques that have been developed in the past to satisfy market demand include oil extraction, biotransformation, and the synthesis of primary or secondary metabolites. Prior to any large-scale metabolite manufacturing, medium optimization remains one of the most studied phenomena. It also presents a number of difficulties. Prior to the 1980s, media optimization was accomplished by traditional techniques that were costly, time-consuming, and required a number of less precise tests. However, since the advent of modern statistical and mathematical methods, media optimization has become more dynamic, efficient, cost-effective, and reliable in delivering outcomes. When designing a production medium, it is necessary to determine and optimize the proper medium components, such as carbon and nitrogen, as well as the optimal fermentation conditions, such as temperature, pH, agitation speed, etc. Additionally, the highest possible product concentration might be attained by enhancing the previously described characteristics. Productivity is one of the primary research areas since it lowers the production and cost of the finished product. To increase production, strain improvement or process parameter adjustments are typically employed. However, optimization and strain improvement are "Catch-22" situations. Researchers from all around the world typically tackle this problem by focusing on a single element at a time. However, if a different medium is chosen, neither method can guarantee that one of the intended strains will be employed. Because of this limitation and Catch-22 scenario, a variety of novel approaches that allow strain improvement and medium design to be applied simultaneously have been proposed and studied.

Medium optimization has been shown to be an effective tool for increasing product yield and process improvement in addition to strain improvement techniques. Screening alternative carbon and nitrogen sources and their various combinations during process development can significantly improve microbial growth, their viability, and overall product output through the evolution of medium compositions. Increasing productivity or replacing more costly components with less expensive ones could lower the cost of fermentation products. These are the goals of a successful optimization strategy. For future advancements and increased accuracy, a few aspects still need to be considered. Because each bacteria have gene-level limitations on producing specific metabolites, for instance, it's always necessary to search for new microbes with higher productivity. Microbes may not always be able to use the less expensive raw material under current conditions, but they may be accepted to ingest low-key substrates with improved performance through mutation. The alternative method of boosting microbial productivity is genetic manipulation. Application of recombinant microorganisms' fermentation in a large-scale has been seriously questioned due to recent worries about genetically modified organisms. Therefore, many researchers and industrial personnel find that using natural bacteria is a terrific option. Since the primary factor influencing the production of secondary metabolites is substrate limitation, planning and refining the chemostat production procedure can increase output and reduce wasteful substrate loss. The alternative for maximizing substrate usage and long-term microbial reuse is the further design of a fluidized bed bioreactor or a mist bioreactor. Understanding the exact nature of microbes, other living systems, and biology is challenging, but as knowledge grows, it will be possible to choose appropriate designs for improved performance. For instance, metabolic flux is the rate at which several enzymes participating in a certain metabolic process regulates any products or molecules produced by that system.

Prior to starting pilot or semi-pilot manufacturing plans, the fermentation media must be optimized for the generation of metabolites. Both the traditional and cutting-edge optimization methods employed in the medium optimization process have been examined and discussed in this critical examination. It was demonstrated that the statistical methods might shorten the experimental period needed for process development and quality enhancement. Optimization methods also aid in lowering the total cost of the product. Applications have been proposed after an analysis of the designs and methods in this review's effectiveness, simplicity, and time consumption. For the media created after using numerous designs, additional testing in real-world production settings and, eventually, full-scale models that accurately depict the manufacturing environment are still required. Anaerobic and aerobic fermentation using the right facultative microorganisms is another sophisticated integrated way of microbe growth that has been investigated and applied recently. Researchers have employed this integrated approach to manufacture ethanol and xylitol using *Candida tropicalis*. Xylitol is the end product that serves as a substrate for the anaerobic production of ethanol under aerobic conditions, they found. It was hypothesized by the researchers that the bacterial community's composition and the subsequent breakdown of chlorophenols in biocathode microbial fuel cells are impacted by the change in anaerobic/aerobic conditions during growing. Among the many benefits of this integrated approach are the elimination of the need for a supplemental operation unit, the generation of zero wastewater, the use of solid pre-treated material for processing, and the elimination of the need for sterilization. These benefits can all contribute to the biotechnology industry becoming more cost-effective and ecologically friendly (Singh et al., 2017).

5 Advances in Fermentation Technology

The metabolic activity of organisms during their entire life cycle, from growth to death, is included in the fermentation process. The fermentation technology employed in the commercial production of food, medications, and alcoholic beverages is powered by these microbes. The basic idea behind industrial fermentation technology is the growth of organisms under ideal conditions. Giving them the nutrients they require to flourish such as carbon, nitrogen, minerals, and vitamins achieves this (Sharma et al., 2020). To guarantee the maximum production of the intended product, the temperature, oxygen, and pH concentrations must also be regulated. The fermentation process is also meticulously watched to make sure the organisms are not exposed to any diseases or poisons that could influence the final result. Recovering fermentation products in such industrial chemicals, organic acids have garnered a lot of attention due to the growing emphasis on environmental preservation and renewable energy. Because of this increase, products other than the conventional low-volume, high-value pharmaceuticals have become more prevalent, and fermentation is now on par with the synthesis of commodity chemicals. For businesses to be competitive in the production of high-volume, low-cost chemicals, they must maximize efficiency and minimize waste byproducts. The biotechnology potential of agro-industrial residues is currently being actively investigated by scientific communities. The food business has undergone numerous revolutions thanks to emerging fermentation-based technology. (Cai et al., 2020) revealed that preheating broccoli florets for three minutes at 65 °C, afterward some lactic acid bacteria fermentation on a laboratory scale, enhanced sulforaphane production by sixteen times. These technologies have also made it possible to produce a pea protein hydrolysate component that improves food's salinity. According to a study, utilizing the probiotic *Lactobacillus gasseri* to ferment carrot juice improved its nutritional profile. The results showed that the sugar content of fermented straight carrot juice was reduced by 27%. These techniques can be used to enhance fermentation procedures and modify the nutritional and sensory qualities of fermented foods (Xu et al., 2020). Many companies are very interested in the benefits of producing lactic acid from renewable resources. This strategy has been adopted by the petrochemical sector because of its high yield and economical productivity using easily accessible substrates. However, a major environmental concern is the disposal of waste materials and biomass from different sources. An integrated biorefinery platform that manages waste and generates high-value bioproducts is the best option. Fermentation-produced optical pure lactic acid is a very attractive choice for packaging materials because of its possible uses in pharmaceutical, cosmetic, textile, and food industries. An eco-friendly substitute for bioplastics derived from petrochemicals is polylactic acid derived from natural resources. Despite the significant benefit of biodegradability, the high expense of lactic acid production has limited the material's widespread application (Kumar et al., 2019; Zhao et al., 2016). Common sense, not scientific knowledge, has enhanced small-scale food fermentation methods in impoverished countries. As a result, many producers need more confidence about the modernization and adjustment of their fermentation processes. While maintaining their uniqueness and cutting manufacturing costs, it is imperative to enhance the safety and quality of fermented foods. The consortium technique is one useful strategy that has greatly enhanced Thailand's small-scale soy sauce fermentation (O'toole, 2019). This strategy makes it easier for academics and industry to continue working together, giving the latter the research emphasis they need to support the industry's success. A new era of customized starter cultures has been brought about by modern molecular biology techniques; however, the small-scale fermentation sector has been reluctant to adopt them because of worries about losing their distinctive flavor. The possibility for personalization was highlighted by recent studies that found differences in sensory quality and composition among bacteria from different product origins. In an effort to facilitate the use of starting cultures in production, a cell bank for Thai fermented pork sausage is presently being developed.

6 An Outlook on Fermentation for the Future

For many years, fermentation techniques have been used to manufacture a variety of goods, like materials, chemicals, and medications. Fermentation is a safe, environmentally friendly, and sustainable process that can provide an increasing number of goods. However, a few problems need to be resolved if this technology is to become more competitive in the future. These include a lack of freshwater, high energy usage, microbial contamination, the difficulty of maintaining sterility, inadequate oxygen uptake in the cultures, the use of food-related components as substrates, and more. Future fermentation procedures should therefore be better and more efficient from the perspective of the aforementioned problems. First, the issues of bacterial contamination should be avoided using innovative fermentation techniques. In fact, the microbes that are now in use are highly vulnerable to pollution from other microbes. For instance, yeast and bacterial strains typically employed in fermentation processes, including *E. Coli*, *Bacillus*, *Corynebacterium* and *Pseudomonas* species, are cultivated in mild environments that also support the growth of the majority of microorganisms found in the soil, water, and air. To avoid microbial contamination, industrial facilities need to be completely sanitized. Because of this, processes become quite complex, which raises the cost of production. Therefore, two key prerequisites for future fermentation process improvement are the employment of more resistant microbes and the installation of contamination prevention methods. To start, innovative fermenting techniques should be employed to avoid bacterial contamination issues. In actuality, the bacteria now in use are highly vulnerable to infection by other microorganisms. For instance, mild conditions that are also favorable for the growth of the majority of soil, water, and airborne microorganisms are employed to cultivate the yeast and bacterial strains frequently utilized in fermentation, such as *E. Coli*, *Bacillus*, *Corynebacterium* and *Pseudomonas* species. Cleaning industrial facilities thoroughly is necessary to avoid microbiological infections. Because of this, processes become quite complex, which raises the cost of production. Therefore, two crucial prerequisites for future fermentation process improvement are the employment of more resistant microbes and the application of contamination control systems. Specifically, substrate-selected bacteria, these germs may grow in conditions that also prevent other microbes from growing. As a result, fermentation processes can be simpler and the need for highly skilled microbial fermentation engineers may be avoided as sterilizing can be avoided (Chen & Jiang, 2018).

Furthermore, the processes can be automated because these extremophile bacteria are more resilient and stable in a variety of growing environments. Reusing the fermentation broth or substituting seawater for fresh water are two potential solutions to the freshwater shortage issue in future fermentation processes. Quorum-sensing molecules, proteins, polysaccharides, lipids, and genetic components are among the chemicals found in the fermentation broth that can be utilized in cellular metabolic activities and, thus, serve as sustenance for the cells. An additional step in the fermentation process's potential is to prioritize anaerobic or microaerobic microbes (Tripathy et al., 2026). Actually, in order to thrive and convert substrates into products, aerobic bacteria utilized in modern fermentation processes need a lot of oxygen. This is especially true in large-scale bioreactors that get their oxygen from air compressors. The majority of these molecules, however, escape into the air and pass through the bioreactors since oxygen is not very soluble in water. Increased oxygen intake can be achieved by overexpressing bacterial hemoglobin, however employing anaerobic/microaerophilic bacteria would also be beneficial and would help save energy consumption, particularly during product creation. Additionally, considerably more energy will be saved by using microorganisms that can grow in a variety of temperatures (Ouyang et al., 2018). Nowadays, a variety of items are mostly made from glucose that is obtained from hydrolyzed starch. Future fermentation will depend on altering and regulating the shape of bacterial cells and their capacity for self-aggregation in order to more easily separate cells and fermented broth using gravity alone, obviating the need for centrifugation and filtration methods. Smaller cells and highly fermented broth are currently separated using these two costly and time-consuming techniques. Wastewater from fermentation industries is also produced in large quantities, and it is primarily made up of inorganic salts and organic cell debris. Waste materials could be transformed into molecules that the cell needs to grow again. By doing this, wastewater production might be prevented and the treated fermented broth could be used again as food for cell growth in the bioreactors. Additionally, new fermentation techniques will emphasize the use of microorganisms that can withstand harsh environments. For example, it was observed that the bacterium *Halomonas campaniensis* sp. LS21 might remain uncontaminated for several days in non-sterilization circumstances (Yue et al., 2014). The majority of fermentation methods used today are fed-batch or batch-operated and heavily rely on manual controls.

7 Conclusion

The scientific studies have shed light on the advantages of fermentation technology for food qualities, human health, and environmental sustainability. Interestingly, the market for functional food has been expanding significantly from the last few years, and fermented functional foods have become a vital subsector of this industry. There are many industrial applications for using different microbial species, including yeasts and different LAB strains, and even more prospective future advantages in creating nutritious cultured foods and drinks. Numerous potential health advantages of fermentation include the production of good lactic acid bacteria, easier metabolism, greater nutrient availability, enhanced mood and behavior, and cardiovascular health advantages. The two most commonly consumed fermented foods are meat and dairy. On the other hand, fermented fruits and vegetables, along with dairy and meat, are widely consumed and recognized for their health advantages. Recently, fermented products with insects and seaweed as primary components have become more popular due to growing awareness of alternative proteins. Furthermore, optimizing food-derived waste and utilizing state-of-the-art biotechnology techniques are essential to maximizing the bioconversion efficiency of food waste. Precision fermentation and other innovative methods allow for the production of molecules with identical constitutions to those of their traditionally farmed counterparts, making them ideal in order to produce desired carbohydrates, lipids, and proteins. Future fermentation industries will run continuously, allowing for the installation of automation, which will reduce human intervention and eliminate errors. Future fermentation industries will run continuously, allowing for the installation of automation, which will reduce human intervention and eliminate errors. In later fermentation processes, inexpensive materials like plastics, cement, or ceramic can be used in place of steel. In exchange, fermentation will be managed by artificial intelligence, resulting in sterile, more effective processes that use less energy, water, and labor.

References

Augustin, M. A., Hartley, C. J., Maloney, G., & Tyndall, S. (2024). Innovation in precision fermentation for food ingredients. *Critical Reviews in Food Science and Nutrition, 64*(18), 6218–6238. https://doi.org/10.1080/10408398.2023.2166014

Ball, P. (2017). Water is an active matrix of life for cell and molecular biology. *Proceedings of the National Academy of Sciences, 114*(51), 13327–13335. https://doi.org/10.1073/pnas.1703781114

Biswal, P., Pal, A., & Das, A. P. (2017). Current trends and future prospective of prebiotics as therapeutic food. *Microbial Production of Food Ingredients and Additives*, 57–88. https://doi.org/10.1016/B978-0-12-811520-6.00003-9

Cai, Y. X., Augustin, M. A., Jegasothy, H., Wang, J. H., & Terefe, N. S. (2020). Mild heat combined with lactic acid fermentation: A novel approach for enhancing sulforaphane yield in broccoli puree. *Food and Function, 11*(1), 779–786. https://doi.org/10.1039/C9FO02089F

Chai, K. F., Ng, K. R., Samarasiri, M., & Chen, W. N. (2022). Precision fermentation to advance fungal food fermentations. *Current Opinion in Food Science, 47*. https://doi.org/10.1016/J.COFS.2022.100881

Chen, G. Q., & Jiang, X. R. (2018). Next generation industrial biotechnology based on extremophilic bacteria. *Current Opinion in Biotechnology, 50*, 94–100. https://doi.org/10.1016/J.COPBIO.2017.11.016

Das, R. K., Brar, S. K., & Verma, M. (2016). Potential use of pulp and paper solid waste for the bio-production of fumaric acid through submerged and solid state fermentation. *Journal of Cleaner Production, 112*, 4435–4444. https://doi.org/10.1016/J.JCLEPRO.2015.08.108

Dulf, F. V., Vodnar, D. C., & Socaciu, C. (2016). Effects of solid-state fermentation with two filamentous fungi on the total phenolic contents, flavonoids, antioxidant activities and lipid fractions of plum fruit (*Prunus domestica* L.) by-products. *Food Chemistry, 209*, 27–36. https://doi.org/10.1016/J.FOODCHEM.2016.04.016

Dursun, D., & Dalgiç, A. C. (2016). Optimization of astaxanthin pigment bioprocessing by four different yeast species using wheat wastes. *Biocatalysis and Agricultural Biotechnology, 7*, 1–6. https://doi.org/10.1016/J.BCAB.2016.04.006

Ghosh, S., Tripathy, B., Dey, S., & Das, A. P. (2023). Comparative investigation of fungal and bacterial manganese biomining mechanisms. *Manganese Mining Microorganisms*, 99–115. https://doi.org/10.1016/b978-0-12-822147-1.00016-9

Gomes, A. M. V., Carmo, T. S., Carvalho, L. S., Bahia, F. M., & Parachin, N. S. (2018). Comparison of yeasts as hosts for recombinant protein production. *Microorganisms, 6*(2). https://doi.org/10.3390/MICROORGANISMS6020038

Gu, B. H., Zheng, P., Yan, Q., & Liu, W. (2014). Aqueous two-phase system: An alternative process for recovery of succinic acid from fermentation broth. *Separation and Purification Technology, 138*, 47–54. https://doi.org/10.1016/J.SEPPUR.2014.09.034

Hong, W., Wu, Y., Fu, X., Chang, Z. (2012). Chaperone-dependent mechanisms for acid resistance in enteric bacteria. *Trends in microbiology, 20*(7), 328–335. https://doi.org/10.1016/j.tim.2012.03.001

Hung, Y.- P., Lee, C.- L. (2017). Higher anti-liver fibrosis effect of *Cordyceps militaris*-fermented product cultured with deep ocean water via inhibiting proinflammatory factors and fibrosis. *Marine Drugs, 15*(6). https://doi.org/10.3390/md15060168

Jin, Y. S., & Cate, J. H. (2017). Metabolic engineering of yeast for lignocellulosic biofuel production. *Current Opinion in Chemical Biology, 41*, 99–106. https://doi.org/10.1016/J.CBPA.2017.10.025

Khatun, R., Singh, S., Dubey, N. K., & Prasad Das, A. (2023). A review on marine-based phytochemicals and their application in biomedical research. *Recent Frontiers of Phytochemicals: Applications in Food, Pharmacy, Cosmetics, and Biotechnology*, 383–395. https://doi.org/10.1016/B978-0-443-19143-5.00013-X

Kumar, A., Thakur, A., & Panesar, P. S. (2019). Recent developments on sustainable solvents for emulsion liquid membrane processes. *Journal of Cleaner Production, 240*. https://doi.org/10.1016/J.JCLEPRO.2019.118250

Kumar, M., Pandit, S., Patel, V., Khanna, N., Nag, M., Lahiri, D., Ray, R. R., Das, A., & Das, D. (2023). Maximization of energy recovery from starch processing wastewater by thermophilic dark fermentation coupled with microbial fuel cell technology. *Geomicrobiology Journal*, 1–12. https://doi.org/10.1080/01490451.2023.2209555

Lian, J., Mishra, S., & Zhao, H. (2018). Recent advances in metabolic engineering of *Saccharomyces cerevisiae*: New tools and their applications. *Metabolic Engineering, 50*, 85–108. https://doi.org/10.1016/J.YMBEN.2018.04.011

Liew, F. M., Martin, M. E., Tappel, R. C., Heijstra, B. D., Mihalcea, C., & Köpke, M. (2016). Gas Fermentation—A flexible platform for commercial scale production of low-carbon-fuels and chemicals from waste and renewable feedstocks. *Frontiers in Microbiology, 7*(May). https://doi.org/10.3389/FMICB.2016.00694/FULL

Mohanty, S., Ghosh, S., Nayak, S., & Das, A. P. (2016). Isolation, identification and screening of manganese solubilizing fungi from low-grade manganese ore deposits. *Geomicrobiological Journal, 34*(4), 309–316. https://doi.org/10.1080/01490451.2016.1189016

Mouritsen, O. G., Duelund, L., Calleja, G., & Frøst, M. B. (2017). Flavour of fermented fish, insect, game, and pea sauces: Garum revisited. *International Journal of Gastronomy and Food Science, 9*, 16–28. https://doi.org/10.1016/J.IJGFS.2017.05.002

Németh, Á., Kaleta, Z. (2015). Complex utilization of dairy waste (whey) in biorefinery. WSEAS *Transactions on Environment and Development*. Retrieved 29 Jan 2025, from http://www.wseas.us/journal/pdf/environment/2015/a1872601-220.pdf

O'toole, D. K. (2019). The role of microorganisms in soy sauce production. *Advances in Applied Microbiology, 108*, 45–113. https://doi.org/10.1016/BS.AAMBS.2019.07.002

Ouyang, P., Wang, H., Hajnal, I., Wu, Q., Guo, Y., & Chen, G. Q. (2018). Increasing oxygen availability for improving poly(3-hydroxybutyrate) production by Halomonas. *Metabolic Engineering, 45*, 20–31. https://doi.org/10.1016/J.YMBEN.2017.11.006

Priyadarsini, S., Alok, Das, P., Journal, G., & Das, A. P. (2025). Isolation and identification of native lithium-resistant bacteria from electronic waste contaminated sites: A focused investigation on tolerance assessment with potential lithium bio-recovery application. *Geomicrobiology Journal*. https://doi.org/10.1080/01490451.2025.2539509

Priyadarsini, S., & Das, A. P. (2024). Analytical and structural characterization of waste lithium-ion batteries for their effective recycling strategy. *Environmental Chemistry and Ecotoxicology*. https://doi.org/10.1016/j.enceco.2024.12.004

Priyadarsini, S., & Das, A. P. (2025). Lithium bioleaching: A review on microbial-assisted sustainable technology for lithium bio-circularity.*Journal of Water Process Engineering, 69*, 106744. https://doi.org/10.1016/j.jwpe.2024.106744

Sadh, P. K., Kumar, S., Chawla, P., Duhan, J. S. (2018). Fermentation: A boon for production of bioactive compounds by processing of food industries wastes (by-products). *Molecules, 23*(10). https://doi.org/10.3390/molecules23102560

Sharma, R., Garg, P., Kumar, P., Bhatia, S. K., & Kulshrestha, S. (2020). Microbial fermentation and its role in quality improvement of fermented foods. *Fermentation, 6*(4). https://doi.org/10.3390/FERMENTATION6040106

Singh, V., Haque, S., Niwas, R., Srivastava, A., Pasupuleti, M., & Tripathi, C. K. M. (2017). Strategies for fermentation medium optimization: An in-depth review. *Frontiers in Microbiology, 7*(Jan). https://doi.org/10.3389/FMICB.2016.02087

Tripathy, B., Singh, S., Behera, I. D., Mishra, S., & Das, A. P. (2026). Metagenomic profiling of diversified marine microbiome across microplastic-contaminated niches of Bay of Bengal, India. *Marine Pollution Bulletin, 222*, 118872. https://doi.org/10.1016/J.MARPOLBUL.2025.118872

Verotta, L., Panzella, L., Antenucci, S., Calvenzani, V., Tomay, F., Petroni, K., Caneva, E., & Napolitano, A. (2018). Fermented pomegranate wastes as sustainable source of ellagic acid:

Antioxidant properties, anti-inflammatory action, and controlled release under simulated digestion conditions. *Food Chemistry, 246*, 129–136. https://doi.org/10.1016/J.FOODCHEM.2017.10.131

Vidhyalakshmi, R., Radhika, R. (2012). Production of xanthan from agro-industrial waste. *Journal of Advanced Scientific Research, 3*(2), 56–59. https://sciensage.info/index.php/JASR/article/view/97

Wu, Y. Z., & Lee, C. L. (2021). Cordyceps cicadae NTTU 868 mycelium with the addition of bioavailable forms of magnesium from deep ocean water prevents the Aβ40 and streptozotocin-induced memory deficit via suppressing alzheimer's disease risk factors and increasing magnesium uptake of brain. *Fermentation, 7*(1). https://doi.org/10.3390/FERMENTATION7010039

Xu, Y., Hlaing, M. M., Glagovskaia, O., Augustin, M. A., & Terefe, N. S. (2020). Fermentation by probiotic *Lactobacillus gasseri* strains enhances the carotenoid and fibre contents of carrot juice. *Foods, 9*(12). https://doi.org/10.3390/FOODS9121803

Yang, R., Chen, Z., Hu, P., Zhang, S., & Luo, G. (2022). Two-stage fermentation enhanced single-cell protein production by *Yarrowia lipolytica* from food waste. *Bioresource Technology, 361*. https://doi.org/10.1016/J.BIORTECH.2022.127677

Yin, X., Li, J., Shin, H., Du, G., Liu, L., Chen, J. (2015). Metabolic engineering in the biotechnological production of organic acids in the tricarboxylic acid cycle of microorganisms: Advances and prospects. *Biotechnology Advances, 33*(6), 830–841. https://doi.org/10.1016/j.biotechadv.2015.04.006

Yue, H., Ling, C., Yang, T., Chen, X., Chen, Y., Deng, H., Wu, Q., Chen, J., & Chen, G. Q. (2014). A seawater-based open and continuous process for polyhydroxyalkanoates production by recombinant *Halomonas campaniensis* LS21 grown in mixed substrates. *Biotechnology for Biofuels, 7*(1). https://doi.org/10.1186/1754-6834-7-108

Zhao, C. J., Schieber, A., & Gänzle, M. G. (2016). Formation of taste-active amino acids, amino acid derivatives and peptides in food fermentations—A review. *Food Research International, 89*, 39–47. https://doi.org/10.1016/J.FOODRES.2016.08.042

Microbial Bioremediation Technology for Conversion of Microfiber Waste to Value-Added Products

Shreyasi Mishra, Sudeshna Dey, and Alok Prasad Das

Abstract Microfibers have emerged as a pressing environmental concern due to their ubiquitous presence, particularly in aquatic ecosystems. Microfibers are defined by a maximum size of less than 5 mm. Aquatic studies predominantly report microfiber abundance exceeding 50%, while annual global microfiber production, estimated at nine million tons, comprises 60% synthetic and 25% natural or non-synthetic fibers. Synthetic fibers account for 14% of global plastic production and can contribute to the formation of synthetic microfibers through processes of fragmentation and degradation. Global consumption of synthetic textiles has surged by about 65% over the past decade with China being the dominant producer, contributing nearly 70% of the total production. In 2022, global fiber production hit a record high of 116 million tons, with projections suggesting it could rise to 147 million tons by 2030 if current trends persist. The rapid expansion of the synthetic textile industry, particularly in Asia, has increased environmental concerns, with studies showing that more than 35% of microplastics are in the oceans. Textile degradation, laundry effluent, and disposable fiber production constitute the primary sources of microfibers. Due to their small diameter, they are more prone to be ingested by organisms that can induce, cytotoxicity, genotoxicity, and oxidative stress, leading to cellular damage and death even at minimal exposures via air and water sources. Microbial metabolic pathways, through a cascade of biochemical reactions, play a significant role in the biodegradation of synthetic microfibers. Microorganisms initiate microplastic degradation by secreting extracellular enzymes that adhere to the plastic surface and hydrolyze into shorter polymer chains. These intermediate products are subsequently ingested and metabolized by bacteria as a carbon source, ultimately yielding carbon dioxide. Notably, several bacterial strains, fungal species, and algae types have been identified with the capacity to degrade synthetic polymers. In conjunction with microbial communities, abiotic factors contribute to the conversion of microplastic waste into

S. Mishra · S. Dey · A. P. Das (✉)
Department of Life Sciences, Rama Devi Women's University, Bhubaneswar, Odisha, India
e-mail: alok1503@gmail.com

S. Dey
Centre for Biotechnology, Siksha 'O' Anusandhan (Deemed to be University), Bhubaneswar, Odisha, India

A. P. Das (ed.), *Bioconversion of Environmental Waste to Value Added Products*, Sustainable Environmental Waste Management Strategies,
https://doi.org/10.1007/978-3-032-14215-3_6

beneficial compounds such as carbon dioxide, nitrogen, methane, water, monomers, dimers, and oligomers. This chapter explores the potential of microorganisms to effectively degrade microfibers and microplastics using the bioremediation principle for sustainable production and generate a novel approach for the future to cultivate more enhanced products.

Keywords Microfibers pollution · Microbial bioremediation · Microfibers treatment · Polymer degradation · Beneficial products

1 Introduction

Since the beginning of the twenty-first century, microplastic and microfiber pollution has become a significant focus in environmental science research. In 2019, an important amendment to the Basel Convention was adopted, formally including plastic waste within the international legal framework governing the movement of hazardous waste. This change marked a pivotal moment in global efforts to address plastic pollution (Reddy Koteswara et al., 2025). In the same year, the G20 summit highlighted microplastic pollution as a crucial issue, with member countries agreeing to cooperate at both international and national levels to prevent and reduce the entry of microplastics into marine environments (Li et al., 2023). The topic of plastic pollution continued to gain prominence at the United Nations Environment Assembly (UNEA) in 2021, where the potential creation of a legally binding international treaty to combat plastic pollution was discussed (Li et al., 2023) The proposed treaty addresses plastic pollution from a life cycle perspective, with a target for adoption in 2024. Microplastics are small plastic particles that form as plastics undergo degradation through physical, chemical, and biological processes. They typically range in size from 1 micron to 5 mm, though more recent studies suggest that microplastics should be defined as particles between 1 micron and 1 mm in size. (Li et al., 2023). Microplastics can be categorized based on their shape into two main groups: fibrous and non-fibrous particles. The non-fibrous group includes fragments, films, and beads. These classifications help in understanding the behavior and impacts of microplastics in the environment, and further refinement of these categories will aid in more precise experimental analysis and mitigation strategies (Mishra et al., 2024c; Priyadarshini et al., 2024b; *RENEWABLE ENERGY GENERATION AND VALUE ADDITION FROM ENVIRONMENTAL*, 2024; Sahu et al., 2024a).

By combining these materials, manufacturers can create fabrics that offer a balance of durability, functionality, and user-friendly qualities. Semi-synthetic or artificial fibers include regenerated cellulose fibers used in clothing, personal care products, and cigarette filters. (Athey & Erdle, 2022) Even at an exceedingly slow rate, plastic debris gradually degrades, breaking down into microplastics (particles smaller than 5 mm) and nano-plastics in soils (Mohanty et al., 2024). As a result, the impacts of microplastics on soil ecosystems have become a growing area of scientific research and concern. (Guo et al., 2021). Estimates suggest that synthetic clothing is responsible for 20–35% of all primary microplastic pollution in the marine environment. Microfibers, in particular, account for the majority (> 85%) of microplastic debris found on shorelines worldwide. While the full scope of primary sources remains under investigation, domestic laundry discharges have long been considered a major contributor to microfiber pollution. However, recent studies have revealed additional insights. For instance, one study found that viscose fibers made up 56.9% of the total microfibers detected in deep sea sediments from the Atlantic Ocean, surpassing polyester—traditionally regarded as the dominant synthetic fiber—by more than twofold. Moreover, research indicates that washing plastic-based fabrics contributes significantly to marine pollution, with laundry discharges responsible for approximately 34.8% of primary microplastics entering the ocean. (Microfibres: The Invisible Pollution from Textile) Among the microfibers found in marine environments, polyester is the most prevalent, both in seawater and sediment samples. This growing body of evidence underscores the significant role of textile fibers in marine pollution and highlights the need for more comprehensive strategies to mitigate their release into the environment. This chapter delves into a comprehensive analysis of microfibers and microplastics, focusing on their sources, and the environmental and health effects they pose. The paper further emphasizes the growing need for effective solutions to mitigate their impact. It highlights the critical role of microbial bioremediation, focusing on how various microorganisms, including algae, fungi, bacteria, and mixed consortia, can break down these pollutants further outlining the potential of these organisms to transform microfibers into valuable byproducts, such as fatty acids, alcohols, and other eco-friendly compounds, offering a promising avenue for sustainable waste management and pollution control.

2 The Origin and Sources of Microfiber Pollution in the Environment

The precise sources of microfibers in the environment remain unclear, yet it is widely recognized that fibrous materials are the primary contributors, released during the production, use, and disposal of textiles. Plastic items degrade and fragment into smaller particles called microplastics as a result of natural environmental factors in these habitats, including ocean currents, sunlight exposure, abrasion, and interactions with vessels and marine organisms. (Chaudhry & Sachdeva, 2021) Primary microfibers are those natural or synthetic fibers used directly in textile manufacturing without further processing. In contrast, secondary microfibers result from the fragmentation of larger textile materials, typically occurring during the life cycle of fabrics, including use, laundering, and disposal. Commonly, microfibers are shed from loosely constructed textiles, such as knitted garments and fleece, especially during daily wear, with pilling being a typical example of this shedding. (Liu et al., 2019). Group of experts on the scientific aspects of Marine Environmental Protection

(GESAMP) scientific findings and recommendations are applied in two primary ways to support policy and decision-making for the effective control and management of marine environmental issues (Bala, et al., 2024b; Prasad. et al., 2024; John & Das, 2024). The classification of nano/microplastics has not yet been standardized, with various categorizations proposed by GESAMP (Watson-Wright et al., 2024). One approach defines microplastics as small if their size is below 1 mm and large if above 1 mm, while another method classifies microplastics based on the resolution level of detected particles, using increments of 100 μm. The presence of microplastics in marine and freshwater environments, such as rivers, oceans, and wastewater, has been widely studied. Among the affected systems, water and wastewater treatment plants are considered significant pathways for microplastics, with individual plants potentially releasing up to 10 billion particles per day in effluent. Despite removal efficiencies of up to 99% in some cases, a substantial number of microplastics are still discharged into the environment through these plants, contributing to the ongoing contamination of aquatic ecosystems. (Enfrin et al., 2019).

In urban environments, domestic wastewater is predominantly routed to wastewater treatment plants, which serve as significant sources of secondary microfibers in aquatic ecosystems, primarily originating from household laundry and textile manufacturing processes. While conventional treatment methods, encompassing primary and secondary stages, are highly effective in removing a substantial portion of microfibers, these plants remain a potential vector for the release of these pollutants. Despite removal efficiencies exceeding 95%, treated effluents may still contain concentrations of up to 10,000 microfibers per cubic meter, constituting a considerable input of microfibers into receiving aquatic environments (Liu et al., 2019). The continuous discharge of these effluents contributes to the ongoing contamination of aquatic systems with microfibers (Priyadarsini & Das, 2025). Furthermore, microfibers have been identified in sewage sludge across all treatment stages, with disposal practices influencing their subsequent environmental distribution. When sewage sludge is repurposed as agricultural fertilizer, it facilitates the dissemination of secondary microfibers into terrestrial and aquatic ecosystems, particularly following the breakdown of larger plastic debris. Consequently, wastewater treatment plants represent a critical source of microfiber pollution within river catchments, and the implementation of advanced treatment technologies holds substantial potential to mitigate microfiber concentrations in both freshwater and marine environments (Mishra et al., 2024b; Ray et al., 2024; Sahu et al., 2024). Microfibers enter aquatic environments via multiple pathways, with wastewater discharge from textile industries and domestic laundering being the most direct and significant sources. Shedding during laundering, driven by mechanical actions such as tumbling and friction, is widely considered a primary route of microfiber release. However, recent studies suggest that the contribution of home laundering to overall microfiber pollution may have been overestimated when compared to discharges from textile factories, particularly those involved in yarn spinning, weaving, dyeing, and finishing processes. This highlights the need to reassess the relative contributions of domestic versus industrial sources to microfiber contamination. (Liu et al., 2019).

3 The Detrimental Impact of Microfiber Pollution on our Ecosystems

Ingestion is considered the primary route of mortal exposure to microplastics. Grounded on estimates of microplastic consumption through food, individuals may ingest between 39,000 and 52,000 microplastic patches per time. These patches can enter the gastrointestinal system through defiled food or potables or reach the gut through mucociliary clearance following inhalation. These changes may also impact the balance of gut microbes, potentially affecting overall metabolic health. Inhalation depends on the microfiber's physical properties, such as size and density (Dey et al., 2024a; Dey et al., 2024b; Tripathy et al., 2024b). Smaller and less dense particles tend to penetrate deeper into the lungs, potentially reaching the alveolar regions where gas exchange occurs. After deposition, the particles may be cleared by macrophages—immune cells responsible for engulfing and removing foreign substances—or they may migrate into the circulatory or lymphatic systems, a process known as particle translocation. However, the large surface area of these small particles can trigger the release of chemotactic factors, which recruit additional immune cells to the site of deposition, preventing proper macrophage migration and increasing vascular permeability. This process can result in chronic inflammation, a condition known as dust overload. As a result, in environments with high concentrations of airborne microplastics or for individuals with heightened susceptibility (e.g., those with pre-existing respiratory conditions), prolonged exposure could lead to damage or lesions in the respiratory epithelium, potentially contributing to conditions such as pulmonary fibrosis or other respiratory diseases. (Prata et al., 2020).

Research has shown that exposure to pollutants such as microplastics and heavy metals significantly inhibits the growth, reproductive success, and overall survival rates of aquatic organisms like water fleas (*Daphnia magna*) and amphipods (*Gammarus* spp.). Studies indicate that the ingestion of synthetic particles by fish leads to gut obstructions, causing reduced nutrient absorption and subsequent nutritional deficiencies, potentially affecting fish health and growth. Evidence suggests that contaminants like pharmaceuticals and heavy metals can reduce the feeding activity of benthic species, including mussels, polychaete worms, and crabs, impairing their ability to obtain adequate nutrition and potentially disrupting local ecosystems. Research indicates that exposure to persistent organic pollutants (POPs) and endocrine-disrupting chemicals (EDCs) can interfere with hormone regulation in marine species, leading to reproductive abnormalities and changes in behavior that compromise their survival and reproduction. Prolonged exposure to airborne pollutants, such as particulate matter from industrial emissions, tobacco smoke, or chemicals like formaldehyde, can lead to chronic respiratory conditions. These conditions include obstructive lung diseases like chronic obstructive pulmonary disease (COPD) and emphysema, which cause inflammation and narrowing of the airways, leading to breathing difficulties and reduced lung function over time. Bisphenol A (BPA), a chemical commonly found in plastics and resins, has been identified as an endocrine-disrupting compound. It can interfere with the normal functioning of hormones in the body, particularly estrogen, leading to reproductive health issues, developmental abnormalities, and an increased risk of hormone-related cancers such as breast and prostate cancer. Studies show that even low levels of BPA exposure may contribute to these adverse health effects. Microfibers, tiny synthetic fibers shed from clothing, textiles, and other materials, are increasingly found in aquatic environments, where they are ingested by marine organisms such as fish, shellfish, and plankton. As these microfibers accumulate in the food web, they are passed along to high trophic levels, including humans, who may unknowingly consume contaminated seafood. The long-term health implications of microplastic ingestion in humans are still under investigation but may include potential toxicological effects, as these fibers can absorb harmful environmental chemicals (Microfibres: The Invisible Pollution from Textile). Table 1 gives an outline of how different microfiber or microplastic particle affects some freshwater and terrestrial organisms.

4 Microbial Bioremediation: A Key to Mitigating Plastic Pollution

Microorganisms and enzymes are pivotal in transforming, cycling, and stabilizing organic matter, directly influencing its storage and availability. Their intricate interactions are fundamental to ecosystem function. (Guo et al., 2021) In the context of microplastic pollution, microbial bioremediation refers to the use of various microorganisms, such as bacteria, fungi, microalgae, and yeast, to break down or modify microplastics in contaminated ecosystems, including soils and aquatic environments. Regarding microbial species, only about 5% of microbial diversity has been characterized to date, leaving an estimated 95% yet to be explored (Srivastava et al., 2014). Effective microbial bioremediation is accomplished when microorganisms interact within their ecological niches under optimal environmental conditions (Bala et al., 2024; Paul et al., 2024; Satapathy et al., 2024). This includes nutrient availability, temperature, pH, and the presence of specific contaminants that the microbes can metabolize. These natural bioremediation processes not only contribute to the detoxification of contaminated environments but also enhance ecosystem resilience, promoting long-term sustainability and environmental recovery. (Niu et al., 2021) Microbial bioremediation represents a natural and highly effective approach to mitigating environmental pollution through the metabolic activities of microorganisms. This process involves both aerobic and anaerobic microbes, each adapted to distinct environmental conditions, yet all fundamentally dependent on organic carbon and an energy source for growth and pollutant degradation (Mishra et al., 2021; Priyadarshini et al., 2024a). Aerobic microorganisms typically require oxygen-rich environments, while anaerobic microorganisms thrive in oxygen-deprived conditions, utilizing alternative electron acceptors like nitrate, sulfate, or carbon dioxide. (Srivastava et al., 2014).

Table 1 Detrimental effect of microfibers and microplastics on various organisms

Serial No.	Microfiber and microplastic type	The affected organisms	Detrimental impact on the organism	References
1	Polyethylene	Daphnia magna	Reduced reproduction and growth	(Besseling et al., 2014)
2	Polyester microfibers	Mytilus edulis (blue mussel)	Tissue inflammation, reduced filtration rates	(Browne, 2007)
3	Polystyrene particles	Danio rerio (zebrafish)	Oxidative stress, behavioral changes	(Lu et al., 2018)
4	Polypropylene fibers	*Crassostrea gigas* (Pacific oyster)	Decreased reproductive success	(Sussarellu et al., 2016)
5	Nylon microfibers	*Carcharhinus limbatus* (blacktip shark)	Bioaccumulation, potential trophic transfer	(Parton, 2020)
6	Polyester and polypropylene fibers	*Canis lupus familiaris* (domestic dog)	Ingestion leading to gastrointestinal distress	(Prata & Dias-Pereira, 2023)
7	Microplastic-laden soil particles	*Rattus norvegicus (Brown rat)*	Oxidative stress and inflammation in organs	(Braeuning, 2018)
8	Synthetic microfibers	*Bos taurus (cattle)*	Presence of microplastics in milk and tissues	(Bahrani et al., 2024)
9	Polystyrene fragments	*Columba livia (pigeon)*	Altered gut microbiome and metabolic changes	(Amélineau et al., 2016)
10	Mixed synthetic fibers	*Oncorhynchus mykiss (rainbow trout)*	Intestinal inflammation, microbiome disruption	(Nelms et al., 2018; Stojanović et al., 2019)

In addition to conventional habitats such as soil, water, and sediments, microbial bioremediation can also occur in diverse and often underexplored ecosystems. These include the rhizosphere, where plant roots foster microbial activity, gravel-associated microbial communities, the highly dynamic aquatic shorelines, subsurface layers of marshlands that experience fluctuating redox conditions, and even weathered rock surfaces that host specialized microbial populations. Each of these environments offers unique physicochemical properties that can support the growth of a wide array of microorganisms capable of degrading a broad spectrum of pollutants, including heavy metals, petroleum hydrocarbons, pesticides, and industrial solvents. (Srivastava et al., 2014) *Pseudomonas* species are known for producing a range of hydrolases, including esterase, lipases, and cutinases, enzymes that significantly contribute to plastic degradation. (Niu et al., 2021) In addition to *Pseudomonas*, other anaerobic or facultative anaerobic bacteria such as *Serratia*, *Bacillus*, *Paraburkholderia*, *Rahnella*, and *Staphylococcus* have also demonstrated the ability to degrade plastics (Niu et al., 2021). These suggest that anaerobic and facultative anaerobic plastic-degrading bacteria, which tend to accumulate on microplastics, play an essential role in the biodegradation of plastics, particularly in deep sediment environments (Dey et al., 2023a; Lahiri et al., 2021; Mallick et al., 2024; Sahu & Das, 2024). The behavior and growth of microbial cells are significantly influenced by both biotic and abiotic factors, which ultimately affect various biological processes occurring within a microbial community. These factors can either promote or hinder the microbial activity necessary for processes like bioremediation, which relies on the natural or enhanced capacity of microorganisms to break down contaminants in the environment.

However, bioremediation is not a straightforward process; it occurs within a multiphasic, heterogeneous environment that often varies in composition, which can greatly affect the rate and efficiency of microbial reactions. Microorganisms, while incredibly adaptable, do have their limitations. Their inherent ability to adjust to environmental changes, such as shifts in nutrient availability or temperature changes, is a remarkable survival trait. However, extreme conditions may overwhelm this adaptive capacity (Mallick et al., 2024; Moharana et al., 2024; Sahu et al., 2024; Singh Jadaun et al., 2022; Tripathy et al., 2024a). For example, while certain bacteria can metabolize hydrocarbons, they might struggle in an environment with extremely low or high temperatures, which can slow down their metabolic processes or even stop them altogether. A deeper understanding of microbial ecology is crucial for enhancing the effectiveness of bioremediation efforts. This review emphasizes that this knowledge is essential for predicting how microbial populations will behave under different environmental stressors and how to design interventions that maximize their potential.

These challenges can be categorized into three main types:

Physicochemical Characteristics

These include temperature, pH, salinity, oxygen levels, and the availability of nutrients. For example, in contaminated soil, the presence of pollutants like oil or heavy metals can drastically affect microbial growth (Priyadarsini et al., 2025; Priyadarsini & Das, 2024, 2025). The pH of the soil can determine the solubility of metals, which may either facilitate or hinder microbial degradation. In such environments, the metabolic rate of microbes is heavily influenced by the ability to adapt to these harsh physicochemical conditions. In oil spills, for example, microbes like *Alcanivorax borkumensis* thrive in the presence of hydrocarbons but are limited by oxygen and nutrient availability. (Srivastava et al., 2014) Thus, understanding these abiotic factors is essential for optimizing microbial bioremediation.

Biological Factors

These factors include interactions between different species of microorganisms, including competition, predation, and symbiosis. For instance, in a microbial community, some bacteria may produce enzymes that break down pollutants, while others may outcompete these pollutant-degrading microbes for nutrients, limiting the overall effectiveness of bioremediation. The success of a microbial community is often dependent on the balance between these different species and their cooperative or competitive interactions. For example, in the case of bioremediation of heavy metals, certain bacteria like *Pseudomonas putida* can resist toxic metals, but their growth may be inhibited by the presence of other microbes that consume available resources or create inhibitory byproducts (Mishra et al., 2024a; Mohanty et al., 2024).

Climatic Conditions

Climatic conditions such as temperature, rainfall, and humidity can also play a significant role in microbial processes. These conditions affect the availability of water, nutrients, and oxygen, all of which are essential for microbial growth and metabolism. For example, in warmer climates, microbial activity generally increases due to enhanced metabolic rates, but extreme heat may also cause stress, leading to a decrease in microbial efficiency. Similarly, in cold environments, the metabolic rates of microbes decrease, slowing down the process of bioremediation. For instance, in cold water environments, microbes may require more time to degrade pollutants due to slower biochemical reactions (Tripathy et al., 2026).

Microorganisms possess remarkable capabilities that make them effective agents for bioremediation. These capabilities include detoxification, degradation, and the accumulation of microplastic components. Over the last two decades, significant advances in bioremediation technologies have focused on enhancing the ability of microorganisms to target and remove pollutants from the environment in a cost-effective manner, contributing to the overall goal of creating an eco-friendly environment (Khatun et al., 2023; Mallick et al., 2023). Despite the development of various methods for bioremediation, it is the microorganisms naturally present in polluted environments that have proven to be central to addressing the challenges associated with the degradation of microplastics.

Microbes are capable of metabolizing a wide variety of substrates, including complex organic pollutants like microplastics. Certain microorganisms possess specialized enzymes that can break down or modify the chemical structure of plastics, making them less harmful to the environment. Many microorganisms can utilize a wide range of organic and inorganic nutrients, allowing them to adapt to environments with limited resources. This flexibility is particularly important when dealing with environments contaminated with microplastics, where nutrient availability may be variable. Microbes can transform microplastics into less toxic or non-toxic compounds through enzymatic activity. This ability is critical in mitigating the environmental hazards associated with plastic waste. Microorganisms used in bioremediation are often referred to as "bioremediators" due to their role in degrading pollutants. The concept of utilizing microorganisms for environmental cleanup dates back to 1981, when the first patent for the biodegradation of petroleum oil using Pseudomonas putida was registered. Since then, the application of microbial bioremediation has expanded to address a wide range of pollutants, including plastic materials. Certain microorganisms produce enzymes such as hydrolases, lignases, and esterases that can break down the polymer chains in microplastics, facilitating their biodegradation. These enzymes enable microorganisms to convert microplastics into simpler, more manageable molecules. The issue of microplastic pollution has been exacerbated by the global COVID-19 pandemic. During this period, the production and consumption of plastic products, particularly personal protective equipment (PPE) such as masks, gloves, and face shields, surged dramatically. According to studies, more than one million tons of plastic waste were generated during the pandemic due to the increased production and use of these plastic-based products. This rapid increase in plastic waste has heightened the urgency for effective bioremediation strategies to manage and mitigate microplastic pollution, and microorganisms have emerged as a viable solution to this growing environmental challenge (Bhattacharjee et al., 2021;

5 The Role of Microbial Consortia

Microplastic degradation typically occurs more effectively when a consortium of microorganisms works together. A single microorganism often lacks the broad range of enzymatic abilities needed to degrade most hazardous microplastics, which are complex synthetic polymers. In a microbial consortium, different microorganisms may have complementary metabolic pathways and enzymatic functions that collectively enhance the degradation process. This sequential degradation, where each species plays a specific role in breaking down various components of microplastics, leads to a more complete and efficient bioremediation process. For example, certain bacteria may initially degrade the polymer chains of plastics into smaller fragments, which fungi or microalgae can further break down. Fungi can secrete enzymes like lignin peroxidases and manganese peroxidases, effectively cleaving the complex bonds in plastic polymers. In turn, the bacterial species can use these smaller fragments as substrates for further metabolism, leading to the complete breakdown of the microplastics into simpler, non-toxic compounds (Ghosh et al., 2018; Ghosh & Das, 2020; Mishra & Das, 2021).

5.1 The Aerobic and Anaerobic Conditions of Bioremediation

Most microbial bioremediation systems operate under aerobic conditions, where oxygen is available, as many microorganisms that degrade organic pollutants, including microplastics, are aerobic organisms. Under these conditions, microorganisms' metabolic processes, such as carbon compounds' oxidation, are more efficient. For example, Pseudomonas and Bacillus species thrive in oxygen-rich environments, effectively degrading various contaminants, including microplastics. However, anaerobic conditions—where oxygen is limited or absent—can also support microbial degradation, although typically less efficiently. Under anaerobic conditions, microorganisms such as *Desulfo vibrio* and other sulfate-reducing bacteria may degrade pollutants via alternative metabolic pathways, such as reduction processes. Although these pathways are slower and less efficient for microplastic degradation, they may still contribute to the breakdown of plastics under specific environmental conditions, such as in deep-sea ecosystems or landfills (Mishra et al., 2020; Mohanty et al., 2018).

5.2 *Environmental Revival Through Microbial Bioremediation*

One of the significant advantages of microbial bioremediation is its ability to restore the health of contaminated environments. As microorganisms degrade microplastics, they break down the toxic plastic particles into simpler, non-toxic substances, such as carbon dioxide, water, and other biodegradable compounds. This process not only mitigates plastic pollution but also revives the ecological balance of affected soil and water systems (Bal et al., 2018; Ghosh & Das, 2017). For example, in polluted marine environments, the bioremediation of microplastics by microbial consortia can lead to the recovery of water quality, benefiting marine life that was previously harmed by the plastic particles. Similarly, in contaminated soils, microbial activity can lead to the restoration of soil fertility, enabling the land to be reused for agricultural purposes. This ecological restoration makes bioremediation a sustainable and cost-effective approach for addressing microplastic pollution.

6 A Multidimensional Approach for Generation of Sustainable Value-Added Products by Microbial Bioremediation

The biofilms form on microplastic surfaces, and microbial activity, supported by the enzymes and metabolites they release, begins the degradation process. This leads to the breakdown of microplastics through molecular fragmentation and depolymerization. In the final stages of biodegradation, the fragmented molecules are assimilated and mineralized, producing CO_2, H_2O, CH_4, and biomass. Various fungi and bacterial species, equipped with powerful enzymes like oxidoreductases and hydrolases, are central to this bioremediation process. To assess biodegradation, a diverse set of analytical techniques is used, including electron microscopy, spectroscopy, mechanical property analysis, molar mass changes, CO_2 evolution, O_2 consumption, radiolabeling, clear-zone assays, enzymatic degradation, and controlled composting tests. Since more than 99% of prokaryotes and some eukaryotic microorganisms are unculturable, culture-independent methods such as metagenomic analysis have become vital tools for identifying microorganisms that can break down plastics (Das & Singh, 2011; Ghosh et al., 2016; Ghosh & Das, 2018; Mishra et al., 2019, 2022; Mishra et al., 2021; Mohanty et al., 2016; Sahoo et al., 2022; Tripathy et al., 2022). This, combined with in silico mining, allows for a deeper investigation into both well-known and unexplored microbial communities, uncovering enzymes and organisms that are effective in microplastic bioremediation. Furthermore, utilizing microbial consortia, engineered microorganisms, and their specific enzymes presents a promising frontier in advancing plastic waste degradation strategies (Hadian-Ghazvini et al., 2022). Table 2 shows various microorganisms for bioremediation of microplastics in freshwater and soil systems, transforming them into valuable byproducts.

6.1 The Bacteria Species Capable of Degrading Microplastics and Microfibers

In recent years, an increasing number of plastic-degrading bacterial strains have been identified and cultured from anoxic and anaerobic environments, including the intestines of insects, activated sludge, and marine sediments. For example, *Halomonas*, typically found in saline habitats and endorheic lakes, has also been detected in urban river sediments, where it plays a role in denitrification and hydrocarbon degradation. This bacterium secretes cytochrome P450 monooxygenases, enzymes capable of catalyzing hydroxylation and demethylation reactions, which are critical for the breakdown of plastics (Niu et al., 2021). Research indicates that natural dissolved organic carbon (DOC) concentrations below 30.7 μmol L − 1 are insufficient to sustain the metabolism of prokaryotic organisms (Arrieta et al., 2015). This value is near the lower boundary of the typical DOC concentrations found in the open ocean, which are estimated to range from 34 to 80 μmol kg − 1. In marine environments, if synthetic polymers, such as plastics, are introduced into the ecosystem, degradation pathways are expected to evolve, leading to the breakdown of these polymers into their constituent monomers or smaller chemical fragments. Once these polymers are hydrolyzed, they can release biodegradable substances that, in theory, should increase the local concentration of organic carbon above approximately 30 μmol C L − 1. This could potentially support microbial communities that rely on these organic carbon sources. A notable example of this phenomenon is the discovery of *Ideonella sakaiensis*, a bacterium capable of degrading polyethylene terephthalate (PET), a commonly used plastic in products like water bottles and clothing fibers. This bacterium was first isolated from outside a plastic recycling facility (Yoshida et al., 2016) and is capable of breaking down PET into its monomeric units, terephthalic acid, and ethylene glycol. (Oberbeckmann & Labrenz, 2020).

Table 2 Different microorganisms for bioremediation of microplastics in freshwater and soil systems and the valuable byproducts generated

Serial No.	Name of the microorganism	Microplastic type	Microbial enzyme	Value-added products generated	References
1	*Bacillus subtilis*	Polyurethane	Esterase	Carbon dioxide, water, organic acids, biomass	(Yuan et al., 2020)
2	*Ideonella sakaiensis*	Polyethylene terephthalate polystyrene	PETase and MHETase	Ethylene glycol, terephthalic acid. Carbon dioxide, water	(Yoshida et al., 2016)
3	*Rhodococcus ruber*	Polystyrene	Dioxygenases and peroxidases	Shorter hydrocarbons, aliphatic compounds, fatty acids, carbon dioxide, water, and biomass.	(Mor & Sivan, 2008)
4	*Aspergillus tubingensis*	High-density polyethylene (HDPE)	Laccases and peroxidases.	Smaller oligomers, fatty acids, aliphatic hydrocarbons, alcohols, carbon dioxide, water, and biomass	(Sangeetha Devi et al., 2015)
5	*Trametes versicolor*	Nylon, polycaprolactone	Laccases, peroxidases (lignin peroxidases and manganese peroxidases), and Esterases.	Smaller fragments, amine groups, aromatic compounds, caproic acid and other short-chain fatty acids, carbon dioxide, water, biomass	(Zhou et al., 2022)
6	*Closterium costatum*	Polyethylene	Laccases, peroxidases, cytochrome P450 enzymes	Smaller polymer fragments, hydrocarbon-based compounds, alcohols, aldehydes, carboxylic acids, carbon dioxide, water, and biomass	(Sarmah & Rout, 2020)
7	*Dunaliella tertiolecta*	Polystyrene	Laccases, peroxidases	Phenolic compounds, aromatic hydrocarbons, carboxylic acids, carbon dioxide, water, biomass	(Priya et al., 2022)

The genus Bacillus includes several strains that have demonstrated significant potential in the biodegradation of microplastics, especially polyethylene and polystyrene. *Bacillus cereus* is particularly notable for its biodegradation capabilities. In a 2018 study, *Bacillus cereus* showed a biodegradation rate of 35.7% for low-density polyethylene after 16 weeks of incubation. This strain is efficient due to its production of a variety of extracellular enzymes that break down plastic polymers, such as esterase and lipases, which target the ester bonds in plastic materials. In laboratory studies, *Pseudomonas* sp. has been shown to degrade up to 20.54% of polyethylene within just one month of anaerobic incubation (Niu et al., 2021). The Pseudomonas genus, including species like *Pseudomonas putida*, is another key contributor to microplastic degradation. *Pseudomonas putida* has shown remarkable abilities to degrade a variety of synthetic plastics, including polyethylene, polystyrene, and polypropylene (Mishra et al., 2019, 2022; Mishra et al., 2021; Sahoo et al., 2022; Tripathy et al., 2022). A study conducted in 2012 over 16 weeks revealed that *Pseudomonas putida* achieved higher degradation rates of low-density polyethylene compared to other bacterial strains, highlighting its efficiency and potential for large-scale bioremediation. (Kyaw et al., 2012) This bacterium is known for its metabolic flexibility and ability to form biofilms on plastic surfaces, which enhances its degradation efficiency. Biofilms are dense communities of microorganisms that can adhere to surfaces and work synergistically, allowing for the sustained breakdown of plastic polymers.

Rhodococcus ruber, a strain known for its biofilm-forming ability, is also significant in the biodegradation of microplastics. (Miloloža et al., 2022) In a study examining biofilm formation on polyethylene (PE) plastic, *Rhodococcus ruber* demonstrated a 7.5% degradation rate over an 8-week exposure period. *Rhodococcus* species are well-known for their ability to degrade complex organic compounds, including hydrocarbons, and have also been observed to degrade plastics such as polyethylene and polypropylene. Their role in forming biofilms helps in enhancing the stability of microbial colonies, making them more effective at breaking down the polymer matrix of plastics over extended periods. A bacterial consortium composed of *Enterobacter* and *Pseudomonas* species has been identified as effective in the biodegradation of polypropylene. This consortium benefits from the complementary metabolic capabilities of each bacterium, where *Pseudomonas* strains can break down larger plastic molecules, while *Enterobacter* species may assist in the mineralization of the resulting smaller molecules into simpler, non-toxic compounds. This synergistic approach is particularly useful in heterogeneous environments where multiple types of plastics may be present. (Skariyachan et al., 2021).

6.2 Biofilm Formation and its Role in Bioremediation

One of the key factors that enhance bacterial effectiveness in microplastic biodegradation is the ability to form biofilms. A biofilm is a structured community of microorganisms attached to a surface, embedded in a self-produced extracellular matrix. This matrix can protect the microorganisms from environmental stressors, such as changes in temperature, pH, or toxic substances, and allows for more efficient interactions between the microorganisms and their environment. In the case of microplastics, the formation of biofilms on plastic surfaces can significantly enhance the bacterial degradation process. The proximity of microbial cells in a biofilm allows for more efficient enzymatic action on the polymer chains of the plastic, leading to increased degradation rates. For example, Rhodococcus ruber forms biofilms on polyethylene, which has been linked to its ability to achieve a 7.5% degradation of microplastics over 8 weeks. This biofilm formation creates a stable environment for the bacteria, increasing their metabolic activity and allowing for more consistent degradation of the plastic material over time (Dey et al., 2023b; Mishra & Das, 2022). As shown in a 2018 study, the biodegradation rate of low-density polyethylene was significantly influenced by its thickness, with thicker plastic requiring more time for complete degradation. This is because thicker plastics provide a larger surface area that bacteria must colonize, and the process of breaking down the polymer chains becomes more challenging due to the reduced availability of surface area for microbial attachment. In these cases, biofilm formation becomes even more crucial as it increases the surface area for bacterial colonization. Furthermore, biofilms can secrete enzymes that help in breaking down plastic polymers, which accelerates the overall degradation process. These findings underscore the importance of optimizing the conditions for bacterial growth, such as nutrient availability and environmental factors, to enhance the biodegradation of thick plastic materials. Figure 1 highlights how the bacterial species breakdown the surface of microfibers and microplastics to generate simpler byproducts.

7 Key Role of Fungi in Microplastic Degradation

The unique biological properties of fungi—such as their ability to degrade complex polymers, form biofilms, and secrete a variety of nonspecific enzymes—make them highly efficient at breaking down hazardous plastic pollutants. The debate regarding whether fungi or bacteria are more effective at microplastic degradation hinges on their respective mechanisms, metabolic flexibility, and environmental adaptability. However, fungi's superior enzymatic capabilities, especially in degrading complex, hydrophobic substances like plastics, place them at the forefront of bioremediation research (Mishra & Das, 2023). Fungi can grow on a variety of substrates, including organic and synthetic materials, which is a critical factor in their ability to degrade microplastics in different environments. Their adaptability allows them to thrive in "plastispheres"—environments dominated by plastic pollutants, such as landfills or marine ecosystems. Fungi can colonize the surface of microplastics and begin the degradation process, often forming dense networks of hyphal growth that increase the surface area available for enzymatic action.

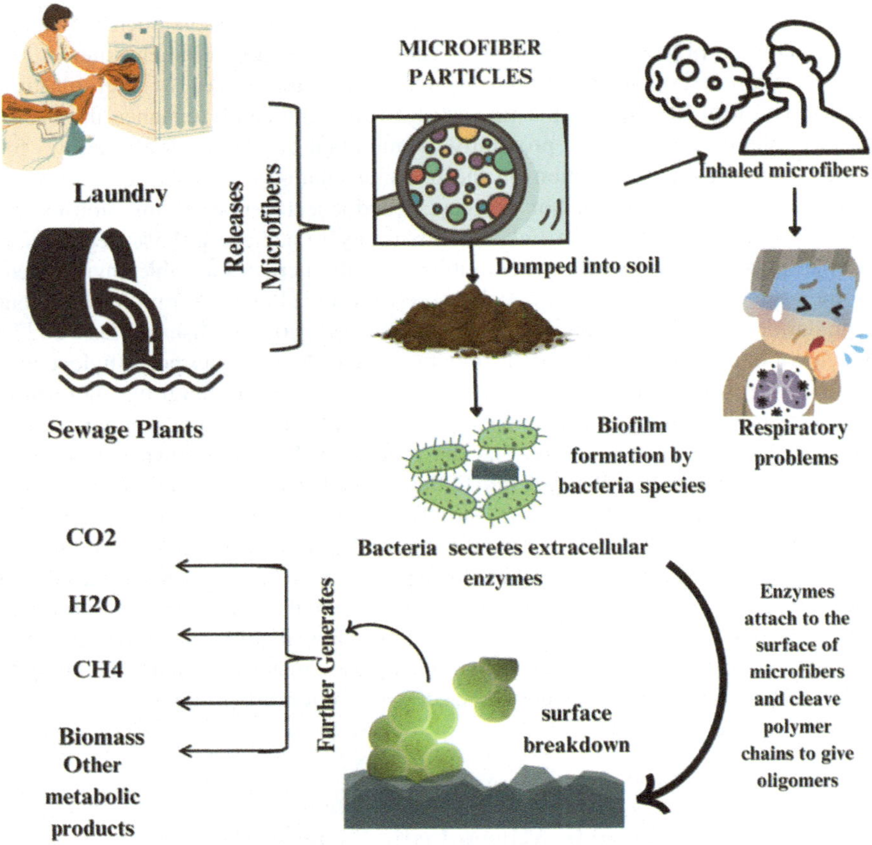

Fig. 1 Microbial bioremediation action of bacteria on microfibers to generate value-added products

Nutrient Absorption and Biofilm Formation by Fungi

Fungi possess an effective nutrient absorption system that allows them to survive in harsh environments with limited resources. This ability is crucial in contaminated environments where plastic pollutants may be the primary carbon source. Additionally, many fungi can form biofilms, which enhance their efficiency in degrading microplastics. Biofilm formation enables fungi to protect themselves from environmental stresses and create a stable environment for sustained biodegradation.

The Enzymatic Systems of Fungi for Degradation

Fungi are equipped with a diverse range of enzymes that are particularly effective at degrading complex polymer structures found in microplastics. These enzymes include:

Hydrolases: Enzymes like esterase, lipases, and proteases, which break down ester and peptide bonds in plastic polymers (Yuan et al., 2020). Oxidoreductases or enzymes like laccases and peroxidases facilitate the oxidation of polymer chains, leading to their breakdown (Sokal et al., 2022). Cytochrome P450 enzymes are part of the intracellular enzymatic system, these enzymes play a role in metabolizing complex hydrocarbons, making fungi particularly effective at degrading hydrophobic and recalcitrant plastics like polyethylene and polypropylene.

Comparison of Rate of Degradation of Microplastics Between Bacteria and Fungi

A pivotal study by Volke Sepúlveda demonstrated the mineralization of thermally oxidized low-density polyethylene by fungal species *Aspergillus niger* and *Penicillium pinophilum*. After 31 months of exposure, A. niger achieved a weight reduction of 0.57% of the thermally oxidized low-density polyethylene, while *Penicillium pinophilum* reduced the plastic weight by 0.37%. (Yuan et al., 2020) This long-term study highlighted the slow but consistent biodegradation capacity of fungi, even on difficult-to-degrade plastics like LDPE. These fungi were able to oxidize the polymer structure, facilitating its breakdown into smaller, more digestible components. While bacteria are often more effective in the initial stages of plastic degradation, fungi, with their ability to oxidize a wider range of compounds and their tolerance to extreme environmental conditions, can provide a more sustained and long-term solution to microplastic contamination 5 (Jagatee et al., 2024; Rathinamoorthy & Raja Balasaraswathi, 2023).

Fungi exhibit remarkable adaptability to challenging environmental conditions, including extremes of temperature, moisture, and plastic concentration. Unlike many bacteria, which may require a specific set of conditions to optimize degradation, fungi can survive and function in a wider range of habitats. They are particularly valuable in environments with higher concentrations of plastic pollutants and lower moisture levels, such as arid soils, landfill sites, and plastic-infested marine environments. This versatility enables fungi to act as effective agents of bioremediation across diverse ecosystems, contributing to the long-term restoration of contaminated areas.

White Rot Fungi in Soil Remediation

White rot fungi, such as *Phanerochaete chrysosporium*, are particularly well-suited for bioremediation of microplastics in soil environments. These fungi are naturally efficient at breaking down lignin, a complex polymer found in plant cell walls, but they also possess the enzymatic machinery to degrade synthetic polymers like microplastics (Abe et al., 2010; Das & Chandran, 2011; Zeenat et al., 2021). White rot fungi can colonize plastic particles in contaminated soil, using their ligninolytic enzymes to break down the polymer chains. This process not only helps reduce plastic waste but also improves soil health by enhancing microbial diversity and restoring soil structure. White rot fungi are particularly advantageous because they can thrive in low-moisture conditions, which is often the case in dry environments where microplastic pollution is a significant concern. (Sokal et al., 2022).

7.1 The Key Algal Species Involved in Microplastic Bioremediation

Algae, particularly microalgae, offer several advantages in microplastic biodegradation, including ease of handling, lower toxicity, and environmental adaptability. While algae have traditionally been used as test organisms to study the toxicity of microplastics, recent studies highlight their active involvement in the degradation process. Algae can colonize the surface of microplastics, and through the exudation of extracellular polymeric substances, they facilitate the biodegradation process. These substances enhance the binding of algae to the microplastic surface, accelerating the degradation rate. Algal colonization of microplastics is influenced by several factors, such as the type of plastic, environmental conditions, and the duration of exposure. Variations in plastic polymer types stimulate different colonization tendencies in algae. For instance, biofilm formation and plastic degradation rates vary between freshwater and soil systems. Environmental parameters like temperature, nutrient availability, and the age of the microplastics also impact the algae's ability to degrade microplastics effectively (Gu, 2003; Jain et al., 2023; Zhang et al., 2021).

Navicula pupula species' ability to exude extracellular polymeric substances helps establish a biofilm on the plastic surface, which accelerates the biodegradation of the microplastics. (Sarmah & Rout, 2020). *Phormidium lucidum* and *Oscillatoria subbrevis* are species of cyanobacteria that are known to efficiently degrade low-density polyethylene. They use it as a carbon and energy source, playing a key role in plastic degradation, especially in aquatic environments where microplastics accumulate (Sarmah & Rout, 2020). *Scenedesmus dimorphus* and *Anabaena spiroides* are freshwater microalgae and have demonstrated high efficacy in breaking down microplastics. *Scenedesmus dimorphus*, in particular, has shown the ability to thrive in polluted environments and can efficiently absorb and metabolize pollutants, including microplastics. (Sarmah & Rout, 2020).

7.2 Mixed Microbial Consortia for Enhanced Biodegradation of Microplastics

While individual microorganisms have shown promise in microplastic biodegradation, mixed microbial consortia—groups of different microorganisms working synergistically—are even more effective in breaking down microplastics. The collaboration between different microbial species enables the secretion of various enzymes that can act on a broader range of microplastic polymers, accelerating the biodegradation process. *Aneurini bacillus* and *Brevi bacillus* species consortium of these two genera has been found to degrade polyester and polypropylene microplastics effectively. The microbial consortium secretes a combination of enzymes that break down the polymer chains, resulting in a significant reduction in the weight and molecular size of the plastics. Bacterial consortium of *Staphylococcus* sp., *Pseudomonas* sp., and *Bacillus* sp. This consortium was studied for its ability to degrade 40-micron polyester sheets over 40 days. The results showed a degradation of 20, 7.5, and 42.5% weight loss, respectively, from these bacteria (Ahmed et al., 2020). The consortium's efficacy was attributed to the combined enzymatic action of the species involved, with *Bacillus* species proving particularly effective in degrading PE. *Paeni bacillus* and *Bacillus* species from municipal solid-waste where a consortium of these bacteria were found to rapidly reduce the molecular weight of PE microplastics. This accelerated degradation occurs due to the combined action of several enzymatic systems, including esterases and hydrolases, which cleave the polymer chains. *Enterobacter and Pseudomonas species* were bacterial consortiums containing these species has been successfully used to degrade polypropylene microplastics. The consortium's ability to work synergistically in degrading the plastic underscores the importance of microbial diversity in bioremediation efforts. (*Microbial Bioremediation and Multiomics Technologies for Sustainable...—Google Books*, n.d.; Sokal et al., 2022).

8 Conclusion

Microfiber pollution represents a growing environmental challenge, particularly due to its widespread presence in aquatic ecosystems and the detrimental effects it has on organisms exposed to these microplastics. However, the potential for microbial bioremediation offers a promising solution to mitigate the harmful impact of microfibers. Through the metabolic pathways of microorganisms, synthetic microfibers can be biodegraded into simpler compounds like simple polymers, digestible compounds, energy sources, carbon dioxide, methane, and water. These microbial processes not only reduce the environmental burden of microfiber waste but can also contribute to the production of value-added products, providing a sustainable alternative for managing plastic waste. Future research into optimizing microbial strains and their enzymatic capabilities will be crucial for enhancing the efficiency of this bioremediation process. By harnessing these natural mechanisms, we can move toward a more sustainable approach to waste management and ecosystem conservation, ultimately advancing the development of novel, eco-friendly solutes for the textile and waste management industries and mediating the microfibrous waste level more efficiently.

References

Abe, M., Kobayashi, K., Honma, N., & Nakasaki, K. (2010). Microbial degradation of poly(butylene succinate) by fusarium solani in soil environments. *Polymer Degradation and Stability, 95*(2), 138–143. https://doi.org/10.1016/J.POLYMDEGRADSTAB.2009.11.042

Ahmed, P. M., De Figueroa, L. I. C., & Pajot, H. F. (2020). Application of microbial consortia in degradation and detoxification of industrial pollutants. *Microorganisms for Sustainable Environment and Health*, 401–418. https://doi.org/10.1016/B978-0-12-819001-2.00020-6

Amélineau, F., Bonnet, D., Heitz, O., Mortreux, V., Harding, A. M. A., Karnovsky, N., Walkusz, W., Fort, J., & Grémillet, D. (2016). Microplastic pollution in the Greenland Sea: Background levels and selective contamination of planktivorous diving seabirds. *Environmental Pollution, 219*, 1131–1139. https://doi.org/10.1016/J.ENVPOL.2016.09.017

Arrieta, J. M., Mayol, E., Hansman, R. L., Herndl, G. J., Dittmar, T., & Duarte, C. M. (2015). Ocean chemistry: Dilution limits dissolved organic carbon utilization in the deep ocean. *Science, 348*(6232), 331–333. https://doi.org/10.1126/SCIENCE.1258955/SUPPL_FILE/ARRIETA-SM.PDF

Athey, S. N., & Erdle, L. M. (2022). Are we underestimating anthropogenic microfiber pollution? A critical review of occurrence, methods, and reporting. *Environmental Toxicology and Chemistry, 41*(4). https://doi.org/10.1002/etc.5173

Bahrani, F., Mohammadi, A., Dobaradaran, S., De-la-Torre, G. E., Arfaeinia, H., Ramavandi, B., Saeedi, R., & Tekle-Röttering, A. (2024). Occurrence of microplastics in edible tissues of livestock (cow and sheep). *Environmental Science and Pollution Research, 31*(14), 22145–22157. https://doi.org/10.1007/S11356-024-32424-9

Bal, B., Ghosh, S., & Das, A. P. (2018). Microbial recovery and recycling of manganese waste and their future application: A review. *Geomicrobiological Journal, 36*(1), 85–96. https://doi.org/10.1080/01490451.2018.1497731

Bala, K., Dogra, A., Lata, P., & Das, A. P. (2024). Discussing sources and biological-chemical remediation approaches for microplastic pollution. *Sustainable Microbial Technology for Synthetic and Cellulosic Microfiber Bioremediation*, 37–59. https://doi.org/10.1007/978-3-031-62600-5_3

Besseling, E., Wang, B., Lürling, M., & Koelmans, A. A. (2014). Nanoplastic affects growth of S. Obliquus and reproduction of D. Magna. *Environmental Science and Technology, 48*(20), 12336–12343. https://doi.org/10.1021/ES503001D/SUPPL_FILE/ES503001D_SI_001.PDF

Bhattacharjee, J., Mishra, S., & Das, A. P. (2021). Recent advances in sensor-based detection of toxic dyes for bioremediation application: A review. *Applied Biochemistry and Biotechnology, 2021*, 1–20. https://doi.org/10.1007/S12010-021-03767-7

Braeuning, A. (2018). Uptake of microplastics and related health effects: a critical discussion of Deng et al., Scientific reports 7:46687, 2017. *Archives of Toxicology, 93*(1), 219–220. https://doi.org/10.1007/S00204-018-2367-9

Browne, M. A. (2007). Environmental and biological consequences of MICROPLASTIC within marine habitats. *Other Faculty of Science and Engineering Theses*.. https://pearl.plymouth.ac.uk/fose-theses-other/314

Chaudhry, A. K., & Sachdeva, P. (2021). Microplastics' origin, distribution, and rising hazard to aquatic organisms and human health: Socio-economic insinuations and management solutions. *Regional Studies in Marine Science, 48*, 102018. https://doi.org/10.1016/J.RSMA.2021.102018

Das, A. P., Ghosh, S., Mohanty, S., & Sukla, L. B. (2015). Advances in manganese pollution and its bioremediation. In L. B. Sukla, N. Pradhan, S. Panda, & B. K. Mishra (Eds.), *Environmental microbe biotechnology* (1st ed., pp. 313–328). Springer. https://doi.org/10.1007/978-3-319-19018-1_16

Das, A. P., & Mishra, S. (2008). Hexavalent chromium (VI): Health hazards and environmental pollutant. *Journal of Environmental Research and Development, 2*(3), 386–392.

Das, A. P., & Singh, S. (2011). Occupational health assessment of chromite toxicity among Indian miners. *Indian Journal of Occupational And Environmental Medicine, 15*(1), 6. https://doi.org/10.4103/0019-5278.82998

Das, N., & Chandran, P. (2011). Microbial degradation of petroleum hydrocarbon contaminants: An overview. *Biotechnology Research International, 2011*, 1–13. https://doi.org/10.4061/2011/941810

Dey, S., Bissoyi, A., Koteswara Reddy, G., & Das, A. P. (2024a). Synthetic microfibers: An urban emerging pollutant demanding for sustainable environmental solutions. *Sustainable Microbial Technology for Synthetic and Cellulosic Microfiber Bioremediation*, 107–121. https://doi.org/10.1007/978-3-031-62600-5_7

Dey, S., Das, A., Mallick, K., Sahu, A., & Das, A. P. (2023a). Environmental Petroleum Waste: Pollution. *Toxicity, Sustainable Remediation.*, 159–175. https://doi.org/10.1007/978-3-031-48220-5_7

Dey, S., Dash, N., Ray, S. K., Priyadarsini, S., Chhotaray, C., & Das, A. P. (2024b). Occurrence and quantification of Tire Wear plastics pollutants from Highway City dust of Bhubaneswar City of Odisha. *Sustainable Microbial Technology for Synthetic and Cellulosic Microfiber Bioremediation,* 77–89. https://doi.org/10.1007/978-3-031-62600-5_5

Dey, S., Tripathy, B., Kumar, M. S., & Das, A. P. (2023b). Ecotoxicological consequences of manganese mining pollutants and their biological remediation. *Environmental Chemistry and Ecotoxicology.* https://doi.org/10.1016/J.ENCECO.2023.01.001

Enfrin, M., Dumée, L. F., & Lee, J. (2019). Nano/microplastics in water and wastewater treatment processes—Origin, impact and potential solutions. *Water Research, 161,* 621–638. https://doi.org/10.1016/J.WATRES.2019.06.049

Ghosh, S., & Das, A. P. (2017). Bioleaching of manganese from mining waste residues using Acinetobacter sp. *Geology, Ecology, and Landscapes, 1*(2), 77–83. https://doi.org/10.1080/24749508.2017.1332847

Ghosh, S., & Das, A. P. (2018). Metagenomic insights into the microbial diversity in manganese-contaminated mine tailings and their role in biogeochemical cycling of manganese. *Scientific Reports 2018 8:1, 8*(1), 1–12. https://doi.org/10.1038/s41598-018-26311-w

Ghosh, S., & Das, A. P. (2020). Microbial metagenomics: Current advances in investigating microbial ecology and population dynamics. In S. K. Nayak & B. B. Mishra (Eds.), *Frontiers in soil and environmental microbiology* (1st ed., pp. 247–254). CRC Press.

Ghosh, S., Kumar, M. S., Bal, B., & Das, A. P. (2018). Application of bioengineering in revamping human health. *Synthetic Biology: Omics Tools and Their Applications,* 21–37. https://doi.org/10.1007/978-981-10-8693-9_2

Ghosh, S., Mohanty, S., Akcil, A., Sukla, L. B., & Das, A. P. (2016). A greener approach for resource recycling: Manganese bioleaching. *Chemosphere, 154,* 628–639. https://doi.org/10.1016/J.CHEMOSPHERE.2016.04.028

Gu, J. D. (2003). Microbiological deterioration and degradation of synthetic polymeric materials: Recent research advances. *International Biodeterioration & Biodegradation, 52*(2), 69–91. https://doi.org/10.1016/S0964-8305(02)00177-4

Guo, Q. Q., Xiao, M. R., Ma, Y., Niu, H., & Zhang, G. S. (2021). Polyester microfiber and natural organic matter impact microbial communities, carbon-degraded enzymes, and carbon accumulation in a clayey soil. *Journal of Hazardous Materials, 405,* 124701. https://doi.org/10.1016/J.JHAZMAT.2020.124701

Hadian-Ghazvini, S., Hooriabad Saboor, F., & Safaee Ardekani, L. (2022). Bioremediation techniques for microplastics removal. *Environmental Footprints and Eco-Design of Products and Processes,* 327–377. https://doi.org/10.1007/978-981-16-8440-1_15

Jagatee, S., Priyadarshini, S., Rath, C. C., & Das, A. P. (2024). Synthetic microfiber: An enduring environmental problem linked to sustainable development. *Environmental Science and Engineering, Part, F2333,* 93–112. https://doi.org/10.1007/978-3-031-51792-1_6

Jain, R., Gaur, A., Suravajhala, R., Chauhan, U., Pant, M., Tripathi, V., & Pant, G. (2023). Microplastic pollution: Understanding microbial degradation and strategies for pollutant reduction. *Science of the Total Environment, 905,* 167098. https://doi.org/10.1016/J.SCITOTENV.2023.167098

John, A., & Das, A. P. (2024). Government strategies for Achieving Circular Economy for plastic waste management in Southeast Asia. *Sustainable Management of Urban Plastic Waste Through Circular Economic Approaches,* 72–82. https://doi.org/10.1201/9781003465447-6/GOVERNMENT-STRATEGIES-ACHIEVING-CIRCULAR-ECONOMY-PLASTIC-WASTE-MANAGEMENT-SOUTHEAST-ASIA-AKBAR-JOHN-ALOK-PRASAD-DAS

Khatun, R., Singh, S., Dubey, N. K., & Prasad Das, A. (2023). A review on marine-based phytochemicals and their application in biomedical research. *Recent Frontiers of Phytochemicals:*

Applications in Food, Pharmacy, Cosmetics, and Biotechnology, 383–395. https://doi.org/10.1016/B978-0-443-19143-5.00013-X

Kyaw, B. M., Champakalakshmi, R., Sakharkar, M. K., Lim, C. S., & Sakharkar, K. R. (2012). Biodegradation of low density polythene (LDPE) by pseudomonas species. *Indian Journal of Microbiology, 52*(3), 411–419. https://doi.org/10.1007/S12088-012-0250-6/METRICS

Lahiri, D., Nag, M., Dey, A., Sarkar, T., Joshi, S., Pandit, S., Das, A. P., Pati, S., Pattanaik, S., Tilak, V. K., & Ray, R. R. (2021). Biofilm mediated degradation of petroleum products. *Geomicrobiology Journal*, 1–10. https://doi.org/10.1080/01490451.2021.1968979

Li, Y., Lu, Q., Xing, Y., Liu, K., Ling, W., Yang, J., Yang, Q., Wu, T., Zhang, J., Pei, Z., Gao, Z., Li, X., Yang, F., Ma, H., Liu, K., & Zhao, D. (2023). Review of research on migration, distribution, biological effects, and analytical methods of microfibers in the environment. *Science of the Total Environment, 855*, 158922. https://doi.org/10.1016/J.SCITOTENV.2022.158922

Liu, J., Yang, Y., Ding, J., Zhu, B., & Gao, W. (2019). Microfibers: A preliminary discussion on their definition and sources. *Environmental Science and Pollution Research, 26*(28). https://doi.org/10.1007/s11356-019-06265-w

Lu, K., Qiao, R., An, H., & Zhang, Y. (2018). Influence of microplastics on the accumulation and chronic toxic effects of cadmium in zebrafish (Danio rerio). *Chemosphere, 202*, 514–520. https://doi.org/10.1016/J.CHEMOSPHERE.2018.03.145

Mallick, K., Sahu, A., Dubey, N. K., & Das, A. P. (2023). Harvesting marine plastic pollutants-derived renewable energy: A comprehensive review on applied energy and sustainable approach. *Journal of Environmental Management, 348*, 119371. https://doi.org/10.1016/J.JENVMAN.2023.119371

Mallick, K., Sahu, S., Sahu, A., Dey, S., & Das, A. P. (2024). A critical review of marine microfiber pollution routes, toxicity, and its sustainable remediation. *Environmental Science and Engineering, Part, F2333*, 189–211. https://doi.org/10.1007/978-3-031-51792-1_10

Microbial Bioremediation *and Multiomics Technologies for Sustainable... - Google Books*. (n.d.). Retrieved 15 January 2025, from https://books.google.co.in/books?hl=en&lr=&id=y-YYEQAAQBAJ&oi=fnd&pg=PA49&dq=microbial+bioremediation+of+microplastics&ots=xfPuMXb7Ve&sig=SQwGgrKQZIop2PQUVQojKMiPvto&redir_esc=y#v=onepage&q&f=true

Miloloža, M., Cvetnić, M., Kučić Grgić, D., Ocelić Bulatović, V., Ukić, Š., Rogošić, M., Dionysiou, D. D., Kušić, H., & Bolanča, T. (2022). Biotreatment strategies for the removal of microplastics from freshwater systems. A review. *Environmental Chemistry Letters, 20*(2), 1377–1402. https://doi.org/10.1007/S10311-021-01370-0

Mishra, S., & Das, A. P. (2021). Current treatment Technologies for Removal of microplastic and microfiber pollutants from wastewater. *Wastewater Treatment: Cutting-Edge Molecular Tools, Techniques and Applied Aspects*, 237–251. https://doi.org/10.1016/B978-0-12-821881-5.00011-8

Mishra, S., & Das, A. P. (2022). Treatment of the wastewater polluted with synthetic microfiber released from washing machine. In B. B. Das, H. Hettiarachchi, P. K. Sahu, & S. Nanda (Eds.), *Lecture notes in civil engineering* (Vol. 207, pp. 109–117). Springer Science and Business Media Deutschland GmbH. https://doi.org/10.1007/978-981-16-7509-6_9/COVER

Mishra, S., & Das, A. P. (2023). Microbial remediation of synthetic microfiber contaminated wastewater. In M. P. Shah (Ed.), *Microbial Technologies in Industrial Wastewater Treatment* (pp. 337–350). Springer. https://doi.org/10.1007/978-981-99-2435-6_17

Mishra, S., Dash, D., & Das, A. P. (2022). Detection, characterization and possible biofragmentation of synthetic microfibers released from domestic laundering wastewater as an emerging source of marine pollution. *Marine Pollution Bulletin, 185*, 114254. https://doi.org/10.1016/J.MARPOLBUL.2022.114254

Mishra, S., Dash, D., & Das, A. P. (2024a). Aquatic microbial diversity on Plastisphere: Colonization and potential role in microplastic biodegradation. *Geomicrobiology Journal, 41*(4), 312–323. https://doi.org/10.1080/01490451.2023.2209750

Mishra, S., Dash, D., & Das, A. P. (2024b). Isolation and molecular characterization of polyester-degrading bacterial strains from municipal wastewater and their application in microplastic degradation in contaminated wastewater. *Geomicrobiology Journal*. https://doi.org/10.1080/01490451.2024.2376101

Mishra, S., Dash, D., & Das, A. P. (2024c). Microbial degradation of polyester microfibers using indigenously isolated bacterial strain Exiguobacterium Sp. *CLEAN–Soil, AirWater, 52*(12), e202300343. https://doi.org/10.1002/CLEN.202300343

Mishra, S., Charan Rath, C., & Das, A. P. (2019). Marine microfiber pollution: A review on present status and future challenges. *Marine Pollution Bulletin, 140*, 188–197. https://doi.org/10.1016/J.MARPOLBUL.2019.01.039

Mishra, S., Rout, P. K., & Das, A. P. (2021). Emerging microfiber pollution and its remediation. In R. Prasad (Ed.), *Environmental and microbial biotechnology* (pp. 247–266). Springer. https://doi.org/10.1007/978-981-15-5499-5_9

Mishra, S., Singh, R. P., Rath, C. C., & Das, A. P. (2020). Synthetic microfibers: Source, transport and their remediation. *Journal of Water Process Engineering, 38*, 101612. https://doi.org/10.1016/J.JWPE.2020.101612

Mohanty, M., Mohanty, J., Dey, S., Dutta, K., Shah, M. P., & Das, A. P. (2024). The face mask: A tale from protection to pollution and demanding sustainable solution. *Emerging Contaminants, 10*(2), 100298. https://doi.org/10.1016/J.EMCON.2023.100298

Mohanty, S., Ghosh, S., Bal, B., & Das, A. P. (2018). A review of biotechnology processes applied for manganese recovery from wastes. *Reviews in Environmental Science and Bio/Technology, 17*(4), 791–811. https://doi.org/10.1007/S11157-018-9482-1

Mohanty, S., Ghosh, S., Nayak, S., & Das, A. P. (2016). Isolation, identification and screening of manganese solubilizing fungi from low-grade manganese ore deposits. *Geomicrobiological Journal, 34*(4), 309–316. https://doi.org/10.1080/01490451.2016.1189016

Moharana, S., Dey, S., Priyadarsini, S., Kumar, M. S., & Das, A. P. (2024). Identification and characterization of microplastic pollutants from the marine sediments of Paradeep coast of bay of Bengal, India for their sustainable management. *Environmental Science and Engineering, Part, F2333*, 57–73. https://doi.org/10.1007/978-3-031-51792-1_4

Mor, R., & Sivan, A. (2008). Biofilm formation and partial biodegradation of polystyrene by the actinomycete Rhodococcus ruber: Biodegradation of polystyrene. *Biodegradation, 19*(6), 851–858. https://doi.org/10.1007/S10532-008-9188-0/METRICS

Nelms, S. E., Galloway, T. S., Godley, B. J., Jarvis, D. S., & Lindeque, P. K. (2018). Investigating microplastic trophic transfer in marine top predators. *Environmental Pollution, 238*, 999–1007. https://doi.org/10.1016/J.ENVPOL.2018.02.016

Niu, L., Li, Y., Li, Y., Hu, Q., Wang, C., Hu, J., Zhang, W., Wang, L., Zhang, C., & Zhang, H. (2021). New insights into the vertical distribution and microbial degradation of microplastics in urban river sediments. *Water Research, 188*, 116449. https://doi.org/10.1016/J.WATRES.2020.116449

Oberbeckmann, S., & Labrenz, M. (2020). Marine microbial assemblages on microplastics: Diversity, adaptation, and role in degradation. *Annual Review of Marine Science, 12*, 209–232. https://doi.org/10.1146/ANNUREV-MARINE-010419-010633/1

Parton, K. (2020). Elasmobranch (sharks and rays) interaction with plastic pollution from global and local perspectives, via entanglement within anthropogenic debris and. https://core.ac.uk/download/pdf/326248134.pdf.

Paul, A., Priyadarsini, S., Karakuş, S., Behera, J. K., Gnanasekaran, R., & Das, A. P. (2024). Fate and accumulation of synthetic microfibers in the environment. *Sustainable Microbial Technology for Synthetic and Cellulosic Microfiber Bioremediation*, 61–76. https://doi.org/10.1007/978-3-031-62600-5_4

Prata, J. C., da Costa, J. P., Lopes, I., Duarte, A. C., & Rocha-Santos, T. (2020). Environmental exposure to microplastics: An overview on possible human health effects. *Science of the Total Environment, 702*, 134455. https://doi.org/10.1016/J.SCITOTENV.2019.134455

Prata, J. C., & Dias-Pereira, P. (2023). Microplastics in terrestrial domestic animals and human health: Implications for food security and food safety and their role as sentinels. *Animals, 13*(4), 661. https://doi.org/10.3390/ANI13040661

Priya, A. K., Jalil, A. A., Dutta, K., Rajendran, S., Vasseghian, Y., Karimi-Maleh, H., & Soto-Moscoso, M. (2022). Algal degradation of microplastic from the environment: Mechanism, challenges, and future prospects. *Algal Research, 67*, 102848. https://doi.org/10.1016/J.ALGAL.2022.102848

Priyadarsini, S., & Das, A. P. (2024). Analytical and structural characterization of waste lithium-ion batteries for their effective recycling strategy. *Environmental Chemistry and Ecotoxicology*. https://doi.org/10.1016/j.enceco.2024.12.004

Priyadarsini, S., & Das, A. P. (2025). Lithium bioleaching: A review on microbial-assisted sustainable technology for lithium bio-circularity. *Journal of Water Process Engineering, 69*, 106744. https://doi.org/10.1016/j.jwpe.2024.106744

Priyadarshini, B., Santhosh Kumar, M., & Das, A. P. (2024a). Microbial degradation of synthetic plastics released from COVID-19 generated single use disposable face masks. *Environmental Science and Engineering, Part, F2827*, 1–16. https://doi.org/10.1007/978-3-031-50840-0_1

Priyadarshini, S., Jagatee, S., & Das, A. P. (2024b). Synthetic fabrics and microfiber pollution–An assessment of their global impact. *Environmental Science and Engineering, Part, F2333*, 137–157. https://doi.org/10.1007/978-3-031-51792-1_8

Priyadarsini, S., Alok, &, Das, P., Journal, G., & Das, A. P. (2025). Isolation and identification of native lithium-resistant bacteria from electronic waste contaminated sites: A focused investigation on tolerance assessment with potential lithium bio-recovery application. *Geomicrobiology Journal*. https://doi.org/10.1080/01490451.2025.2539509

Rathinamoorthy, R., & Raja Balasaraswathi, S. (2023). Characterization of microfibers released from chemically modified polyester fabrics — A step towards mitigation. *Science of the Total Environment, 866*. https://doi.org/10.1016/j.scitotenv.2022.161317

Ray, S. K., Dey, S., Das, A. P., & John, A. (2024). Microplastic waste management through greener technologies: A review. *Sustainable Management of Urban Plastic Waste Through Circular Economic Approaches*, 237–261. https://doi.org/10.1201/9781003465447-14

Reddy Koteswara, G., Bhargavi, S., Yamuna, D., Anjani Devi, M., Jeevana Lakshmi Kumar, G., Manas, T., Adarsh, S., Bhavya Sai, V., Nath, H., Alreshidi, M. A., Togun, H., Ram, D. K., & Das, A. P. (2025). Recent advancements in the remediation of marine hydrocarbon and plastic pollutants. *Environmental Science and Engineering, Part, F218*, 127–142. https://doi.org/10.1007/978-3-031-81615-4_7

RENEWABLE ENERGY *GENERATION AND VALUE ADDITION FROM ENVIRONMENTAL*. (2024).

Sahoo, P. P., Singh, S., Rout, P. K., Mishra, S., & Das, A. P. (2022). Microbial remediation of plastic pollutants generated from discarded and abandoned marine fishing nets. *Biotechnology and Gnetic Engineering Reviews*. https://doi.org/10.1080/02648725.2022.2152629

Sahu, A., Mallick, K., & Das, A. P. (2024). Environmental impact of single-use synthetic face mask and its recycling: A sustainable approach. 197–212. doi:https://doi.org/10.1007/978-3-031-50840-0_11.

Sahu, S., & Das, A. P. (2024). Microplastics degradation mechanism by aquatic Plastiomes: A review. *Environmental Science and Engineering, Part, F2827*, 213–226. https://doi.org/10.1007/978-3-031-50840-0_12

Sangeetha Devi, R., Rajesh Kannan, V., Nivas, D., Kannan, K., Chandru, S., & Robert Antony, A. (2015). Biodegradation of HDPE by aspergillus spp. from marine ecosystem of gulf of Mannar, India. *Marine Pollution Bulletin, 96*(1–2), 32–40. https://doi.org/10.1016/J.MARPOLBUL.2015.05.050

Sarmah, P., & Rout, J. (2020). Role of algae and cyanobacteria in bioremediation: Prospects in polyethylene biodegradation. *Advances in Cyanobacterial Biology, 333–349*. https://doi.org/10.1016/B978-0-12-819311-2.00022-X

Satapathy, S., Kar, D., Priyadarsini, S., Das, A. P., & Kuanar, A. (2024). Application of microorganisms in emerging waste management strategies and remediation of synthetic and cellulosic microfibers. *Sustainable Microbial Technology for Synthetic and Cellulosic Microfiber Bioremediation, 183–206*. https://doi.org/10.1007/978-3-031-62600-5_10

Singh Jadaun, J., Bansal, S., Sonthalia, A., Rai, A. K., & Singh, S. P. (2022). Biodegradation of plastics for sustainable environment. *Bioresource Technology, 347*, 126697. https://doi.org/10.1016/J.BIORTECH.2022.126697

Skariyachan, S., Taskeen, N., Kishore, A. P., Krishna, B. V., & Naidu, G. (2021). Novel consortia of Enterobacter and pseudomonas formulated from cow dung exhibited enhanced biodegradation of polyethylene and polypropylene. *Journal of Environmental Management, 284*, 112030. https://doi.org/10.1016/J.JENVMAN.2021.112030

Sokal, S., Palsania, P., & Kaushik, G. (2022). Bioremediation and functional metagenomics: Advances, challenges, and opportunities. *Omics Insights in Environmental Bioremediation, 3–36*. https://doi.org/10.1007/978-981-19-4320-1_1

Srivastava, J., Naraian, R., Kalra, S. J. S., & Chandra, H. (2014). Advances in microbial bioremediation and the factors influencing the process. *International journal of Environmental Science and Technology, 11*(6), 1787–1800. https://doi.org/10.1007/S13762-013-0412-Z/METRICS

Stojanović, K., Živić, M., Marković, Z., Đorđević, J., Jovanović, J., & Živić, I. (2019). How changes in water quality under the influence of land-based trout farms shape chemism of the recipient streams—Case study from Serbia. *Aquaculture International, 27*(6), 1625–1641. https://doi.org/10.1007/S10499-019-00414-1/METRICS

Sussarellu, R., Suquet, M., Thomas, Y., Lambert, C., Fabioux, C., Pernet, M. E. J., Le Goïc, N., Quillien, V., Mingant, C., Epelboin, Y., Corporeau, C., Guyomarch, J., Robbens, J., Paul-Pont, I., Soudant, P., & Huvet, A. (2016). Oyster reproduction is affected by exposure to polystyrene microplastics. *Proceedings of the National Academy of Sciences of the United States of America, 113*(9), 2430–2435. https://doi.org/10.1073/PNAS.1519019113/SUPPL_FILE/PNAS.1519019113.SD03.XLS

Tripathy, B., Dash, A., & Das, A. P. (2022). Detection of environmental microfiber pollutants through vibrational spectroscopic techniques: Recent advances of environmental monitoring and future prospects. *Critical Reviews of Analytical Chemistry*, 1–11. https://doi.org/10.1080/10408347.2022.2144994

Tripathy, B., Dey, S., Dubey, N. K., Lo, W.-C., & Das, A. P. (2024a). *An omics-based approach for managing the risk of COVID-19-generated wastes*. 51–66. https://doi.org/10.1007/978-3-031-50840-0_4

Tripathy, B., Sahoo, P. P., Sundaray, H., & Das, A. P. (2024b). In-vitro biodegradation of discarded marine microplastics across the eastern coast of the bay of Bengal, India using Exiguobacterium sp. *Environmental Chemistry and Ecotoxicology, 6*, 236–247. https://doi.org/10.1016/J.ENCECO.2024.07.003

Tripathy, B., Singh, S., Behera, I. D., Mishra, S., & Das, A. P. (2026). Metagenomic profiling of diversified marine microbiome across microplastic-contaminated niches of Bay of Bengal, India. *Marine Pollution Bulletin, 222*, 118872. https://doi.org/10.1016/J.MARPOLBUL.2025.118872

Watson-Wright, W. M., Wells, P. G., Duce, R. A., Gilardi, K. V., Girvan, A. S. T., Huber, M. E., Kershaw, P. J., Linders, J. B. H. J., Luit, R. J., Vivian, C. M. G., & Vousden, D. H. (2024). The UN joint Group of Experts on the scientific aspects of marine environmental protection (GESAMP) — An ocean science-policy interface standing the test of time. *Marine Pollution Bulletin, 199*, 115917. https://doi.org/10.1016/J.MARPOLBUL.2023.115917

Yoshida, S., Hiraga, K., Takehana, T., Taniguchi, I., Yamaji, H., Maeda, Y., Toyohara, K., Miyamoto, K., Kimura, Y., & Oda, K. (2016). A bacterium that degrades and assimilates poly(ethylene terephthalate). *Science, 351*(6278), 1196–1199. https://doi.org/10.1126/SCIENCE.AAD6359/SUPPL_FILE/AAD6359-YOSHIDA-SM.PDF

Yuan, J., Ma, J., Sun, Y., Zhou, T., Zhao, Y., & Yu, F. (2020). Microbial degradation and other environmental aspects of microplastics/plastics. *Science of the Total Environment, 715*, 136968. https://doi.org/10.1016/J.SCITOTENV.2020.136968

Zeenat, E. A., Bukhari, D. A., Shamim, S., & Rehman, A. (2021). Plastics degradation by microbes: A sustainable approach. *Journal of King Saud University-Science, 33*(6), 101538. https://doi.org/10.1016/J.JKSUS.2021.101538

Zhang, K., Hamidian, A. H., Tubić, A., Zhang, Y., Fang, J. K. H., Wu, C., & Lam, P. K. S. (2021). Understanding plastic degradation and microplastic formation in the environment: A review. *Environmental Pollution, 274*. https://doi.org/10.1016/J.ENVPOL.2021.116554

Zhou, Y., Kumar, M., Sarsaiya, S., Sirohi, R., Awasthi, S. K., Sindhu, R., Binod, P., Pandey, A., Bolan, N. S., Zhang, Z., Singh, L., Kumar, S., & Awasthi, M. K. (2022). Challenges and opportunities in bioremediation of micro-nano plastics: A review. *Science of the Total Environment, 802*, 149823. https://doi.org/10.1016/J.SCITOTENV.2021.149823

Advanced Microbial Technology for Sustainable Management of Environmental Waste

H. S. Rita, Swarna M. Patra, Glen Cletus DSouza, P. Ramya, and R. P. Vijayakumar

Abstract Anthropogenic activities have caused significant ecological harm, making microbial technologies a promising solution for waste management and environmental sustainability. This chapter explores the use of microbial consortia, genetically modified microorganisms, and bioremediation techniques to effectively degrade pollutants such as heavy metals, plastics, and organic contaminants. By harnessing microbial metabolic pathways, these processes not only detoxify hazardous waste but also convert it into valuable resources, supporting a circular economy. Case studies highlight the successful application of these technologies across diverse settings, showcasing their effectiveness and adaptability. The chapter emphasizes the importance of integrating microbial methods into existing waste management frameworks to foster sustainable practices, reduce environmental impact, and enhance ecosystem health.

Keywords Microbial technology · Waste management · Environmental sustainability · Bioremediation · Circular bioeconomy

---

H. S. Rita · S. M. Patra (✉)
Department of Chemistry, RV College of Engineering, Bengaluru, India
e-mail: swarnamp@rvce.edu.in

G. C. DSouza (✉)
Department of Chemical and Biochemical Engineering, Western University, London, ON, Canada
e-mail: gdsouza6@uwo.ca

P. Ramya
Department of Physics, RV College of Engineering, Bengaluru, India

R. P. Vijayakumar
Department of Chemical Engineering, Visvesvaraya National Institute of Technology, Nagpur, India

A. P. Das (ed.), *Bioconversion of Environmental Waste to Value Added Products*, Sustainable Environmental Waste Management Strategies,
https://doi.org/10.1007/978-3-032-14215-3_7

1 Introduction to Microbial Technology

Microbial technology is a field of science and engineering that leverages the abilities of microorganisms—such as bacteria, fungi, and algae—for various industrial, environmental, and biotechnological applications. These microorganisms possess natural metabolic capabilities to break down, synthesize, or transform substances, making them invaluable tools for solving real-world problems (Priyadarsini & Das, 2024, 2025). This technology capitalizes on microorganisms' roles in ecosystems, where they act as decomposers, nutrient cyclers, and environmental cleansers (Fouad et al., 2023; Mirza et al., 2024; Timmis & Hallsworth, 2024). By understanding and optimizing these processes, microbial technology provides innovative solutions in areas like agriculture (Prasad et al., 2023; Singh et al., 2024), medicine (Zhang et al., 2023), energy production (Lal, 2013), and waste management (Adebayo & Smart, 2018; Kannan et al., 2024; Kaur & Vyas, 2024). Microbial technology has become a cornerstone in waste management, addressing challenges such as pollution, resource depletion, and energy scarcity. It offers eco-friendly alternatives to traditional methods, ensuring the sustainable handling of organic and inorganic waste while contributing to the circular economy. This field continues to expand with advancements in genetic engineering, synthetic biology, and bioprocessing, unlocking new possibilities for harnessing the power of microorganisms .

1.1 *Overview of Microbial Technology and its Significance in Waste Management*

Microbial technology utilizes microorganisms such as bacteria, fungi, and algae to perform various biological processes, including the breakdown of complex substances into simpler compounds, the transformation of waste into valuable resources, and the mitigation of environmental pollution (; Ghosh et al., 2023; Nag et al., 2023a). In waste management, this technology leverages the natural abilities of microorganisms to degrade, detoxify, and recycle organic and inorganic waste, offering an environmentally friendly, cost-effective, and sustainable alternative to conventional methods (Kaur & Vyas, 2024). Its significance lies in numerous aspects: it reduces reliance on chemical treatments and fossil fuels while minimizing the release of toxic substances, thus delivering substantial environmental benefits. Additionally, it supports resource recovery by converting waste into valuable products like biogas, compost, and biofertilizers, promoting circular economy principles (. Microbial processes are energy-efficient, generating renewable energy sources such as biogas and electricity, which contribute to sustainable energy goals. They are also cost-effective, with lower operational expenses compared to traditional waste management methods, reducing the economic burden of waste disposal. Furthermore, microbial technology fosters long-term sustainability by aligning with global environmental goals, reducing greenhouse gas emissions, and encouraging biodiversity (A. Prasad. Das & Ghosh, 2023f, 2023 g).

1.2 Historical Context and Evolution of Microbial Applications in Environmental Sustainability

Microbial technology, the application of microorganisms to solve practical problems, has its roots in ancient practices but has significantly evolved with scientific advancements. From early uses in fermentation to modern biotechnological innovations, microorganisms have played a vital role in human progress, particularly in promoting environmental sustainability (). The evolution of microbial applications in environmental sustainability has been transformative, leveraging the natural capabilities of microorganisms to address critical environmental challenges (Table 1).

Microbial technology has revolutionized the way we address environmental challenges, offering eco-friendly, efficient, and scalable solutions. Its applications reduce reliance on chemical treatments, minimize greenhouse gas emissions, and recover resources from waste streams. The field continues to evolve, bridging ancient wisdom with cutting-edge science to create a sustainable future.

2 Microbial Bioremediation Techniques

2.1 Definitions and Principles of Bioremediation

Microbial bioremediation is an advanced and sustainable process that utilizes microorganisms to detoxify, degrade, or remove pollutants from the environment. By exploiting the metabolic pathways of microbes, this technique effectively addresses contamination in soil, water, and air (Bhowmick et al., 2024; Mokrani et al., 2024; Saravanan et al., 2023). It stands out as an eco-friendly alternative to traditional remediation methods, offering a cost-effective and efficient means of managing environmental pollution. The choice of bioremediation approach depends on several factors, including the type of contaminant, site-specific conditions, and the desired remediation outcomes.

During microbial metabolism, microorganisms utilize contaminants as a source of carbon or energy, breaking them down into simpler, non-toxic substances such as water, carbon dioxide, or benign metabolites. The effectiveness of this process is influenced by environmental factors like pH, temperature, oxygen availability, and nutrient levels, which collectively determine microbial activity. Furthermore, the bioavailability of pollutants plays a critical role; contaminants must be accessible

Table 1 Key stages and advancements

Time period	Key developments	Applications/impact
Ancient times	Fermentation (example: Bread, beer, wine, yogurt)	Food preservation and beverage production through microbial activity
	Composting and manure use	Natural recycling of organic waste and enhancing soil fertility
Seventeenth century	Antonie van Leeuwenhoek observes microbes using simple microscope (1676)	First discovery of microorganisms
Nineteenth century	Louis Pasteur develops '*germ theory*'	Understanding microbial roles in disease and fermentation
	Robert Koch identifies pathogens and develops postulates	Establishment of microbiology as a scientific field
	Development of microbial ecology by Sergei Winograd sky	Recognition of microbial roles in nutrient cycles
Early twentieth century	Discovery of nitrogen fixing bacteria (example- rhizobia)	Use in sustainable agriculture through biofertilizers
	Development of activated sludge process (1914)	Efficient wastewater treatment
Mid-twentieth century	Identification of oil-degrading microbes and early bioremediation practices	Cleaning oil spills and pollutants using microbes
	Discovery of bacillus thurigiensis (Bt) as a biopesticide	Biological pest control in agriculture
Late twentieth century	Emergence of genetic engineering (1970s) for microbial modifications	Development of microbes for tailored applications like pollutant degradation and biofuel production
	Introduction of recombinant DNA technology	Creation of bioengineered microbes for sustainability
	Biofertilizers and bioremediation gain traction	Reducing dependence on chemical fertilizers
	Focus on bioremediation	Pollution control and ecosystem restoration
	Phytoremediation with microbial plant partnerships	Soil and water decontamination using microbial activity
	Microbial consortia studied synergistic applications	Enhancing efficiency of bioremediation and waste treatment
Twenty-first century	Advancements in metagenomics and synthetic biology	Enhanced understanding of microbial communities
	Development of microbial fuel cells	Bioenergy production from organic waste

(continued)

Table 1 (continued)

Time period	Key developments	Applications/impact
	Exploration of microbes for carbon sequestration	Mitigating greenhouse gas emissions
	Synthetic biology for designing microbes	Tailored solutions for environmental problems like plastic degradation
	CRISPR and AI-driven microbial innovations	Precision microbial engineering for sustainability
Future directions	Integration with circular economy: Bioengineered microbes for specific challenges (example, microplastic degradation)	Revolutionizing waste management and resource recovery
	Precision environmental monitoring: AI-driven biosensors	Using microbes track environmental health in real time

to microbes for degradation, which depends on their solubility, chemical form, and interaction with soil or water matrices.

Site-specific conditions and microbial adaptability further determine the feasibility of bioremediation. In some cases, native microorganisms adapt to degrade pollutants, while engineered microbes with enhanced degradative abilities may be introduced (Curiel-Alegre et al., 2022). Regular monitoring and adjustments, such as nutrient addition or aeration, ensure optimal conditions for microbial activity. This process effectively addresses various contaminants, including hydrocarbons, pesticides, heavy metals, and industrial chemicals, making it a sustainable and versatile solution for environmental cleanup.

2.2 *Types of Bioremediations: In Situ Versus Ex Situ Methods*

Bioremediation involves two main approaches: in situ and ex situ methods (Perez-Vazquez et al., 2024). In situ bioremediation treats contaminants directly at the site without excavation using techniques like bioventing, biosparging, and natural attenuation (Kuppan et al., 2024; Philp & Atlas, 2005). Bioventing supplies oxygen to stimulate aerobic degradation in soil, while biosparging injects air and nutrients into contaminated groundwater or soil to enhance microbial activity. Natural attenuation relies on existing microbial processes to degrade pollutants without external intervention. In situ methods are commonly applied to petroleum hydrocarbons and volatile organic compounds (VOCs) and offer advantages like minimal environmental disturbance, cost-effectiveness, and suitability for large-scale applications. However, they are slower and require favorable site conditions for microbial growth.

Ex situ bioremediation involves excavating contaminated material for treatment in a controlled environment (Balseiro-Romero et al., 2019; Tomei & Daugulis, 2013). Techniques include biopiles, where soil is piled and aerated to enhance biodegradation; landfarming, which spreads contaminated soil over prepared beds for periodic tilling; and bioreactors, engineered vessels where contaminants are degraded under controlled conditions (Paul et al., 2023; Pradhan et al., 2023). Ex situ methods are faster and more efficient, with greater control over environmental factors, making them suitable for addressing oil spills, industrial effluents, and heavy metals. However, they are more expensive due to excavation and transportation and are less effective for deeply contaminated sites (; Kanungo et al., 2024).

2.3 Case Studies Showcasing Successful Bioremediation Efforts

Bioremediation has been successfully applied in various environmental restoration efforts, showcasing its versatility and effectiveness. In the Exxon Valdez oil spill in Alaska (1989), biostimulation was employed by applying fertilizers containing nitrogen and phosphorus to stimulate indigenous oil-degrading bacteria. This approach significantly degraded hydrocarbons, mitigating the environmental impact. Similarly, in California, groundwater contaminated with trichloroethylene (TCE) and other volatile organic compounds was treated through in situ bioremediation using bioaugmentation with anaerobic bacteria, leading to successful detoxification and safe groundwater use. Following the Chernobyl disaster in Ukraine (1986), fungi like *Cladosporium sphaerospermum* were applied to remediate radioactive waste, reducing contamination in the surrounding area (Bhattacharjee et al., 2021; Dey et al., 2024; Nag et al., 2023b).

Another notable example is the remediation of polychlorinated biphenyls (PCBs) in the Hudson River, USA. Ex situ bioremediation techniques involving biopiles treated with PCB-degrading microbes effectively reduced PCB levels. These case studies underscore the potential of bioremediation as a sustainable and evolving solution for managing environmental pollution. Advances in microbial engineering, genomics, and environmental monitoring continue to enhance its scope and efficiency, addressing diverse pollutants across various ecosystems.

3 Microbial Consortia and their Role in Enhancing Pollutant Degradation

Microbial consortia are communities of interacting microorganisms that work synergistically to perform complex metabolic processes, particularly in environmental applications like pollutant degradation (Lü et al., 2024a). By combining the diverse metabolic capabilities of different species, consortia can efficiently break down complex and recalcitrant pollutants. Each microbial species within the consortium contributes to the process by dividing metabolic tasks—one species may transform a pollutant into intermediate compounds, which another species further metabolizes into harmless by-products. This collaborative approach makes microbial consortia highly effective for addressing a wide range of pollutants, including hydrocarbons, plastics, heavy metals, and persistent compounds like polychlorinated biphenyls (PCBs) and dioxins, especially in mixed-contaminant environments such as industrial effluents.

In addition to their efficiency, microbial consortia demonstrate remarkable resilience and adaptability to changing environmental conditions, such as variations in pH, temperature, and nutrient availability. They can simultaneously target multiple pollutants, improving the overall effectiveness of bioremediation. Processes like co-metabolism, where certain microbes degrade pollutants incidentally while metabolizing other compounds, further enhance their utility. Techniques such as bioaugmentation—introducing engineered consortia to contaminated sites—and biostimulation—enhancing native consortia through nutrient addition—accelerate degradation processes. Applications of microbial consortia include oil spill cleanup, pesticide degradation, wastewater treatment, and solid waste management (Lü et al., 2024b; Matilda & Samuel, 2024). Their collaborative nature and versatility make them indispensable tools in bioremediation and environmental restoration strategies.

3.1 Mechanisms of Interaction among Different Microbial Species

Microbial consortia rely on intricate interactions to perform their metabolic functions effectively, particularly in environmental applications like pollutant degradation. A key mechanism is metabolic cross-feeding (syntrophy), where one species produces metabolic by-products that serve as substrates for another. For example, in anaerobic digestion, fermentative bacteria produce short-chain fatty acids, which methanogens convert into methane. Nutrient sharing also enhances the collective growth and activity of consortia, with different species recycling essential nutrients such as nitrogen, phosphorus, and sulfur (Kumar et al., 2022). An example includes nitrifying bacteria converting ammonia into nitrites, which are further oxidized to nitrates by other species.

Additional mechanisms include quorum sensing, where microbes communicate using signaling molecules to coordinate their activities like enzyme production or biofilm formation, as seen in the cooperative degradation of pollutants. Spatial organization within consortia optimizes metabolic efficiency by creating microenvironments; for instance, anaerobic and aerobic microbes coexist in bioreactors, utilizing oxygen gradients to support diverse pathways (Zuroff & Curtis, 2012). Competitive and mutualistic interactions further shape consortia dynamics. While competition for resources may occur, mutualism—such as lactic acid bacteria lowering pH to create favorable conditions for other microbes—ensures their success.

These complex interactions make microbial consortia indispensable for environmental applications. Their ability to synergistically degrade complex substances, adapt to diverse conditions, and efficiently utilize resources positions them as powerful tools for bioremediation and waste management. Advancing our understanding and engineering of these interactions will further unlock their potential for sustainable environmental solutions.

4 Genetically Engineered Microorganisms (GEMs) and Their Applications in Waste Management and Pollutant Degradation

Genetically engineered microorganisms (GEMs) are microbes whose genetic material has been modified to enhance their natural abilities for specific tasks. These modifications allow GEMs to degrade pollutants, process waste, and produce valuable by-products, making them a powerful tool in environmental management. With advancements in genetic engineering, particularly CRISPR/Cas9, scientists can make precise changes to microbial genomes, improving their efficiency for targeted applications (Fatima et al., 2024; Rafeeq et al., 2023; Wu et al., 2021). CRISPR/Cas9 has enabled tasks such as gene knockouts and insertions, creating strains that are better equipped for environmental challenges (Hamdan & Tan, 2024).

Recent breakthroughs in genetic engineering have expanded the potential of GEMs beyond traditional methods. Technologies like base editing and prime editing refine genetic modifications with high precision, minimizing unintended changes. Multiplex genome editing allows researchers to alter multiple genes at once, tailoring GEMs for complex environmental problems. Additionally, synthetic biology has opened new possibilities by designing custom metabolic pathways, enabling microbes to break down difficult pollutants or convert waste materials into useful products. For example, microbes engineered to produce enzymes like PETase can degrade plastics into biodegradable components, offering solutions for the growing problem of plastic waste (DSouza et al., 2021; Tamoor et al., 2021).

GEMs play a crucial role in managing waste and reducing environmental pollutants. Engineered strains such as *Pseudomonas* and *Sphingomonas* are capable of breaking down petroleum hydrocarbons, polychlorinated biphenyls (PCBs), and industrial dyes. In wastewater treatment, modified strains of *E. coli* and *Bacillus* improve the removal of nitrogen compounds, phosphates, and organic waste (Sharma et al., 2024). GEMs are also used to detoxify heavy metals like mercury and cadmium by expressing proteins such as metallothioneins that bind and neutralize these toxins (P. Sharma et al., 2021).

Beyond waste treatment, GEMs contribute to landfill management by boosting methane production during anaerobic digestion or reducing harmful gas emissions.

In air pollution control, GEMs are deployed in biofilters and bioreactors to remove volatile organic compounds (VOCs) from industrial emissions (Arora et al., 2023). GEMs also aid in recycling e-waste by bioleaching valuable metals and degrading synthetic drugs and other persistent pollutants. Additionally, they serve as biosensors to detect contaminants such as heavy metals, providing quick and effective environmental monitoring. The versatility and efficiency of GEMs make them an essential part of modern strategies for waste management and environmental conservation.

5 Microbial Processes for Waste Valorization

Waste valorization refers to the process of converting waste materials into valuable products, such as biofuels, bioplastics, and other high-value chemicals. Microbial processes play a central role in this, as microorganisms can convert organic waste into energy, biodegradable plastics, and other useful bioproducts through natural metabolic pathways. These processes are not only environmentally sustainable but also economically beneficial, offering new avenues for waste recycling and resource recovery (Ansari et al., 2024; Kapoor et al., 2024; Sharma et al., 2024; Vishwas et al., 2024).

5.1 Conversion of Waste into Biofuels Through Microbial Processes

Microorganisms play a vital role in transforming organic and inorganic waste into valuable biofuels, offering a sustainable approach to waste management and renewable energy production. In the production of bioethanol, microbes like *Saccharomyces cerevisiae* (yeast) and *Zymomonas mobilis* ferment organic waste, including agricultural residues, food waste, and lignocellulosic biomass, into ethanol (Prasad et al., 2019). For example, sugars extracted from corn stover, a residual biomass, are metabolized by these microorganisms to produce ethanol and carbon dioxide, providing a renewable energy source.

Similarly, microbial processes are employed to produce biodiesel from waste oils, such as used cooking oil and animal fats. Lipid-producing microorganisms like *Yarrowia lipolytica* convert these waste oils into biodiesel through fermentation (Carsanba et al., 2018). This renewable fuel serves as an alternative to fossil fuels, supporting energy generation and vehicle operation while recycling waste materials. These microbial pathways not only reduce waste but also contribute to the development of a circular economy by converting discarded resources into clean energy.

5.2 *Conversion of Waste into Bioplastics Through Microbial Processes*

The production of bioplastics through microbial processes offers a sustainable solution to waste management while reducing dependency on petroleum-based plastics. Microorganisms such as *Cupriavidus necator* and *Pseudomonas* species are capable of synthesizing polyhydroxyalkanoates (PHAs), a family of biodegradable plastics, from various organic waste streams (Getino et al., 2024; Morlino et al., 2023). These microbes metabolize carbon-rich sources such as glucose, fatty acids, or waste oils, converting them into PHAs through intracellular accumulation during nutrient-limited growth conditions. The resulting PHAs are versatile materials suitable for applications ranging from packaging to medical devices. An exemplary pathway involves the utilization of waste glycerol, a by-product of biodiesel production, as a feedstock for engineered bacteria (Thamarai et al., 2024). By optimizing metabolic pathways, these bacteria efficiently convert glycerol into PHAs, showcasing an integrated approach to waste valorization and bioplastics production.

5.3 *Conversion of Waste into Valuable Biochemicals Through Microbial Processes*

Microbial processes enable the transformation of waste materials into valuable biochemicals, offering sustainable alternatives for various industrial applications. One significant example is the production of lactic acid, a key precursor for biodegradable plastics like polylactic acid (PLA). Bacteria such as *Lactobacillus* and *Enterococcus* species ferment waste sugars or agricultural residues, including sugarcane bagasse, to produce lactic acid (Katepogu et al., 2023). This bioconversion not only reduces waste but also supports the growing demand for eco-friendly plastics.

In addition to lactic acid, microbial fermentation of organic and agricultural waste can yield other valuable chemicals such as succinic acid, acetic acid, and butanol. Succinic acid, derived from waste biomass, is a versatile compound used in the production of plastics, solvents, and detergents, as well as in pharmaceuticals and food production (Sadare et al., 2021). Furthermore, microbial processes can produce bio-phenols from lignin-rich agricultural residues, which can replace petroleum-based phenols in resins, adhesives, and other industrial products (DSouza et al., 2024, 2025).

Volatile fatty acids (VFAs) are another class of valuable intermediates generated through microbial fermentation of organic waste materials, such as food waste and manure. Anaerobic digestion processes convert these waste streams into VFAs, which serve as precursors for bioplastics, solvents, and other industrial chemicals (Strazzera et al., 2018). VFAs like acetic acid, propionic acid, and butyric acid can be further refined into chemicals such as acetone, butanol, and ethanol, expanding their utility in industrial applications.

5.4 Conversion of Waste into Proteins and Animal Feed Through Microbial Processes

Single-cell protein (SCP) production involves the use of microorganisms like *Saccharomyces cerevisiae*, *Candida* species, and *Trichoderma* to convert organic waste, such as agricultural residues and food scraps, into protein-rich biomass suitable for animal feed or human consumption (Koukoumaki et al., 2024). Through microbial fermentation, waste biomass is metabolized to produce high-quality protein, rich in essential amino acids. For example, food waste or brewery by-products can be transformed into SCP, providing a sustainable and cost-effective protein source for livestock.

The SCP production process leverages advanced fermentation systems to optimize microbial growth and protein yield. Parameters such as pH, temperature, and nutrient availability are tightly controlled to enhance the efficiency of conversion (Rajput et al., 2024). Additionally, waste streams rich in carbohydrates, such as lignocellulosic hydrolysates or starchy residues, can serve as feedstock, maximizing resource utilization. Advances in microbial engineering have further improved the amino acid profile and digestibility of SCP, making it an increasingly viable alternative to conventional protein sources like soymeal and fishmeal in animal nutrition (Jones et al., 2020).

5.5 Economic and Environmental Benefits of Waste Valorization Through Microbial Processes

Waste valorization through microbial processes provides transformative economic advantages by converting waste into valuable resources. Microbial techniques, such as anaerobic digestion and fermentation, enable the production of biofuels, bioplastics, and biochemicals. For instance, agricultural residues can be transformed into bioethanol by microbes like *Saccharomyces cerevisiae*, while waste glycerol from biodiesel production is processed into bioplastics like polyhydroxyalkanoates (PHAs) through engineered bacteria such as *Cupriavidus necator*. These processes not only reduce waste disposal costs but also provide additional revenue streams by creating high-value products. Moreover, microbial valorization techniques consume less energy compared to traditional incineration or chemical treatment, offering cost-effective alternatives for waste-heavy industries such as agriculture and food processing (Geetha et al., 2024).

Beyond cost savings, microbial valorization supports energy production and resource efficiency. Biogas, a renewable energy source generated through anaerobic digestion, offers a sustainable alternative to fossil fuels while providing grid-compatible energy (Kumar & Sharma, 2024). Processes involving volatile fatty acids (VFAs), derived from food waste and manure, produce intermediates for the synthesis of valuable chemicals like acetone, butanol, and ethanol. These outputs further diversify the economic benefits by enabling biorefineries to expand their portfolios. Additionally, investments in microbial processing foster employment opportunities in waste collection, fermentation systems, and bioproduct commercialization, supporting regional economic growth. The modular scalability of microbial technologies also enables their integration across industries, from municipal solid waste management to industrial effluent treatment.

Environmentally, microbial waste valorization significantly reduces landfill reliance, diverting organic and biodegradable waste streams, and mitigating methane emissions from unmanaged anaerobic decomposition (Devi et al., 2024). The microbial processing of heavy metals, such as using metallothionein-expressing microbes for mercury or arsenic detoxification, helps neutralize industrial pollutants that would otherwise contaminate soil and water ecosystems. Processes like bioaugmentation with genetically engineered microorganisms (GEMs) have demonstrated efficiency in breaking down persistent organic pollutants such as pesticides and PCBs, further reducing environmental harm. Additionally, microbial bioplastics production, such as PHAs, offers an eco-friendly alternative to petroleum-based plastics, cutting down non-biodegradable waste accumulation in land and marine ecosystems.

Microbial processes also enrich soil health and agricultural sustainability. For example, biofertilizers and compost generated through microbial treatment of organic waste reduce dependency on chemical fertilizers while restoring soil nutrients. VFAs produced from anaerobic digestion serve as precursors for bio-based solvents and plastics, illustrating a closed-loop system that minimizes resource wastage. Moreover, microbial methods efficiently manage mixed waste streams, including municipal solid waste and agricultural residues, ensuring broader applicability. By leveraging these processes, industries not only achieve waste management goals but also contribute to renewable resource creation and long-term environmental resilience.

6 Challenges and Limitations of Microbial Technologies

Microbial technologies have great potential for environmental sustainability, including waste management, pollutant degradation, and resource recovery. However, there are several technical, regulatory, and social challenges that need to be addressed to ensure the successful implementation and widespread adoption of these technologies (Eissa et al., 2024; Stanton et al., 2021; Weisenfeld et al., 2023).

6.1 Technical Challenges in Implementing Microbial Solutions

Microbial solutions face significant challenges in addressing complex pollutants, many of which resist degradation due to their chemical stability and structural complexity. Persistent organic pollutants (POPs), heavy metals, and plastics often require engineered microbial strains or consortia capable of breaking down these recalcitrant compounds (Gaur et al., 2018). Genetic engineering and co-culturing techniques have been employed to enhance microbial metabolic pathways, enabling efficient conversion of pollutants into non-toxic by-products. For example, engineered microbes expressing enzymes like laccases have demonstrated potential in degrading aromatic hydrocarbons.

Environmental variability further complicates microbial efficacy, as factors like temperature, pH, oxygen levels, and nutrient availability greatly influence microbial activity. Fluctuations in these parameters, particularly in outdoor or natural settings, can significantly affect treatment outcomes. Solutions include the development of resilient microbial strains and the use of bioreactors that allow precise control over environmental conditions. For instance, thermophilic microbes optimized for high-temperature settings ensure consistent pollutant degradation despite external variability.

Scaling microbial technologies from lab to industry introduces challenges related to maintaining optimal growth conditions, controlling microbial populations, and managing nutrient distribution on a large scale. Modular bioreactor designs and multi-stage systems are being explored to address these issues, providing localized optimal conditions within larger setups. Coupled with advanced process control technologies, these systems enhance scalability while maintaining efficiency. At the same time, cost remains a significant hurdle, with high initial investments in infrastructure and operational expenses for growth media and energy inputs. Strategies like using waste feedstocks as substrates, recycling nutrients, and optimizing bioreactor energy efficiency aim to make microbial solutions more economically viable.

Microbial stability and contamination are additional concerns in large-scale applications (Hou et al., 2020; Wehrs et al., 2019). Non-target microbes can infiltrate systems, disrupting treatment processes and reducing efficacy. Maintaining the functional stability of microbial strains over extended periods is also critical. Robust microbial strains engineered for resistance to contaminants, encapsulation technologies, and optimized system designs help mitigate these issues, ensuring consistent performance in industrial-scale implementations. Together, these approaches address the technical challenges of microbial technologies, paving the way for their broader adoption.

6.2 Regulatory and Public Acceptance Issues Related to Genetically Modified Organisms (GMOs)

The implementation of microbial technologies involving genetically modified organisms (GMOs) faces significant challenges stemming from stringent regulatory frameworks and public skepticism. Regulatory oversight varies widely across regions, creating inconsistencies that hinder global applications. For example, the European

Union enforces strict GMO regulations under Directive 2001/18/EC, often leading to bans, while the USA adopts a case-by-case approach overseen by agencies such as the FDA, USDA, and EPA. These discrepancies result in trade barriers, as GMOs approved in one jurisdiction may face restrictions elsewhere. The approval process for GMOs is lengthy and costly, further impeding commercialization. Biosafety concerns, including the risks of gene flow to wild populations or unintended ecological impacts like horizontal gene transfer, necessitate robust containment strategies and post-release monitoring (Anyshchenko & Yarnold, 2021). These requirements significantly increase operational costs. Ethical debates over the morality of genetic modification and disputes regarding intellectual property rights add another layer of complexity, particularly for applications in food production and environmental restoration (Wani & Aftab, 2022).

Public acceptance issues further hinder the deployment of GMO-based microbial technologies. As shown in a GMOs awareness survey conducted on US citizens (Fig. 1), a significant portion of the population cited a lack of information about GMOs, contributing to uncertainty and resistance. While some individuals perceive GMOs as safe, a notable percentage considers them unsafe, reflecting public fears about unknown risks, including potential ecological or health consequences. Misinformation about GMOs as "unnatural" organisms exacerbates these concerns, alongside fears that GM microbes could mutate, spread uncontrollably, or cause unforeseen harm (McFadden, 2016). Negative associations with genetically modified crops have also spilled over into skepticism about non-food applications, such as waste management. Activist groups and cultural or religious beliefs further amplify these concerns, fostering public distrust of biotech companies and regulatory agencies, often perceived as profit-driven. Addressing these challenges requires clear, accessible communication of the benefits, safety measures, and rigorous oversight of GMOs to foster public trust and support.

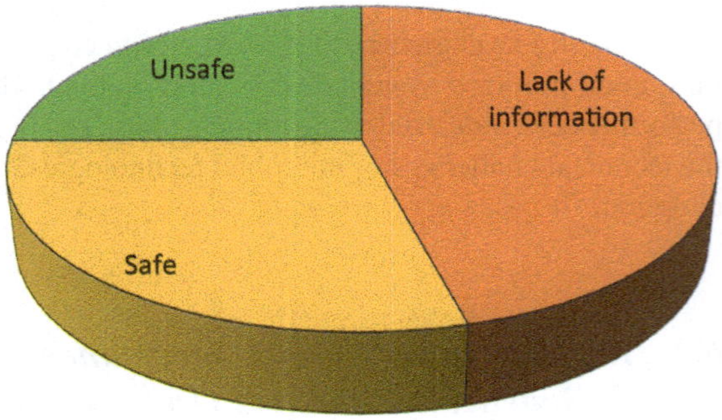

Fig. 1 Results of a GMO awareness survey conducted on US citizens in 2001 (Wunderlich & Gatto, 2015)

7 Future Directions in Microbial Technology

7.1 Emerging Trends and Innovations in Microbial Applications for Waste Management

One of the most promising areas of innovation is the use of multi-microbial consortia, which involve diverse microbial communities working synergistically to degrade complex waste streams. Unlike single-strain systems, consortia can simultaneously target multiple pollutants, making them ideal for applications such as degrading plastics, remediating oil spills, and treating industrial waste. For example, mixed microbial consortia have been employed to break down plastics and hydrocarbons with enhanced efficiency (Nesterov et al., 2024; Ye et al., 2024).

The development of engineered microbial strains through advanced tools like CRISPR/Cas9 and synthetic biology is another transformative innovation. These microbes are designed to degrade recalcitrant materials such as polyethylene, polystyrene, and toxic chemicals. Additionally, engineered strains can convert waste into high-value products, such as bioplastics, biofuels, or even pharmaceuticals, enhancing the economic feasibility of microbial technologies (Balakrishnan et al., 2024; Swaminaathan et al., 2024).

Microbial fuel cells (MFCs) are gaining traction for their ability to generate electricity from organic waste while simultaneously treating it (Nawaz et al., 2022; Sonawane et al., 2024). Electroactive bacteria in MFCs can metabolize organic compounds and transfer electrons to electrodes, enabling waste-to-energy conversion. These systems are particularly useful in decentralized wastewater treatment facilities, addressing both waste management and energy needs. Similarly, microbial CO_2 conversion technologies are being developed to capture and transform carbon dioxide into valuable products like ethanol, methane, or biodegradable polymers, offering a dual benefit of greenhouse gas mitigation and resource recovery (Huo & Wang, 2024).

Efforts in plastic biodegradation and upcycling are advancing rapidly, with microbes engineered to produce enzymes such as PETase and MHETase that break down PET plastics into reusable raw materials. This not only addresses plastic pollution but also facilitates the upcycling of waste plastics into new products. Complementing these efforts are circular bioeconomy solutions, which integrate microbial technologies into waste valorization processes. Food waste can be transformed into biofuels, bioplastics, or animal feed, while bioleaching techniques recover rare earth elements and metals from e-waste, closing resource loops, and reducing environmental impact (Nesterov et al., 2024; Swaminaathan et al., 2024; Ye et al., 2024).

The integration of artificial intelligence (AI) and machine learning is further revolutionizing microbial technology. AI tools are being employed to predict microbial behavior, optimize bioprocess conditions, and design site-specific microbial strains. For instance, AI-driven modeling of microbial consortia has enhanced the efficiency of waste remediation processes by tailoring them to local environmental conditions

(Balakrishnan et al., 2024; Janga et al., 2023). Finally, decentralized waste management systems featuring small-scale, modular microbial bioreactors are emerging as a practical solution for local waste management. These systems reduce transportation costs and emissions associated with centralized waste treatment, making microbial technologies accessible to remote and underdeveloped areas while promoting resource recovery (Sonawane et al., 2024).

7.2 Impact of Synthetic Biology on Environmental Sustainability

Synthetic biology, which involves designing and engineering new biological systems, offers transformative potential for enhancing microbial applications in environmental sustainability. It enables the creation of microbes with enhanced pollutant-degrading abilities, such as breaking down persistent organic pollutants (POPs) like dioxins and PCBs or detoxifying heavy metals and radioactive waste. Synthetic biology also facilitates the precise design of custom microbial consortia, optimizing interactions between species to efficiently target specific pollutants and improve degradation in diverse environments (Tang et al., 2021).

Additionally, synthetic biology drives sustainable production by enabling microbes to convert waste carbon sources into bioplastics like polyhydroxyalkanoates (PHA), reducing reliance on petroleum-based plastics and addressing plastic pollution. In carbon management, engineered microbes are being used to fix CO_2 and convert it into valuable products such as biofuels, chemicals, and synthetic fibers. For example, engineered cyanobacteria produce ethanol and biodegradable materials, mitigating climate change by reducing atmospheric CO_2 levels. Synthetic biology further enhances environmental monitoring through the development of biosensors that detect pollutants such as heavy metals and hydrocarbons in real time, enabling rapid responses to contamination events. Importantly, safety measures like "kill-switch" systems and synthetic nutrient dependencies ensure that engineered microbes cannot survive outside controlled environments, minimizing ecological risks.

8 Conclusion and Recommendations

8.1 Summary of Key Findings and Implications for Future Research

Microbial technologies have emerged as highly effective solutions for waste management, demonstrating their ability to degrade organic waste, reduce environmental pollution, and recover valuable by-products such as bioenergy, bioplastics, and compost. These technologies contribute significantly to lowering greenhouse gas emissions, mitigating landfill dependency, and advancing resource recovery. Despite these benefits, barriers remain, such as the high initial setup costs, limited technical expertise, and challenges associated with scaling microbial systems to industrial levels. Knowledge gaps persist in areas like long-term efficacy, microbial strain optimization, and the integration of microbial technologies with existing waste management systems. Furthermore, supportive policies and financial incentives, such as subsidies and streamlined regulatory frameworks, are critical to promoting their adoption and expanding their application.

8.2 Recommendations for Policy and Practice to Enhance Adoption of Microbial Technologies in Waste Management

To address these challenges, policy recommendations should focus on financial incentives and regulatory support. Providing grants, tax breaks, and subsidies can help offset initial costs, while funding pilot projects can encourage innovation and adoption. Clear regulatory standards and streamlined approval processes are essential to ensure safety, efficiency, and compliance while reducing administrative hurdles. Government investment in research and development can accelerate advancements in microbial strain optimization, scalability, and integration with conventional waste management systems. Nationwide public awareness campaigns can build trust by educating stakeholders about the environmental and economic benefits of microbial technologies and highlighting successful case studies. Policies mandating waste separation at the source can also maximize the efficiency of microbial processing and promote circular economy practices.

In terms of practical measures, capacity building and infrastructure development are paramount. Training programs and certification schemes can enhance the technical expertise of waste management professionals, ensuring high-quality implementation and operation of microbial systems. Investment in modular and scalable waste treatment facilities can expand accessibility, particularly in small-scale and rural applications. Partnerships between private companies, research institutions, and municipal governments should be encouraged to share resources, knowledge, and risk, while fostering international collaborations to adopt best practices. Monitoring and evaluation systems should be established to assess the performance of microbial technologies, facilitating continuous improvement and adaptation. Additionally, the co-benefits of microbial technologies, such as bioenergy production and organic fertilizer generation, should be emphasized to attract investment and community support. Aligning these initiatives with broader sustainability goals, including climate change mitigation and resource recovery, can further enhance their impact.

References

Adebayo, F., & Smart, O. (2018). Microorganisms in Waste Management. *Research Journal of Science and Technology, 10*. https://doi.org/10.5958/2349-2988.2018.00005.0

Ansari, S. A., Kumar, T., Sawarkar, R., Gobade, M., Khan, D., & Singh, L. (2024). Valorization of food waste: A comprehensive review of individual technologies for producing bio-based products. *Journal of Environmental Management, 364*, 121439. https://doi.org/10.1016/j.jenvman.2024.121439

Anyshchenko, A., & Yarnold, J. (2021). From 'mad cow' crisis to synthetic biology: Challenges to EU regulation of GMOs beyond the European context. *International Environmental Agreements: Politics, Law and Economics, 21*(3), 391–404. https://doi.org/10.1007/s10784-020-09516-1

Arora, K., Kumar, V., Nayak, S. J., Surya, A., & Sugumar, S. (2023). Recent advances in biological remediation of volatile organic compounds (VOCs) and heavy metals. *Bio-Nano Filtration in Industrial Effluent Treatment*, 49–68. https://doi.org/10.1201/9781003165149

Balakrishnan, D., Sharma, P., Bora, B. J., & Dizge, N. (2024). Harnessing biomass energy: Advancements through machine learning and AI applications for sustainability and efficiency. *Process Safety and Environmental Protection, 191*, 193–205. https://doi.org/10.1016/j.psep.2024.08.084

Balseiro-Romero, M., Monterroso, C., Kidd, P. S., Lu-Chau, T. A., Gkorezis, P., Vangronsveld, J., & Casares, J. J. (2019). Modelling the ex situ bioremediation of diesel-contaminated soil in a slurry bioreactor using a hydrocarbon-degrading inoculant. *Journal of Environmental Management, 246*, 840–848. https://doi.org/10.1016/j.jenvman.2019.06.034

Bhattacharjee, J., Mishra, S., & Das, A. P. (2021). Recent advances in sensor-based detection of toxic dyes for bioremediation application: A review. *Applied Biochemistry and Biotechnology, 2021*, 1–20. https://doi.org/10.1007/S12010-021-03767-7

Bhowmick, K., Roy, D., Rana, D., Ghosh, A., Sadhukhan, S., Chakraborty, M., Chattopadhyay, D., & Ghosh, T. K. (2024). Potential microbes in bioremediation: A review. *Materials Today Sustainability, 28*, 101032. https://doi.org/10.1016/j.mtsust.2024.101032

Carsanba, E., Papanikolaou, S., & Erten, H. (2018). Production of oils and fats by oleaginous microorganisms with an emphasis given to the potential of the nonconventional yeast Yarrowia lipolytica. *Critical Reviews in Biotechnology, 38*(8), 1230–1243. https://doi.org/10.1080/07388551.2018.1472065

Curiel-Alegre, S., Velasco-Arroyo, B., Rumbo, C., Khan, A. H. A., Tamayo-Ramos, J. A., Rad, C., Gallego, J. L. R., & Barros, R. (2022). Evaluation of biostimulation, bioaugmentation, and organic amendments application on the bioremediation of recalcitrant hydrocarbons of soil. *Chemosphere, 307*, 135638. https://doi.org/10.1016/j.chemosphere.2022.135638

Devi, R., Singh, A. K., Kumar, A., Kumar, R., Rani, S., & Chandra, R. (2024). Development of Technologies for Municipal Solid Waste Management: Current status, challenges, and future perspectives. In A. Gupta, R. Kumar, & V. Kumar (Eds.), *Integrated waste management: A sustainable approach from waste to wealth* (pp. 37–62). Springer Nature Singapore. https://doi.org/10.1007/978-981-97-0823-9_3

Dey, S., Dash, A., & Das, A. P. (2024). Bioremediation as a sustainable and circular solution for mining waste management and bioenergy generation. *Sustainable Management of Mining Waste and Tailings: A Circular Economy Approach*, 107–119. https://doi.org/10.1201/9781003442455-6/BIOREMEDIATION-SUSTAINABLE-CIRCULAR-SOLUTION-MINING-WASTE-MANAGEMENT-BIOENERGY-GENERATION-SUDESHNA-DEY-AKANSHYA-DASH-ALOK-PRASAD-DAS.

DSouza, G. C., Chio, C., Ray, M. B., Prakash, A., Qin, W., & Xu, C. (2024). Synthesis and characterization of biodegradable Kraft lignin-based hydrophilic phenol formaldehyde foams. *Sustainable Materials and Technologies, 41*, e01064. https://doi.org/10.1016/j.susmat.2024.e01064

DSouza, G. C., Dodangeh, F., Venkata, G. B., Ray, M. B., Prakash, A., & Xu, C. (2025). A comprehensive review of biobased polyurethane and phenol formaldehyde hydrophilic foams for environmental remediation, floral, and hydroponics applications. *Biomass and Bioenergy, 192*, 107493. https://doi.org/10.1016/j.biombioe.2024.107493

DSouza, G. C., Sheriff, R. S., Ullanat, V., Shrikrishna, A., Joshi, A. V., Hiremath, L., & Entoori, K. (2021). Fungal biodegradation of low-density polyethylene using consortium of aspergillus species under controlled conditions. *Heliyon, 7*(5), e07008. https://doi.org/10.1016/j.heliyon.2021.e07008

Eissa, F., Zidan, N. E.-H., & El-Banna, A. (2024). Analysis of EU RASFF notifications on genetically modified food and feed from 2002 to 2023. *Journal of Food Composition and Analysis, 136*, 106801. https://doi.org/10.1016/j.jfca.2024.106801

Fatima, Z., Azam, A., Iqbal, M. Z., Badar, R., & Muhammad, G. (2024). A comprehensive review on effective removal of toxic heavy metals from water using genetically modified microorganisms. *Desalination and Water Treatment, 319*, 100553. https://doi.org/10.1016/j.dwt.2024.100553

Fouad, F. A., Youssef, D. G., Shahat, F. M., & Abd El-Ghany, M. N. (2023). Role of microorganisms in biodegradation of pollutants. In G. A. M. Ali & A. S. H. Makhlouf (Eds.), *Handbook of biodegradable materials* (pp. 221–260). Springer International Publishing. https://doi.org/10.1007/978-3-031-09710-2_11

Gaur, N., Narasimhulu, K., & Y, P. (2018). Recent advances in the bio-remediation of persistent organic pollutants and its effect on environment. *Journal of Cleaner Production, 198*, 1602–1631. https://doi.org/10.1016/j.jclepro.2018.07.076

Geetha, B. V., Shreenidhi, K. S., Nethra, L., & Padhmalakshmi, S. (2024). Microbial communities: A promising tool for treatment of wastewater and solid waste. In M. P. Shah (Ed.), *Microbial remediation of hazardous Chemicals from Water & Industrial Wastewater Treatment Plant* (pp. 197–218). Springer International Publishing. https://doi.org/10.1007/978-3-031-62898-6_9

Getino, L., Martín, J. L., & Chamizo-Ampudia, A. (2024). A review of Polyhydroxyalkanoates: Characterization, production, and application from waste. *Microorganisms, 12*(10). https://doi.org/10.3390/microorganisms12102028

Ghosh, S., Tripathy, B., Dey, S., & Das, A. P. (2023). Comparative investigation of fungal and bacterial manganese biomining mechanisms. *Manganese Mining Microorganisms*, 99–115. doi:https://doi.org/10.1016/b978-0-12-822147-1.00016-9.

Hamdan, M. F., & Tan, B. C. (2024). Genetic modification techniques in plant breeding: A comparative review of CRISPR/Cas and GM technologies. *Horticultural Plant Journal.*. https://doi.org/10.1016/j.hpj.2024.02.012

Hou, D., O'Connor, D., Igalavithana, A. D., Alessi, D. S., Luo, J., Tsang, D. C. W., Sparks, D. L., Yamauchi, Y., Rinklebe, J., & Ok, Y. S. (2020). Metal contamination and bioremediation of agricultural soils for food safety and sustainability. *Nature Reviews Earth & Environment, 1*(7), 366–381. https://doi.org/10.1038/s43017-020-0061-y

Huo, D., & Wang, X. (2024). A new era in healthcare: The integration of artificial intelligence and microbial. *Medicine in Novel Technology and Devices, 23*, 100319. https://doi.org/10.1016/j.medntd.2024.100319

Janga, J. K., Reddy, K. R., & Raviteja, K. V. N. S. (2023). Integrating artificial intelligence, machine learning, and deep learning approaches into remediation of contaminated sites: A review. *Chemosphere, 345*, 140476. https://doi.org/10.1016/j.chemosphere.2023.140476

Jones, S. W., Karpol, A., Friedman, S., Maru, B. T., & Tracy, B. P. (2020). Recent advances in single cell protein use as a feed ingredient in aquaculture. *Current Opinion in Biotechnology, 61*, 189–197. https://doi.org/10.1016/j.copbio.2019.12.026

Kannan, U., Pullangott, G., & Ramachandran, J. (2024). Leveraging microbes for waste management. In P. Verma (Ed.), *Industrial microbiology and biotechnology: A new horizon of the microbial world* (pp. 979–997). Springer Nature Singapore. https://doi.org/10.1007/978-981-97-6270-5_32

Kanungo, J., Sahoo, T., Bal, M., & Behera, I. D. (2024). Performance of bioremediation strategy in waste lubricating oil pollutants: A review. *Geomicrobiology Journal, 41*(4), 360–373. https://doi.org/10.1080/01490451.2023.2245395

Kapoor, A., Raghunathan, M., Lal, B., Kumar, P., Srivastava, N., Devnani, G. L., & Pal, D. B. (2024). Sustainable valorization of waste plastic into nanostructured materials for environmental, energy, catalytic and biomedical applications: A review. *Chemosphere, 364*, 143279. https://doi.org/10.1016/j.chemosphere.2024.143279

Katepogu, H., Wee, Y. J., Chinni, S. V., Gopinath, S. C. B., Syed, A., Bahkali, A. H., Elgorban, A. M., & Lebaka, V. R. (2023). Lactic acid production from sugarcane field residue as renewable and economical bioresource by newly isolated Pediococcus pentosaceus HLV1. *Biomass Conversion and Biorefinery, 13*(16), 14927–14937. https://doi.org/10.1007/s13399-022-03267-6

Kaur, K., & Vyas, P. (2024). Microbes for sustainable development of environmental issues. In A. Karnwal & A. R. Mohammad Said Al-Tawaha (Eds.), Microbial applications for environmental sustainability (pp. 1–20). Springer Nature Singapore. https://doi.org/10.1007/978-981-97-0676-1_1.

Koukoumaki, D. I., Tsouko, E., Papanikolaou, S., Ioannou, Z., Diamantopoulou, P., & Sarris, D. (2024). Recent advances in the production of single cell protein from renewable resources and applications. *Carbon Resources Conversion, 7*(2), 100195. https://doi.org/10.1016/j.crcon.2023.07.004

Kumar, M., & Sharma, S. (2024). Renewable energy and sustainable transportation. In R. C. Sobti (Ed.), *Role of science and Technology for Sustainable Future: Volume 1: Sustainable development: A primary goal* (pp. 375–414). Springer Nature Singapore. https://doi.org/10.1007/978-981-97-0710-2_22

Kumar, S., Diksha, S., & S. S., & Kumar, R. (2022). Biofertilizers: An ecofriendly technology for nutrient recycling and environmental sustainability. *Current Research in Microbial Sciences, 3*, 100094. https://doi.org/10.1016/j.crmicr.2021.100094

Kuppan, N., Padman, M., Mahadeva, M., Srinivasan, S., & Devarajan, R. (2024). A comprehensive review of sustainable bioremediation techniques: Eco friendly solutions for waste and pollution management. *Waste Management Bulletin, 2*(3), 154–171. https://doi.org/10.1016/j.wmb.2024.07.005

Lal, D. (2013). Microbes to generate electricity. *Indian Journal of Microbiology, 53*(1), 120–122. https://doi.org/10.1007/s12088-012-0343-2

Lü, H., Wei, J.-L., Tang, G.-X., Chen, Y.-S., Huang, Y.-H., Hu, R., Mo, C.-H., Zhao, H.-M., Xiang, L., Li, Y.-W., Cai, Q.-Y., & Li, Q. X. (2024). Microbial consortium degrading of organic pollutants: Source, degradation efficiency, pathway, mechanism and application. *Journal of Cleaner Production, 451*, 141913. https://doi.org/10.1016/j.jclepro.2024.141913

Matilda, M. I., & Samuel, H. S. (2024). Bioremediation of oil spill: Concept, methods and applications. *Discover Chemistry, 1*(1), 42. https://doi.org/10.1007/s44371-024-00038-2

McFadden, B. R. (2016). Examining the gap between science and public opinion about genetically modified food and global warming. *PLoS One, 11*(11), e0166140. https://doi.org/10.1371/journal.pone.0166140

Mirza, A., Kumar, A., Singh, G., Arya, S. K., Bhalla, A., & Singh, J. (2024). Microbial technology: Tools for waste management; environmental sustainability and environmental safety. In A. Karnwal & A. R. Mohammad Said Al-Tawaha (Eds.), Microbial applications for environmental sustainability (pp. 41–70). Springer Nature Singapore. doi:https://doi.org/10.1007/978-981-97-0676-1_3.

Mokrani, S., Houali, K., Yadav, K. K., Arabi, A. I. A., Eltayeb, L. B., Awjan Alreshidi, M., Benguerba, Y., Cabral-Pinto, M. M. S., & Nabti, E. (2024). Bioremediation techniques for soil organic pollution: Mechanisms, microorganisms, and technologies—a comprehensive review. *Ecological Engineering, 207*, 107338. https://doi.org/10.1016/j.ecoleng.2024.107338

Morlino, M. S., Serna García, R., Savio, F., Zampieri, G., Morosinotto, T., Treu, L., & Campanaro, S. (2023). Cupriavidus necator as a platform for polyhydroxyalkanoate production: An overview of strains, metabolism, and modeling approaches. *Biotechnology Advances, 69*, 108264. https://doi.org/10.1016/j.biotechadv.2023.108264

Nag, M., Lahiri, D., Ghosh, S., Sarkar, T., Pati, S., Das, A. P., Ram, D. K., Bhattacharya, D., & Ray, R. R. (2023). Application of microorganisms in biotransformation and bioremediation of environmental contaminant: A review. *Geomicrobiology Journal*. https://doi.org/10.1080/01490451.2023.2261443

Nawaz, A., ul Haq, I., Qaisar, K., Gunes, B., Raja, S. I., Mohyuddin, K., & Amin, H. (2022). Microbial fuel cells: Insight into simultaneous wastewater treatment and bioelectricity generation. *Process Safety and Environmental Protection, 161*, 357–373. https://doi.org/10.1016/j.psep.2022.03.039

Nesterov, D., Barrera-Martínez, I., Martínez-Sánchez, C., Sandoval-González, A., & Bustos, E. (2024). Approaching the circular economy: Biological, physicochemical, and electrochemical methods to valorize agro-industrial residues, wastewater, and industrial wastes. *Journal of Environmental Chemical Engineering, 12*(5), 113335. https://doi.org/10.1016/j.jece.2024.113335

Paul, A., Dey, S., Ram, D. K., & Das, A. P. (2023). Hexavalent chromium pollution and its sustainable management through bioremediation. Geomicrobiology Journal, 1–11. https://doi.org/10.1080/01490451.2023.2218377.

Perez-Vazquez, A., Barciela, P., & Prieto, M. A. (2024). In situ and ex situ bioremediation of different persistent soil pollutants as Agroecology tool. *PRO, 12*(10). https://doi.org/10.3390/pr12102223

Philp, J. C., & Atlas, R. M. (2005). Bioremediation of contaminated soils and aquifers. In *bioremediation* (pp. 139–236). https://doi.org/10.1128/9781555817596.ch5

Pradhan, G., Tripathy, B., Ram, D. K., Digal, A. K., & Das, A. P. (2023). Bauxite mining waste pollution and its sustainable management through bioremediation. Geomicrobiology Journal, 1–10. doi:https://doi.org/10.1080/01490451.2023.2235353.

Prasad, M., Brar, B., Bala, K., & Singh, N. (2023). Emerging microbial technologies. *Indian Journal of Microbiology, 63*(3), 231–234. https://doi.org/10.1007/s12088-023-01103-7

Prasad, R. K., Chatterjee, S., Mazumder, P. B., Gupta, S. K., Sharma, S., Vairale, M. G., Datta, S., Dwivedi, S. K., & Gupta, D. K. (2019). Bioethanol production from waste lignocelluloses: A review on microbial degradation potential. *Chemosphere, 231*, 588–606. https://doi.org/10.1016/j.chemosphere.2019.05.142

Priyadarsini, S., & Das, A. P. (2024). Analytical and structural characterization of waste lithium-ion batteries for their effective recycling strategy. *Environmental Chemistry and Ecotoxicology.* https://doi.org/10.1016/j.enceco.2024.12.004

Priyadarsini, S., & Das, A. P. (2025). Lithium bioleaching: A review on microbial-assisted sustainable technology for lithium bio-circularity. *Journal of Water Process Engineering, 69*, 106744. https://doi.org/10.1016/j.jwpe.2024.106744

Rafeeq, H., Afsheen, N., Rafique, S., Arshad, A., Intisar, M., Hussain, A., Bilal, M., & Iqbal, H. M. N. (2023). Genetically engineered microorganisms for environmental remediation. *Chemosphere, 310*, 136751. https://doi.org/10.1016/j.chemosphere.2022.136751

Rajput, S. D., Pandey, N., & Keshavkant, S. (2024). Optimization strategies for enhanced production of single cell protein: Recent advances and perspectives. *Reviews in Environmental Science and Bio/Technology, 23*(4), 1015–1040. https://doi.org/10.1007/s11157-024-09706-2

Sadare, O. O., Ejekwu, O., Moshokoa, M. F., Jimoh, M. O., & Daramola, M. O. (2021). Membrane purification techniques for recovery of succinic acid obtained from fermentation broth during bioconversion of Lignocellulosic biomass: Current advances and future perspectives. *Sustainability, 13*(12). https://doi.org/10.3390/su13126794

Saravanan, A., Kumar, P. S., Duc, P. A., & Rangasamy, G. (2023). Strategies for microbial bioremediation of environmental pollutants from industrial wastewater: A sustainable approach. *Chemosphere, 313*, 137323. https://doi.org/10.1016/j.chemosphere.2022.137323

Sharma, M., Bains, A., Goksen, G., Dhull, S. B., Ali, N., Rashid, S., Elossaily, G. M., & Chawla, P. (2024). A review of valorization of agricultural waste for the synthesis of cellulose membranes: Separation of organic, inorganic, and microbial pollutants. *International Journal of Biological Macromolecules, 277*, 134170. https://doi.org/10.1016/j.ijbiomac.2024.134170

Sharma, P., Sirohi, R., Tong, Y. W., Kim, S. H., & Pandey, A. (2021). Metal and metal(loids) removal efficiency using genetically engineered microbes: Applications and challenges. *Journal of Hazardous Materials, 416*, 125855. https://doi.org/10.1016/j.jhazmat.2021.125855

Singh, P., Singh, R., Singh, S., Chauhan, R. S., Bala, S., Pathak, N., Singh, K., & Tripathi, M. (2024). Microbial engineering for a greener ecosystem and agriculture: Recent advances and challenges. *Journal of Pure and Applied Microbiology, 18*, 797–807. https://doi.org/10.22207/JPAM.18.2.23

Sonawane, A. V., Rikame, S., Sonawane, S. H., Gaikwad, M., Bhanvase, B., Sonawane, S. S., Mungray, A. K., & Gaikwad, R. (2024). A review of microbial fuel cell and its diversification in the development of green energy technology. *Chemosphere, 350*, 141127. https://doi.org/10.1016/j.chemosphere.2024.141127

Stanton, J., Rezai, G., & Baglione, S. (2021). The effect of persuasive/possessing information regarding GMOs on consumer attitudes. *Future Foods, 4*, 100076. https://doi.org/10.1016/j.fufo.2021.100076

Strazzera, G., Battista, F., Garcia, N. H., Frison, N., & Bolzonella, D. (2018). Volatile fatty acids production from food wastes for biorefinery platforms: A review. *Journal of Environmental Management, 226*, 278–288. https://doi.org/10.1016/j.jenvman.2018.08.039

Swaminaathan, P., Saravanan, A., & Thamarai, P. (2024). Utilization of bioresources for high-value bioproducts production: Sustainability and perspectives in circular bioeconomy. *Sustainable Energy Technologies and Assessments, 63*, 103672. https://doi.org/10.1016/j.seta.2024.103672

Tamoor, M., Samak, N. A., Jia, Y., Mushtaq, M. U., Sher, H., Bibi, M., & Xing, J. (2021). Potential use of microbial enzymes for the conversion of plastic waste into value-added products: A viable solution. *Frontiers in Microbiology, 12*. https://www.frontiersin.org/journals/microbiology/articles/10.3389/fmicb.2021.777727

Tang, T.-C., An, B., Huang, Y., Vasikaran, S., Wang, Y., Jiang, X., Lu, T. K., & Zhong, C. (2021). Materials design by synthetic biology. *Nature Reviews Materials, 6*(4), 332–350. https://doi.org/10.1038/s41578-020-00265-w

Thamarai, P., Vickram, A. S., Saravanan, A., Deivayanai, V. C., & Evangeline, S. (2024). Recent advancements in biosynthesis, industrial production, and environmental applications of polyhydroxyalkanoates (PHAs): A review. *Bioresource Technology Reports, 27*, 101957. https://doi.org/10.1016/j.biteb.2024.101957

Timmis, K., & Hallsworth, J. E. (2024). This is the age of microbial technology: Crucial roles of learned societies and academies. *Microbial Biotechnology, 17*(5), e14450. https://doi.org/10.1111/1751-7915.14450

Tomei, M. C., & Daugulis, A. J. (2013). Ex situ bioremediation of contaminated soils: An overview of conventional and innovative technologies. *Critical Reviews in Environmental Science and Technology, 43*(20), 2107–2139. https://doi.org/10.1080/10643389.2012.672056

Vishwas, P. A., Kumar, A., & Meena, K. R. (2024). Valorization of Agri-wastes through microbial consortia and its conversion into bioethanol by co-fermentation method using Saccharomyces cerevisiae NCIM 3594 and Pichia stipitis NCIM 3497 strains. *Bioresource Technology Reports, 28*, 101991. https://doi.org/10.1016/j.biteb.2024.101991

Wani, K. I., & Aftab, T. (2022). Limitations, biosafety, ethics, regulatory issues in molecular farming in plants. In K. I. Wani & T. Aftab (Eds.), *Plant molecular farming: Applications and new directions* (pp. 61–74). Springer International Publishing. https://doi.org/10.1007/978-3-031-12794-6_5

Wehrs, M., Tanjore, D., Eng, T., Lievense, J., Pray, T. R., & Mukhopadhyay, A. (2019). Engineering robust production microbes for large-scale cultivation. *Trends in Microbiology, 27*(6), 524–537. https://doi.org/10.1016/j.tim.2019.01.006

Weisenfeld, U., Hauerwaas, A., Elshiewy, O., Halder, P., Wesseler, J., Cingiz, K., & Broer, I. (2023). Beyond plastic—Consumers prefer food packaging derived from genetically modified plants. *Research Policy, 52*(10), 104883. https://doi.org/10.1016/j.respol.2023.104883

Wu, C., Li, F., Yi, S., & Ge, F. (2021). Genetically engineered microbial remediation of soils co-contaminated by heavy metals and polycyclic aromatic hydrocarbons: Advances and ecological risk assessment. *Journal of Environmental Management, 296*, 113185. https://doi.org/10.1016/j.jenvman.2021.113185

Wunderlich, S., & Gatto, K. (2015). Consumer perception of genetically modified organisms and sources of information. *Advances in Nutrition: An International Review Journal, 6*, 842–851. https://doi.org/10.3945/an.115.008870

Ye, Y., Guo, W., Ngo, H. H., Wei, W., Cheng, D., Bui, X. T., Hoang, N. B., & Zhang, H. (2024). Biofuel production for circular bioeconomy: Present scenario and future scope. *Science of the Total Environment, 935*, 172863. https://doi.org/10.1016/j.scitotenv.2024.172863

Zhang, F., Wang, W., Nie, Y., Li, J., & He, X. (2023). From microbial technology to microbiota medicine as a clinical discipline: Sustainable development goal. *Microbial Biotechnology, 16*(9), 1705–1708. https://doi.org/10.1111/1751-7915.14317

Zuroff, T. R., & Curtis, W. R. (2012). Developing symbiotic consortia for lignocellulosic biofuel production. *Applied Microbiology and Biotechnology, 93*(4), 1423–1435. https://doi.org/10.1007/s00253-011-3762-9

Smart and Innovative Technologies for Waste Disposal: Tools for Sustainable Environmental Management

Ankita Mohanty, Aditya Kishore Dash, Akshaya Kumar Verma, and Alok Prasad Das

Abstract The world's trash problem is a long-term issue and is not expected to solve away in the near future. The traditional garbage management systems like incineration, landfill, composting, recycling, waste compaction, underground disposal have its own drawback and are not equipped to manage with the hues waste generated by the current day industrialization and population growth. The environment and human health can be harmed by improper waste collection, treatment, and also disposal. The types of wastes that are generated such as agricultural waste, municipal solid wastes, hazardous waste, biomedical waste, industrial waste, and chemical waste need to be disposed in a non-detrimental way. Waste containing hazardous substances can endanger human health and harm natural ecosystem. To solve this problem, efficient and sustainable waste management is becoming more and more crucial. Hence, to overcome the issue, it is required to adopt smart waste management technologies that can increase the management efficiency, minimum collection cost, and can divert more and more waste away from the sanitary landfills. The smart innovative technologies for management of wastes are like 'smart waste bins, waste level sensors, e-waste kiosks, pneumatic waste pipes, artificial intelligence (AI) recycling robots, garbage truck weighing mechanisms, solar-powered trash compactors, recycling apps', etc. It has been studied that start-ups have received grants of multimillion dollars during the year 2021 in order to develop smart environmentally friendly technologies. It is expected that the smart waste collection market to reach revenue of about $4.5 billion in the year 2027. Further, the state governments are looking to adopt stringent waste recycling regulations. The present study aims to discuss the various smart and innovative waste management techniques for a sustainable and greener environment.

A. Mohanty · A. K. Dash (✉)
Chemistry Department, ITER, Siksha 'O' Anusandhan (Deemed to be University), Bhubaneswar, India
e-mail: adityadash@soa.ac.in

A. K. Verma
Department of Civil Engineering, ITER, Siksha 'O' Anusandhan (Deemed to be University), Bhubaneswar, India

A. P. Das
Department of Life Sciences, Rama Devi Women's University, Bhubaneswar, India

A. P. Das (ed.), *Bioconversion of Environmental Waste to Value Added Products*, Sustainable Environmental Waste Management Strategies,
https://doi.org/10.1007/978-3-032-14215-3_8

Keywords Waste management · Public health · Waste level sensors · AI recycling robots · Pneumatic waste pipes

1 Introduction

Change in lifestyle on top of increasing commercial and industrial growth in many countries world wide in the past few decades has resulted a rapid accumulation of both industrial and municipal solid wastes (Renou et al., 2008). Improper management of solid wastes results in serious environmental, ecological as well and health issues (Biswal et al., 2021; Singh Jadaun et al., 2022). It is reported that up to 95% of the MSW collected were disposed of through landfill methods worldwide (Kurniawan et al., 2006). Landfill generates the polluted liquid called leachate (Parida et al., 2023). In developing nations, waste management sustainability is found to be more complex due to rapid urbanization and industrialization, particularly change in waste generation and composition (Agamuthu et al., 2007). Waste collection rate mostly linked with the area, socioeconomic level of the region, and the waste management becomes more challenging due to complex process of waste collection and disposal, which is further influenced by the political, legal, environmental as well as economic factors of the area (Mishra et al., 2022; Paul et al., 2024; Ghosh et al., 2023).

The rapid increase in waste generation has posed a great challenge for the management of wastes through traditional methods (Behera et al., 2022; Raj & Das, 2023). Hence, there is a need to assess and explore the present advancement in different computational tools for management of wastes. In this context, the cutting-edge waste management technologies like IoT, artificial intelligence, big data analysis, etc., can be adopted. This can help the governments and municipalities to optimize the waste collection, for enhancing resource efficiency, and protection of the environment, there by improving the quality of urban lifestyle (Abdallah, 2020). The different stages for waste management with respect to innovative computational tools are waste generation, waste accumulation, waste collection, and waste disposal (Dey et al., 2023a, b). The various components of waste generation are source classification, waste reduction, prediction for waste generation, and smart charging (Jagatee et al., 2024). Smart bins, wireless communication, capacity design are the components of waste accumulation. Waste collection involves route optimization and dynamic routing. Similarly, final waste disposal includes smart industry technology, separation mechanism, circular economy, and waste to energy (Nesmachnow, 2025).

2 Impact of Waste Disposal on Environment

Waste disposal plays a critical role in environmental sustainability, particularly in mitigating the adverse effects of industrial production. Without effective waste management, excessive waste accumulation causes significant environmental degradation, including soil contamination, water pollution, air quality deterioration, and adverse effects on biodiversity, as well as human and animal health (Chaudhary et al., 2024). According to United Nations Environment Programme (UNEP 2024), global municipal solid waste generation exceeds 1.3 billion tonnes annually, contributing 25% of greenhouse gas emissions. Only a small fraction of this waste is recycled, while the remainder is landfilled or incinerated, exacerbating pollution. Recycling conserves energy, and composting reduces water demand, yet waste treatment methods vary in effectiveness and environmental impact (Parida et al., 2023). Improper waste disposal refers to the careless and unsystematic discharge of waste, including municipal, recyclable, and hazardous materials, which can result in significant environmental and public health hazards. Such practice contaminates air, soil, and water, posing serious health risks to humans and numerous wildlife species (Fig. 1). As reported by Siddiqua et al. (2022), the collection of non-biodegradable and poisonous materials further intensifies the already existing pollution, accelerates the loss of biodiversity, and elevates emissions of greenhouse gases (Dey et al., 2024; Nag et al., 2023). These factors also pose health hazards to the inhabitants in the vicinity of polluted groundwater and gas emissions with carcinogenic and non-carcinogenic effects. These impacts highlight the need for proper planning and strategies focused on the efficient disposal of waste in order to diminish the damage inflicted on nature and promote healthier industrial activities 2025;.

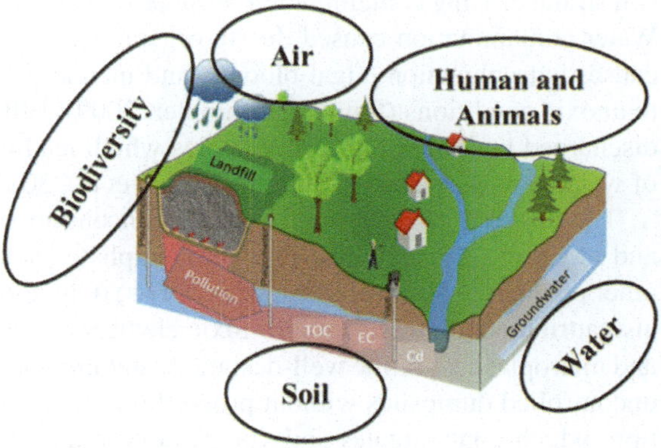

Fig. 1 Various environmental effects related to improper waste disposal (Siddiqua et al., 2022)

2.1 Environmental Pollution and Climate Change

Atmospheric pollution is a problem affecting everyone, and careless waste management practices contribute heavily to the accumulation of pollutants in the environment. The biological breakdown of waste in landfills and open dumping sites emit greenhouse gases like methane, which worsens climate change and environmental degradation (; Priyadarsini & Das, 2024). Landfills account for around 5% of the total emissions of greenhouse gases, primarily due to methane emissions from the anaerobic decomposition of organic wastes (Zhang et al., 2019). According to reports, the amount of carbon dioxide and other pollutants created by burning garbage frequently exceeds the rates of greenhouse gas emissions from a number of conventional energy sources per unit of electricity generated (Tangri, 2023). These practices put both human and animal health at risk while deteriorating air quality due to pollution. Waste that is not adequately handled contributes to climate change resulting ice cap melting leads to sea level rise, putting ecosystems and human infrastructure at risk. Additionally, it makes extreme weather events like hurricanes and floods more frequent. To reduce the negative effects, sustainable waste management techniques must be put into place, such as recycling, trash minimization, and enhanced treatment technologies (Abubakar et al., 2022).

The harmful waste like plastics and chemicals can pollute the water bodies, and there would be any serious impact on the ecosystem as well as on human population. The decomposition of these pollutants generates toxic chemicals, raises water toxicity levels, and endangers biodiversity. Contaminant accumulation in freshwater systems makes them unsuitable for drinking and recreational uses, which has significant ecological and societal health threats (Jambeck et al., 2015). Landfills may ooze hazardous chemicals (leachate) to the ground's surface and can carry through soil strata, causing contamination of subterranean water reserves (Parida et al. 2023). Water contamination caused due to organic waste, pathogens, and nutrient runoff, causes eutrophication, algal blooms, and increased biodepletion processes leading to anoxic conditions (Smith & Schindler, 2009). Effluents from industries are often discharged undiluted into water bodies which lead to the pollution and destruction of water resources and ecosystems (Singh et al., 2023).

The disposal of waste increases the concentration of dissolved organic material and inorganic elements like iron (Fe), sulphate (SO_4^{2-}), chloride (Cl^-) as well as other potentially harmful substances (PTEs) (Cittadino et al., 2020). Soil pollution is also attributed to the disposal of toxic elements, polycyclic aromatic hydrocarbons, and microplastics. While well-designed sanitary landfills solve these problems, other uncontrolled dumpsites without proper liners tend to allow leachate and gas migration, which contaminates soil via infiltration and diffusion (Rouhani & Hejcman, 2024; Mishra et al., 2023). Open burning of waste leads to the formation of ash with highly concentrated deposits of potentially toxic elements (PTEs). These undergo oxidation and the subsequent mobilization by rainwater percolation aids in encapsulating the soil's vegetation, thus entering the food chain manner (Kanmani & Gandhimathi, 2013).

2.2 Biodiversity, Human, and Animal Health

Improper waste disposal practices cause irreparable harm to ecological systems and biodiversity by disrupting the balance of the ecosystem and damaging natural habitats. Poor waste disposal practices cause habitat destruction by polluting air, water, and soil while (Mishra & Das, 2023; Mohanty et al., 2018) posing risks to animals in terms of drowning or choking. Harmful food chain pollutants change the composition of species that comprise food chains and, consequently, jeopardize the balance in the ecosystem (Dubey, 2024). With the accumulation of hazardous waste and plastic pollution, both marine and terrestrial ecosystems suffer a loss of biodiversity. For example, Fig. 2 shows plastic marine pollution as a multifaceted problem impacting coastal and marine resources far beyond ecology, leading to socioeconomic decline in tourism and fisheries-reliant communities. Plastic debris causes loss of biodiversity which alters the trophic structure, biological productivity, and population maintenance of various organisms in different ecosystems (Mishra et al., 2022; Mishra & Das, 2021). Eco-system services crucial for climate moderation, like carbon storage, are further endangered due to this decline (Vergara et al., 2025). Addressing these issues requires more comprehensive approaches such as better education about public policy enforcement, public promotion, effective legislation, and improvement of recycling programmes. Incorporating sustainable methods of waste management is critical to prevent further deterioration of the environment and protect biodiversity for future generations. Figure 2 depicts the multifaceted impacts of plastics on marine environment.

The breakdown of municipal solid waste in landfills leads to the emission of bioaerosols rich in bacteria, fungi, and enteric pathogens, which can significantly impede occupational safety and public health (Bala et al., 2024; Kumar et al., 2023; Mishra & Das, 2022). They are easily inhaled and remain airborne during waste transferring, and prolonged exposure to this increases the risk of respiratory diseases such as chronic obstructive pulmonary disease (Nair, 2021). A study by Ithnin et al. (2013) observed high lead concentrations and challenging respiratory symptoms among students living near an abandoned landfill, exemplifying how pollution affects waste as well as air quality. Adana, uncontrolled urbanization along with the progressive increase in population has raised serious problems related to the collection and disposition of municipal solid waste (MSW) (Unal et al., 2019). The uncontrolled placement of MSW facilities has led to worsening health and humanitarian issues (Unal et al., 2019). Some of the world's most densely populated areas in Africa, particularly Nigeria, Kenya, and Sub-Saharan Africa, face simultaneous increases in resource consumption and waste production, as well as extremely deficient waste collection, treatment, and disposal systems (Kumar & Agrawal, 2020). Poor waste management practices aggravate pollution and increase the risk of various respiratory diseases (Ray et al., 2024). Attaining these objectives requires the adoption of environmentally responsible waste management techniques. Various sources of wastes to garbage open dumping and burning with possible environmental and health hazards are depicted in Fig. 3.

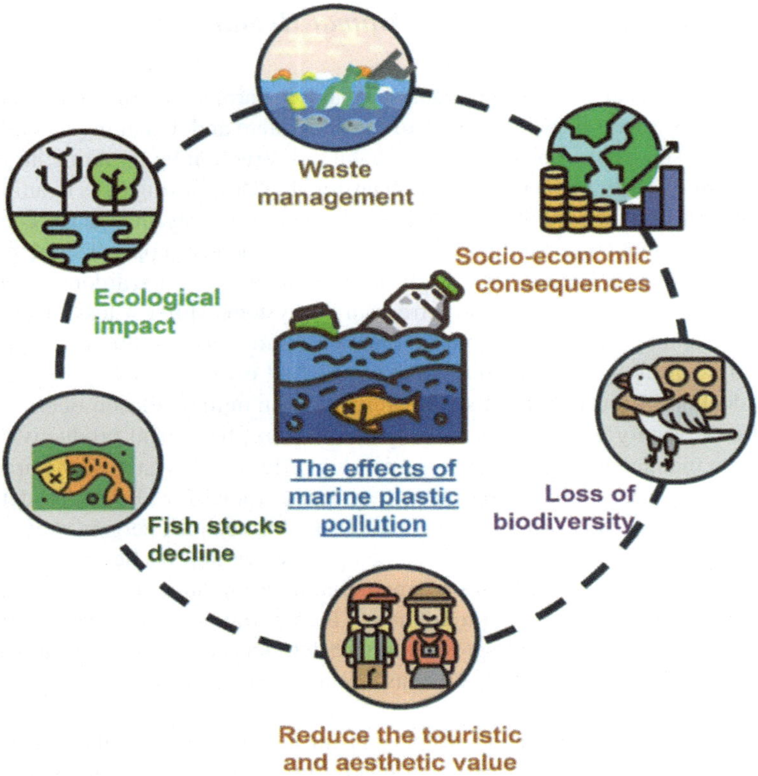

Fig. 2 Multifaceted impacts of marine plastic pollution (Vergara et al., 2025)

3 Innovative Solutions for Waste Management for a Clean and Green Future

Traditional waste disposal methods were unscientific, due to which it leads to environmental degradation causing harm to ecosystem and human health. Innovative solutions for waste management promote a clean and green future. In the process of waste-to-energy technology, the electricity is generated from the wastes. Advanced recycling techniques help to promote the environmental purification. Other innovative techniques for better management of wastes are smart waste bin and sensor, data analysis system for smart waste management, waste level sensor, AI recycling robots, garbage truck weighing machines, pneumatic waste pipes, solar-powered trash compactors, e-waste kiosks.

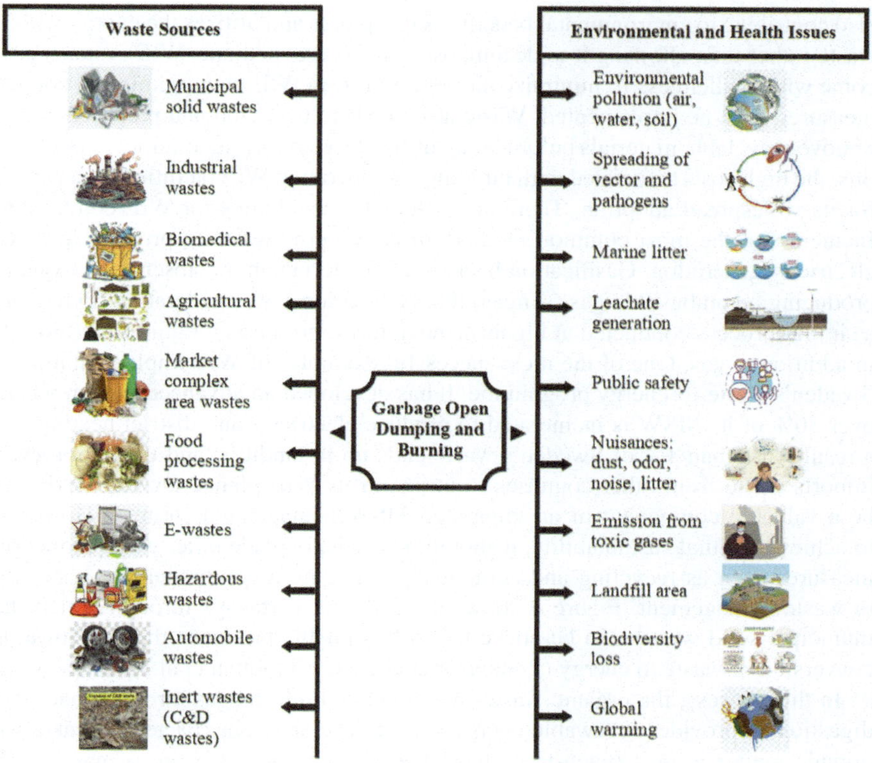

Fig. 3 Sources of wastes to garbage open dumping and burning with possible environmental and health hazards

3.1 *Waste-To-Energy Conversion*

By converting waste into usable energy, primarily electricity, through processes like incineration, waste-to-energy (WtE) plays a major role in municipal solid waste (MSW) management. This approach not only reduces waste volume sent to landfills but also provides an alternative energy source, thus serving as a dual-purpose solution for waste disposal and energy generation (Ghosh & Das, 2017; Sahu et al., 2024). For example, India has the potential to generate 5690 MW of electricity from municipal solid waste and industrial waste, but as of May 2023, only 556 MW of this capacity had been realized (Times of India, 2023). WtE offers several advantages in MSW management. These facilities significantly reduce the amount of solid waste that requires land filling through incinerating waste in a controlled environment, thus conserving space and potentially lowering landfill gas emissions. This WtE technology offers an alternative source of renewable energy as the heat is produced from waste combustion that produces electricity. When properly managed, by reducing greenhouse gas emissions compared to traditional land filling, WtE can also contribute to environmental benefits as it captures and utilizes the energy within waste rather than allowing it to decompose and release methane. WtE systems also come with challenges. To minimize air pollution from WtE plants, emission control measures must be implemented. Waste also needs to be sorted and preprocessed to remove recyclable materials before being utilized for energy generation. Apart from this, the high costs associated with building and operating WtE facilities are a barrier for its widespread adoption. There are different technologies for WtE conversion. Incineration, the most common method, directly burns waste to produce heat for electricity generation. Gasification heats waste materials in the absence of oxygen, producing a combustible gas (syngas) that is used for power generation. Pyrolysis, a similar process conducted at higher temperatures, produces a liquid fuel (bio-oil) in addition to gas. One of the most successful examples of WtE implementation is Sweden's waste-to-energy programme. It has developed an advanced system where over 50% of its MSW is incinerated to produce electricity and district heating. As a result, less than 1% of Sweden's waste ends up in landfills, and the nation even imports waste from other countries to utilize in its WtE plants. Overall, WtE can be a valuable component of an integrated MSW management strategy. However, to achieve optimal sustainability, it should be used alongside other waste reduction measures such as recycling and composting, ensuring a comprehensive approach to waste management. Figure 4 shows the different pathways through which the municipal solid wastes can be converted in to valuable products, the step towards conversion of waste to energy (Prabhakar et al., 2019; Pradhan et al., 2023).

In this process, the organic wastes are converted into biogas through anaerobic digestion. It provides renewable energy source. The anaerobic bacteria breakdown organic matter in an oxygen-free environment releasing a mixture of gases such as CH_4 and CO_2. In this process, the organic wastes such as food scraps, animal manure, and agricultural residues are placed inside a digester chamber. In the presence of anaerobic bacteria, these organic wastes are digested to release biogas. In waste-to-energy conversion, there are three vortices at the waste triangle, i.e. plastics, pollutants, and green gases. Household waste contains hydrocarbons that can be transformed into electricity, which is used in thermoelectric anaerobic digestion and thermoelectric plants (Fan et al., 2021). It damages the infrastructures of buildings and roads, which is transferred into heat energy storage materials (Ho et al., 2021). This process involves in different stages. These are hydrolysis, acidogenesis, acetogenesis, and finally methanogenesis. In each stage different types of bacteria digest to form biogas which is a renewable energy and can be used for heating, electricity generation as a fuel. Some new waste-to-energy conversion technology includes gasification, hydrothermal carbonization, and dendro liquid energy.

Gasification: It is a thermochemical process by which the waste is converted into a combustible synthetic gas in high temperature called syngas. The syngas is used as fuel for transportation, which is alternative to natural gas. It produces less pollution. Gasification also has many benefits and uses that reliable technology used to produce transportation fuels, electricity, chemicals, etc.

Hydrothermal Carbonization: In this process, the waste is broken down by industrial-sized pressure cooker by applying heat and pressure. It is considered as

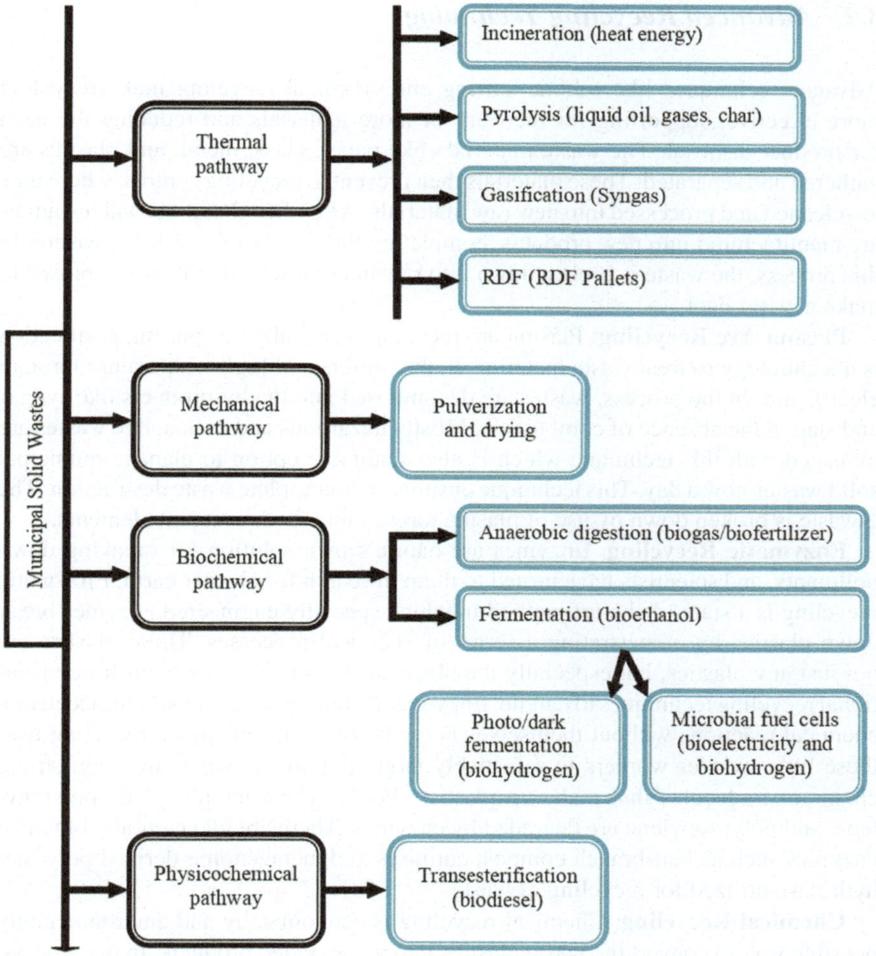

Fig. 4 Various pathways showing the conversion of municipal solid wastes in value-added products

an emerging waste-to-energy technology. It is a thermochemical process that uses biomass and organic waste to convert from heat and pressure. It also takes place in the presence of water, and it forms three products, like solid, gas, and liquid.

Dendro Liquid Energy: It is also a WtE conversion technology by which the waste is broken down into fuel by moderate temperature. The plants can work with both wet and dry wastes. It can handle various types of waste products, i.e. wood, plastic, and other organic material. It helps municipalities to manage their waste in a sustainable or sustainable future.

3.2 Advanced Recycling Technology

Advance techniques like robotic sorting and chemical recycling make recycling more effective, supporting for recovery of more materials and reducing the need for product disposal. The waste materials like paper, glass, metal, and plastics are gathered and separated. These materials then prevent to recycling centres, where they are cleaned and processed into new raw materials. After that, these renewal resources are manufactured into new products, completing the cycle from trash to treasure. In this process, the waste is broken down into raw materials, so that it can be reused to make new product.

Plasma Arc Recycling: Plasma arc recycling also called as plasma gasification is a technology to treat waste materials in the presence of high temperature through electric arc. In the process, wastes are decomposed into the by-products like syngas and slag in the absence of combustion. Mostly hazardous and radioactive wastes are managed with this technique which is also a suitable option to manage municipal solid wastes now a day. This technique ensures near-complete waste destruction. The e-waste is broken down by use of plasma torches into its constituent elements.

Enzymatic Recycling: Enzymes are nature's own solution for breaking down pollutants, and scientists have turned to them in search for cleaner earth. Enzymatic recycling is a state-of-the-art method in which specially engineered enzymes break down plastics by orchestrating a series of chemical processes. These plastics are not just any plastics, but especially the obstinate ones that are difficult for conventional recycling techniques to handle. Enzymes are biological catalysts that accelerate chemical reactions without themselves being consumed. Enzymatic recycling uses these little wonder workers to selectively target polymers, which are long, strong chains of molecules that make up plastics. Polyethylene terephthalate, polyethylene, and polypropylene are degraded by enzymes. There are two naturally obtained enzymes such as leaf-branch compost cutinises and metagenome-derived polyester hydrolase utilized for recycling of waste.

Chemical Recycling: Chemical recycling is economically and environmentally possible way to convert the waste plastics into value-added products. In the process, liquid oils with high heating values, different carbon materials, hydrogen, syngas, and other chemicals are recovered from plastic waste, by applying the techniques such as pyrolysis, reforming, gasification, and chemolysis. The plastic waste streams which are not recycled mechanically can recycle by chemical recycling. In this process, the plastic waste streams can be converted back into feedstock for the chemical industry and allocated to the products manufactured in BASF'S integrated chemical production network through a mass balance approach. In the process of chemical recycling, the plastic wastes are broken down by heat or chemicals. Depending on the type of plastic and desired product, the reaction is carried out involving heat, pressure, solvents, or catalysts. The end goal is to decompose long, stubborn polymer chains into simpler and versatile monomers. Then it can be purified and re-polymerized into new, high-quality plastics.

Mobile Recycling Units: Mobile recycling units are convenient for people who don't have access to a recycling centre and cannot transport their recyclables. In this service, the unit comes to home to collect recyclable materials like e-wastes. Types of mobile recycling units are:

Trashpresso: It is a solar-powered and AI-powered recycling machine which converts plastic waste into construction tiles. It involves three steps that are: size reduction, reshaping, and purification. It protects the environmental discharge from the process with air and water filtration.

Mobile C&D Waste Recycling Plant: It is a small, self-contained system mint for waste material processing on the spot which minimizes the transportation cost. It has several units such as crushing unit, screening unit, and dust control system. They help for conversion of waste into valuable products.

Hook-Type Mobile Recycling Machines: It can be loaded and transported on a truck and easy to handle. These units can be moved to different locations to respond the waste generation. Many mobile machines are equipped with hybrid technology.

Product Distribution Service: Product distribution service for waste management describes to the logistical process of collecting, transporting, and delivering different types of waste materials. It involves in following such as waste collection, waste segregation, transportation, and disposal. It is also made from natural materials that are grown without chemicals and fertilizers. It is also more sustainable because they don't produce harmful toxins.

Zero Waste Initiatives: It is a holistic step for addressing the issues of unsustainable resource flow. The aim of zero waste is to reduce waste, promote reuse, and keep materials out of landfills. Adopting zero waste policy can significantly minimize the environmental footprint of a business which can lower down the emission of greenhouse gases, reduces pollution load, and also conserves natural resources. It includes policies, programmes, and every day practices. Zero waste initiatives have several aspects like:

Policy Development: So many policies are developed to create strategies to reduce waste, promote recycling, and reuse.

Closed-Loop System: The systems are designed to keep product and materials in use for as long as possible.

Reduce Consumption: To create zero waste, we have to reduce consumption and avoiding single use plastics.

3.3 Smart Waste Bin and Sensor

Utilizing advanced sensors and a data analysis system is a clever and sophisticated method of keeping rubbish in the bins. This is to control the bins' capacity to fill, in which the sensor determines the waste kind and filling level. When the bin reaches its limit, an alert message is send to the data analysis system by the sensor. It is very easy to manage the waste, and ultimately it improves the waste management system by managing real-time data of type of waste and level or quantity of waste in the bin.

The smart waste bin sensors are various types, i.e. ultrasonic sensor for measure filled level, infrared sensors for distinguish the type of waste, weight sensors for weighing of waste, capacitive sensors for detection of changes in waste composition, and RFID sensors for tracking different types of bins and waste types. It is used to first detect the type of waste, i.e. e-waste or domestic waste or biomedical waste or industrial waste, etc., and then the waste automatically stored in the separate containers which are already under the conveyor belt que. After that, when the container filled up to the upper level. Then it sends the alert to the system to stop the filling process of container. The sensors data is transmitted by the help of Bluetooth or Wi-Fi. Because of this, it prevents from overflow and reducing environmental impact from the hazardous wastes.

3.4 Data Analysing System for Smart Waste Management

For storing and managing the waste related informations, a data analysis system is used. It also tracks the number and waste disposal habits such as recycling, reusing, and composting in a realtime sensor. The smart waste bins data can be easily accessed from the control panel, and due to the stored data, the control panel easily understands the waste trends and habits of the waste. The various types of analysers used are image-based analysers, weight-based analysers, sensor-based analysers (including near infrared, Raman spectroscopy), and combined systems.

3.5 Waste Level Sensor

Businesses centres and houses over the nation depend on the routine waste collection service to discard of their garbage. Although weekly services have existed for many years, they aren't always the best choice. Waste level sensor also called as fill-level sensor or ultrasonic sensors tracks the amount of waste in bins and containers. It can be water-resistant and can accurately transfer data in different conditions. It is most beneficial for waste management and can help cities and businesses centres to understand the waste production and optimize collection routes. This device is most efficient among other devices. It can help in maximizing efficiencies by monitoring waste capacity demands. These are different types of waste level sensors like:

Recycle Bin Ultra Eye Sensor: It can be mounted on bin lids, poles, or other customized arrangements.

Sensoneo Smart Sensor: It is an ultrasonic sensor for monitoring the fill levels in different waste containers.

Smart Bin Sensor: Sensor transmits data in real time to a smart waste management platform.

3.6 *AI Recycling Robots*

Recycling equipments are necessary to lowering the quantity of waste that reaches landfills and waterways annually. Many facilities are finding it difficult to meet demand as a result of a decreased workforce during the COVID-19 pandemic. Tool like artificial intelligence (AI)-powered recycling robots can help and make up part of the difference. Designed to precisely find and sort recyclable items, these robots boost efficiency and lower the need for human labour. Long-term financial savings for recycling facilities are also achieved, as materials that would otherwise wind up in landfills are diverted.

3.7 *Garbage Truck Weighing Mechanisms*

The waste and recycling industry is dependent on careful measurement for monitoring the volume transported through trucks. In order to weight the collected waste in real time, an 'onboard weighing system' is installed on the garbage tank. By measuring and documenting garbage containers, they then use the data to project fill levels over time. By adopting this technology, cities may more precisely project how often they need to send their trucks and reduce their annual collecting costs.

3.8 *Pneumatic Waste Pipes*

These are underground pipes that use air pressure to transport waste to a central collection point. It is an alternative to the traditional waste collection systems. In this process, the fans generate air that creates negative pressure in the pipes. Then the air enters into the pipes at atmospheric pressure and the solid wastes are trapped into it. The waste is transported to collection stations and into sealed containers. It improves sanitation and environmental conditions. These pipes are used in residential, roadside, and other locations. The pipes carry trash to a waste collection facility, where it is either sorted or removed. This system lowers energy expenses, boosts overall efficiency, and does away with the need for conventional waste collection.

3.9 *Solar-Powered Trash Compactors*

A solar-powered trash compactor can hold up to five times as much as conventional trash cans. It includes solar panel, battery storage, a compaction machine, a fill-level sensor, and a control unit trash. The bin can hold significantly more waste before needing to be emptied, minimizing collecting trips. It also helps to reduce environmental pollution, and it is a most cost-saving device.

3.10 E-Waste Kiosks

An e-waste kiosk is a designated drop-off point usually present in public places like shopping malls, community centres, etc. The e-waste is disposed properly and it undergoes recycling, minimizing the harmful substances properly. These kiosks collect items like old phones, computers, TVs, and small appliances. These devices reduce the amount of hazardous materials entering landfills. It also helps for extraction of valuable metals like copper, gold, and silver by the extraction process. It provides a user-friendly way for people to dispose e-waste without the need to actively search for recycling facilities. The important references that are reviewed in this article under different heads like route optimization, application of computational intelligence with respect to waste classification and identification, role and functioning of smart bins, waste management applying IoT and communications are reflected in Table 1 (Nesmachnow et al., 2025). The table also includes the study concept with their key limitations and the corresponding references under each head.

4 Conclusion

Effective municipal solid waste management has become more critical than ever with rapid urbanization, population growth, and increasing waste generation. In order to minimize environmental impact, the future of waste management must focus on circular economy principles, sustainability, and technological advancements. Integration of new technologies which convert low-calorific waste into energy at a reasonable cost is essential, as they can aid in the cleanup of ageing landfill sites while simultaneously generating energy. Waste-to-energy technologies have redefined waste as a resource rather than a burden. However, the composition of waste is quite crucial in energy recovery, as the existing WtE systems function best with high-calorific-value waste. Garbage-to-energy solutions are emerging as a viable alternative to traditional disposal methods in India, as increasing amounts of high-calorie waste are ending up in landfill site. The future in MSW management involves a multi-pronged approach that integrates technology, policy, community participation, and sustainability principles. In order to establish a cleaner, greener, and more livable environments, cities must prioritize resource recovery, waste minimization, and eco-friendly disposal techniques. Municipalities can enhance waste management efficiency while reducing their environmental footprint through adoption of strategies such as improved waste segregation, circular economy-based recycling programmes, and innovative waste processing methods in addition to the emerging technologies like IoT, artificial intelligence, and big data analysis.

Table 1 Study concept and their key limitations for different aspects of waste management like smart waste management, computational intelligence for waste classification and waste identification, IoT system and communications for waste management and smart bins

Study concept	Key limitations	References
Smart waste management with respect to route optimization techniques		
IoT-based WMS for optimizing the schedule and routes for waste collection	There is no comparison with similar methods	Idwan et al. (2020)
Real-time bin status for improving collection efficiency of the waste and to reduce cost of collection and emission	Did not presented any real evaluation	Hannan et al. (2020)
Metaheuristic optimization for e-waste vehicle routing	No report on validation over actual scenarios	Nowakowski et al. (2020)
Metaheuristic for optimizing routs and schedule for vehicles to collect from waste bins	Real validation in the area was reported	Bin et al. (2021)
Used real data for waste collection routing	There is no description on real implementation or validation	Burduk et al. (2022)
Sustainable waste collection routes identification framework	There is no real implementation	Hu et al. (2023)
Autonomous emptying process approach and to predict the filling levels of the bin	No methods for route optimization have been studied for efficient time and energy for the robots	Pollak et al. (2024)
Computational intelligence applied for waste classification and waste identification		
IoT for detection of human activity and classification of waste management	No implication or validation	Alqahtani et al. (2020)
ANN for waste classification and identification	No real evaluation, accurate baseline methods	Alsubaei et al. (2022)
Automatic visual warning and identification for illegal waste disposal	Performed with less dataset, real implementation was not presented	Liu et al. (2023)
Waste classification by using IoT, edge/cloud computing, and deep learning	No real implementation/ validation	Cheema et al. (2022)
Waste classification applying deep learning	No state-of-the-art classification, only ad hoc database	Kumar et al. (2023)
IoT system and communications for management of wastes		
IoT application as a multilevel architecture for smart city waste management	Communication metrics evaluation only	Marques et al. (2019)
Real-time smart bin monitoring applying IoT and AI	No real evaluation	Bano et al. (2020)

(continued)

Table 1 (continued)

Study concept	Key limitations	References
Smart bins	Chances of theft and damage	Jhaveri et al. (2021)
Smart waste removal system based on rapid exploration and random trees	Validated only in few case studies	Zhang et al. (2020)
Smart bin for communal WMS	No deployment or real evaluation	Azyze et al. (2022)
GA-fuzzy method for smart waste management	No real scenario, only done by simulation	Thaseen et al. (2023)
IoT-based household WMS	No real scenarios, simple evaluation	Wong et al. (2022)
Smart bins		
Household e-waste collection smart bin	No real validation/ implementation	Kang et al. (2020)
Underground structure for landfill for waste collection and separation	Proposed system has no real evaluation	Majidi (2021)
Waste disposal prediction and monitoring with smart bin using IoT	Real implementation was not reported	John et al. (2021)
Deep learning and IoT for improving collection rate and minimizing cost	Need few training data, GPU, lifetime of the batteries	Sallang et al. (2021)
Smart cities self-powered WMS, smart bin using IoT sensors	No real implementation	Thamarai et al. (2023)

References

Abdallah, M., Talib, M., Feroz, S., Nasir, Q., Abdalla, H., & Mahfood, B. (2020). Artificial intelligence applications in solid waste management: A systematic research review. *Waste Management, 109*, 231–246.

Abubakar, I. R., Maniruzzaman, K. M., Dano, U. L., AlShihri, F. S., AlShammari, M. S., Ahmed, S. M. S., Al-Gehlani, W. A. G., & Alrawaf, T. I. (2022). Environmental sustainability impacts of solid waste management practices in the Global South. *International Journal of Environmental Research and Public Health, 19*(19), 12717. https://doi.org/10.3390/ijerph191912717

Agamuthu, P., Fauziah, S.H., Khidzir, K.M., & Noorazamimah Aiza, A. (2007). Sustainable waste management—Asian perspectives. In *Proceedings of the International Conference on Sustainable Solid Waste Management* (pp. 15–26).

Alqahtani, F., Al-Makhadmeh, Z., Tolba, A., & Said, W. (2020). Internet of Things-based urban waste management system for smart cities using a cuckoo search algorithm. *Cluster Computing, 23*, 1769–1780.

Alsubaei, F., Al-Wesabi, F., & Hilal, A. (2022). Deep learning-based small object detection and classification model for garbage waste management in smart cities and IoT environment. *Applied Sciences, 12*, 2281.

Azyze, N., Isa, I., & Chin, T. (2022). IoT-based communal garbage monitoring system for smart cities. *Indonesian Journal of Electrical Engineering and Computer Science, 27*, 37.

Bala, K., Anshul, Verma, S., Dogra, A., Das, A. P., Kumar, R., & Nandini. (2024). Sustainable management for urban plastic waste generation. In *Sustainable management of urban plastic waste through circular economic approaches* (pp. 211–236). https://doi.org/10.1201/9781003465447-13

Bano, A., Ud Din, I., & Al-Huqail, A. (2020). A IoT-based smart bin for real-time monitoring and management of solid waste. *Scientific Programming, 2020*, Article 613263.

Behera, I. D., Nayak, M., Mishra, A., Meikap, B. C., & Sen, R. (2022). Strategic implementation of integrated bioaugmentation and biostimulation for efficient mitigation of petroleum hydrocarbon pollutants from terrestrial and aquatic environment. *Marine Pollution Bulletin, 177*, Article 113492. https://doi.org/10.1016/J.MARPOLBUL.2022.113492

Bin, C., Xinghan, C., Zhihan, L., Ruichang, L., & Shanshan, F. (2021). Optimization of classified municipal waste collection based on the internet of connected vehicles. *IEEE Transactions on Intelligent Transportation Systems, 22*, 5364–5373.

Biswal, P., Ghosh, S., Pal, A., & Das, A. P. (2021). Exploration of probiotic microbial biodiversity in acidic environments (curd) and their futuristic pharmaceutical applications. *Geomicrobiology Journal*. https://doi.org/10.1080/01490451.2021.1956020

Burduk, A., Łapczyn´ska, D., Kochan´ska, J., Musiał, K., Wie˛cek, D., & Kuric, I. (2022). Waste management with the use of heuristic algorithms and Internet of Things technology. *Sensors, 22*, 8786.

Cheema, S., Hannan, A., & Pires, I. (2022). Smart waste management and classification systems using cutting edge approach. *Sustainability, 14*, 10226.

Choudhary, M., Singh, D., Parihar, M., Choudhary, K. B., Nogia, M., Samal, S. K., & Mishra, R. (2024). Impact of municipal solid waste on the environment, soil, and human health. In *Agricultural soil degradation and restoration. Waste management for sustainable and restored agricultural soil* (pp. 33–58). Academic Press. ISBN 9780443184864.

Cittadino, A., Ocello, N., Majul, M. V., Ajhuacho, R., Dietrich, P., & Igarzabal, M. A. (2020). Heavy metal pollution and health risk assessment of soils from open dumps in the Metropolitan Area of Buenos Aires, Argentina. *Environmental Monitoring and Assessment, 192*(5), 291. https://doi.org/10.1007/s10661-020-8246-x

Das, A., Dey, S., Gacem, A., Yadav, K. K., Nath, H., Ram, D. K., & Das, A. P. (2025). Revolutionizing solid hydrocarbon sourcing: Plastic's journey from waste to upcycled treasure. *Environmental Science and Engineering, Part F218*, 331–351. https://doi.org/10.1007/978-3-031-81615-4_14

Dey, S., Bissoyi, A., Koteswara Reddy, G., & Das, A. P. (2024). Synthetic microfibers: An urban emerging pollutant demanding for sustainable environmental solutions. *Sustainable Microbial Technology for Synthetic and Cellulosic Microfiber Bioremediation*, 107–121. https://doi.org/10.1007/978-3-031-62600-5_7

Dey, S., Das, A., Mallick, K., Sahu, A., & Das, A. P. (2023a). *Environmental petroleum waste: Pollution, toxicity, sustainable remediation* (pp. 159–175). https://doi.org/10.1007/978-3-031-48220-5_7

Dey, S., Das, A., Mallick, K., Sahu, A., & Das, A. P. (2023b). *Environmental petroleum waste: Pollution, toxicity, sustainable remediation* (pp. 159–175). https://doi.org/10.1007/978-3-031-48220-5_7

Dubey, D. K., Agarwal, S., Yadav, M. P., Goswami, A., & Ravat, A. (2024). The study of the effects of improper hazardous waste disposal on ecosystems. *Indian Journal of Science and Research, 4*(4), 130–136. https://www.researchgate.net/publication/385241984

Fan, Z., Zhang, Y., Pan, L., Ouyang, J., & Zhang, Q. (2021). Recent developments in flexible thermoelectric: From materials to devices. *Renew. Sustain. Energy Revs., 137*, Article 110448. https://doi.org/10.1016/j.rser.2020.110448

Ghosh, S., & Das, A. P. (2017). Bioleaching of manganese from mining waste residues using *Acinetobacter* sp. *Geology, Ecology, and Landscapes, 1*(2), 77–83. https://doi.org/10.1080/24749508.2017.1332847

Ghosh, S., Sinha, J. K., Ghosh, S., Vashisth, K., Han, S., & Bhaskar, R. (2023). Microplastics as an emerging threat to the global environment and human health. *Sustainability, 15*(14), 10821. https://doi.org/10.3390/SU151410821

Hannan, M., Begum, R., Al-Shetwi, A., Ker, P., Al Mamun, M., Hussain, A., Basri, H., & Mahlia, T. (2020). Waste collection route optimisation model for linking cost saving and emission reduction to achieve sustainable development goals. *Sustainable Cities and Society, 62*, Article 102393.

Ho, H.-J., Iizuka, A., & Shibata, E. (2021). Chemical recycling and use of various types of concrete waste: A review. *Journal of Cleaner Production* 124785. https://doi.org/10.1016/j.jclepro.2020.124785

Hu, S., An, L., & Shen, L. (2023). A multi-objective modeling and optimization approach to municipal solid waste collection for classified treatment in China towards sustainable development. *Sustainable Cities and Society, 98*, Article 104846.

Idwan, S., Mahmood, I., Zubairi, J., & Matar, I. (2020). Optimal management of solid waste in smart Cities using Internet of Things. *Wireless Personal Communications, 110*, 485–501.

Ithnin, A., Abd Rahman, M. S., Awang, N., Mohd Yusuf, N., Abdullah, R., & Ariffin, F. D. (2013). Study on air quality in a school located near a former landfill site and its influences on students' respiratory health. *Middle-East Journal of Scientific Research, 14*, 371–374.

Jagatee, S., Priyadarshini, S., Rath, C. C., & Das, A. P. (2024). Synthetic microfiber: An enduring environmental problem linked to sustainable development. *Environmental Science and Engineering, Part F2333*, 93–112. https://doi.org/10.1007/978-3-031-51792-1_6

Jambeck, J. R., Geyer, R., Wilcox, C., Siegler, T. R., Perryman, M., Andrady, A., & Law, K. L. (2015). Plastic waste inputs from land into the ocean. *Science, 347*(6223), 768–771.

Jhaveri, R., Shah, P., Patel, N., Patel, D., Thanki, S., Bhoi, A., & Bhatia, J. (2021). An ICT-based solid waste management system for smart cities: A case of municipality in India. *International Journal of Hoc and Ubiquitous Computing, 38*, 158.

John, J., Varkey, M., Podder, R., Sensarma, N., Selvi, M., Santhosh, S., & Kannan, A. (2021). Smart prediction and monitoring of waste disposal system using IoT and cloud for IoT based smart cities. *Wireless Personal Communications, 122*, 243–275.

Kang, K., Kang, H., Ilankoon, I., & Chong, C. (2020). Electronic waste collection systems using Internet of Things (IoT): Household electronic waste management in Malaysia. *Journal of Cleaner Production, 252*, Article 119801.

Kanmani, S., & Gandhimathi, R. (2013). Assessment of heavy metal contamination in soil due to leachate migration from an open dumping site. *Applied Water Science, 3*, 193–205. https://doi.org/10.1007/s13201-012-0072-z

Kumar, A., & Agrawal, A. (2020). Recent trends in solid waste management status, challenges, and potential for the future Indian cities—A review. *Current Research in Environmental Sustainability, 2*, Article 100011. https://doi.org/10.1016/j.crsust.2020.100011

Das, D. (2023). Maximization of energy recovery from starch processing wastewater by thermophilic dark fermentation coupled with microbial fuel cell technology. *Geomicrobiology Journal*, 1–12. https://doi.org/10.1080/01490451.2023.2209555

Kurniawan, T. A., Lo, W. H., & Chan, G. Y. S. (2006). Physicochemical treatments for removal of recalcitrant contaminants from landfill leachate. *Journal of Hazardous Materials, 129*(1–3), 80–100.

Liu, J., Prabuwono, A., Abulfaraj, A., Miniaoui, S., & Taheri, N. (2023). Cognitive cloud framework for waste dumping analysis using deep learning vision computing in healthy environment. *Computers & Electrical Engineering, 110*, Article 108814.

Majidi, A. (2021). City with smart waste management: Introducing waste bin technologies. *Nexo Rev. Científica., 34*, 1032–1038.

Marques, P., Manfroi, D., Deitos, E., Cegoni, J., Castilhos, R., Rochol, J., Pignaton, E., & Kunst, R. (2019). An IoT-based smart cities infrastructure architecture applied to a waste management scenario. *Ad Hoc Networks, 87*, 200–208.

Mishra, S., & Das, A. P. (2021). Current treatment technologies for removal of microplastic and microfiber pollutants from wastewater. In *Wastewater treatment: Cutting-edge molecular tools, techniques and applied aspects* (pp. 237–251). https://doi.org/10.1016/B978-0-12-821881-5.00011-8

Mishra, S., & Das, A. P. (2022). Treatment of the wastewater polluted with synthetic microfiber released from washing machine. In B. B. Das, H. Hettiarachchi, P. K. Sahu, & S. Nanda (Eds.), *Lecture Notes in Civil Engineering* (Vol. 207, pp. 109–117). Springer Science and Business Media Deutschland GmbH. https://doi.org/10.1007/978-981-16-7509-6_9/COVER

Mishra, S., & Das, A. P. (2023). Microbial remediation of synthetic microfiber contaminated wastewater. In M. P. Shah (Ed.), *Microbial technologies in industrial wastewater treatment* (pp. 337–350). Springer. https://doi.org/10.1007/978-981-99-2435-6_17

Mishra, S., Dash, D., Al-Tawaha, A. R. M. S., & Das, A. P. (2022). A review on heavy metal ion adsorption on synthetic microfiber surface in aquatic environments. *Applied Biochemistry and Biotechnology, 194*(10), 4639–4654. https://doi.org/10.1007/S12010-022-04029-W/METRICS

Mishra, S., Ghosh, S., van Hullebusch, E. D., Singh, S., & Das, A. P. (2023). A critical review on the recovery of base and critical elements from electronic waste-contaminated streams using microbial biotechnology. *Applied Biochemistry and Biotechnology, 2023*, 1–30. https://doi.org/10.1007/S12010-023-04440-X

Mohanty, S., Ghosh, S., Bal, B., & Das, A. P. (2018). A review of biotechnology processes applied for manganese recovery from wastes. *Reviews in Environmental Science and Bio/Technology 2018, 17*(4), 791–811. https://doi.org/10.1007/S11157-018-9482-1

Nag, M., Lahiri, D., Ghosh, S., Sarkar, T., Pati, S., Das, A. P., Ram, D. K., Bhattacharya, D., & Ray, R. R. (2023). Application of microorganisms in biotransformation and bioremediation of environmental contaminant: A review. *Geomicrobiology Journal.* https://doi.org/10.1080/01490451.2023.2261443

Nair, A. T. (2021). Bioaerosols in the landfill environment: An overview of microbial diversity and potential health hazards. *Aerobiologia, 37*, 185–203.

Nesmachnow, S., Rossit, D., & Moreno-Bernal, P. (2025). A literature review of recent advances on innovative computational tools for waste management in smart cities. *Urban Science, 9*(16), 1–61. https://doi.org/10.3390/urbansci9010016

Nesmachnow, S., Rossit, D., & Moreno-Bernal, P. (2025). A literature review of recent advances on innovative computational tools for waste management in smart cities. *Urban Science, 9*, 16. https://doi.org/10.3390/urbansci9010016

Nowakowski, P., Szwarc, K., & Boryczka, U. (2020). Combining an artificial intelligence algorithm and a novel vehicle for sustainable e-waste collection. *Science of the Total Environment, 730*, Article 138726.

Parida, B., Dash, A. K., & Verma, A. K. (2023a). Technologies for treatment of landfill leachate: A brief review. *Recent Innovations in Chemical Engineering, 16*(4), 260–284. https://doi.org/10.2174/0124055204276847231013040437

Parida, B., Dash, A. K., & Verma, A. K. (2023b). Technologies for treatment of landfill leachate: A brief review. *Recent Innovations in Chemical Engineering, 16*(4), 260–284. https://doi.org/10.2174/0124055204276847231013040437

Paul, A., Priyadarsini, S., Karakuş, S., Behera, J. K., Gnanasekaran, R., & Das, A. P. (2024). Fate and accumulation of synthetic microfibers in the environment. In *Sustainable microbial technology for synthetic and cellulosic microfiber bioremediation* (pp. 61–76). https://doi.org/10.1007/978-3-031-62600-5_4

Pollak, A., Gupta, A., & Göhlich, D. (2024). Optimized operation management with predicted filling levels of the litter bins for a fleet of autonomous urban service robots. *IEEE Access, 12*, 7689–7703.

Prabhakar, A., Mishra, S., & Das, A. P. (2019). Isolation and identification of lead (Pb) solubilizing bacteria from automobile waste and its potential for recovery of lead from end of life waste batteries. *Geomicrobiological Journal, 36*(10), 894–903. https://doi.org/10.1080/01490451.2019.1654044

Pradhan, G., Tripathy, B., Ram, D. K., Digal, A. K., & Das, A. P. (2023). Bauxite mining waste pollution and its sustainable management through bioremediation. *Geomicrobiology Journal* 1–10. https://doi.org/10.1080/01490451.2023.2235353

Priyadarsini, S., & Das, A. P. (2024). Analytical and structural characterization of waste lithium-ion batteries for their effective recycling strategy. In *Environmental chemistry and ecotoxicology*. https://doi.org/10.1016/j.enceco.2024.12.004

Raj, K., & Das, A. P. (2023). Lead pollution: Impact on environment and human health and approach for a sustainable solution. *Environmental Chemistry and Ecotoxicology, 5*, 79–85. https://doi.org/10.1016/J.ENCECO.2023.02.001

Ray, S. K., Dey, S., Das, A. P., & John, A. (2024). Microplastic waste management through greener technologies: A review. In *Sustainable management of urban plastic waste through circular economic approaches* (pp. 237–261). https://doi.org/10.1201/9781003465447-14

Renou, S., Givaudan, J. G., Poulain, S., Dirassouyan, F., & Moulin, P. (2008). Landfill leachate treatment: Review and opportunity. *Journal of Hazardous Materials, 150*(3), 468–493. https://doi.org/10.1016/j.jhazmat.2007.09.077

Rouhani, A., & Hejcman, M. (2024). A review of soil pollution around municipal solid waste landfills in Iran and comparable instances from other parts of the world. *International Journal of Environmental Science and Technology*. https://doi.org/10.1007/s13762-024-05728-z

Sahu, A., Dey, S., Mallick, K., Priyadarsini, S., Behera, I. D., & Das, A. P. (2024). A review on advanced technology for sustainable management of synthetic microplastic waste. In *Sustainable management of urban plastic waste through circular economic approaches* (pp. 262–279). https://doi.org/10.1201/9781003465447-15

Sallang, N., Islam, M., & Arshad, H. (2021). A CNN-based smart waste management system using tensorflow lite and LoRa-GPS shield in Internet of Things environment. *IEEE Access, 9*, 153560–153574.

Siddiqua, A., Hahladakis, J. N., & Al-Attiya, W. A. K. A. (2022). An overview of the environmental pollution and health effects associated with waste landfilling and open dumping. *Environmental Science and Pollution Research, 29*, 58514–58536. https://doi.org/10.1007/s11356-022-21578-z

Singh Jadaun, J., Bansal, S., Sonthalia, A., Rai, A. K., & Singh, S. P. (2022). Biodegradation of plastics for sustainable environment. *Bioresource Technology, 347*, Article 126697. https://doi.org/10.1016/J.BIORTECH.2022.126697

Singh, B. J., Chakraborty, A., & Sehgal, R. (2023). A systematic review of industrial wastewater management: Evaluating challenges and enablers. *Journal of Environmental Management, 348*, Article 119230. https://doi.org/10.1016/j.jenvman.2023.119230

Smith, V. H., & Schindler, D. W. (2009). Eutrophication science: Where do we go from here? *Trends in Ecology & Evolution, 24*(4), 201–207. https://doi.org/10.1016/j.tree.2008.11.009

Tangri, N. (2023). Waste incinerators undermine clean energy goals. *PLOS Climate, 2*(6), Article e0000100. https://doi.org/10.1371/journal.pclm.0000100

Thamarai, M., & Naresh, V. (2023). Smart self-power generating garbage management system using deep learning for smart cities. *Microprocessors and Microsystems, 98*, Article 104816.

Thaseen, S., Mohanraj, V., Ramachandran, S., & Balakrishnan, A. (2023). An intelligent waste management application using IoT and a genetic algorithm–fuzzy inference system. *Applied Sciences, 13*, 3943.

Unal, M., Cilek, A., & Guner, E. D. (2019). Implementation of fuzzy, Simos and strengths, weaknesses, opportunities and threats analysis for municipal solid waste landfill site selection: Adana City case study. *Waste Management Research, 38*, 45–64.

United Nations Environment Programme. (2024). *Global waste management outlook 2024*. https://www.unep.org/resources/global-waste-management-outlook-2024. Accessed April 03, 2025

Vergara, D., de la Hoz-M, J., Ariza-Echeverri, E. A., Fernández-Arias, P., & Antón-Sancho, Á. (2025). Evaluating solutions to marine plastic pollution. *Environments, 12*(3), 86. https://doi.org/10.3390/environments12030086

Wong, S., Han, H., Cheng, K., Koo, A., & Yussof, S. (2022). ESS-IoT: The smart waste management system for general household. *Pertanika Journal of Science and Technology, 31*, 311–325.

Zhang, C., Xu, T., Feng, H., & Chen, S. (2019). Greenhouse gas emissions from landfills: A review and bibliometric analysis. *Sustainability, 11*(8), 2282. https://doi.org/10.3390/su11082282

Zhang, Q., Li, H., Wan, X., Skitmore, M., & Sun, H. (2020). An intelligent waste removal system for smarter communities. *Sustainability, 12*, 6829.

Microbial Strategies for Sustainable Management of Environmental Waste

Abhinashi Singh Sodhi, Sonu Bhatia, and Navneet Batra

Abstract Waste generated from different domestic and industrial sources has become a major global concern with pollutants such as pharmaceutical compounds, heavy metals, radioactive substances, hydrocarbons, microplastics posing serious threat to environment as well as public health. Microorganisms-based strategies have emerged as an innovative and sustainable alternative for environmental waste treatment over conventional techniques due to environment friendly nature and cost effectiveness. Methodologies including microbial fuel cells, biofilm reactors, bioremediation, bioaugmentation, denitrification presents immense potential toward a more sustainable and efficient way of treatment. Use of microbial consortium has also gained attention in degradation of persistent environmental pollutants due to their diverse metabolic capabilities and synergistic interactions. Recent breakthroughs in technologies such as CRISPR/Cas, metabolic engineering, and metagenomics resulted in obtaining novel microbes with tailored abilities for enhanced degradation of pollutants and resource recovery. Additionally, microbe-assisted nanotechnology is another area offering excellent wastewater treatment opportunities. However, several challenges including waste specific process optimization and scale up, maintaining stable microbial communities, and integrating new microbial technologies with existing treatment infrastructure needs to be addressed for successful implementation of these strategies. These methods represent an efficient approach to managing and mitigating environmental pollution, contributing to sustainable development.

Keywords Environmental waste · Bioremediation · Microbial fuel cell · Metagenomics · Microbe-assisted nanotechnology

A. S. Sodhi (✉) · S. Bhatia · N. Batra
Department of Biotechnology, Goswami Ganesh Dutta Sanatan Dharma College, Chandigarh, India
e-mail: abhinashisingh90@gmail.com

A. P. Das (ed.), *Bioconversion of Environmental Waste to Value Added Products*, Sustainable Environmental Waste Management Strategies,
https://doi.org/10.1007/978-3-032-14215-3_9

1 Introduction

An enormous amount of solid and liquid waste is generated through different domestic and industrial sources. As per the Global Waste Management Outlook report of 2024, solid waste generation is expected to reach 3.8 billion tonnes by 2050. These waste materials are often left untreated and dumped in open or discharged into aquatic bodies, thus posing a severe threat to biodiversity and the environment (Mishra et al., 2023a, 2023b). These techniques are simple to employ, but considering the amount of waste being generated worldwide, they are ineffective in managing such a large volume of waste, leading to contamination in groundwater, open fires, and contributing to greenhouse gases (Global waste management outlook 2024; Sodhi et al., 2022).

Considering the limitations associated with the present waste management technologies, researchers explored and put efforts into developing biological methods with attributes such as sustainability, cost-effectiveness, and ability to mitigate environmental damage caused by hazardous substances (Dey et al., 2023; Mishra et al., 2019). Microorganism-based bioremediation is widely considered as an innovative and effective strategy to degrade various pollutants such as hydrocarbons, heavy metals, and pharmaceutical compounds (, Mohanty et al., 2018). This approach can be applied both in situ and ex situ, depending on the type of pollutant and specific conditions of the remediation site. Many microorganisms, including bacteria, fungi, algae, have been identified with promising pollutant degradation abilities that can be employed for environment remediation (Mishra et al., 2024a; Tripathy et al., 2024). Over the years, a variety of microorganism-based methodologies have been successfully developed exhibiting significant potential to remove the persistent pollutants and restore the health of soil and water-based environment (Saeed et al., 2022). New age technologies namely, microbial fuel cells, biofilm reactors, bioremediation, bioaugmentation, denitrification have gained attention for enhanced degradation of pollutants and environment restoration (Fig. 1). Microbial consortiums have also gained attention in the degradation of persistent environmental pollutants due to their diverse metabolic capabilities and synergistic interactions (Ghosh & Das, 2020; Priyadarsini & Das, 2025). Cutting edge technologies CRISPR/Cas, metabolic engineering and metagenomics are breaking the barriers of microbial capabilities and help in developing microbes with targeted modifications to enhance their resistance to pollutants, improve their degradation capabilities, and increase their efficiency in bioremediation processes (, Mishra & Das, 2023). Microbe-assisted nanotechnology offer promising waste treatment approach by combining unique characteristics of nanomaterials with the metabolic capabilities of microorganisms enabling removal of toxic pollutants in a sustainable manner.

Together, these technologies are paving the way for the development of custom-designed microbes with enhanced abilities to address a wide range of environmental challenges, including the degradation of hazardous pollutants and the restoration of contaminated ecosystems. The present chapter will comprehensively discuss the fundamental concept of microbial bioremediation, different strategies involved and

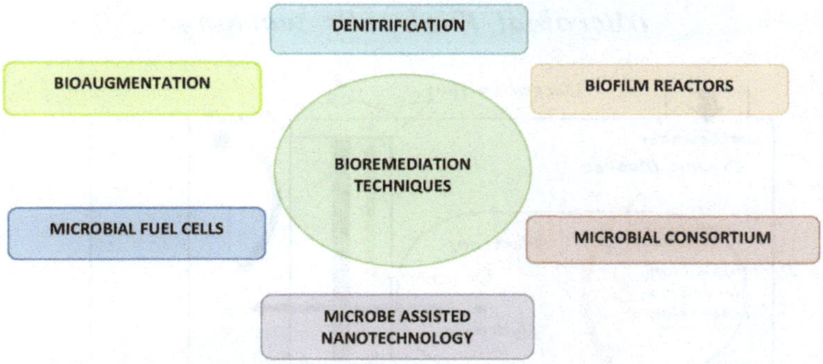

Fig. 1 Various microbe-based techniques for bioremediation

their applications in degradation of a wide range of pollutants. Further, we will also describe the role of modern day biotechnology methods including CRISPR/Cas, metabolic engineering, nanotechnology in improving the microbial abilities in bioremediation.

2 Microbial-Based Treatment Strategies

2.1 Microbial Fuel Cells

Microbial fuel cells (MFCs) are a promising technology wherein microbial cells are used to convert the inorganic or organic matter into electrical energy by catalyzing a series of electrochemical reactions (Pandya et al., 2024). The key components of MFCs includes an anode, a cathode, and proton exchange membrane to allow the flow of ions between anode and cathode (Sharma et al., 2025). Electrons are directed to the cathodic chamber from anode side having electrochemically active microbe (catalyzing organic matter degradation) through the proton exchange membrane, leading to the successful reduction reaction and producing water (Fig. 2). A variety of substrates, also known as anolyte can be potentially used for electricity production including pure compounds such as glucose, acetate, butyrate, citrate, glycerol, malate, lignocellulosic biomass, and organic matter present in wastewater (Pant et al., 2010; Obileke et al., 2021; Nawaz et al., 2022).

MFCs offer advantages including reduced energy consumption, resource recovery, ability to be coupled with constructed wetlands, less sensitive to temperature variations, thus enhancing the removal of pollutants while generating electricity (He et al., 2017). For instance, Xu et al. (2023) demonstrated the removal of antibiotic (sulfamethoxazole) and nitrogen from microbial fuel cell constructed wetlands packed with coke and quartz sand as substrate. High percentage of sulfamethoxazole (94%), chemical oxygen demand (COD) (78%), ammonium (80%), nitrate

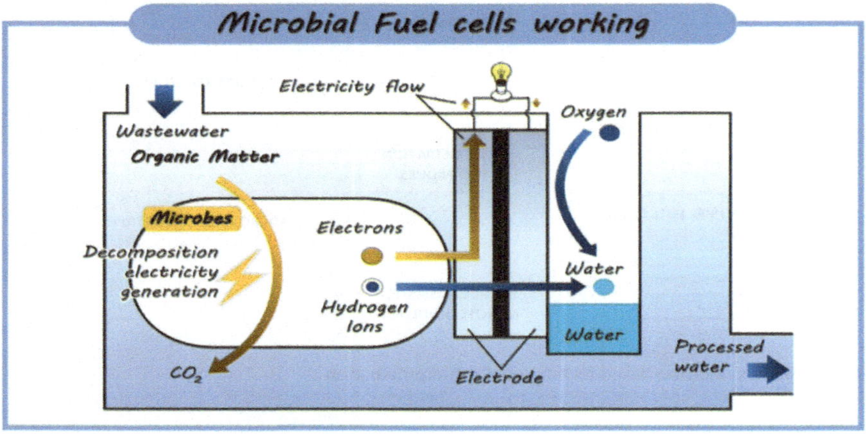

Fig. 2 Schematic representation of workflow of a typical microbial fuel cell (Reproduced from Nawaz et al., 2022)

(83%), and total nitrogen (70%) was removed when coke used as substrate. The coke substrate improved microbial activity, enhancing pathways related to cellular transport, amino acid, and carbohydrate metabolism. Additionally, MFC-constructed wetlands generated more electricity. Dominant microbial groups in the system included *Firmicutes, Proteobacteria, and Bacteroidetes,* which contributed to both antibiotic and nitrogen transformation as well as electricity production. Similarly, simultaneous removal of sulfamethazine (70.3–85.6%) and Chromium (VI) (>98%) was observed by using microbial fuel cell coupled constructed wetland. Additionally, efficiency for N, P, and COD ranged from 74.0% to 96.1%, 83.6% to 94.1%, and 91.1% to 95.3%, respectively (Chen et al., 2025).

Microbial fuel cells has been potentially evaluated for their ability to remove heavy metals from wastewater. For instance, significant removal of gold (95%) from a solution of pure tetrachloroaurate ions has been observed after 5 h of contact time and using 80 g/L of yeast biomass. Further incubation for 48 h under the same conditions increased the removal efficiency to 98.86% (Al-Asheh et al., 2022).

Similarly, strategy based upon the coupling of microbial fuel cell with constructed wetland were used to remove zinc and nickel from sludge and simultaneous electricity generation. Significant removal rates of zinc (76.88%) and nickel (66.02%) were observed. Moreover, the system also resulted in the generation of 534.30 mV of electricity (Wang et al., 2022a, 2022b). On a similar pattern, MFC-CW were used to achieve significant removal of Cu, Pb, Zn, and Cd upto 98.83%, 95.74%, 92.91%, and 90.75% respectively. Further, maximum voltage, current, and power densities reached 687 mV, 4000 mA/m^3, and 785.86 mW/m^3, respectively, were also obtained in microbial fuel cell constructed wetland (Kahrizi et al., 2024).

In a study conducted at University of Dar es Salaam (UDSM), MFC prototype was developed and fed with sewage as substrate with aluminum /carbon electrodes

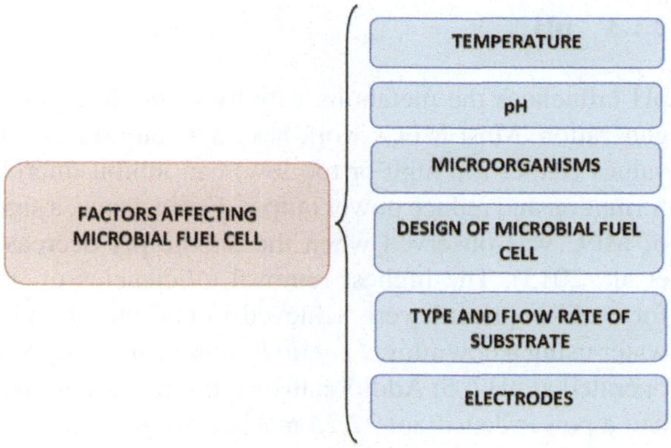

Fig. 3 Important factors affecting microbial fuel cell performance

for electricity generation as well as waste treatment resulting in generation of 1.73 V of stable voltage and 0.07 mA of current for a period of one hour. As per the study, electricity generated was sufficient to light up a LED bulb of 6.25 mW (Said et al., 2023).

2.1.1 Factors Affecting Microbial Fuel Cells

Several factors influence the overall efficiency of MFC system and need to be addressed. Key variables, including the type of electrodes, proton exchange membrane material, substrate type, pH, temperature have a profound affect on the system's functioning (Fig. 3) (Obileke et al., 2021; Nawaz et al., 2022).

2.1.2 Temperature

The performance of MFCs is affected by variation in the operating temperature by affecting microbial kinetics, mass transfer and thermodynamic. Temperature range of 25–30 °C has been found to be optimum for efficient working of MFCs (Amend, 2001).

Song et al. (2017) demonstrated the temperature of 30 °C to be more beneficial to obtain higher power generation. The reason behind this could be production of biofilms exhibiting maximum catalyst activity at this temperature. Similarly, maximum power output of 54mW/m^2 and current of 348 mA/m^2 was obtained when inoculated with sludge and incubated at a temperature of 25 °C for 2 weeks. However, a significant decrease in electricity generation was witnessed with subsequent increase in temperature (Song et al., 2017).

2.1.3 pH

pH influences the metabolic activity of the microorganisms involved in electricity generation. Most MFCs work best at a neutral to slightly acidic pH, as extreme pH values (either too high or too low) can inhibit microbial activity, affecting biofilm formation and reduce power output. For instance, a significant impact on the working of MFC was observed when the anodic pH decreased from 6.23 to 4.35 (Zhang et al., 2013). The highest removal efficiencies of 99% for chromium and 78.6% for p-chlorophenol were achieved in chromium and p-chlorophenol-contaminated water using a downflow *Leersia hexandra* microbial fuel cell (DLCW-MFC) system operated at pH 6.5. Additionally, electricity generation of 543 mV in output voltage and a power density of 72.25 mW/m^2 were obtained at this pH. Notably, the results were significantly improved at pH 6.5 in comparison to when the system operated at pH levels of 5.8 and 7.4, highlighting the importance of pH for optimal performance (Wang et al., 2024).

2.1.4 Microorganisms

Microorganisms are essential components of microbial fuel cells (MFCs), driving the biochemical processes that convert organic matter into electrical energy. They either act as mediators or directly facilitate electron transfer across the electrodes, catalyzing the breakdown of organic matter (Mishra et al., 2021, 2024b). This process completes the electrical circuit while simultaneously generating electricity. Bacterial species belonging to archaebacteria, acidobacteria, cyanobacteria, proteobacteria constitutes the major microorganisms being used in MFCs (Nawaz et al., 2022). *Actinobacillus succinogenes*, *Proteus mirabilis*, *Proteus vulgaris*, *Shewanella putrefaciens*, *Geobacter* are common examples of bacteria employed directly or as mediators for electron transfer between the electrodes and generate electrochemical process (Park et al., 1999; Kim et al., 2002; Choi et al., 2003; Aghababaie et al., 2015). A comprehensive list of various microorganisms used in microbial fuel cells and their power density is given in Table 1.

2.1.5 Design of Microbial Fuel Cell

The overall working and performance of microbial fuel cell is also dependent on its design. Depending upon the design of the reactor a variety of MFCs are available including single chamber MFC, double chamber MFC, upflow MFC, stacked MFC, cylindrical MFC and flat plate MFC (Fig. 4) (Nawaz et al., 2022).

Table 1 Various microbes used in microbial fuel cell and their respective power density

Microbe	Power density	References
Geobacter sulfurreducens	3147 mA/m^2	Inoue et al. (2010)
Desulfovibrio desulfuricans	233 mA/m^2	Kang et al. (2014)
Klebsiella pneumonia	199 mA/m^2	Deng et al. (2010)
Thermincola ferriacetica	12,000 mA/m^2	Prathap et al. (2013)
Ochrobactrum sp	2625 mW/m^3	Li et al. (2014)
Shewanella xiamenensis	38.89 W/m^3	Truong et al. (2021)
Exiguobacterium, Desulfovibrio, Macellibacteroide	197.94 mW/m^2	Patel et al. (2021)
Shewanella algae	143.3 mA/m^2	Choudhury et al. (2021)

2.1.6 Type of Substrate and Loading Rate

A variety of substrate can be employed for power generation in microbial fuel cell starting from simple organic substrates such as glucose, acetate, butyrate, complex organic substrates (wastewater, lignocellulosic biomass) to persistent pollutants, heavy metals, and pharmaceutical products (Bal et al., 2018; Mishra et al., 2023a, 2023b). These substrate act as source of nutrients for microbes and electrons that can significantly affect the microbial community structure in the anode biofilm as well as power density and Coulombic efficiency (Chae et al., 2009; Tan et al., 2020). High amount of power generation and COD removal efficiency of 84% was achieved by using organic substrate at a concentration 2000 mg/L (Rahmani et al., 2020).

Substrate loading rate refers to the amount of substrate (or organic matter) supplied to the MFC per unit time. Optimizing the substrate flow rate can greatly enhance degradation of organic matter and simultaneous power generation (Abdallah et al., 2019). For instance, increasing the flow rate of organic substrate from 0.7 g COD/L d to 1.9 g COD/L/d significantly improves power yield from 42 mW/m^2 to 116.03 mW/m^2 (Tamilarasan et al., 2017). In another study, maximum chemical oxygen demand, power and current density of 84.64%, 162.59 mW/m^2 and 468.74 mA/m^2 was obtained at a substrate concentration of 0.810 g/l and flow rate of 0.266 ml/min (Tan et al., 2020).

2.1.7 Electrodes

The type of electrode used in a microbial fuel cell (MFC) can significantly impact its performance by influencing factors such as electricity generation, microbial activity, and the efficiency of electron transfer between microbes and the electrode surface. Anode is the site responsible for electron and proton generation along with organic pollutant oxidation, biofilm formation whereas cathode is the site where electrons are received during the electrochemical process in a microbial fuel cell. Further, the material for anode and cathode must be stable, have large surface

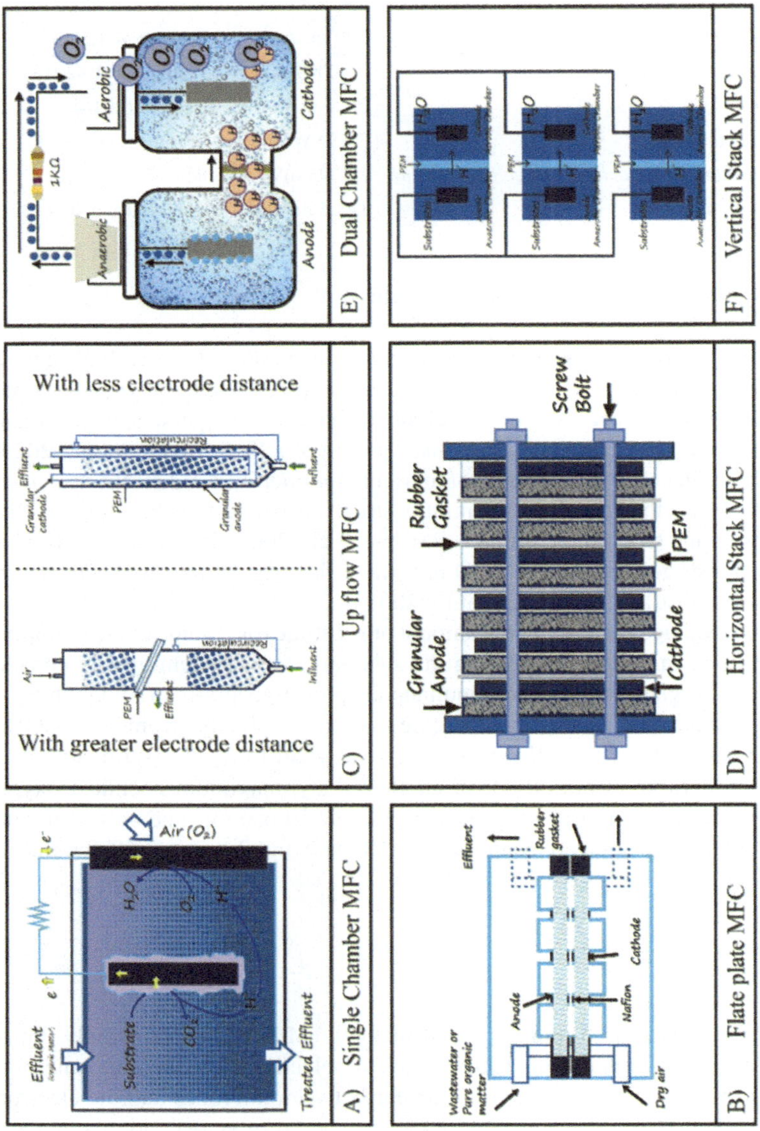

Fig. 4 Schematic representation of various types of microbial fuel cells (Reproduced from Nawaz et al., 2022)

area, non-corrosive, suitable to support high and continuous conductivity, should be compatible with microorganisms (Jaiswal et al., 2020; Nawaz et al., 2022). Electrodes made up of materials such as platinum and gold show high conductivity but they are often more expensive and may not be as suitable for long-term operation due to antimicrobial activity affecting microbial growth and biofilm formation. Carbon-based electrodes comprised of graphite, activated carbon, carbon cloth, carbon fiber, and carbon nanotube are common because of their high conductivity, corrosion resistance, chemical stability, large surface area ability to support biofilm growth and low cost. The porosity of these materials also allows for better microbial attachment (Mustakeem, 2015; Singh et al., 2019; Slate et al., 2019; Munoz-Cupa et al., 2020; Jatoi et al., 2020; Nawaz et al., 2022).

After 115 days of operation, soil microbial fuel cells exhibited optimal performance with graphite felt as the anode, yielding a stable voltage of 345 mV and a power density of 24.0 mW/m^2. Additionally, the removal rates for pollutants were 59.14% for total petroleum hydrocarbons, 61.65% for anthracene, and 55.92% for pyrene, respectively (Yu et al., 2021).

Microbial fuel cells (MFCs) have emerged as an innovative approach to waste treatment, harnessing the metabolic activity of microorganisms to generate electricity. Despite their potential, the technology faces significant challenges, primarily the insufficient power output, limited life span and the high construction cost of key components, such as electrode materials and proton exchange membranes. Overcoming these limitations is crucial to making MFCs more economically viable and scalable for widespread application.

2.2 *Bioaugmentation*

Bioaugmentation refers to the process of introducing specific type of microbes with the ability to degrade harmful pollutants to soil or water and accelerating the degradation rate of contaminants without harming the native microflora (Herrero & Stuckey, 2015; Chettri et al., 2024). To increase the bioremediation process, bioaugmentation can be combined with methods such as biostimulation and phytoremediation. The process of bioaugmentation is influenced by several environmental factors including the nature of pollutant, selection of microbial strain and remediation cycles (Gao et al., 2022).

Microbial strain should be selected based on its stability and ability to withstand toxic environment of the contaminated site and operating conditions, including temperature, pH, and indigenous microbes (Bitton, 2011). Bioaugmentation being a site-specific process, the microbes are selected and isolated from the polluted site itself as it enhances the effectiveness of bioremediation efforts by using microbes that are naturally adapted to the specific environmental conditions and contaminants present at the site (Raper et al., 2018).

Several studies have proven the potential of bioaugmentation as a significant method in the treatment of soil or water contaminated by various pollutants, including heavy metals, pesticides and petroleum compounds. For instance, printing and dyeing wastewater treated using *Ochrobactrum anthropi* S1 combined with anaerobic and aerobic technology resulted in final removal efficiencies of reactive black 5, Cr (VI), chemical oxygen demand, and ammonia nitrogen of 95%, 65%, 90%, and 85%, respectively (Sun et al., 2025). In another study, microbial consortium HT comprising of thermophilic petroleum degrading organisms was employed to treat petroleum contaminated soils. The application resulted in significant degradation of hydrocarbons including carbon chains of C16–C21 (87.1%) and C21–C40 (67.2%) within 140 days under high temperature (Wang et al., 2025). In a similar study, conducted to bioremediate spent engine oil-contaminated soil involves biostimulation using sodium dodecyl sulfate followed by introduction of indigenous hydrocarbonoclastic bacterial isolates. Results showed a significant reduction in the initial total petroleum hydrocarbon value by 88.28% (Onwujekwe et al., 2024).

In addition to the type of microbial species being used for remediation, another factor that plays a significant role in the functioning of microbes and efficiency of the technique is dosing ratio and dosing time. In a study, Chen et al. (2022) investigated the impact of dosing ratios and timing on the efficiency of bioaugmentation in seafood processing wastewater (WW) treatment using a sequencing batch bioreactor (SBBR). The reactor followed an 8-h cycle, with phases for settling, decanting, filling, anaerobic reaction, and aerobic reaction. The bioaugmentation process used a mixed bacterial system, with bacterial agents dosed at 2.5% on day 1 and day 10 of the experiment. The results showed that these dosing strategies significantly enhanced the removal of total nitrogen in winter, increasing removal efficiency from 66.89% and 52.77% to 79.02% and 69.97%, respectively.

The above-mentioned studies highlighted that both the timing and ratio of bacterial dosing are crucial factors in optimizing the bioaugmentation process for treating seafood processing wastewater. The improved nitrogen removal achieved through the strategic dosing of bacterial agents demonstrates the effectiveness of these dosing strategies in enhancing the coordination and adaptive potential of functional bacteria within the bioaugmentation system (Mishra et al., 2009).

Bioaugmentation has numerous advantages including easy process, low cost, and sustainable way to handle contaminants. However, maintaining a stable microbial community for longer time period to carry out effective bioremediation in a highly toxic environment is one of the major limitations of bioaugmentation (Behera et al., 2021, 2022) (Table 2).

2.3 Biofilm-Based Bioreactors

Biofilm-based bioreactors are a type of bioreactor where microorganisms form a biofilm on a surface, allowing them to interact with the substrate (such as wastewater, gases, or liquids) that is being treated or processed. These bioreactors can be coupled with a variety of different processes such as metal catalysts and electrochemistry, increasing their efficiency in removing pollutants and reducing carbon

Table 2 Microorganisms used in bioaugmentation of pollutants and their degradation efficiency

Microbe	Target environment	Pollutant	Exposure	Efficiency	References
Sphingobium SA2	Soil	Mercury	28 days	30–50%	Mahbub et al. (2017)
Serratia marcescens, Pseudomonas pyogenes Erwnia amylovora, Enterobacter cloacae	Soil	Paper mill effluent containing heavy metals	180 days	Pb (74%) Cd (51.6%) Zn (71.6%) As (72.6%) Cr (78.2%) Cu (55.9%) Ni (78.1%)	Nwaehiri et al. (2020)
Pannonibacter phragmitetus BB	Contaminated slag	Chromium (VI)	2 days	99%	Wang et al. (2015)
Micrococcus sp.	Soil	Crude oil	60 days	36%	Giwa and Ibitoye (2017)
Staphylococcus aureus	Soil	Crude oil	60 days	89.5%	
Penicillium sp. *Ulocladium sp.* *Aspergillus sp.* *Fusarium sp.*	Soil	Hydrocarbons	120 days	39.90%	Medaura et al. (2021)
Micrococcus luteus, Rhizopus arrhizus	Soil	Hydrocarbon	8 weeks	75.70%	Nwankwegu and Onwosi (2017)
Sphingobium sp. strain YBL2-gfp	Soil	Isoproturon	21 days	–	Zhang et al. (2024)
Actinobacteria, Proteobacteria, Firmicutes, Halobacteria	Soil	Crude oil	12 months	82.7–93%	Lee et al. (2024)
Pseudomonas stutzeri M3	Soil	Paraffins, iso-paraffins, olefins, naphthenes, aromatics fractions	60 days	96.5%	Li et al. (2024)

emissions (Mei et al., 2025). Different types of biofilm reactors have been developed namely, trickling filter reactor, moving bed biofilm reactor, rotating biological contactor, membrane biofilm reactor, passive aeration simultaneous nitrification, and denitrification (PASND) biofilm reactor (Abdelfattah et al., 2020). One of the most widely accepted biofilm reactors is membrane biofilm reactor that combines the technology of biological treatment with membrane filtration. This hybrid approach is used primarily for the removal of contaminants, particularly those that are difficult to treat using traditional methods, such as nitrogen (nitrates, ammonium), phosphorus, or carbon-based pollutants (Dominic & Baidurah, 2022).

2.4 Denitrification

Nitrogen in the form of compounds such as nitrates and ammonium can be a significant pollutant in aquatic ecosystems. Discharge from agricultural lands, industrial waste and sewage lead toward excessive nitrogen deposition in water bodies resulting in eutrophication causing threat to the biodiversity and environment. Denitrification is a widely used method for removing excess of nitrogen from wastewater and improving the overall water quality before discharging them into aquatic bodies (Yang et al., 2021). Advantages including high efficiency of the process, minimum by-product generation, cost effectiveness and environment friendly makes it a method of choice (Liu et al., 2018).

It is a sequential process involving the reduction of nitrate to volatile compounds such as dinitrogen (Zumft, 1997). The whole process is enzymatically controlled and bacteria utilizes a variety of enzyme systems to perform denitrification. The process begins with the reduction of nitrate to nitrite, catalyzed by nitrate reductase. Further, nitrite is converted to nitric oxide through the action of nitrite reductase. Finally, nitric oxide reductase further reduces nitric oxide to nitrous oxide (Fig. 5) (Brozinčević et al., 2024).

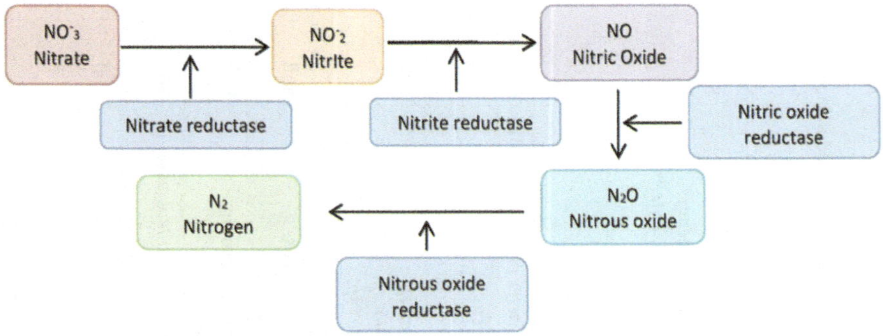

Fig. 5 Process of bacterial denitrification

Nitrate/nitrite reduction process utilizes organic compounds in wastewater or external carbon source as an electron donor. Various bacteria utilize these modules in diverse contexts and enzyme systems, offering a level of metabolic flexibility that is gradually being uncovered through the increasing availability of genomic data. Each stage of this metabolic pathway is driven by intricate metalloenzymes, which have distinctive structural and spectroscopic properties (Einsle & Kroneck, 2004).

Several bacterial species have been explored and tested for their denitrification ability such as *Bacillus, Enterobacter, Micrococcus, Pseudomonas,* and *Spirillum* (Meng et al., 2014). The feasibility of the technique has been validated by a number of studies for instance, employing denitrifying *Bacillus subtilis* JD-014 demonstrated nitrogen pollution depletion with maximum removal efficiency of NO-3-N and NO-2-N of 98.91% and 81.99%, respectively (Yang et al., 2021). Similarly, employing algal–bacterial consortium and *Bacillus subtilis, Pseudomonas stutzeri, Rhodococcus* sp. resulted in total nitrogen removal to 74.6% and 86% respectively (Li et al., 2023; Zhang et al., 2019).

Denitrification technique have certain limitations such as substantial amount of land requirements, longer time period, high energy consumption, and high initial investment and operational costs. However, technological advances are leading the way in developing methods to reduce the processing time and lower down the operational as well as maintenance costs (Huang et al., 2023).

2.5 *Microbial Consortia for Degradation of Persistent Pollutants*

Over the years, numerous microbial strains have been identified and utilized in the development of various bioremediation strategies. However, often individual strains have certain limitations in terms of pollutant degradation efficiency, substrate specificity and production of limited type of enzymes. In contrast, developing microbial consortium can offer distinct advantages over single isolated microbes, as the consortia facilitate more synchronous cellular interactions resulting in enhanced physiological and biochemical reactions leading to more efficient degradation of complex compounds such as pharmaceutical waste, azo dyes, petroleum hydrocarbons. Further, optimized metabolic pathways, reduction in cross reactions and intermediate product accumulation can improve the microbial action (Cao et al., 2022; Zhang & Zhang, 2022).

For instance, remediation of oil polluted soil was carried out using two oil degrading strains namely, Strain A and Strain B in a combination (ratio of 7:3) resulting in significant degradation rate of 68.27% after 40 days with wheat bran (158 g/kg) and swine wastewater (232 mL/kg) being used as substrate material (Zhang et al., 2020). Similarly, employing microbial consortia of *Ochrobactrum pseudintermedium* C1 and *Bacillus cereus* K1 exhibited crude oil degradation rate upto 70.54% after 72 h. Moreover, the consortia also demonstrated good emulsification activity of 80.5% toward crude oil (Bhattacharya et al., 2019). In another study, degradation of polycyclic aromatic hydrocarbons benzo[a]pyrene, benzo[b]fluoranthene, phenanthrene, pyrene, was performed using seven different bacterial consortia comprising of *Mycobacterium gilvum* MI, *Mycobacterium* sp. ZL7 and*Rhodococcus rhodochrous* Q3 achieving a degradation efficiency of 59% after 8 days (Zhou et al., 2023). Inoculating acephate contaminated soil with microbial consortium, ZQ01, resulted in degradation of the pollutant upto 89.5% within 32 h (Lin et al., 2022).

The utilization of different microbial strains in a mixture prove to be a promising technique in bioremediation of polluted sites offering an eco-friendly approach to land restoration and opening new avenues for sustainable development. However, the mechanisms of intercellular communication within large microbial consortia remain unclear and regulatory processes are still not fully understood. As research in this area progresses, the enhanced metabolic capacity and resilience of these consortia are expected to drive their broader application in degrading complex compounds.

A list of various microbial consortia and pollutants degraded by them is given in Table 3.

Table 3 Various microbial consortia and their targeted pollutants

Microbial consortia	Targeted pollutant	References
Bacillus sp. *Pseudomonas sp.*	Heavy metals	Liu et al. (2023)
Bacillus pumilus KS2 *Bacillus cereus R2*	Petroleum hydrocarbons	Patowary et al. (2016)
Bacillus cereus MG257494.1 *Alcaligenes faecalis MG966440.1* *Alcaligenes faecalis MG257493.1*	Heavy metals	Abou-Aly et al. (2021)
Algal–bacterial biomass	Cu, Ni, Cd, Zn Salicylate	Muñoz et al. (2006)
Cellulomonas iranensis *Pseudomonas brassicacearum*	Cd Pb	Che et al. (2024)
γ-Proteobacteria, Firmicutes	Zinc	Carpio et al. (2016)
Pseudomonas aeruginosa REBP5 *Pseudomonas alloputida* REBP7 *Castellaniella denitrificans* REBF6 *Debaryomyces hansenii* RELF8	Plastic	Salinas et al. (2023)

3 Innovative Approaches to Enhance Bioremediation Potential of Microbes

3.1 CRISPR/Cas9

Genome editing tools have paved the way for improving the functionality of biological systems. Rapid advancements in the field resulted in efficient editing of the microbial genome and incorporate tailored abilities in them. CRISPR/Cas systems have revolutionized genetic engineering and proved to be an effective strategy for enabling precise and targeted modifications in microbial genomes.

In a study, CRISPR-ddAsCpf1 system was employed to enhance the extracellular electron transfer capacity of *Shewanella oneidensis* MR-1 by repressing seven putative genes. The resulted modification increased the rate of methyl orange and chromium removal (Li et al., 2020). Similarly, gene knockdown in *Aeromonas hydrophila* done using CRISPRi resulted in transcriptional repression of various genes. leading toward arsenic detoxification and methyl orange reduction (Wu et al., 2020).

CRISPR-dCas9 system was used to genetically alter *Acidithiobacillus ferridurans* JAGS by transcriptional repression of the gene HdrB3, promoter substitution and gene insertion involved in sulfur oxidation resulting in lower production of sulfuric acid in leachate and reducing groundwater contamination (Chen et al., 2023).

3.2 Metabolic Engineering

Metabolic engineering for bioremediation is a promising field that combines genetic engineering, systems biology, and microbiology to enhance the capabilities of microorganisms in degrading, transforming, or removing pollutants from the environment. Native strains possess great ability to degrade a variety of pollutants such as heavy metals, azo dyes and hydrocarbons, however by modifying the metabolic pathways involved in their degradation, the rate of reaction can be significantly enhanced thus completing the process in short time period. Further, engineered strains have the potential to degrade a broader spectrum of persistent pollutants which cannot be processed by native strains. (Yaashikaa et al., 2022). In a study, metabolic engineering tools have been applied to specifically modify the catabolic repertoire of *Pseudomonas putida* in order to enhance the enzymatic mechanisms of the microbe for in situ bioremediation (Nikel & Lorenzo, 2021). In another study, *E.coli* transformant (BL-MP) was produced by introducing six synthetic genes *opdS, pnpAS, pnpBS, pnpCS, pnpDS*, and *pnpES*. The resulting BL-MP transformant completed degraded methyl parathion, an organophosphate within 2 h (Xu et al., 2022). Similarly, *E. coli* strain reconstructed after introducing 4-fluorophenol degrading genes from Arthrobacter sp. namely, *fpdA2*, *fpdB*, *fpdC*, and *fpdD* resulted in complete degradation of the 4-flurophenol from industrial wastewater in 3 h (Wang et al., 2022a, 2022b).

3.3 Metagenomics

Conventionally, to study the microbiome of a particular environment was completely dependent on culture dependent approaches wherein the microbial cells were isolated on specific growth medium followed by their desired characterization and analysis for potential applications. However such approaches have their limitation such as not all microorganisms are culturable and therefore it was close to impossible to study the complete microbial community of an environment. In recent times, metagenomics has transformed the methodology to study the microbial communities by employing next generation sequencing techniques and retrieve information about the complete metagenome (DNA of the complete microbiota) of environmental microbes (Fig. 6).

Metagenomics is a scientific methodology to study the whole genomic material recovered directly from environmental samples, enabling the analysis of microbial communities without the need for isolating and culturing individual organisms. It provides insights into the diversity, structure, and function of microorganisms in a given environment whether from soil and water to the human gut (Bharagava et al., 2019; Wani et al., 2022).

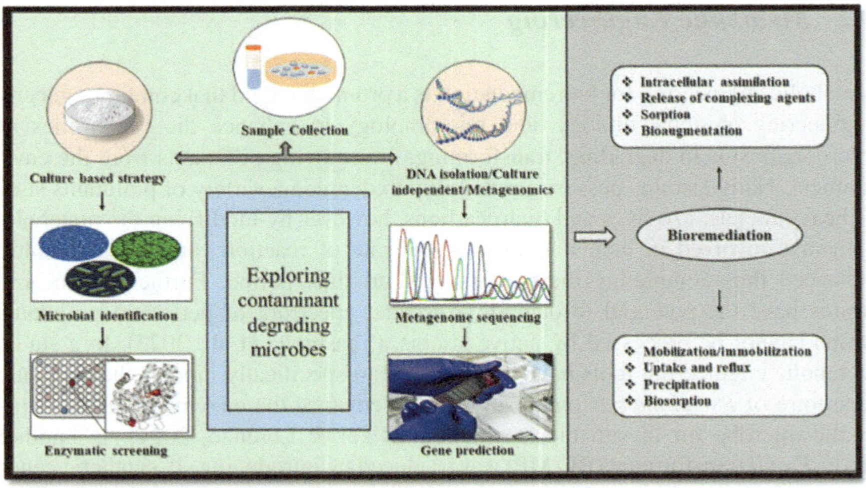

Fig. 6 Basic workflow of metagenomics based analysis and its application in bioremediation (Reproduced from Wani et al., 2022)

Metagenomics can be applied to identify novel genes or gene clusters, their functional aspects in pollutant removal, degradation and monitoring. Further, metagenomic libraries can be used to study dynamic changes in microbial communities during the bioremediation process and provide insights about functional roles and their association to biodegradation process (Datta et al., 2020; Zhang et al., 2021). The basic workflow of the metagenomic analysis includes isolation of complete genome of the microbes directly from the environmental sample, followed by construction of metagenomic libraries and sequencing based analysis (Fig. 7). High throughput next generation sequencing techniques such as Illumina, shotgun sequencing, Roche 454, Sequencing by oligonucleotide ligation and detection (SOLiD), PacBio, Nanopore sequencing have rapidly made significant advancements in functional metagenomics by providing unparalleled platform for large scale data preparation and analysis.

Microbial communities profiling using Illumina MiSeq revealed the presence of *Acidobacteria, Actinobacteria, Bacteroidetes, Chloroflexi, Firmicutes, Patesibacteria, Proteobacteria, Spirochetes* in the pulp and paper industry wastewater containing toxic metals. The identified microbial communities possess essential properties required for degradation of contaminants (Sharma et al., 2021).

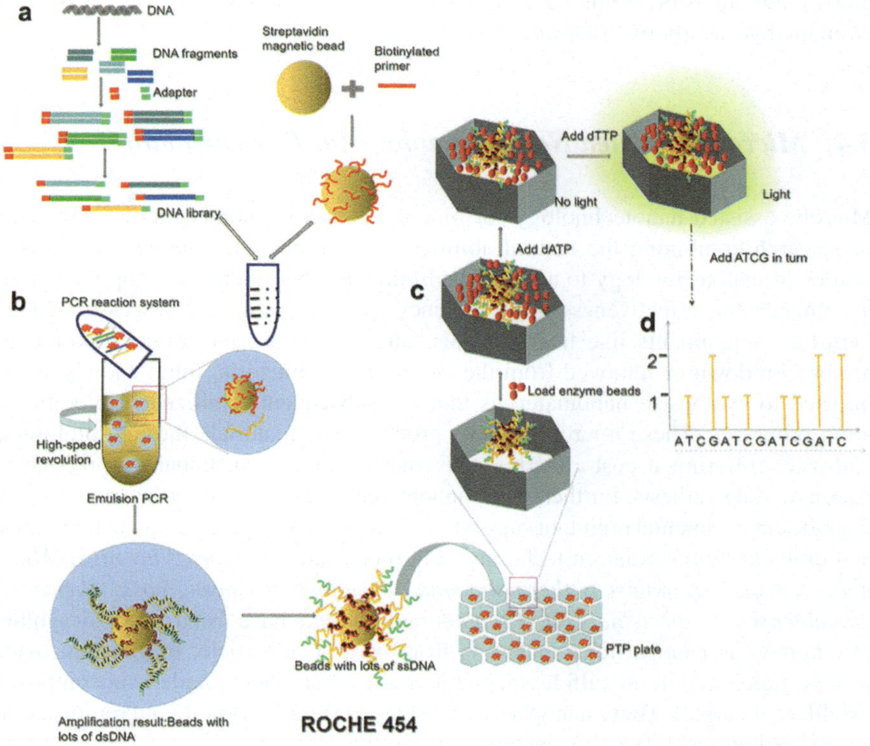

Fig. 7 Stepwise procedure of ROCHE 454 sequencing (Reproduced from Zhang et al., 2021)

Metagenomics can be combined with other prominent omics approaches including transcriptomics, proteomics and metabolomics providing greater information about genetic and functional diversity of the microbial community present in any environmental sample (Kumar et al., 2020). In another study, microbial characterization of sediment samples from Kinnickinnic, Milwaukee, and Menomonee Rivers near Milwaukee, Wisconsin, USA using shotgun sequencing approach demonstrated the presence of high amount of bacteria and archaea with respect to elevated pollutant concentrations, suggesting their property of metabolic adaptation and their potential to degrade complex compounds such as polycyclic aromatic hydrocarbons. Further, low population of fungi, viruses and protozoa at higher concentration of pollutant indicates the vulnerability of these groups and underscore the importance of protecting sensitive taxa to maintain ecological stability. Moreover, identification of microbial populations and functional genes associated to pollutant degradation lead toward a promising application of microbiome diversity as a biomarker for environmental monitoring (Muhie et al., 2025).

Similarly, MinION shotgun sequencing was employed to study biodegradation related genes in the polluted urban riverine system of Western India. The study revealed the presence of large number of genes associated with degradation of xenobiotic compounds including benzoate, nitrotoluene, aminobenzoate, and polycyclic aromatic hydrocarbons (Yadav et al., 2021).

3.4 Microbe-Assisted Nanotechnology for Bioremediation

Microbe-assisted nanotechnology for bioremediation is another cutting-edge area of research combining the natural abilities of microbes with the advanced capabilities of nanotechnology to address environmental pollution. This approach aims to enhance the effectiveness and efficiency of bioremediation processes, where harmful contaminants like heavy metals, oils, pesticides, and organic pollutants are broken down or removed from the environment. Bacteria, fungi and algae can be used to synthesize nanomaterials that are subsequently utilized in bioremediation processes. These microorganisms produce nanomaterials through biological pathways, offering a cost-effective and environmentally sustainable approach to nanomaterial synthesis. Further, these nanomaterials exhibit properties to remove or degrade environmental pollutant suggesting their promising potential as an innovative and efficient bioremediation technique. Microbes such as *Aspergillus niger, Marinobacter sp., Streptomyces thermolineatus, Pseudomonas aeruginosa, Shewanellaoneidensis, Escherichia coli, Pseudonomas putida* have been used to synthesize numerous nanoparticles including iron oxide nanoparticle, magnesium oxide nanoparticles, cadmium sulfide, magnetite/reduced graphene oxide nanocomposite (Malik et al., 2022). These nanoparticles displayed significant pollutant removal efficiency as high as 100%. For instance, iron oxide nanoparticles synthesized using

Aspergillus tubingensis demonstrated significant removal of heavy metals lead, nickel, copper and zinc upto 90% with a reuse potential of five cycles (Mahanty et al., 2020).

Govarthan et al. (2020) demonstrated the removal of 91% of phosphate and 85% of ammonia using iron oxide nanoparticles co-precipitated with exopolysaccharides from *Chlorella vulgaris.* Copper nanoparticles synthesized from *Escherichia* sp. SINT7 and *Bacillus flexus* exhibited azo dye degradation properties and resulted in reduction of reactive black-5, congo red, direct blue-1, and malachite green by 83.61, 97.07, 88.42, 90.55% and 88.16, 88.45, 67.274, 90.433% respectively (Batool et al., 2024; Noman et al., 2020). Complete removal of malachite green dye from wastewater was obtained using silver nanoparticles synthesized from *Bacillus paralicheniformis* strain KJ-16 (Allam et al., 2019).

The above studies actively demonstrates the potential of microbe-assisted nanotechnology in removing toxic pollutants from environment in a fast and efficient manner, reducing secondary pollutant generation. However further research is required to uncover the large scale application of this technique. Nanotechnology based microbial remediation are generally accompanied with few limitations including their in situ use, loss of reactivity, transportation, and effect on microorganisms (Das et al., 2018; Shahi et al., 2021). Further, the commercialization of these technologies on a large scale are quite difficult with only 1% of these are commercialized so far and require support from government funding in terms of incentives or grants thus making their production sustainable (Dwevedi et al., 2019).

4 Conclusion

1. Microorganism-based bioremediation offers a promising and sustainable approach to remediating environments contaminated by pollutants. The use of microorganisms has enabled the degradation of complex pollutants that conventional physical or chemical methods could not address. Rapid advancements in innovative technologies, such as CRISPR/Cas, metabolic engineering, and metagenomics, have significantly enhanced the biodegradation capabilities of native microbial strains. Metagenomics combined with omics based techniques provides a comprehensive genetic profile of the entire microbial community within an environmental sample, offering real-time insights into the functional capabilities of microbes. Such information is invaluable for identifying new genes with bioremediating potential and for designing novel remediation strategies. Additionally, the synergistic combination of microbiology and nanotechnology has created an efficient platform for remediating toxic contaminants, including heavy metals and organic pollutants. However, most of these techniques still need to be optimized for their better large scale application. Further, maintaining stable microbial communities under toxic environments is a complex process requiring more deep studies to avoid any environmental fluctuations. While traditional methods might not be considered highly efficient and effective in bioremediation

and waste management, they remain the most widely used methods. Therefore, successful implementation of new technologies will require careful work for their integration with the existing infrastructure. Collaborative efforts between researchers, engineers, and policymakers are essential for bridging the gap and enhancing the overall effectiveness of bioremediation.

References

Abdallah, M., Feroz, S., Alani, S., Sayed, E. T., & Shanableh, A. (2019). Continuous and scalable applications of microbial fuel cells: a critical review. *Reviews in Environmental Science and Bio/Technology, 18*(3), 543–578.

Abdelfattah, A., Hossain, M. I., & Cheng, L. (2020). High-strength wastewater treatment using microbial biofilm reactor: A critical review. *World Journal of Microbiology and Biotechnology, 36*(5), 75.

Abou-Aly, H. E., Youssef, A. M., Tewfike, T. A., El-Alkshar, E. A., & El-Meihy, R. M. (2021). Reduction of heavy metals bioaccumulation in sorghum and its rhizosphere by heavy metals-tolerant bacterial consortium. *Biocatalysis and Agricultural Biotechnology, 31*, Article 101911.

Al-Asheh, S., Bagheri, M., & Aidan, A. (2022). Removal of heavy metals from industrial wastewater using microbial fuel cell. Engineering in Life Sciences, 22(8), 535–549.

Allam, N. G., Ismail, G. A., El-Gemizy, W. M., & Salem, M. A. (2019). Biosynthesis of silver nanoparticles by cell-free extracts from some bacteria species for dye removal from wastewater. *Biotechnology Letters, 41*, 379–389.

Amend, J. P., & Shock, E. L. (2001). Energetics of overall metabolic reactions of thermophilic and hyperthermophilic archaea and bacteria. *FEMS Microbiology Reviews, 25*(2), 175–243.

Arzate-Cárdenas, M. A., Carbajal-Hernández, A. L., Ramos-Garza, J., Arreguin-Rebolledo, U., & Rico-Martínez, R. (2024). Study of microbial communities in degrading toxic pollutants in the wastewater and solid waste treatment industries. In *Microbial remediation of hazardous chemicals from water & industrial wastewater treatment plant* (pp. 71–94). Cham: Springer International Publishing.

Bal, B., Ghosh, S., & Das, A. P. (2018). Microbial recovery and recycling of manganese waste and their future application: A review. *Geomicrobiological Journal, 36*(1), 85–96. https://doi.org/10.1080/01490451.2018.1497731

Batool, F., Shahid, M., Mahmood, F., Shahzad, T., Azeem, F., Hussain, S., Algarni, T. S., Elshikh, M. S., Onazi, W. A. A., & Mustafa, S. (2024). Biosynthesis of copper nanoparticles using Bacillus flexus and estimation of their potential for decolorization of azo dyes and textile wastewater treatment. *Journal of King Saud University-Science, 36*(8), Article 103309.

Behera, I. D., Nayak, M., Biswas, S., Meikap, B. C., & Sen, R. (2021). Enhanced biodegradation of total petroleum hydrocarbons by implementing a novel two-step bioaugmentation strategy using indigenous bacterial consortium. *Journal of Environmental Management, 292*, Article 112746. https://doi.org/10.1016/J.JENVMAN.2021.112746

Behera, I. D., Nayak, M., Mishra, A., Meikap, B. C., & Sen, R. (2022). Strategic implementation of integrated bioaugmentation and biostimulation for efficient mitigation of petroleum hydrocarbon pollutants from terrestrial and aquatic environment. *Marine Pollution Bulletin, 177*, Article 113492. https://doi.org/10.1016/J.MARPOLBUL.2022.113492

Bharagava, R. N., Purchase, D., Saxena, G., & Mulla, S. I. (2019). Applications of metagenomics in microbial bioremediation of pollutants: from genomics to environmental cleanup. In *Microbial diversity in the genomic era* (pp. 459–477). Academic Press.

Bhattacharya, M., Guchhait, S., Biswas, D., & Singh, R. (2019). Evaluation of a microbial consortium for crude oil spill bioremediation and its potential uses in enhanced oil recovery. *Biocatalysis and Agricultural Biotechnology, 18*, Article 101034.

Bitton, G. (2011). *Wastewater microbiology*, 4th edn. Wiley-Blackwell, New York

Brozinčević, A., Grgas, D., Štefanac, T., Habuda-Stanić, M., Zelić, B., & Landeka Dragičević, T. (2024). Cost reduction in the process of biological denitrification by choosing traditional or alternative carbon sources. *Energies, 17*(15), 3660.

Cao, Z., Yan, W., Ding, M., & Yuan, Y. (2022). Construction of microbial consortia for microbial degradation of complex compounds. *Frontiers in Bioengineering and Biotechnology, 10*, 1051233.

Carpio, I. E. M., Franco, D. C., Sato, M. I. Z., Sakata, S., Pellizari, V. H., Ferreira Filho, S. S., & Rodrigues, D. F. (2016). Biostimulation of metal-resistant microbial consortium to remove zinc from contaminated environments. *Science of the Total Environment, 550*, 670–675.

Chae, K., Choi, M., Lee, J., Kim, K., & Kim, I. S. (2009). Bioresource technology effect of different substrates on the performance, bacterial diversity, and bacterial viability in microbial fuel cells. *Bioresource Technology, 100*, 3518–3525.

Che, S., Wang, J., Zhou, Y., Yue, C., Zhou, X., Xu, Y.,... & Yang, M. (2024). The adsorption and fixation of Cd and Pb by the microbial consortium weakened the toxic effect of heavy metal-contaminated soil on rice. Chemical Engineering Journal, 497, 154684.

Chen, J., Liu, Y., & Mahadevan, R. (2023). Genetic engineering of *Acidithiobacillus ferridurans* using CRISPR systems to mitigate toxic release in biomining. *Environmental Science & Technology, 57*(33), 12315–12324.

Chen, X., Wang, Y., Mamathaxim, N., Habibul, N., & Hu, Y. (2025). Simultaneous sulfamethazine and Cr (VI) removal in lab-scale microbial fuel cell-constructed wetland. *Journal of Environmental Sciences, 154*, 91–101.

Chen, Y., Wang, S., Geng, N., Wu, Z., Xiong, W., & Su, H. (2022). Artificially constructing mixed bacteria system for bioaugmentation of nitrogen removal from saline wastewater at low temperature. *Journal of Environmental Management, 324*, Article 116351.

Chettri, D., Verma, A. K., & Verma, A. K. (2024). Bioaugmentation: an approach to biological treatment of pollutants. Biodegradation, 35(2), 117–135.

Choi, Y., Jung, E., Kim, S., & Jung, S. (2003). Membrane fluidity sensoring microbial fuel cell. *Bioelectrochemistry, 59*, 121–127.

Choudhury, P., Ray, R. N., Bandyopadhyay, T. K., Basak, B., Muthuraj, M., & Bhunia, B. (2021). Process engineering for stable power recovery from dairy wastewater using microbial fuel cell. *International Journal of Hydrogen Energy, 46*(4), 3171–3182.

Das, S., Chakraborty, J., Chatterjee, S., & Kumar, H. (2018). Prospects of biosynthesized nanomaterials for the remediation of organic and inorganic environmental contaminants. *Environmental Science: Nano, 5*, 2784–2808.

Datta, S., Rajnish, K. N., Samuel, M. S., Pugazlendhi, A., & Selvarajan, E. (2020). Metagenomic applications in microbial diversity, bioremediation, pollution monitoring, enzyme and drug discovery. A review. *Environmental Chemistry Letters, 18*, 1229–1241.

Deng, L., Li, F., Zhou, S., Huang, D., & Ni, J. (2010). A study of electron-shuttle mechanism in Klebsiella pneumoniae based-microbial fuel cells. *Chinese Science Bulletin, 55*(1), 99–104.

Dey, S., Das, A., Mallick, K., Sahu, A., & Das, A. P. (2023). *Environmental petroleum waste: Pollution, toxicity, sustainable remediation* (pp. 159–175). https://doi.org/10.1007/978-3-031-48220-5_7

Dominic, D., & Baidurah, S. (2022). Recent developments in biological processing technology for palm oil mill effluent treatment—A review. *Biology, 11*(4), 525.

Dwevedi, A. (2019). *Solutions to environmental problems involving nanotechnology and enzyme technology*. Academic Press.

Einsle, O., & Kroneck, P. M. (2004). Structural basis of denitrification. *Biological chemistry, 385*(10).

Gao, D., Zhao, H., Wang, L., Li, Y., Tang, T., Bai, Y., & Liang, H. (2022). Current and emerging trends in bioaugmentation of organic contaminated soils: A review. *Journal of Environmental Management, 320*, Article 115799.

Ghosh, S., & Das, A. P. (2020). Microbial metagenomics: current advances in investigating microbial ecology and population dynamics. In S. K. Nayak, & B. B. Mishra (Eds.), *Frontiers in soil and environmental microbiology* (1st ed., pp. 247–254). CRC Press, London

Giwa, O. E., & Ibitoye, F. O. (2017). Bioremediation of heavy metal in crude oil contaminated soil using isolated Indigenous microorganism cultured with *E. coli* DE3 BL21. *International Journal of Engineering and Applied Sciences, 4*(6), Article 257436.

Govarthanan, M., Jeon, C. H., Jeon, Y. H., Kwon, J. H., Bae, H., & Kim, W. (2020). Non-toxic nano approach for wastewater treatment using *Chlorella vulgaris* exopolysaccharides immobilized in iron-magnetic nanoparticles. *International Journal of Biological Macromolecules, 162*, 1241–1249.

He, L., Du, P., Chen, Y., Lu, H., Cheng, X., Chang, B., & Wang, Z. (2017). Advances in microbial fuel cells for wastewater treatment. Renewable and Sustainable Energy Reviews, 71, 388–403.

Herrero, M., & Stuckey, D. C. (2015). Bioaugmentation and its application in wastewater treatment: A review. *Chemosphere, 140*, 119–128.

Huang, S., Fu, Y., Zhang, H., Wang, C., Zou, C., & Lu, X. (2023). Research progress of novel bio-denitrification technology in deep wastewater treatment. *Frontiers in Microbiology, 14*, 1284369.

Inoue, K., Leang, C., Franks, A. E., Woodard, T. L., Nevin, K. P., & Lovley, D. R. (2010). Specific localization of the c-type cytochrome OmcZ at the anode surface in current- producing biofilms of Geobacter sulfurreducens. *Environmental Microbiology Reports, 3*, 211–217.

International Solid Waste Association. (2024). *Global waste management outlook 2024: Beyond an age of waste, turning rubbish into a resource.*

Jaiswal, K. K., Kumar, V., Vlaskin, M. S., Sharma, N., Rautela, I., Nanda, M.,... & Chauhan, P. K. (2020). Microalgae fuel cell for wastewater treatment: Recent advances and challenges. Journal of Water Process Engineering, 38, 101549.

Jatoi, A. S., Akhter, F., Mazari, S. A., Sabzoi, N., Aziz, S., Soomro, S. A., Mubarak, N. M., Baloch, H., Memon, A. Q., Ahmed, S. (2020). Advanced microbial fuel cell for waste water treatment—A review. *Environmental Science and Pollution Research* 1–15.

Kahrizi, H., Garmdareh, S. E. H., & Abbassi, R. (2024). Simultaneous removal of heavy metals and electricity generation from wastewater in constructed wetland-microbial fuel cells. *Process Safety and Environmental Protection, 190*, 921–929.

Kang, C. S., Eaktasang, N., Kwon, D. Y., & Kim, H. S. (2014). Enhanced current production by Desulfovibrio desulfuricans biofilm in a mediator-less microbial fuel cell. Bioresource technology, 165, 27–30.

Kannan, P., Verma, I., Banerjee, B., & Saleena, L. M. (2024). Unveiling bacterial consortium for xenobiotic biodegradation from Pichavaram mangrove forest soil: A metagenomic approach. *Archives of Microbiology, 206*(1), 27.

Kim, H. J., Park, H. S., Hyun, M. S., Chang, I. S., Kim, M., & Kim, B. H. (2002). A mediator-less microbial fuel cell using a metal reducing bacterium, Shewanella putrefaciens. Enzyme and Microbial technology, 30(2), 145–152.

Kim, H. J., Park, H. S., Hyun, M. S., Chang, I. S., Kim, M., & Kim, B. H. (2002). A mediatorless microbial fuel cell using a metal reducing bacterium, Shewanella putrefaciens. *Enzyme and Microbial Technology, 30*, 145–152

Lee, K. C., Archer, S. D., Kansour, M. K., & Al-Mailem, D. M. (2024). Bioremediation of oily hyper-saline soil via autochthonous bioaugmentation with halophilic bacteria and archaea. *Science of the Total Environment, 922*, Article 171279.

Li, J., Tang, Q., Li, Y., Fan, Y. Y., Li, F. H., Wu, J. H., Min, D., Li, W. W., Lam, P. K., & Yu, H. Q. (2020). Rediverting electron flux with an engineered CRISPR-ddAsCpf1 system to enhance the pollutant degradation capacity of Shewanella oneidensis. *Environmental Science & Technology, 54*(6), 3599–3608.

Li, L., Liu, Y., Zhang, Y., Wang, X., Liu, H., Liu, Q., Chen, S., & Zhou, G. (2024). Dynamic responses in Pseudomonas stutzeri M3 bioaugmentation of crude-oil-contaminated soil: Hydrocarbons, microbial community structures, and functional genes. *Journal of Environmental Chemical Engineering, 12*(1), Article 111863.

Li, Q., Xu, Y., Liang, C., Peng, L., & Zhou, Y. (2023). Nitrogen removal by algal-bacterial consortium during mainstream wastewater treatment: Transformation mechanisms and potential N_2O mitigation. *Water Research, 235*, Article 119890.

Li, X., Zhong, G. Z., Qiao, Y., Huang, J., Hu, W. H., Wang, X. G., & Li, C. M. (2014). A high performance xylose microbial fuel cell enabled by Ochrobactrum sp. 575 cells. RSC Advances, 4(75), 39839–39843.

Lin, Z., Pang, S., Zhou, Z., Wu, X., Li, J., Huang, Y., Zhang, W., Lei, Q., Bhatt, P., Mishra, S., & Chen, S. (2022). Novel pathway of acephate degradation by the microbial consortium ZQ01 and its potential for environmental bioremediation. *Journal of Hazardous Materials, 426*, Article 127841.

Liu, S., Chen, Q., Ma, T., Wang, M., & Ni, J. (2018). Genomic insights into metabolic potentials of two simultaneous aerobic denitrification and phosphorus removal bacteria, *Achromobacter* sp. GAD3 and *Agrobacterium* sp. LAD9. *FEMS Microbiology Ecology, 94*(4).

Liu, X., Mei, S., & Salles, J. F. (2023). Inoculated microbial consortia perform better than single strains in living soil: A meta-analysis. *Applied Soil Ecology, 190*, Article 105011.

Mahanty, S., Chatterjee, S., Ghosh, S., Tudu, P., Gaine, T., Bakshi, M., et al. (2020). Synergistic approach towards the sustainable management of heavy metals in wastewater using mycosynthesized iron oxide nanoparticles: Biofabrication, adsorptive dynamics and chemometric modeling study. *Journal of Water Process Engineering, 37*, Article 101426.

Mahbub, K. R., Krishnan, K., Andrews, S., Venter, H., Naidu, R., & Megharaj, M. (2017). Bioaugmentation and nutrient amendment decrease concentration of mercury in contaminated soil. *Science of the Total Environment, 576*, 303–309.

Mandeep, & Shukla, P. (2020). Microbial nanotechnology for bioremediation of industrial wastewater. *Frontiers in Microbiology, 11*, 590631.

Mathuriya, A. S., & Yakhmi, J. V. (2014). Microbial fuel cells to recover heavy metals. *Environmental Chemistry Letters, 12*, 483–494.

Medaura, M. C., Guivernau, M., Moreno-Ventas, X., Prenafeta-Boldú, F. X., & Viñas, M. (2021). Bioaugmentation of native fungi, an efficient strategy for the bioremediation of an aged industrially polluted soil with heavy hydrocarbons. *Frontiers in Microbiology, 12*, Article 626436.

Mei, C., Zhang, M., Chen, Y., Dong, K., Sun, R., Zhang, X., & Li, H. (2025). Exploring the applications and carbon reduction of multi-technology-coupled membrane biofilm reactors for sustainable wastewater treatment: A review. *Environmental Science: Water Research & Technology, 11*, 793–808.

Meng, P., Pei, H., Hu, W., Shao, Y., & Li, Z. (2014). How to increase microbial degradation in constructed wetlands: Influencing factors and improvement measures. *Bioresource Technology, 157*, 316–326.

Mishra, S., & Das, A. P. (2023). Microbial remediation of synthetic microfiber contaminated wastewater. In M. P. Shah (Ed.), *Microbial technologies in industrial wastewater treatment* (pp. 337–350). Springer. https://doi.org/10.1007/978-981-99-2435-6_17

Mishra, S., Das, A. P., & Seragadam, P. (2009). Microbial remediation of hexavalent chromium from chromite contaminated mines of Sukinda Valley, Orissa (India). *J Environ Res Dev, 3*, 1122–1127.

Mishra, S., Dash, D., & Das, A. P. (2023). Aquatic microbial diversity on plastisphere: Colonization and potential role in microplastic biodegradation. *Geomicrobiology Journal*, 1–12. https://doi.org/10.1080/01490451.2023.2209750

Mishra, S., Dash, D., & Das, A. P. (2024a). Isolation and molecular characterization of polyester-degrading bacterial strains from municipal wastewater and their application in microplastic degradation in contaminated wastewater. *Geomicrobiology Journal, 41*(7), 732–741.

Mishra, S., Dash, D., & Das, A. P. (2024b). Microbial degradation of polyester microfibers using indigenously isolated bacterial strain *Exiguobacterium* Sp. *CLEAN – Soil, Air, Water, 52*(12), e202300343. https://doi.org/10.1002/CLEN.202300343

Mishra, S., Ghosh, S., van Hullebusch, E. D., Singh, S., & Das, A. P. (2023b). A critical review on the recovery of base and critical elements from electronic waste-contaminated streams using microbial biotechnology. *Applied Biochemistry and Biotechnology, 2023*, 1–30. https://doi.org/10.1007/S12010-023-04440-X

Mishra, S., Rout, P. K., & Das, A. P. (2019). Solar photovoltaic panels as next generation waste: A review. *Biointerface Research in Applied Chemistry, 9*(6).

Mishra, S., Swain, S., Sahoo, M., Mishra, S., & Das, A. P. (2021). Microbial colonization and degradation of microplastics in aquatic ecosystem: A review. *Geomicrobiology Journal*. https://doi.org/10.1080/01490451.2021.1983670

Mohanty, S., Ghosh, S., Bal, B., & Das, A. P. (2018). A review of biotechnology processes applied for manganese recovery from wastes. *Reviews in Environmental Science and Bio/Technology, 17*(4), 791–811. https://doi.org/10.1007/S11157-018-9482-1

Muhie, S., Gautam, A., Mylroie, J., Sowe, B., Campbell, R., Perkins, E. J., Hammamieh, R., & Garcia-Reyero, N. (2025). Effects of environmental chemical pollutants on microbiome diversity: Insights from shotgun metagenomics. *Toxics, 13*(2), 142.

Munoz-Cupa, C., Hu, Y., Xu, C. C., & Bassi, A. (2020). An overview of microbial fuel cell usage in wastewater treatment, resource recovery and energy production. *Science of the Total Environment, 754*, 142429

Muñoz, R., Alvarez, M. T., Muñoz, A., Terrazas, E., Guieysse, B., & Mattiasson, B. (2006). Sequential removal of heavy metals ions and organic pollutants using an algal-bacterial consortium. *Chemosphere, 63*(6), 903–911.

Mustakeem, M. (2015). Electrode materials for microbial fuel cells: nanomaterial approach. *Materials for Renewable and Sustainable Energy, 4*, 22.

Nawaz, A., ul Haq, I., Qaisar, K., Gunes, B., Raja, S. I., Mohyuddin, K., & Amin, H. (2022). Microbial fuel cells: Insight into simultaneous wastewater treatment and bioelectricity generation. Process Safety and Environmental Protection, 161, 357–373.

Nikel, P. I., & de Lorenzo, V. (2021). Metabolic engineering for large-scale environmental bioremediation. *Metabolic Engineering: Concepts and Applications, 13*, 859–890.

Noman, M., Shahid, M., Ahmed, T., Niazi, M. B. K., Hussain, S., Song, F., & Manzoor, I. (2020). Use of biogenic copper nanoparticles synthesized from a native *Escherichia* sp. as photocatalysts for azo dye degradation and treatment of textile effluents. *Environmental Pollution, 257*, 113514.

Nwaehiri, U. L., Akwukwaegbu, P. I., & Nwoke, B. E. B. (2020). Bacterial remediation of heavy metal polluted soil and effluent from paper mill industry. *Environmental Analysis Health and Toxicology, 35*, 1–10.

Nwankwegu, A. S., & Onwosi, C. O. (2017). Bioremediation of gasoline contaminated agricultural soil by bioaugmentation. *Environmental Technology & Innovation, 7*, 1–11.

Obileke, K., Onyeaka, H., Meyer, E. L., & Nwokolo, N. (2021). Microbial fuel cells, a renewable energy technology for bio-electricity generation: A mini-review. Electrochemistry Communications, 125, 107003.

Onwujekwe, E. C., Ogbonna, J. C., & Eze, C. N. (2024). Remediation of spent engine oil-contaminated soil through biostimulation and bioaugmentation with sodium dodecyl sulphate (SDS) and indigenous hydrocarbonoclastic bacterial isolates. *World Journal of Microbiology and Biotechnology, 40*(2), 50.

Pandya, R. S., Kaur, T., Bhattacharya, R., Bose, D., & Saraf, D. (2024). Harnessing microorganisms for bioenergy with microbial fuel cells: Powering the future. Water-Energy Nexus, 7, 1–12.

Pant, D., Van Bogaert, G., Diels, L., & Vanbroekhoven, K. (2010). A review of the substrates used in microbial fuel cells (MFCs) for sustainable energy production. *Bioresource Technology, 101*(6), 1533–1543.

Parameswaran, P., Bry, T., Popat, S. C., Lusk, B. G., Rittmann, B. E., & Torres, C. I. (2013). Kinetic, electrochemical, and microscopic characterization of the thermophilic, anode-respiring bacterium Thermincola ferriacetica. Environmental science & technology, 47(9), 4934–4940.

Park, D. H., & Zeikus, J. G. (1999). Utilization of electrically reduced neutral red by Actinobacillus succinogenes: Physiological function of neutral red in membrane driven fumarate reduction and energy conservation. *Journal of Bacteriology, 181*, 2403–2410.

Patel, D., Bapodra, S. L., Madamwar, D., & Desai, C. (2021). Electroactive bacterial community augmentation enhances the performance of a pilot scale constructed wetland microbial fuel cell for treatment of textile dye wastewater. Bioresource Technology, 332, 125088.

Patowary, K., Patowary, R., Kalita, M. C., & Deka, S. (2016). Development of an efficient bacterial consortium for the potential remediation of hydrocarbons from contaminated sites. *Frontiers in Microbiology, 7*, 1092.

Priyadarsini, S., & Das, A. P. (2025). Lithium bioleaching: A review on microbial-assisted sustainable technology for lithium bio-circularity. *Journal of Water Process Engineering, 69*, Article 106744. https://doi.org/10.1016/j.jwpe.2024.106744

Rahmani, A. R., Navidjouy, N., Rahimnejad, M., Alizadeh, S., Samarghandi, M. R., & Nematollahi, D. (2022). Effect of different concentrations of substrate in microbial fuel cells toward bioenergy recovery and simultaneous wastewater treatment. *Environmental Technology, 43*(1), 1–9.

Raper, E., Stephenson, T., Anderson, D. R., Fisher, R., & Soares, A. (2018). Industrial wastewater treatment through bioaugmentation. *Process Safety and Environmental Protection, 118*, 178–187.

Saeed, M. U., Hussain, N., Sumrin, A., Shahbaz, A., Noor, S., Bilal, M., Aleya, L., & Iqbal, H. M. (2022). Microbial bioremediation strategies with wastewater treatment potentialities—A review. *Science of the Total Environment, 818*, Article 151754.

Said, M. M., Saad, A. A., John, G. R., & Mushi, A. T. (2023). Generation of electricity by using microbial fuel cell prototype fed by sewage: Case study at the University of Dar es Salaam. *Tanzania Journal of Science, 49*(1), 240–249.

Salinas, J., Carpena, V., Martínez-Gallardo, M. R., Segado, M., Estrella-González, M. J., Toribio, A. J., Jurado, M. M., López-González, J. A., Suárez-Estrella, F., & López, M. J. (2023). Development of plastic-degrading microbial consortia by induced selection in microcosms. *Frontiers in Microbiology, 14*, 1143769.

Shahi, M. P., Kumari, P., Mahobiya, D., & Shahi, S. K. (2021). Nano-bioremediation of environmental contaminants: Applications, challenges, and future prospects. In *Bioremediation for environmental sustainability* (pp. 83–98).

Sharma, M., Sharma, S., Alkhanjaf, A. A. M., Arora, N. K., Saxena, B., Umar, A.,... & Baskoutas, S. (2025). Microbial fuel cells for azo dye degradation: A perspective review. *Journal of Industrial and Engineering Chemistry, 142*, 45–67.

Sharma, P., Tripathi, S., & Chandra, R. (2021). Metagenomic analysis for profiling of microbial communities and tolerance in metal-polluted pulp and paper industry wastewater. *Bioresource Technology, 324*, Article 124681.

Singh, H. M., Pathak, A. K., Chopra, K., Tyagi, V. V., Anand, S., & Kothari, R. (2019). Microbial fuel cells: A sustainable solution for bioelectricity generation and wastewater treatment. *Biofuels, 10*(1), 11–31.

Slate, A. J., Whitehead, K. A., Brownson, D. A., & Banks, C. E. (2019). Microbial fuel cells: An overview of current technology. *Renewable and Sustainable Energy Reviews, 101*, 60–81.

Sodhi, A. S., Sharma, N., Bhatia, S., Verma, A., Soni, S., & Batra, N. (2022). Insights on sustainable approaches for production and applications of value added products. *Chemosphere, 286*, Article 131623.

Song, Y., An, J., & Chae, K. J. (2017). Effect of temperature variation on the performance of microbial fuel cells. *Energy Technology, 5*(12), 2163–2167.

Sun, J., Yuan, M., Zhou, H., Chen, Z., Wang, Y., & Cheng, H. (2025). Enhanced printing and dyeing wastewater treatment using anaerobic-aerobic systems with bioaugmentation. *Journal of Hazardous Materials, 486*, Article 136982.

Tamilarasan, K., Banu, J. R., Jayashree, C., Yogalakshmi, K. N., & Gokulakrishnan, K. (2017). Effect of organic loading rate on electricity generating potential of upflow anaerobic microbial fuel cell treating surgical cotton industry wastewater. *Journal of Environmental Chemical Engineering, 30*, 145–152.

Tamilarasan, K., Banu, J. R., Jayashree, C., Yogalakshmi, K. N., & Gokulakrishnan, K. (2017). Effect of organic loading rate on electricity generating potential of upflow anaerobic microbial fuel cell treating surgical cotton industry wastewater. *Journal of environmental chemical engineering, 5*(1), 1021–1026.

Tan, S. M., Ong, S. A., Ho, L. N., Wong, Y. S., Thung, W. E., & Teoh, T. P. (2020). The reaction of wastewater treatment and power generation of single chamber microbial fuel cell against substrate concentration and anode distributions. Journal of Environmental Health Science and Engineering, 18(2), 793–807.

Tripathy, B., Dey, S., Dubey, N. K., Lo, W.-C., & Das, A. P. (2024). *An omics-based approach for managing the risk of COVID-19-generated wastes* (pp. 51–66). https://doi.org/10.1007/978-3-031-50840-0_4

Truong, D. H., Dam, M. S., Bujna, E., Rezessy-Szabo, J., Farkas, C., Vi, V. N. H., Csernus, O., Nguyen, V. D., Gathergood, N., Friedrich, L., & Hafidi, M. (2021). In situ fabrication of electrically conducting bacterial cellulose-polyaniline-titanium-dioxide composites with the immobilization of Shewanella xiamenensis and its application as bioanode in microbial fuel cell. *Fuel, 285*, Article 119259.

Wang, L., Peng, R., Tian, Y., Xu, J., Wang, B., Han, H., Fu, X., Gao, J., & Yao, Q. (2022a). Metabolic engineering of *Escherichia coli* for efficient degradation of 4-fluorophenol. *AMB Express, 12*(1), 55.

Wang, L., Xu, D., Zhang, Q., Liu, T., & Tao, Z. (2022b). Simultaneous removal of heavy metals and bioelectricity generation in microbial fuel cell coupled with constructed wetland: An optimization study on substrate and plant types. *Environmental Science and Pollution Research, 29*, 768–778.

Wang, Q., Hou, J., Peng, L., Liu, W., & Luo, Y. (2025). Dynamic responses in bioaugmentation of petroleum-contaminated soils using thermophilic degrading consortium HT: Hydrocarbons, microbial communities, and functional genes. *Journal of Hazardous Materials, 487*, Article 137222.

Wang, Y., Peng, B., Yang, Z., Chai, L., Liao, Q., Zhang, Z., & Li, C. (2015). Bacterial community dynamics during bioremediation of Cr(VI)-contaminated soil. *Applied Soil Ecology, 85*, 50–55.

Wang, Y., Zhang, X., & Lin, H. (2024). Effects of pH on simultaneous Cr (VI) and p-chlorophenol removal and electrochemical performance in Leersia hexandra constructed wetland-microbial fuel cell. *Environmental Technology, 45*(3), 483–494.

Wani, A. K., Akhtar, N., Naqash, N., Chopra, C., Singh, R., Kumar, V., Kumar, S., Mulla, S. I., & Américo-Pinheiro, J. H. P. (2022). Bioprospecting culturable and unculturable microbial consortia through metagenomics for bioremediation. *Cleaner Chemical Engineering, 2*, Article 100017.

Wu, J., Cheng, Z. H., Min, D., Cheng, L., He, R. L., Liu, D. F., & Li, W. W. (2020). CRISPRi system as an efficient, simple platform for rapid identification of genes involved in pollutant transformation by Aeromonas hydrophila. *Environmental Science & Technology, 54*(6), 3306–3315.

Xu, J., Wang, B., Wang, M. Q., Gao, J. J., Li, Z. J., Tian, Y. S., Peng, R. H., & Yao, Q. H. (2022). Metabolic engineering of *Escherichia coli* for methyl parathion degradation. *Frontiers in Microbiology, 13*, Article 679126.

Xu, W., Yang, B., Wang, H., Zhang, L., Dong, J., & Liu, C. (2023). Simultaneous removal of antibiotics and nitrogen by microbial fuel cell-constructed wetlands: Microbial response and carbon–nitrogen metabolism pathways. *Science of the Total Environment, 893*, Article 164855.

Yaashikaa, P. R., Devi, M. K., & Kumar, P. S. (2022). Engineering microbes for enhancing the degradation of environmental pollutants: A detailed review on synthetic biology. *Environmental Research, 214*, Article 113868.

Yadav, R., Rajput, V., & Dharne, M. (2021). Functional metagenomic landscape of polluted river reveals potential genes involved in degradation of xenobiotic pollutants. *Environmental Research, 192*, Article 110332.

Yang, T., Yang, Q., Shi, Y., Xin, Y., Zhang, L., Gu, Z., & Shi, G. (2021). Insight into the denitrification mechanism of *Bacillus subtilis* JD-014 and its application potential in bioremediation of nitrogen wastewater. *Process Biochemistry, 103*, 78–86.

Yu, B., Feng, L., He, Y., Yang, L., & Xun, Y. (2021). Effects of anode materials on the performance and anode microbial community of soil microbial fuel cell. *Journal of Hazardous Materials, 401*, Article 123394.

Zhang, C., Wu, D., & Ren, H. (2020). Bioremediation of oil contaminated soil using agricultural wastes via microbial consortium. *Scientific Reports, 10*(1), 9188.

Zhang, E. R., Liu, L., & Cui, Y. Y. (2013). Effect of pH on the performance of the anode in microbial fuel cells. *Advanced Materials Research, 608*, 884–888.

Zhang, H., Zhao, Z., Li, S., Chen, S., Huang, T., Li, N., Yang, S., Wang, Y., Kou, L., & Zhang, X. (2019). Nitrogen removal by mix-cultured aerobic denitrifying bacteria isolated by ultrasound: Performance, co-occurrence pattern and wastewater treatment. *Chemical Engineering Journal, 372*, 26–36.

Zhang, L., Chen, F., Zeng, Z., Xu, M., Sun, F., Yang, L., Bi, X., Lin, Y., Gao, Y., Hao, H., Yi, W., Li, M, Xie, Y. (2021). Advances in metagenomics and its application in environmental microorganisms. *Frontiers in microbiology, 12*, 766364.

Zhang, M., Liu, X., Zhu, W., Hu, S., Yan, X., & Hong, Q. (2024). Remediation of isoproturon-contaminated soil by *Sphingobium* sp. strain YBL2: Bioaugmentation, detoxification and community structure. *Journal of Hazardous Materials, 476*, 134968.

Zhang, T., & Zhang, H. (2022). Microbial consortia are needed to degrade soil pollutants. *Microorganisms, 10*(2), 261.

Zhou, H., Gao, X., Wang, S., Zhang, Y., Coulon, F., & Cai, C. (2023). Enhanced bioremediation of aged polycyclic aromatic hydrocarbons in soil using immobilized microbial consortia combined with strengthening remediation strategies. *International Journal of Environmental Research and Public Health, 20*(3), 1766.

Zumft, W. G. (1997). Cell biology and molecular basis of denitrification. *Microbiology and Molecular Biology Reviews, 61*, 533–546.

Biotechnological Exploitation of Microbes for Sustainable Waste Management

Rahul Gogoi and Sudipta Sankar Bora

Abstract Recent decades have seen a rapid rise in industrialization and urbanization, thereby intensifying the process of environmental pollution and waste generation, which are today some of the single largest threats to sustainable development. Microbial technology provides an insightful road for waste management practices, utilizing the capabilities of microorganisms to degrade, detoxify, and recycle various environmental pollutants. Thus, this chapter discusses a range of advanced microbial technologies including, but certainly not limited to, bioaugmentation, bioremediation, and genetically modified microorganisms to preferentially attack waste. Special attention is paid to the role of microbial consortia in the biodegradation of complex pollutants where synergetic interactions have been documented that enhance degradation efficiency. Alongside, various case studies are given on the application of these technologies toward wastewater, solid waste, and heavy metals. It extends the idea of microbial fuel cells and biosensors as new vectors in waste-to-energy conversion and real-time environmental monitoring. Challenges of scale-up, regarding its regulatory framework and public perception, are thereby additionally considered, alongside future perspectives on integrating microbial technology within the circular economy models. In conclusion, this latter chapter discusses that advanced microbial approaches hold promise toward the establishment of a sustainable and eco-friendlier framework for environmental waste management.

Keywords Microbial technology · Sustainable waste management · Genetically modified microorganisms (GMOs) · Microbial consortia · Microbial fuel cells · Waste-to-energy conversion · Circular economy

---

R. Gogoi
Department of Agricultural Biotechnology, Assam Agricultural University, Jorhat, Assam, India

S. S. Bora (✉)
Silchar Regional Forensic Science Laboratory, Assam Under the Directorate of Forensic Science, Kahilipara, Guwahati, Assam, India
e-mail: sudip.asm@gmail.com

A. P. Das (ed.), *Bioconversion of Environmental Waste to Value Added Products*, Sustainable Environmental Waste Management Strategies,
https://doi.org/10.1007/978-3-032-14215-3_10

1 Introduction

Pollution and waste are among the largest problems in the world today. The speed at which industry, urbanization, and population have expanded over the last hundred years has been catastrophic for the environment (Ahmed et al., 2022). Heavy metals, chemical leaks and spills, agricultural runoff, and others of this sort pollute air, water, and soil and pose a "serious danger to human beings and ecological systems" (Mishra et al., 2022). As an example, the United Nations Environmental Program, UNEP, indicated that more than 8 million tons of plastic waste yearly reached our oceans by 2019, damaging marine biodiversity and affecting coastal communities (Mishra et al., 2023). In addition, polluted waters from industrial waste and sewage still affect many water sources, posing access to clean water as a problem in several areas of the world (Benya et al., 2023). This continuous generating of waste has considerably overstepped ordinary waste management capabilities and thus indicates the rising necessity for ways to confront pollution in sustainable and effective manners (Anekwe & Isa, 2023).

Microbial technology therefore represents an attractive option under such circumstances. Microorganisms, including bacteria, fungi, and algae, have developed an incredible ability to break down a wide range of pollutants (Maglione et al., 2024). These tiny organisms naturally contribute to nutrient cycling by decomposing organic matter and converting harmful substances into less toxic forms. The use of microbial technology for environmental cleanup commonly known as bioremediation takes advantage of the metabolic versatility of microbes to detoxify or degrade pollutants. Microbial biotechnology, which has been developed in recent decades to employ natural processes more efficiently and specifically to remediate environmental waste, is now being used more broadly (Hyde et al., 2019).

Microbial technology offers numerous benefits compared to traditional waste management methods. Widespread methodologies such as landfills, burning, or chemical processes generally expend a large amount of energy and create toxic remnants that can be an environmental hazard for the long term (Singh et al., 2024). Microbial processes, on the other hand are more sustainable, economical, and environmentally benign. Depending on the type of waste, microbes may be used to treat waste "in situ," or at the place of contamination, or "ex situ," or in a controlled environment (Nag et al., 2023). Nonetheless, these microbial approaches reduce dependence on toxic chemicals, reduce pollution, and are applicable at different scales and to various pollutants, whether organic matter or heavy metals (Siddiqua et al., 2022).

The field of advanced microbial approaches in waste management is rapidly expanding as researchers discover new uses and refine existing techniques. One of the principal drivers of this growth, is the capacity of microorganisms to persist and grow in varied and extreme conditions. Some microbes can break down the complex hydrocarbons found in oil spills and some heavy metal-resistant microbes which bioaccumulate heavy metals, making them an essential step in remediating contaminated sites (Gonzalez & Aranda, 2023). Additionally, the potential of genetically modified microorganisms (GMOs) has emerged, enhancing their biodegradation abilities to tackle pollutants that are tough to break down. The ability to engineer microbes to perform specific tasks like degrading plastic polymers or lowering toxic nitrate levels in groundwater opens up exciting possibilities for tackling environmental problems (Han et al., 2022).

In addition, what a microbial technological approach can offer in a sustainable waste management is highly correlated with the philosophy of the circular economy. Within this economic form, it encourages zero waste as well as continual resource reuse such as recycling, upcycling, and recovery (Ghosh et al., 2023). Microorganisms play an important role in this vision as they are responsible for recycling organic waste and turning it into valuable resources such as biogas, compost, or bio-based chemicals. Microbial fuel cells, for instance, can transform organic waste into electricity, thus transforming waste into a renewable energy source. Microbial processes can also recycle materials such as plastics and metals and contribute in this way to closing the loop of industry.

2 Microbial Technologies for Waste Degradation

Microbial technologies have proved effective solutions for remediating environmental waste. The ability for bacteria to digest complex organic waste and inorganic pollutants has been employed for a range of biotechnological procedures. Advanced technologies including bioaugmentation, bioremediation, and genetically modified organisms (GMOs) have collaboratively boosted the efficiency of waste treatment systems. Research suggests that these.

2.1 Bioaugmentation

Bioaugmentation is a technique used in bioremediation that includes purposely adding specific strains of bacteria to contaminated areas to enhance the breakdown of certain contaminants. This approach works especially effectively in areas where the indigenous microbial populations aren't strong enough or lack the ability to rapidly digest these dangerous compounds. Bioaugmentation generally utilizes either native (autochthonous) or foreign (allochthonous) microbial strains. These strains can assist directly break down pollutants or boost the general metabolic capacities of the existing microbial community (Nzila et al., 2016).

For bioaugmentation to be successful, numerous aspects come into play, such as how compatible the microorganisms are with the current microbial population, ambient conditions like pH, temperature, and nutrient levels, as well as the properties of the pollutants involved. Recent breakthroughs in molecular biology and genetic engineering have increased bioaugmentation approaches by generating microbial strains that can digest pollutants more effectively, adapt better to severe environments, and survive longer in contaminated settings (Muter, 2023).

A famous example of bioaugmentation in action is its application in clearing up hydrocarbon contamination, particularly from marine oil spills. A study indicated how bioaugmentation could speed up the process of oil spill restoration. In this work, researchers injected hydrocarbon-eating bacteria, such as *Pseudomonas aeruginosa* and *Alcanivorax borkumensis*, to oil-polluted sea areas (Gao et al., 2023). These bacteria are notable for their capacity to exploit hydrocarbons as a food supply, breaking down complicated petroleum molecules into less hazardous chemicals. Over four weeks, the researchers documented a considerable decline in total petroleum hydrocarbons (TPH), confirming the effectiveness of bioaugmentation in reducing oil pollution (Xu et al., 2018).These findings underline the promise of bioaugmentation as a sustainable and effective option for addressing marine oil spills, especially when employed alongside other bioremediation strategies like biostimulation, which involves balancing nutrients and oxygen to stimulate microbial growth (Fallahi et al., 2023).

However, despite its benefits, bioaugmentation does come with problems that need to be overcome for it to be generally utilized. Factors such as environmental stress, competition from local bacteria, and possible dangers from protozoan predation can affect the survival and efficacy of the introduced microbes. Moreover, thorough evaluation of regulatory problems and the ecological implications of introducing non-native microbial strains is required (Behera et al., 2021).

2.2 *Bioremediation*

A natural cleaning method employing microorganisms to degrade, neutralize, or transform pollutants into benign forms is bioremediation. Bioremediation falls into two categories: one is in situthat is, right at the contaminated location; the second is ex situ, whereby the contaminants are removed for treatment in a controlled environment (Romantschuk et al., 2023). In situ methods tend to be cheaper and less intrusive, employing things like bioventing, biosparging, and phytoremediation (Paul et al., 2023; Pradhan et al., 2023). One widely established application of in situ bioremediation is the treatment of nitrate contamination in groundwater by nitrate-reducing bacteria through the process of denitrification, which converts nitrates to nitrogen gas (Safonov et al., 2018). Ex situ remediation requires the excavation of contaminated materials for treatment in bioreactors, composting systems, or land farming techniques. A popular method of composting is to use municipal solid wastes along with useful bacteria to speed up the degradation of organic matters and recycle nutrients (Thirumalaivasan et al., 2024).

However, bioremediation is suitable for the cleaning of soils contaminated by heavy metals. Tang et al. (2024) studied the abilities of metal-resistant bacteria, including *Bacillus subtilis* and *Pseudomonas fluorescens*, to remediate agricultural soil polluted by lead and cadmium (Tang et al., 2024). Such bacteria were extraordinarily capable of absorbing heavy metals and could reduce their concentration by more than 60% within 90 days. This says that bioremediation stands to become a sustainable option for the treatment of heavy metal contamination and thus a more environmentally friendly alternative to physicochemical methods. While the advantages of bioremediation are many, among which cost-effectiveness and eco-friendliness have been emphasized, there remain issues that must be addressed by future research to enhance its success. These include the survival of the microbes themselves, site-specific differences, and increased bioavailability of metals (Abo-Alkasem et al., 2023).

2.3 *Genetically Modified Microorganisms (GMOs)*

Genetically modified microorganisms (GMOs) have been produced to improve their ability to break down contaminants, making them a useful resource for cleaning up the environment. Modifications to their DNA could enable us to produce microbes that do their jobs more effectively and are able to withstand a wider range of conditions (Deckers et al., 2020). They are highly capable of conquering various forms of environmental pollution. Overall, using GMOs for waste degradation is advantageous because more enzymes increase treatment speed, they can metabolize items that other treatments cannot, they handle tough circumstances well and waste treatment costs less and finishes faster (Yang et al., 2024).

There are highly capable genetically modified microbes that are used to manage waste. For example, *Escherichia coli* has been engineered to more effectively break-down polychlorinated biphenyls (PCBs), whereas *Pseudomonas putida* has been modified to break down aromatic hydrocarbons like benzene and toluene (Rafeeq et al., 2023). Moreover, scientists have modified *Saccharomyces cerevisiae* to remove heavy metals from industrial waste which turns out to be more sustainable than traditional wastewater treatment methods. Managing waste becomes easier when GMOs are used to process plastic. A significant study by Kalathil et al. (2022) found *Ideonella sakaiensis*, a bacterium that can destroy polyethylene terephthalate (PET) plastic (Kalathil et al., 2022). Modifying the genes in these bacteria markedly improved how fast it decomposed PET. This innovation solved the issue of plastic pollution, as it offers a possible natural way to handle the plastic we have stored in landfills globally. Yet, using genetically modified microorganisms to help the environment raises concerns about security, their effects on nature and the restrictions placed by regulations.

2.4 *Role of Microbial Consortia*

It has been demonstrated that microbial consortia, which are made up of different interacting microbial species, are more efficient at decomposing pollutants than single-strain techniques. They are therefore a viable choice for bioremediation. Through the use of many metabolic pathways, the cooperative interactions within these consortia enable a more effective breakdown of complex contaminants. These communities' function through a number of primary mechanisms, including redox cycling, in which certain microbes alter environmental conditions to benefit others, horizontal gene transfer, which increases microbial capabilities by exchanging genes linked to degradation, and syntrophic interactions, in which various species cooperate to break down complex compounds (Li et al., 2022).

The treatment of industrial wastewater is one important use for microbial consortia. *Pseudomonas*, *Bacillus*, and *Rhodococcus* species worked well together in this situation (Chattopadhyay et al., 2022). The corporation demonstrated its potential for large-scale waste treatment by successfully breaking down phenolic chemicals and reducing chemical oxygen demand (COD) by 90%. The consortium's microbial species diversity improved overall treatment efficiency by enabling a greater variety of substrates and increased resistance to environmental changes.

Microbial consortiums have applications outside wastewater treatment, such as hydrocarbon degradation, heavy metal bioremediation, and the breakdown of plastic waste. However, maintaining these consortia's stability and functionality under various environmental circumstances continues to provide difficulties. Developments in synthetic biology, metagenomics, and microbial ecology are assisting in the development of optimized microbial communities designed to combat certain contaminants (Lokesh et al., 2023).

2.5 Microbial Fuel Cells (MFCs) for Waste-to-Energy Conversion

Microbial fuel cells (MFCs) are a viable alternative for wastewater treatment and renewable energy since they use the natural mechanisms of electrogenic bacteria to convert organic waste into power (Jalili et al., 2024). These systems rely on microorganisms such as *Shewanella oneidensis* and *Geobacter sulfurreducens*, which produce bioelectricity by breaking down organic materials and passing electrons to an electrode. MFCs effectively remove organic pollutants and provide clean energy by utilizing the processes of microbial respiration (Garbini et al., 2023).

However, MFCs' potential uses extend beyond wastewater treatment; they may also be used in bioelectric sensors, desalination systems, and the bioremediation of toxic materials. To enable larger-scale deployments, however, a number of obstacles must be overcome, including low power density, electrode fouling problems, and expensive material prices. Thankfully, advancements in synthetic biology, system integration, and nanomaterials are improving MFC performance and opening the door for their eventual commercialization (Fatehbasharzad et al., 2022).

3 Role of Microbial Consortia in Waste Degradation

Microbial consortia, made up of many interacting microbial species, are crucial when it comes to breaking down trash. Unlike individual microbial strains, these consortia exhibit synergistic interactions that dramatically accelerate the breakdown of complex contaminants. To fully mineralize a variety of pollution types microbial communities, work together metabolically, share resources, and feed on one another. With the present challenges with waste disposal, microbial consortia are an acceptable, sustainable bioremediation solution (Zhang & Zhang, 2022).

3.1 Mechanisms of Microbial Consortia in Waste Degradation

Microbial consortia are diverse groups of microbial species that synergistically communities that utilize diverse pathways of degradation to foster waste partitioning (Fig. 1). With respect to bioremediation, microbial systems are integral to the improvement of the bioremediation process, which supports microbial consortia as a viable intervention for environmental management (Cao et al., 2022).

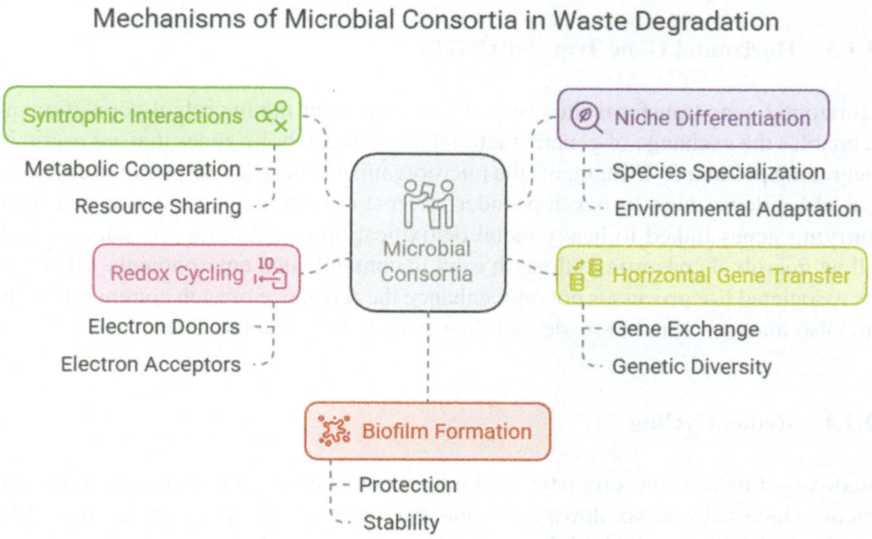

Fig. 1 Overview of microbial consortia in waste water treatment

3.1.1 Syntrophic Interactions

Microbial communities can have species that cooperate to degrade complex contaminants to simpler ones. The simpler ones can be taken up by other members of the community. This collaboration enables the efficient breakdown of large chemical molecules. Anaerobic digesting systems show this mechanism. The first stage of the method is to convert organic waste to fatty acids using fermentative bacteria. The second stage is achieved with methanogenic archaea which convert those fatty acids to methane. The breakdown of organic waste occurs much faster when it is accomplished in two stages (Cai et al., 2023).

3.1.2 Niche Differentiation

By focusing on distinct metabolic pathways, microbial consortia are able to differentiate their habitats and flourish. The breakdown of different contaminants is accelerated and competition is lessened because to this division of labor. Anaerobic microorganisms work with aromatic hydrocarbons in low-oxygen environments, whereas aerobic bacteria primarily deal with alkanes in the decomposition of petroleum hydrocarbons. By working together this way, the microbial community can more effectively target a wider variety of contaminants (Castro et al., 2022).

3.1.3 Horizontal Gene Transfer (HGT)

Horizontal gene transfer is thus very advantageous for the microbial population, as it enables the exchange of genetic material, such as catabolic genes that are useful in degrading pollutants. It augments the microorganisms' destructive potential capacity, but also allows them to develop under adverse conditions. Considering plasmids carrying genes linked to heavy-metal detoxification or antibiotic resistance which allow microbial consortia to flourish even in contaminated environments. HGT and its associated life processes not only enhance the resilience of such communities but are also major factors in broadening their potential (Sitaraman, 2018).

3.1.4 Redox Cycling

Redox cycling is when certain types of microbes change the redox conditions around them, which helps break down pollutants that need specific electron acceptors like oxygen, nitrate, or sulfate. Take iron-reducing bacteria, for example; they change the redox potential in contaminated sediments, making it easier to break down organic pollutants. In the cleaning up of polluted sites, this process is very important as it creates these conducive conditions which help in shedding these persevering pollutants (Dong et al., 2022).

3.1.5 Biofilm Formation

Many groups of microorganisms create biofilms, which help give them structural support and safeguard them from environmental stress like extreme temperatures or changes in pH. Additionally, these biofilms boost how effectively pollutants are broken down. In wastewater treatment systems, biofilm bacteria actually eliminate organic matter and heavy metals more efficiently than those found in free planktonic conditions. The biofilm matrix allows microbes easier contact with pollutants while guarding the community from outside dangers, thus making the overall degradation process stronger and more sustainable (Muhammad et al., 2020).

3.2 Applications of Microbial Consortia in Waste Management

Microbial consortia are commonly used in a variety of waste degradation processes, providing an effective and robust approach to treating wastewater, breaking down solid waste, and cleaning up hazardous materials (Fig. 2). Their capacity to collaborate enables them to tackle complex pollutants, making them an important asset in managing environmental issues (Kuppan et al., 2024).

Fig. 2 Microbial consortia for sustainable waste management

3.2.1 Wastewater Treatment

Microbial consortia are essential for performing the breakdown of organic pollutants, nitrogenous compounds, and toxic materials in the treatment of wastewater including industrial waste. The diversity of microbial communities that can exist in activated sludge systems (*Pseudomonas*, *Bacillus*, and *Nitrosomonas*) contribute to the effective mineralization of organic matter and the uptake and destruction of nutrients such as nitrogen and phosphorus (Ziembińska-Buczyńska et al., 2019).The first step in eliminating water pollution is managing and cleaning out its organic and industrial contaminants. Li et al. (2023) pointed out the significance of microbial consortia in their study about the biodegradation of crude oil by these three microorganisms. In 30 days, the partnership was able to decompose 75% of the total petroleum hydrocarbons (TPH), according to the research (Li et al., 2023).

3.2.2 Solid Waste Decomposition

Microbial communities are essential for reducing solid waste because they promote organics' anaerobic breakdown and composting. Examples of microbial communities are the various bacteria and fungi of species *Aspergillus niger* and *B. subtilis* that aid in the breakdown of lignocellulosic materials in composting and produce compost with nutrients that are beneficial to soil quality (Rastogi et al., 2020). Likewise, these microbial communities are important for the anaerobic digestion process that converts organic waste to biogas, a renewable source of energy. A study assessed the functions of bacterial and fungal groups when composting municipal solid waste and found that a consortium of *A. niger*, *Trichoderma harzianum*, and *B. subtilis* increased the degradation rate of cellulose and lignin and reduced composting time by 40%. This strongly indicates the potential for microbial consortia in improving efficiencies of waste composting (Yao et al., 2023).

3.2.3 Bioremediation of Heavy Metals and Xenobiotics

Microbial consortia play an important role in remediating heavy metals and persistent organic pollutants (POPs).These microbial consortia can detoxify all kinds of pollution through biosorption, bioaccumulation, and enzymatic degradation and remediation. For instance, a group of intricate microorganisms (*Bacillus*, *Pseudomonas*, and sulfate-reducing bacteria) effectively removed toxic metals (arsenic and cadmium) from industrial waste, which is an eco-friendly method of industrial waste treatment (Pande et al., 2022). Zulfiqar et al. (2023) explored the ability of metal-resistant microorganisms in rehabilitating lead and cadmium polluted soil. The research indicated that the specific consortium (*B. subtilis* and *P. fluorescens*) exhibited excellent biosorption efficiency with more than 60% removal of heavy metals within about 90 days. The research emphasizes the viability of microbial consortia for the treatment of soil heavy metal contamination (Zulfiqar et al., 2023).

3.2.4 Degradation of Plastic Waste

Some groups of microbes are impressively capable of breaking down synthetic plastics like polyethylene, polypropylene, and polystyrene. For instance, *I. sakaiensis* and other bacteria have shown promise in degrading polyethylene terephthalate (PET), a prevalent type of plastic waste (Choi et al., 2024). Global plastic waste management could undergo a radical change if a biological process that breaks down plastics is discovered. Sharma et al. (2023a, 2023b) investigated how well specific microbial consortia treated wastewater from textile production operations (Sharma et al., 2023a, 2023b).Through successful decolorization and 90% chemical oxygen demand (COD) reduction, their study demonstrated the effectiveness of such microbial consortia in treating complicated industrial wastewaters. *Rhodococcus*, *Sphingomonas*, and *Pseudomonas* comprised the microbial consortium they employed.

4 Microbial Approaches for Specific Waste Types

Microbial technology is increasingly being accepted as green technology for waste treatment. Different microbial groups, such as bacteria, fungi, and archaea, have certain enzymes that allow them to break down and transform complex waste into less toxic or even valuable by-products. With the variety of waste types that need special microbial techniques for maximum efficiency, this section discusses distinct microbial techniques for different waste classes, illustrated through relevant case studies.

4.1 Microbial Degradation of Organic Waste

Degradation of various types of organic waste by microorganisms were presented in (Table 1).

4.2 Microbial Treatment of Plastic Waste

Treatment of various types of plastic waste by microorganisms were presented in (Table 2).

4.3 Microbial Bioremediation of Heavy Metals

Bioremediation of heavy metals by microorganisms were presented in (Table 3).

Table 1 Different microorganisms involved in organic waste degradation, types of organic waste they act on, their mechanisms of action

Microorganism	Type of organic waste	Mechanism of action	References
Bacillus spp.	Agricultural residues, food waste, manure	Generates extracellular enzymes (such as cellulases, amylases, and proteases) that degrade starch, cellulose, and proteins	Rowińska et al. (2024)
Pseudomonas spp.	Petroleum waste, industrial waste, food waste	Produces biosurfactants to accelerate hydrocarbon breakdown; oxidizes organic contaminants	Pérez-Armendáriz et al. (2019)
Trichoderma spp.	Lignocellulosic biomass (wood, crop residues)	Breaks down lignocellulose by secreting ligninases, cellulases, and hemicellulases	Organo et al. (2022)
Aspergillus spp.	Agricultural waste, municipal waste, food waste	Enables the breakdown of carbohydrates by producing cellulases, xylanases, and pectinases	Heidarzadeh et al. (2019)
Lactobacillus spp.	Food waste, composting materials	Organic matter ferments to produce lactic acid, which lowers pH and promotes breakdown	Tran et al. (2019)
Methanobacterium spp.	Organic sludge, animal waste, wastewater	Biogas (methane) is produced by anaerobic digestion and methanogenesis	Yang et al. (2025)
Actinomycetes (e.g., *Streptomyces* spp.)	Compost, soil organic matter, plant residues	To break down complex organic matter, it produces cellulolytic and ligninolytic enzymes	Farda et al. (2022)

Table 2 Different microorganisms involved in plastic waste treatment, the types of plastic they act on, their mechanisms of action

Microorganism	Type of plastic waste	Mechanism of action	References
Ideonella sakaiensis	Polyethylene terephthalate (PET)	Produces the enzymes PETase and MHETase, which convert PET into ethylene glycol and terephthalic acid	Sevilla et al. (2023)
Pseudomonas aeruginosa	Polyurethane (PU)	Releases hydrolases and esterases that break down PU into monomers and oligomers	Su et al. (2023)
Aspergillus niger	Polyethylene (PE), Polypropylene (PP)	Creates the oxidizing enzymes (peroxidases, laccases) that start the breakdown of polymers	Temporiti et al. (2022)
Bacillus subtilis	Polycaprolactone (PCL), Polylactic acid (PLA)	Produces lipases and proteases that hydrolyze biodegradable plastics' ester bonds	Bonifer et al. (2019)
Rhodococcus ruber	Polyethylene (PE)	Releases oxidative enzymes and forms a biofilm, which reduces PE to smaller pieces	Zampolli et al. (2022)
Penicillium spp.	Polyvinyl chloride (PVC), Polyethylene (PE)	Releases extracellular enzymes that cause the oxidation and fragmentation of polymers	El-Dash et al. (2023)
Comamonas acidovorans	Polyhydroxyalkanoates (PHA)	PHA biopolymers are hydrolyzed to produce monomers for microbial metabolism	Vermeer et al. (2022)

4.4 *Microbial Approaches to Electronic Waste (E-Waste) Recycling*

E-Waste Recycling by microorganisms were presented in (Table 4).

5 Emerging Microbial Technologies

With more attention on the environment and better waste control, scientists are focusing on using microorganisms. With the help of bacteria, those new techniques are able to treat water, recycle waste and make useful goods. Artificial intelligence

Table 3 Different microorganisms involved in heavy metal bioremediation, the types of heavy metals they act on, their mechanisms of action

Microorganism	Heavy metal	Mechanism of action	References
P. aeruginosa	Lead (Pb), Cadmium (Cd), Chromium (Cr)	Using metal-binding proteins and extracellular polysaccharides for biosorption and bioaccumulation	Tasleem et al. (2023)
B. subtilis	Arsenic (As), Mercury (Hg), Lead (Pb)	uses precipitation, bioaccumulation, and enzymatic reduction to lessen metal toxicity	Wróbel et al. (2023)
Escherichia coli	Mercury (Hg), Arsenic (As)	Arsenate reduction to arsenite; mercuric reductase conversion of Hg^{2+} to volatile Hg^0	Tseng et al. (2023)
Rhizobium spp.	Copper (Cu), Zinc (Zn), Lead (Pb)	Root nodule chelation and sequestration, soil metal immobilization	Riseh et al. (2023)
A. niger	Lead (Pb), Cadmium (Cd), Zinc (Zn)	Increases bioavailability by producing organic acids that solubilize heavy metals	Pande et al. (2022)
Saccharomyces cerevisiae	Chromium (Cr), Lead (Pb), Copper (Cu)	Biosorption involves surface adsorption and metal binding through cell wall proteins	Igiri et al. (2018)
Desulfovibrio desulfuricans	Uranium (U), Chromium (Cr), Manganese (Mn)	Metals are precipitated by sulfate-reducing bacteria (SRB) using bio reduction	Kapahi and Sachdeva (2019)
Shewanella oneidensis	Iron (Fe), Manganese (Mn), Chromium (Cr)	As electron acceptors, metals are used in dissimilatory metal reduction	Ng et al. (2015)

(AI)-assisted microbial optimization, microbiome engineering, nanobiotechnology, and synthetic biology are among the techniques opening the door to greater economic viability and environmental sustainability.

5.1 *Synthetic Biology for Enhanced Bioremediation*

With the use of synthetic biology, researchers can now modify microorganisms so they cause less trash waste. They choose organisms from various bacteria that are

Table 4 Different microorganisms involved in electronic waste (E-waste) recycling, the types of E-waste they act on, their mechanisms of action

Microorganism	Type of e-waste	Mechanism of action	References
Acidithiobacillus ferrooxidans	Printed circuit boards (PCBs), gold-containing waste	To make metals like Cu, Au, and Zn soluble, bioleaching oxidizes Fe^{2+} and S^{2+}	Tapia et al. (2022)
Leptospirillum ferrooxidans	PCBs, copper wires	Dissolves Cu and other metals by producing ferric ions (Fe^{3+})	Wu et al. (2018)
P. aeruginosa	PCBs, mobile phone waste	Enhances metal recovery by producing organic acids and biosurfactants	Hossain and Sahajwalla (2024)
Fusarium oxysporum	E-waste containing gold (Au)	Gold ions are bound and reduced to nanoparticles through biomineralization	Kitching et al. (2015)
A. niger	Cathode ray tubes (CRTs), PCBs	Produces oxalic and citric organic acids, which dissolve heavy metals	Alias et al. (2021)
Cupriavidus metallidurans	Gold-containing waste, electronic chips	Toxic metal ions are transformed into innocuous nanoparticles via bioprecipitation	Golzar-Ahmadi et al. (2024)
Bacillus cereus	E-waste contaminated soil	Heavy metals such as Pb, Cd, and Ni are accumulated from waste sites through biosorption	Priyashantha et al. (2022)

able to uptake and decompose harmful compounds better, faster and more cheaply. Dealing with both plastic waste and heavy metal contamination is one of the key problems that this strategy can solve (Bala et al., 2022).

Synthetic biology has expanded what's possible for managing wastes with the help of microbes: scientists are now able to genetically modify microorganisms to more efficiently process waste. And by selectively culling through microbial genomes and metabolic pathways, researchers are enlisting microbes that can tackle complex pollutants more quickly and cheaply, with less energy. The method can clearly solve some significant environmental problems, like plastic waste and heavy metal pollution (Sadanov et al., 2025).

Another significant application of synthetic biology is the design of microorganisms that can help clean up heavy metal contamination. Microbes like *E. coli* and *P. putida* have been genetically enhanced to improve their ability to capture and neutralize heavy metals from polluted environments. By expressing metal-binding proteins and transporters, these modified bacteria can effectively sequester toxic metals such as cadmium and lead, making them less available and toxic (Thai et al., 2023). Moreover, these synthesized microbes have the potential to employ methods such as bioaccumulation, biomineralization and likewise, extracellular precipitation to sequester the heavy metals and prevent their spread throughout the environment. According to a study by Lu et al. (2023), 90% of the lead and cadmium in wastewater was removed by genetically modified *E. coli* strains that expressed more metallothioneins and phytochelatins (Lu et al., 2023). Consequently, these bacteria could work at a high level and save a lot of money when it comes to big-scale heavy metal cleanup.

5.2 Nanobiotechnology in Microbial Waste Management

The fascinating field of nanobiotechnology combines the unique properties of nanomaterials with microbial activities to combat pollution, recover resources, and keep an eye on the environment. This field increases the efficacy and efficiency of bioremediation techniques by fusing nanoparticles with microbial systems, particularly in challenging environmental settings. The collaboration between nanomaterials and microorganisms paves the way for innovative methods to manage pollution and enhance ecosystem health (Koul et al., 2021).

Nanoparticles like iron oxide, silver, and carbon-based materials are vital in improving how well microbes break down hydrocarbons during oil spills. These tiny particles help make hydrocarbons more bioavailable, meaning they're easier for microbes to access. They contribute to microbial metabolism, which speeds up the breakdown of contaminants in both terrestrial and marine environments. Iron oxide nanoparticles, for instance, dramatically accelerated the biodegradation rate of the hydrocarbon-degrading bacterium *A. borkumensis*, and after 30 days, crude oil contamination decreased by 75% (Patowary et al., 2023). This highlights the potential that nanoparticles may have for influencing microbial processes, and offers a new approach for remediation of oil spills in affected locations.

An unusual and complex method in nanobiotechnology involves nano-encapsulation which encases microorganisms in nanomaterials to shield them from environmental stress while preserving their biodegradation functions. This technique protects microorganisms against extreme pH values, elevated temperatures, and toxic contaminants to maintain their survival and activity in difficult environments (Pateiro et al., 2021). The system enables a temporally controlled microbial discharge which enhances their bioremediation performance. Research showed that encapsulating *P. fluorescens* in silica nanoparticles increased pesticide degradation rates by 60% compared to free-living bacteria (Singh & Saxena, 2022). The improved stability and efficiency of these encapsulated microbes make this technology a valuable option for bioremediation in areas with changing environmental conditions.

5.3 *Microbiome Engineering for Waste Recycling*

Technological advancements are essential in improving biofuel production methods from organic waste to generate biogas, bioethanol, and biodiesel. By precisely adjusting microbial populations, the decomposition process for waste materials becomes more efficient, leading to increased biofuel production. Microbiome engineering aims to modify microbial communities to enhance their natural functions which transform waste into valuable products including biofuels, bioplastics, and biofertilizers. Researchers can manipulate microbial interactions within communities to boost waste conversion processes, which then deliver sustainable waste management and resource recovery solutions (Banu et al., 2021).

The relationships among different microbes in these engineered communities are carefully managed to maximize substrate use and improve conversion efficiency. For instance, a study found that engineered anaerobic consortia boosted methane yields by 40% in food waste digestion systems (Gong et al., 2021). This increase not only makes biofuel production more efficient but also offers a cleaner alternative to traditional fossil fuels, paving the way for greener energy solutions.

Utilizing bacterial metabolic processes, microbial electrochemical technologies such as microbial fuel cells (MFCs) and microbial electrolysis cells (MECs) produce hydrogen and energy from organic waste products. By cleaning wastewater and generating renewable energy, these systems offer two benefits. Microbial populations break down organic matter inside these cells, releasing electrons that are then trapped and used to generate power. MFCs based on *S. oneidensis* converted wastewater to power with a 70% conversion efficiency (Chakma et al., 2025). Such technologies present a scalable, sustainable approach to both waste treatment and energy generation, making them an exciting option for decentralized wastewater management and clean energy production.

On another note, pharmaceutical waste, particularly antibiotics and analgesics, poses serious environmental and health risks due to the persistence of these substances in water systems. A study by Gong et al. (2018) targeted the rational design of *P. putida* for biodegradation of pharmaceutically active compounds in water purification works (Gong et al., 2018). The engineered strains managed to break down 95% of these pollutants, working to mitigate the effect of the pharmaceutical residues and slow the proliferation of antibiotic-resistant bacteria. This work highlights how synthetic biology can fight the widespread problem of pharmaceutical pollution in our lakes and rivers.

Biosensor technology also plays a critical role by using genetically engineered microbes to detect and assess environmental contaminants in real-time. Mendes et al. (2025) conducted a field study employing bioluminescent *E. coli* strains to spot arsenic contamination in groundwater (Mendes et al., 2025). This biosensor system provided a quick, accurate, and cost-effective solution for pollution monitoring, especially in regions with limited resources. Rapid detection of pollutants is essential for protecting public health and ensuring timely responses to environmental contamination.

Lastly, the advent of CRISPR-Cas9 genome editing has made it possible to execute targeted changes in microbial genomes, enhancing their ability to break down environmental pollutants. Chen et al., (2022) led a pioneering study where *S. oneidensis* was engineered with CRISPR to improve its capacity for chromium reduction (Chen et al., 2022). The modified microbial strain demonstrated an incredible threefold increase in chromium reduction capacity, providing a scalable environmentally friendly way to treat industrial wastewater that contains dangerous heavy metals. The study shows that gene editing methods increase the efficacy of microbial bioremediation in industrial settings.

6 Challenges and Limitations

Microbial processes are appealing alternatives for sustainable waste treatments. They rely on the ability of microorganisms to metabolize and/or transform pollutants, recycle waste, and/or produce economically valuable commodities. However, problems and limitations will prevent the widespread application or the practical realization of this capacity.

6.1 Scale-Up Difficulties

One of the greatest challenges of using microbial technologies for waste management purposes is the translation of lab results into industrial or field applications. Microbial processes are sometimes more illustrious in the lab for waste breakdown purposes, and once the lab scale increases to industrial or field capacities, many more variables need to be monitored and maintained so the microbes are effective at metabolizing that waste (Crater & Lievense, 2018). Environmental and operational factors may fluctuate and affect microbial performance, such as temperature, pH, nutrient levels, and quantities of contaminant. Accordingly, the fluctuations present in larger environments for metabolism of waste can also affect microbial viability and growth, through nutrient nutrition. In addition to this, when the scale of application increases, the microbial population can easily go out of balance, which can dramatically limit the utility of bioremediation (Ashfaq et al., 2023).

In wastewater treatment, methods like microbial biofilms and activated sludge systems tend to perform really well in controlled environments. However, when it comes to scaling up these approaches for handling larger quantities of wastewater, particularly in industrial contexts, things can get tricky. Additionally, keeping the right mix of microbes for large-scale operations means you need to constantly monitor and fine-tune the process, which can make everything more complicated and expensive (Godzieba et al., 2022).

6.2 *Economic Feasibility*

Microbial process has been considered as cost-effective alternatives to chemical and physical treatment. The actual application of the microbial technologies at full scale typically proves difficult due to lack of technical/economic feasibility. These costs rapidly increase when growing specific microorganisms in addition to costs for controlling optimal environmental parameters, and for (large scale) bioreactors (Sharma et al., 2023a, 2023b). The microbially mediated treatment times in one or more of these microbial treatment processes can exceed conventional treatment times, thereby increasing the cost of operation of the processes. Processes of bioremediation degrade heavy metals and pesticides and do so more slowly than whenever chemicals are used, hence they are not suitable for those fields in which it is required a rapid decontamination. The cost of investment in the development of infrastructure and indigenous knowledge to scale a microbial process becomes a significant issue when industries are situated in economically backward regions (Yamin et al., 2023).

6.3 *Regulatory Frameworks and Biosafety Concerns*

Applying genetically modified organisms (GMO) to waste disposal have enormous regulatory and biosafety issues. These creatures may be built by scientists to rid a site of contaminants, but when released into the wild, they can be harmful and disrupt local insect populations. Most countries have regulations that require testing and approval before the release of modified organisms into the environment (Abdul Aziz et al., 2022).People also worry about horizontal gene transfer where changed genes from GMOs could spread to native microbial populations producing unforeseen changes in the ecosystem. This implies having safety rules and containment measures is crucial. Yet, these safety precautions make the process more complex and potentially slow down the wider usage of GMOs in environmental activities (Mitchell & Bartsch, 2020).

6.4 *Public Perception and Ethical Considerations*

Adoption and usage of microbe-based technologies (those associated with synthetic biology and genetically modified organisms) may be hampered by public perceptions. Many people have ethical and safety concerns about releasing modified microbes into the environment. This hesitation toward GMOs driven by concerns about losing biodiversity harming the environment, and risking health, can slow down regulatory approvals and make it tougher for these breakthroughs to gain traction (Ou & Guo, 2023).Ethics also play a significant role when we talk about altering microbial genomes for environmental purposes.

6.5 *Environmental Factors and Uncertainty*

The treatment process in microbial waste degradation is subject to various environmental factors, i.e., temperature, moisture content, inhibitors (organic solvents and heavy metals), and contamination (pharmaceuticals or transgenic hybrids) that can greatly interfere with the process. Bacteria respond to changes in their environment, and any environmental change can hinder bacteria's ability to break down waste. For instance, temperatures that are too high or too low, or pollutant levels that are too high can trigger stress responses or even kill cells (Pande et al., 2022). Also, microbial communities are not necessarily capable of degrading all pollutants effectively. Therefore, in some cases, certain toxic pollutants, may require certain microbial species, or enzymatic pathways that are not available in natural communities, and in certain cases genetic modifications will need to occur for these processes to become effective (Miglani et al., 2022).

7 Future Prospects and Circular Economy Integration

Considering their part in a circular economy which helps to limit our environmental impact by means of reducing, reusing, and recycling materials microbial technologies seems to have a bright future in waste management. By reducing pollutants, recycling trash, and producing valuable byproducts, microbial technologies are best suited to improve this situation.

7.1 *Microbial Technologies for Waste-to-Resource Conversion*

Key elements of the solution to turn waste into a resource, microbial technologies can help with this paradigm change. For example, microbial fuel cells (MFCs). By producing power and treating organic trash, this innovative approach lowers waste and offers an energy source (Roy et al., 2023).Moreover, as synthetic biology and genetic engineering advance, we will be able to modify these microorganisms to create bio-based products that could replace petroleum-based plastic, therefore fostering a permanent material economy. From organic trash, microbes can also create bioplastics, biofuels, and several other bioproducts (Folino et al., 2023). These technologies fit perfectly into current garbage disposal systems to assist waste-to-energy projects and support the sustainable energy model of a circular economy.

7.2 *Integration with Renewable Energy Systems*

Nowadays, there is a global trend toward renewable sources of energy, which can be used for establishing integral waste-treatment platforms that combine microbial and solar, wind and other forms of alternative energy. For example, microbially mediated processes to form biogas by anaerobic digestion can be employed in the context of renewable energy systems constituting a stable and clean energy supply (Kumar et al., 2023). These types of relationships may increase resilience, generate sustainability and enable decentralized community energy production and waste management.

7.3 *Microbial Approaches in Resource Recovery*

Microbial communities have a potential involvement in the recovery of valuable metals and minerals from wastes, including mining tailings and electronic waste. These bacteria are able to leach and absorb the copper, gold and vital rare earths from waste electronics (Newsome & Falagán, 2021). The approach will reduce dependence on traditional mining means that are resource depleting and an environmental hazardous supporting the concept of circular economy.

8 Conclusion

Microbial technologies offer a novel approach to efficient waste treatment, providing environmentally benign substitutes for traditional chemical and physical processes. These technologies can recycle waste, break down pollutants, and produce electricity, biofuels, and bioplastics by harnessing the natural abilities of bacteria. Microbial biotechnologies find their place in a circular economy as an effective means of waste minimization and resource recovery while contributing positively to sustainability. Nevertheless, the challenge of the scale of operation, the economic viability, and the addressing of concerns such as regulatory compliance and social acceptance are the problems that must be surmounted for these technologies to be efficiently integrated into the market. These core issues have another set that appear with changing skies: climatic variation, long-lived biodegradation, and the safe handling of gene-engineered microbes. The development of synthetic biology, bioengineering, and microbial consortium studies has been paving the way toward stronger and more workable microbial systems able to tackle many waste types: industrial by-product wastes and e-waste as cases.

Also, microbial technologies in a circular economy are based on renewable energy and recovering resources technologies, like microbes' technologies for biogas and microbial fuel cells. Ultimately, these projects make the ecological and monetary benefits tangible, the junk can be harvested for rare metals and energy can be generated from garbage. Garbage disposal has been a huge problem, and ultimately, microbial technologies could change that into a massive opportunity. Instead, it is important that further research, supportive legislation, and technology advancements come into play to harness their potential for global waste challenges, and the transition to a more sustainable and circular economy.

References

Abdul Aziz, M., Brini, F., Rouached, H., & Masmoudi, K. (2022). Genetically engineered crops for sustainably enhanced food production systems. *Frontiers in Plant Science, 13*. https://doi.org/10.3389/fpls.2022.1027828

Abo-Alkasem, M. I., Hassan, N. H., & Abo Elsoud, M. M. (2023). Microbial bioremediation as a tool for the removal of heavy metals. *Bulletin of the National Research Centre, 47*(1), 31. https://doi.org/10.1186/s42269-023-01006-z

Ahmed, F., Ali, I., Kousar, S., & Ahmed, S. (2022). The environmental impact of industrialization and foreign direct investment: Empirical evidence from Asia-Pacific region. *Environmental Science and Pollution Research, 29*(20), 29778–29792. https://doi.org/10.1007/s11356-021-17560-w

Alias, C., Bulgari, D., Bilo, F., Borgese, L., Gianoncelli, A., Ribaudo, G., Gobbi, E., & Alessandri, I. (2021). Food waste-assisted metal extraction from printed circuit boards: The *Aspergillus niger* route. *Microorganisms, 9*(5), 895. https://doi.org/10.3390/microorganisms9050895

Anekwe, I. M. S., & Isa, Y. M. (2023). Bioremediation of acid mine drainage—Review. *Alexandria Engineering Journal, 65*, 1047–1075. https://doi.org/10.1016/j.aej.2022.09.053

Ashfaq, R., Rasul, A., Asghar, S., Kovács, A., Berkó, S., & Budai-Szűcs, M. (2023). Lipid nanoparticles: An effective tool to improve the bioavailability of nutraceuticals. *International Journal of Molecular Sciences, 24*(21), 15764. https://doi.org/10.3390/ijms242115764

Bala, S., Garg, D., Thirumalesh, B. V., Sharma, M., Sridhar, K., Inbaraj, B. S., & Tripathi, M. (2022). Recent strategies for bioremediation of emerging pollutants: A review for a green and sustainable environment. *Toxics, 10*(8), 484. https://doi.org/10.3390/toxics10080484

Banu, J. R., Kumar, G., & Chattopadhyay, I. (2021). Management of microbial enzymes for biofuels and biogas production by using metagenomic and genome editing approaches. *3 Biotech, 11*(10), 429. https://doi.org/10.1007/s13205-021-02962-x

Behera, I. D., Nayak, M., Biswas, S., Meikap, B. C., & Sen, R. (2021). Enhanced biodegradation of total petroleum hydrocarbons by implementing a novel two-step bioaugmentation strategy using indigenous bacterial consortium. *Journal of Environmental Management, 292*, 112746. https://doi.org/10.1016/J.JENVMAN.2021.112746

Benya, A., Mohanty, S., Hota, S., Das, A. P., Rath, C. C., Achary, K. G., & Singh, S. (2023). Endangered Curcuma caesia Roxb.: Qualitative and quantitative analysis for identification of industrially important elite genotypes. *Industrial Crops and Products, 195*, 116363. https://doi.org/10.1016/J.INDCROP.2023.116363

Bonifer, K. S., Wen, X., Hasim, S., Phillips, E. K., Dunlap, R. N., Gann, E. R., DeBruyn, J. M., & Reynolds, T. B. (2019). *Bacillus pumilus* B12 degrades polylactic acid and degradation is affected by changing nutrient conditions. *Frontiers in Microbiology, 10*. https://doi.org/10.3389/fmicb.2019.02548

Cai, Z., Li, M., Zhu, Z., Wang, X., Huang, Y., Li, T., Gong, H., & Yan, M. (2023). Biological degradation of plastics and microplastics: A recent perspective on associated mechanisms and influencing factors. *Microorganisms, 11*(7), 1661. https://doi.org/10.3390/microorganisms11071661

Cao, Z., Yan, W., Ding, M., & Yuan, Y. (2022). Construction of microbial consortia for microbial degradation of complex compounds. *Frontiers in Bioengineering and Biotechnology, 10*. https://doi.org/10.3389/fbioe.2022.1051233

Castro, A. R., Martins, G., Salvador, A. F., & Cavaleiro, A. J. (2022). Iron compounds in anaerobic degradation of petroleum hydrocarbons: A review. *Microorganisms, 10*(11), 2142. https://doi.org/10.3390/microorganisms10112142

Chakma, R., Hossain, M. K., Paramasivam, P., Bousbih, R., Amami, M., Toki, G. F. I., Haldhar, R., & Karmaker, A. K. (2025). Recent applications, challenges, and future prospects of microbial fuel cells: A review. *Global Challenges, 9*(5). https://doi.org/10.1002/gch2.202500004

Chattopadhyay, I., J, R. B., Usman, T. M. M., & Varjani, S. (2022). Exploring the role of microbial biofilm for industrial effluents treatment. *Bioengineered, 13*(3), 6420–6440. https://doi.org/10.1080/21655979.2022.2044250

Chen, L., Wu, Y., Shen, Q., Zheng, X., & Chen, Y. (2022). Enhancement of hexavalent chromium reduction by *Shewanella oneidensis* MR-1 in presence of copper nanoparticles via stimulating bacterial extracellular electron transfer and environmental adaptability. *Bioresource Technology, 361*, 127686. https://doi.org/10.1016/j.biortech.2022.127686

Choi, J., Kim, H., Ahn, Y.-R., Kim, M., Yu, S., Kim, N., Lim, S. Y., Park, J.-A., Ha, S.-J., Lim, K. S., & Kim, H.-O. (2024). Recent advances in microbial and enzymatic engineering for the biodegradation of micro- and nanoplastics. *RSC Advances, 14*(14), 9943–9966. https://doi.org/10.1039/D4RA00844H

Crater, J. S., & Lievense, J. C. (2018). Scale-up of industrial microbial processes. *FEMS Microbiology Letters, 365*(13). https://doi.org/10.1093/femsle/fny138

Deckers, M., Deforce, D., Fraiture, M.-A., & Roosens, N. H. C. (2020). Genetically modified microorganisms for industrial food enzyme production: An overview. *Foods, 9*(3), 326. https://doi.org/10.3390/foods9030326

Dong, H., Huang, L., Zhao, L., Zeng, Q., Liu, X., Sheng, Y., Shi, L., Wu, G., Jiang, H., Li, F., Zhang, L., Guo, D., Li, G., Hou, W., & Chen, H. (2022). A critical review of mineral–microbe interaction and co-evolution: mechanisms and applications. *National Science Review, 9*(10). https://doi.org/10.1093/nsr/nwac128

El-Dash, H. A., Yousef, N. E., Aboelazm, A. A., Awan, Z. A., Yahya, G., & El-Ganiny, A. M. (2023). Optimizing eco-friendly degradation of polyvinyl chloride (PVC) plastic using environmental strains of malassezia species and *Aspergillus fumigatus*. *International Journal of Molecular Sciences, 24*(20), 15452. https://doi.org/10.3390/ijms242015452

Fallahi, M., Sarempour, M., & Mirzadi Gohari, A. (2023). Potential biodegradation of polycyclic aromatic hydrocarbons (PAHs) and petroleum hydrocarbons by indigenous fungi recovered from crude oil-contaminated soil in Iran. *Scientific Reports, 13*(1), 22153. https://doi.org/10.1038/s41598-023-49630-z

Farda, B., Djebaili, R., Vaccarelli, I., Del Gallo, M., & Pellegrini, M. (2022). Actinomycetes from caves: An overview of their diversity, biotechnological properties, and insights for their use in soil environments. *Microorganisms, 10*(2), 453. https://doi.org/10.3390/microorganisms10020453

Fatehbasharzad, P., Aliasghari, S., Shaterzadeh Tabrizi, I., Khan, J. A., & Boczkaj, G. (2022). Microbial fuel cell applications for removal of petroleum hydrocarbon pollutants: A review. *Water Resources and Industry, 28*, 100178. https://doi.org/10.1016/j.wri.2022.100178

Folino, A., Pangallo, D., & Calabrò, P. S. (2023). Assessing bioplastics biodegradability by standard and research methods: Current trends and open issues. *Journal of Environmental Chemical Engineering, 11*(2), 109424. https://doi.org/10.1016/j.jece.2023.109424

Gao, Y., Cai, M., Shi, K., Sun, R., Liu, S., Li, Q., Wang, X., Hua, W., Qiao, Y., Xue, J., & Xiao, X. (2023). Bioaugmentation enhance the bioremediation of marine crude oil pollution: Microbial communities and metabolic pathways. *Water Science and Technology, 87*(1), 228–238. https://doi.org/10.2166/wst.2022.406

Garbini, G. L., Barra Caracciolo, A., & Grenni, P. (2023). Electroactive bacteria in natural ecosystems and their applications in microbial fuel cells for bioremediation: A review. *Microorganisms, 11*(5), 1255. https://doi.org/10.3390/microorganisms11051255

Ghosh, S., Tripathy, B., Dey, S., & Das, A. P. (2023). Comparative investigation of fungal and bacterial manganese biomining mechanisms. *Manganese Mining Microorganisms*, 99–115. https://doi.org/10.1016/b978-0-12-822147-1.00016-9

Godzieba, M., Zubrowska-Sudol, M., Walczak, J., & Ciesielski, S. (2022). Development of microbial communities in biofilm and activated sludge in a hybrid reactor. *Scientific Reports, 12*(1), 12558. https://doi.org/10.1038/s41598-022-16570-z

Golzar-Ahmadi, M., Bahaloo-Horeh, N., Pourhossein, F., Norouzi, F., Schoenberger, N., Hintersatz, C., Chakankar, M., Holuszko, M., & Kaksonen, A. H. (2024). Pathway to industrial application of heterotrophic organisms in critical metals recycling from e-waste. *Biotechnology Advances, 77*, 108438. https://doi.org/10.1016/j.biotechadv.2024.108438

Gong, C., Singh, A., Singh, P., & Singh, A. (2021). Anaerobic digestion of agri-food wastes for generating biofuels. *Indian Journal of Microbiology, 61*(4), 427–440. https://doi.org/10.1007/s12088-021-00977-9

Gong, T., Xu, X., Dang, Y., Kong, A., Wu, Y., Liang, P., Wang, S., Yu, H., Xu, P., & Yang, C. (2018). An engineered *Pseudomonas putida* can simultaneously degrade organophosphates, pyrethroids and carbamates. *Science of the Total Environment, 628–629*, 1258–1265. https://doi.org/10.1016/j.scitotenv.2018.02.143

Gonzalez, J. M., & Aranda, B. (2023). Microbial growth under limiting conditions-future perspectives. *Microorganisms, 11*(7), 1641. https://doi.org/10.3390/microorganisms11071641

Han, P., Teo, W. Z., & Yew, W. S. (2022). Biologically engineered microbes for bioremediation of electronic waste: Wayposts, challenges and future directions. *Engineering Biology, 6*(1), 23–34. https://doi.org/10.1049/enb2.12020

Heidarzadeh, M. H., Amani, H., & Javadian, B. (2019). Improving municipal solid waste compost process by cycle time reduction through inoculation of *Aspergillus niger*. *Journal of Environmental Health Science and Engineering, 17*(1), 295–303. https://doi.org/10.1007/s40201-019-00348-z

Hossain, R., & Sahajwalla, V. (2024). Current recycling innovations to utilize e-waste in sustainable green metal manufacturing. *Philosophical Transactions of the Royal Society A: Mathematical, Physical and Engineering Sciences, 382*(2284). https://doi.org/10.1098/rsta.2023.0239

Hyde, K. D., Xu, J., Rapior, S., Jeewon, R., Lumyong, S., Niego, A. G. T., Abeywickrama, P. D., Aluthmuhandiram, J. V. S., Brahamanage, R. S., Brooks, S., Chaiyasen, A., Chethana, K. W. T., Chomnunti, P., Chepkirui, C., Chuankid, B., de Silva, N. I., Doilom, M., Faulds, C., Gentekaki, E., … Stadler, M. (2019). The amazing potential of fungi: 50 ways we can exploit fungi industrially. *Fungal Diversity, 97*(1), 1–136. https://doi.org/10.1007/s13225-019-00430-9

Igiri, B. E., Okoduwa, S. I. R., Idoko, G. O., Akabuogu, E. P., Adeyi, A. O., & Ejiogu, I. K. (2018). Toxicity and bioremediation of heavy metals contaminated ecosystem from tannery wastewater: A review. *Journal of Toxicology, 2018*, 1–16. https://doi.org/10.1155/2018/2568038

Jalili, P., Ala, A., Nazari, P., Jalili, B., & Ganji, D. D. (2024). A comprehensive review of microbial fuel cells considering materials, methods, structures, and microorganisms. *Heliyon, 10*(3), e25439. https://doi.org/10.1016/j.heliyon.2024.e25439

Kalathil, S., Miller, M., & Reisner, E. (2022). microbial fermentation of polyethylene terephthalate (PET) plastic waste for the production of chemicals or electricity**. *Angewandte Chemie International Edition, 61*(45). https://doi.org/10.1002/anie.202211057

Kapahi, M., & Sachdeva, S. (2019). Bioremediation options for heavy metal pollution. *Journal of Health and Pollution, 9*(24). https://doi.org/10.5696/2156-9614-9.24.191203

Kitching, M., Ramani, M., & Marsili, E. (2015). Fungal biosynthesis of gold nanoparticles: Mechanism and scale up. *Microbial Biotechnology, 8*(6), 904–917. https://doi.org/10.1111/1751-7915.12151

Koul, B., Poonia, A. K., Yadav, D., & Jin, J.-O. (2021). Microbe-mediated biosynthesis of nanoparticles: Applications and future prospects. *Biomolecules, 11*(6), 886. https://doi.org/10.3390/biom11060886

Kumar, A., Verma, L. M., Sharma, S., & Singh, N. (2023). Overview on agricultural potentials of biogas slurry (BGS): Applications, challenges, and solutions. *Biomass Conversion and Biorefinery, 13*(15), 13729–13769. https://doi.org/10.1007/s13399-021-02215-0

Kuppan, N., Padman, M., Mahadeva, M., Srinivasan, S., & Devarajan, R. (2024). A comprehensive review of sustainable bioremediation techniques: Eco friendly solutions for waste and pollution management. *Waste Management Bulletin, 2*(3), 154–171. https://doi.org/10.1016/j.wmb.2024.07.005

Li, P., Liang, X., Shi, R., Wang, Y., Han, S., & Zhang, Y. (2023). Unraveling the functional instability of bacterial consortia in crude oil degradation via integrated co-occurrence networks. *Frontiers in Microbiology, 14*. https://doi.org/10.3389/fmicb.2023.1270916

Li, X., Lu, C., Dai, Y., Yu, Z., Gu, W., Li, T., Li, X., Li, X., Wang, X., Su, Z., Xu, M., & Zhang, H. (2022). Characterizing the microbial consortium L1 capable of efficiently degrading chlorimuron-ethyl via metagenome combining 16S rDNA sequencing. *Frontiers in Microbiology, 13*. https://doi.org/10.3389/fmicb.2022.912312

Lokesh, P., Shobika, R., Omer, S., Reddy, M., Saravanan, P., Rajeshkannan, R., Saravanan, V., & Venkatkumar, S. (2023). Bioremediation of plastics by the help of microbial tool: A way for control of plastic pollution. *Sustainable Chemistry for the Environment, 3*, 100027. https://doi.org/10.1016/j.scenv.2023.100027

Lu, C.-W., Ho, H.-C., Yao, C.-L., Tseng, T.-Y., Kao, C.-M., & Chen, S.-C. (2023). Bioremediation potential of cadmium by recombinant *Escherichia coli* surface expressing metallothionein MTT5 from *Tetrahymena thermophila*. *Chemosphere, 310*, 136850. https://doi.org/10.1016/j.chemosphere.2022.136850

Maglione, G., Zinno, P., Tropea, A., Mussagy, C. U., Dufossé, L., Giuffrida, D., & Mondello, A. (2024). Microbes' role in environmental pollution and remediation: A bioeconomy focus approach. *AIMS Microbiology, 10*(3), 723–755. https://doi.org/10.3934/microbiol.2024033

Mendes, F., Machado, B. O., Castro, B. B., Sousa, M. J., & Chaves, S. R. (2025). Harnessing the power of biosensors for environmental monitoring of pesticides in water. *Applied Microbiology and Biotechnology, 109*(1), 92. https://doi.org/10.1007/s00253-025-13461-x

Miglani, R., Parveen, N., Kumar, A., Ansari, Mohd. A., Khanna, S., Rawat, G., Panda, A. K., Bisht, S. S., Upadhyay, J., & Ansari, M. N. (2022). Degradation of xenobiotic pollutants: An environmentally sustainable approach. *Metabolites, 12*(9), 818. https://doi.org/10.3390/metabo12090818

Mishra, R. K., Mentha, S. S., Misra, Y., & Dwivedi, N. (2023). Emerging pollutants of severe environmental concern in water and wastewater: A comprehensive review on current developments and future research. *Water-Energy Nexus, 6*, 74–95. https://doi.org/10.1016/j.wen.2023.08.002

Mishra, S., Dash, D., Al-Tawaha, A. R. M. S., & Das, A. P. (2022). A review on heavy metal ion adsorption on synthetic microfiber surface in aquatic environments. *Applied Biochemistry and Biotechnology, 194*(10), 4639–4654. https://doi.org/10.1007/S12010-022-04029-W/METRICS

Mitchell, H. J., & Bartsch, D. (2020). Regulation of GM organisms for invasive species control. *Frontiers in Bioengineering and Biotechnology, 7*. https://doi.org/10.3389/fbioe.2019.00454

Muhammad, M. H., Idris, A. L., Fan, X., Guo, Y., Yu, Y., Jin, X., Qiu, J., Guan, X., & Huang, T. (2020). Beyond risk: Bacterial biofilms and their regulating approaches. *Frontiers in Microbiology, 11*. https://doi.org/10.3389/fmicb.2020.00928

Muter, O. (2023). Current trends in bioaugmentation tools for bioremediation: A critical review of advances and knowledge gaps. *Microorganisms, 11*(3), 710. https://doi.org/10.3390/microorganisms11030710

Nag, M., Lahiri, D., Ghosh, S., Sarkar, T., Pati, S., Das, A. P., Ram, D. K., Bhattacharya, D., & Ray, R. R. (2023). Application of microorganisms in biotransformation and bioremediation of environmental contaminant: A review. *Geomicrobiology Journal*, 1–18. https://doi.org/10.1080/01490451.2023.2261443

Newsome, L., & Falagán, C. (2021). The microbiology of metal mine waste: Bioremediation applications and implications for planetary health. *GeoHealth, 5*(10). https://doi.org/10.1029/2020GH000380

Ng, I.-S., Ndive, C. I., Zhou, Y., & Wu, X. (2015). Cultural optimization and metal effects of Shewanella xiamenensis BC01 growth and swarming motility. *Bioresources and Bioprocessing, 2*(1), 28. https://doi.org/10.1186/s40643-015-0055-7

Nzila, A., Razzak, S., & Zhu, J. (2016). Bioaugmentation: An emerging strategy of industrial wastewater treatment for reuse and discharge. *International Journal of Environmental Research and Public Health, 13*(9), 846. https://doi.org/10.3390/ijerph13090846

Organo, N. D., Granada, S. M. J. M., Pineda, H. G. S., Sandro, J. M., Nguyen, V. H., & Gummert, M. (2022). Assessing the potential of a Trichoderma-based compost activator to hasten the decomposition of incorporated rice straw. *Scientific Reports, 12*(1), 448. https://doi.org/10.1038/s41598-021-03828-1

Ou, Y., & Guo, S. (2023). Safety risks and ethical governance of biomedical applications of synthetic biology. *Frontiers in Bioengineering and Biotechnology, 11*. https://doi.org/10.3389/fbioe.2023.1292029

Pande, V., Pandey, S. C., Sati, D., Bhatt, P., & Samant, M. (2022). Microbial interventions in bioremediation of heavy metal contaminants in agroecosystem. *Frontiers in Microbiology, 13*. https://doi.org/10.3389/fmicb.2022.824084

Pateiro, M., Gómez, B., Munekata, P. E. S., Barba, F. J., Putnik, P., Kovačević, D. B., & Lorenzo, J. M. (2021). Nanoencapsulation of promising bioactive compounds to improve their absorption, stability, functionality and the appearance of the final food products. *Molecules, 26*(6), 1547. https://doi.org/10.3390/molecules26061547

Patowary, R., Devi, A., & Mukherjee, A. K. (2023). Advanced bioremediation by an amalgamation of nanotechnology and modern artificial intelligence for efficient restoration of crude petroleum oil-contaminated sites: A prospective study. *Environmental Science and Pollution Research, 30*(30), 74459–74484. https://doi.org/10.1007/s11356-023-27698-4

Paul, A., Dey, S., Ram, D. K., & Das, A. P. (2023). Hexavalent chromium pollution and its sustainable management through bioremediation. *Geomicrobiology Journal*, 1–11. https://doi.org/10.1080/01490451.2023.2218377

Pérez-Armendáriz, B., Cal-y-Mayor-Luna, C., El-Kassis, E. G., & Ortega-Martínez, L. D. (2019). Use of waste canola oil as a low-cost substrate for rhamnolipid production using *Pseudomonas aeruginosa. AMB Express, 9*(1), 61. https://doi.org/10.1186/s13568-019-0784-7

Pradhan, G., Tripathy, B., Ram, D. K., Digal, A. K., & Das, A. P. (2023). Bauxite mining waste pollution and its sustainable management through bioremediation. *Geomicrobiology Journal*, 1–10. https://doi.org/10.1080/01490451.2023.2235353

Priyashantha, A. K. H., Pratheesh, N., & Pretheeba, P. (2022). E-waste scenario in South-Asia: An emerging risk to environment and public health. *Environmental Analysis Health and Toxicology, 37*(3), e2022022. https://doi.org/10.5620/eaht.2022022

Rafeeq, H., Afsheen, N., Rafique, S., Arshad, A., Intisar, M., Hussain, A., Bilal, M., & Iqbal, H. M. N. (2023). Genetically engineered microorganisms for environmental remediation. *Chemosphere, 310*, 136751. https://doi.org/10.1016/j.chemosphere.2022.136751

Rastogi, M., Nandal, M., & Khosla, B. (2020). Microbes as vital additives for solid waste composting. *Heliyon, 6*(2), e03343. https://doi.org/10.1016/j.heliyon.2020.e03343

Riseh, R. S., Vazvani, M. G., Hajabdollahi, N., & Thakur, V. K. (2023). Bioremediation of heavy metals by rhizobacteria. *Applied Biochemistry and Biotechnology, 195*(8), 4689–4711. https://doi.org/10.1007/s12010-022-04177-z

Romantschuk, M., Lahti-Leikas, K., Kontro, M., Galitskaya, P., Talvenmäki, H., Simpanen, S., Allen, J. A., & Sinkkonen, A. (2023). Bioremediation of contaminated soil and groundwater by in situ biostimulation. *Frontiers in Microbiology, 14*. https://doi.org/10.3389/fmicb.2023.1258148

Rowińska, P., Gutarowska, B., Janas, R., & Szulc, J. (2024). Biopreparations for the decomposition of crop residues. *Microbial Biotechnology, 17*(8). https://doi.org/10.1111/1751-7915.14534

Roy, H., Rahman, T. U., Tasnim, N., Arju, J., Rafid, Md. M., Islam, Md. R., Pervez, Md. N., Cai, Y., Naddeo, V., & Islam, Md. S. (2023). Microbial fuel cell construction features and application for sustainable wastewater treatment. *Membranes, 13*(5), 490. https://doi.org/10.3390/membranes13050490

Sadanov, A. K., Baimakhanova, B. B., Orasymbet, S. E., Ratnikova, I. A., Turlybaeva, Z. Z., Baimakhanova, G. B., Amitova, A. A., Omirbekova, A. A., Aitkaliyeva, G. S., Kossalbayev, B. D., & Belkozhayev, A. M. (2025). Engineering useful microbial species for pharmaceutical applications. *Microorganisms, 13*(3), 599. https://doi.org/10.3390/microorganisms13030599

Safonov, A. V., Babich, T. L., Sokolova, D. S., Grouzdev, D. S., Tourova, T. P., Poltaraus, A. B., Zakharova, E. V., Merkel, A. Y., Novikov, A. P., & Nazina, T. N. (2018). Microbial community and in situ bioremediation of groundwater by nitrate removal in the zone of a radioactive waste surface repository. *Frontiers in Microbiology, 9*. https://doi.org/10.3389/fmicb.2018.01985

Sevilla, M. E., Garcia, M. D., Perez-Castillo, Y., Armijos-Jaramillo, V., Casado, S., Vizuete, K., Debut, A., & Cerda-Mejía, L. (2023). Degradation of PET bottles by an engineered *Ideonella sakaiensis* PETase. *Polymers, 15*(7), 1779. https://doi.org/10.3390/polym15071779

Sharma, M., Agarwal, S., Agarwal Malik, R., Kumar, G., Pal, D. B., Mandal, M., Sarkar, A., Bantun, F., Haque, S., Singh, P., Srivastava, N., & Gupta, V. K. (2023a). Recent advances in microbial engineering approaches for wastewater treatment: a review. *Bioengineered, 14*(1). https://doi.org/10.1080/21655979.2023.2184518

Sharma, S., Mohler, J., Mahajan, S. D., Schwartz, S. A., Bruggemann, L., & Aalinkeel, R. (2023b). Microbial biofilm: A review on formation, infection, antibiotic resistance, control measures, and innovative treatment. *Microorganisms, 11*(6), 1614. https://doi.org/10.3390/microorganisms11061614

Siddiqua, A., Hahladakis, J. N., & Al-Attiya, W. A. K. A. (2022). An overview of the environmental pollution and health effects associated with waste landfilling and open dumping. *Environmental Science and Pollution Research, 29*(39), 58514–58536. https://doi.org/10.1007/s11356-022-21578-z

Singh, H., YT, K., Mishra, A. K., Singh, M., Mohanto, S., Ghumra, S., Seelan, A., Mishra, A., Kumar, A., Pallavi, J., Ahmed, M. G., Sangeetha, J., & Thangadurai, D. (2024). Harnessing the foundation of biomedical waste management for fostering public health: strategies and policies for a clean and safer environment. *Discover Applied Sciences, 6*(3), 89. https://doi.org/10.1007/s42452-024-05735-2

Singh, Y., & Saxena, M. K. (2022). Insights into the recent advances in nano-bioremediation of pesticides from the contaminated soil. *Frontiers in Microbiology, 13*. https://doi.org/10.3389/fmicb.2022.982611

Sitaraman, R. (2018). Prokaryotic horizontal gene transfer within the human holobiont: Ecological-evolutionary inferences, implications and possibilities. *Microbiome, 6*(1), 163. https://doi.org/10.1186/s40168-018-0551-z

Su, T., Zhang, T., Liu, P., Bian, J., Zheng, Y., Yuan, Y., Li, Q., Liang, Q., & Qi, Q. (2023). Biodegradation of polyurethane by the microbial consortia enriched from landfill. *Applied Microbiology and Biotechnology, 107*(5–6), 1983–1995. https://doi.org/10.1007/s00253-023-12418-2

Tang, H., Xiang, G., Xiao, W., Yang, Z., & Zhao, B. (2024). Microbial mediated remediation of heavy metals toxicity: mechanisms and future prospects. *Frontiers in Plant Science, 15*. https://doi.org/10.3389/fpls.2024.1420408

Tapia, J., Dueñas, A., Cheje, N., Soclle, G., Patiño, N., Ancalla, W., Tenorio, S., Denos, J., Taco, H., Cao, W., Alexandrino, D. A. M., Jia, Z., Vasconcelos, V., Carvalho, M. de F., & Lazarte, A. (2022). Bioleaching of heavy metals from printed circuit boards with an acidophilic iron-oxidizing microbial consortium in stirred tank reactors. *Bioengineering, 9*(2), 79. https://doi.org/10.3390/bioengineering9020079

Tasleem, M., Hussein, W. M., El-Sayed, A.-A.A.A., & Alrehaily, A. (2023). An in silico bioremediation study to identify essential residues of metallothionein enhancing the bioaccumulation of heavy metals in *Pseudomonas aeruginosa*. *Microorganisms, 11*(9), 2262. https://doi.org/10.3390/microorganisms11092262

Temporiti, M. E. E., Nicola, L., Nielsen, E., & Tosi, S. (2022). Fungal enzymes involved in plastics biodegradation. *Microorganisms, 10*(6), 1180. https://doi.org/10.3390/microorganisms10061180

Thai, T. D., Lim, W., & Na, D. (2023). Synthetic bacteria for the detection and bioremediation of heavy metals. *Frontiers in Bioengineering and Biotechnology, 11*. https://doi.org/10.3389/fbioe.2023.1178680

Thirumalaivasan, N., Gnanasekaran, L., Kumar, S., Durvasulu, R., Sundaram, T., Rajendran, S., Nangan, S., & Kanagaraj, K. (2024). Utilization of fungal and bacterial bioremediation techniques for the treatment of toxic waste and biowaste. *Frontiers in Materials, 11*. https://doi.org/10.3389/fmats.2024.1416445

Tran, Q. N. M., Mimoto, H., Koyama, M., & Nakasaki, K. (2019). Lactic acid bacteria modulate organic acid production during early stages of food waste composting. *Science of the Total Environment, 687*, 341–347. https://doi.org/10.1016/j.scitotenv.2019.06.113

Tseng, A. S., Roberts, M. C., Weissman, S. J., & Rabinowitz, P. M. (2023). Study of heavy metal resistance genes in *Escherichia coli* isolates from a marine ecosystem with a history of environmental pollution (arsenic, cadmium, copper, and mercury). *PLoS ONE, 18*(11), e0294565. https://doi.org/10.1371/journal.pone.0294565

Vermeer, C. M., Bons, L. J., & Kleerebezem, R. (2022). Production of a newly discovered PHA family member with an isobutyrate-fed enrichment culture. *Applied Microbiology and Biotechnology, 106*(2), 605–618. https://doi.org/10.1007/s00253-021-11742-9

Wróbel, M., Śliwakowski, W., Kowalczyk, P., Kramkowski, K., & Dobrzyński, J. (2023). Bioremediation of heavy metals by the Genus bacillus. *International Journal of Environmental Research and Public Health, 20*(6), 4964. https://doi.org/10.3390/ijerph20064964

Wu, W., Liu, X., Zhang, X., Zhu, M., & Tan, W. (2018). Bioleaching of copper from waste printed circuit boards by bacteria-free cultural supernatant of iron–sulfur-oxidizing bacteria. *Bioresources and Bioprocessing, 5*(1), 10. https://doi.org/10.1186/s40643-018-0196-6

Xu, X., Liu, W., Tian, S., Wang, W., Qi, Q., Jiang, P., Gao, X., Li, F., Li, H., & Yu, H. (2018). Petroleum hydrocarbon-degrading bacteria for the remediation of oil pollution under aerobic conditions: A perspective analysis. *Frontiers in Microbiology, 9*, 2885. https://doi.org/10.3389/fmicb.2018.02885

Yamin, D., Uskoković, V., Wakil, A., Goni, M., Shamsuddin, S., Mustafa, F., Alfouzan, W., Alissa, M., Alshengeti, A., Almaghrabi, R., Fares, M., Garout, M., Al Kaabi, N., Alshehri, A., Ali, H., Rabaan, A., Aldubisi, F., Yean, C., & Yusof, N. (2023). Current and future technologies for the detection of antibiotic-resistant bacteria. *Diagnostics, 13*(20), 3246. https://doi.org/10.3390/diagnostics13203246

Yang, J., Zhang, J., Du, X., Gao, T., Cheng, Z., Fu, W., & Wang, S. (2025). Ammonia inhibition in anaerobic digestion of organic waste: A review. *International Journal of Environmental Science and Technology, 22*(5), 3927–3942. https://doi.org/10.1007/s13762-024-06029-1

Yang, P., Condrich, A., Lu, L., Scranton, S., Hebner, C., Sheykhhasan, M., & Ali, M. A. (2024). Genetic engineering in bacteria, fungi, and oomycetes, Taking advantage of CRISPR. *DNA, 4*(4), 427–454. https://doi.org/10.3390/dna4040030

Yao, X., Guo, H., Zhang, K., Zhao, M., Ruan, J., & Chen, J. (2023). Trichoderma and its role in biological control of plant fungal and nematode disease. *Frontiers in Microbiology, 14*. https://doi.org/10.3389/fmicb.2023.1160551

Zampolli, J., Orro, A., Vezzini, D., & Di Gennaro, P. (2022). Genome-based exploration of rhodococcus species for plastic-degrading genetic determinants using bioinformatic analysis. *Microorganisms, 10*(9), 1846. https://doi.org/10.3390/microorganisms10091846

Zhang, T., & Zhang, H. (2022). Microbial consortia are needed to degrade soil pollutants. *Microorganisms, 10*(2), 261. https://doi.org/10.3390/microorganisms10020261

Ziembińska-Buczyńska, A., Ciesielski, S., Żabczyński, S., & Cema, G. (2019). Bacterial community structure in rotating biological contactor treating coke wastewater in relation to medium composition. *Environmental Science and Pollution Research, 26*(19), 19171–19179. https://doi.org/10.1007/s11356-019-05087-0

Zulfiqar, U., Haider, F. U., Maqsood, M. F., Mohy-Ud-Din, W., Shabaan, M., Ahmad, M., Kaleem, M., Ishfaq, M., Aslam, Z., & Shahzad, B. (2023). Recent advances in microbial-assisted remediation of cadmium-contaminated soil. *Plants, 12*(17), 3147. https://doi.org/10.3390/plants12173147

Recovery of Industrial Carbon Fiber Waste in Sustainable Green Transformation: Pros and Cons

Nuray Beköz Üllen, Gizem Karabulut Şevk, Mustafa Akoğlu, İbrahim Halil Kalkan, and Okan Otuz

Abstract Carbon fiber has become an indispensable material for many industries with its high-performance properties and wide applications. It offers significant technical and economic advantages and is expected to become more widespread in the future. These materials, particularly carbon fiber-reinforced composites, are widely used in aerospace, automotive, sports equipment, and renewable energy sectors. Carbon fiber-reinforced composites combine carbon fibers with a polymer, metallic, ceramic, or carbon matrix to create composites that leverage the strengths of both the fibers and the matrix, resulting in superior mechanical properties and design flexibility. The recycling of carbon fiber materials is a critical aspect of their lifecycle, addressing environmental and economic concerns. Although recycling carbon fiber composites is challenging due to their durability and the strong bonding between fibers and matrices, advancements in recycling technologies are gradually making it more feasible. Effective recycling not only reduces waste but also conserves resources and lowers the environmental impact of producing new carbon fibers. Looking ahead, the future of carbon fiber materials appears promising. Compared to virgin carbon fibers, recycled carbon fibers retain many of their properties, providing an environmentally friendly and economical alternative, creating excellent usage potential. Ongoing research and development aim to reduce production costs and improve recycling methods, making these materials more accessible and sustainable. As these advancements continue, the adoption of carbon fiber composites is expected to expand, driving innovation and efficiency across multiple industries.

Keywords Recycling · Carbon fiber composites · Sustainability

N. Beköz Üllen (✉) · G. Karabulut Şevk
Department of Metallurgical and Materials Engineering, Faculty of Engineering, İstanbul University-Cerrahpaşa, Istanbul, Türkiye
e-mail: nbekoz@iuc.edu.tr

M. Akoğlu
VIMANSYS Digital Solutions and Smart Manufacturing, Bursa, Türkiye

İ. H. Kalkan · O. Otuz
Daimler Mercedes-Benz Turk, Istanbul, Türkiye

A. P. Das (ed.), *Bioconversion of Environmental Waste to Value Added Products*, Sustainable Environmental Waste Management Strategies,
https://doi.org/10.1007/978-3-032-14215-3_11

1 Introduction

Carbon fiber (CF) is a material known for its high-strength and low-weight properties and found in thin strands where carbon atoms are tightly bonded to each other (De et al., 2024). It usually forms composite materials by being contained in a matrix (Ateeq et al., 2023). These composite materials are widely used in applications that require particularly high performance and durability. Although CFs are much lighter than steel, they can have the same or higher strength (Ahmad et al., 2020). Because of its low density, it is ideal for applications requiring weight reduction. CFs have high resistance to bending and tensile forces. It is resistant to many chemicals and environmental influences (Jang et al., 2022). It is used in the construction of aircraft and space vehicles, performance cars, and racing cars due to its lightness and high strength, increasing fuel efficiency and performance (Das et al., 2019; Muhammad et al., 2021; Seydibeyoğlu et al., 2020; Sreejith & Rajeev, 2021). Due to its strength and lightness, it is preferred in items such as bicycles, tennis rackets, golf clubs, and ski equipment (Lina & Zhen, 2012; Zhang, 2013). They are preferred in bridges, buildings, and other structural elements due to their durability and long life (Carolin, 2003; Hegde et al., 2019; Tuakta & Whittle, 2005). CF has become an indispensable material for many industries with its high-performance properties and wide applications (Lin et al., 2024; Ye et al., 2024). It offers significant technical and economic advantages and is expected to become more widespread in the future. Historically important developments of CF from the past to the present are shown in Fig. 1. Although it is known that the first studies on CF were made by Sir Joseph Wilson Swan in 1860, its first commercial use was performed in 1958 by American chemist Dr. Roger Bacon. While working at Union Carbide, Bacon conducted a series of experiments to produce graphite fibers. In these experiments, it was observed that the fibers obtained by heating polymers at high temperatures had a structure similar to graphite crystals (Peijs et al., 2022). Later, as a reflection of the developing technology, the CF market has gradually increased over the years with the developments in CF production processes and products. Nowadays, due to the increasing amount of CF products, the product cost has been reduced considerably and the competition in the market has increased considerably. With the replacement of traditional materials, their increasing use brings with it many environmental problems. In addition to the use of petroleum-based raw materials, CF transformation has come to the fore in recent years due to the increasing amount of carbon emissions and the increasing prevalence of concepts such as sustainable living and circular economy.

CFs are mainly used in composite materials as a dopant material. CF-reinforced composite materials are advanced engineering materials created by embedding CFs in polymer, metallic, ceramic, and carbon matrices (Hu et al., 2024; Newcomb, 2016; Park, 2018). Polymer matrix CF composites are favored in the automotive and aerospace industries for their lightweight and flexibility (Alshammari et al., 2021; Zhang et al., 2023). However, metallic matrix composites provide high-temperature

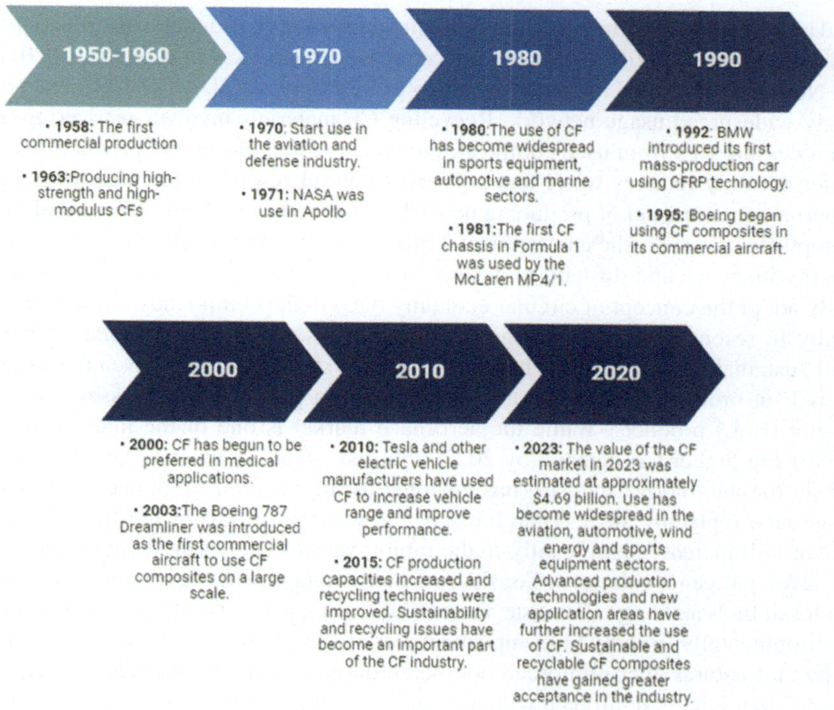

Fig. 1 Historical development of CF technology

resistance and strength, making them suitable for aerospace and defense applications (Shirvanimoghaddam et al., 2017). Ceramic matrix CF composites offer exceptional thermal and chemical resistance, which is valuable in rocket engines and thermal protection systems (Dhanasekar et al., 2022; Tang & Hu, 2017). On the other hand, carbon matrix composites deliver outstanding thermal conductivity and corrosion resistance, making them ideal for high-performance electronics and energy storage systems (Chawla, 2019; Chung, 2016). By combining CFs with various matrix materials, these composites offer unique advantages in their respective fields, providing high-performance solutions tailored to specific engineering challenges (Koumoulos et al., 2019; Sayam et al., 2022). CF holds significant strategic importance as a material due to its exceptional strength-to-weight ratio (Ahmad et al., 2020), high stiffness, and corrosion resistance (Ahmad et al., 2020). These properties make it indispensable in aerospace (Barile & Casavola, 2019; Tang & Hu, 2017), defense (Kacan & Elaldi, 2020), automotive (Ahmad et al., 2020), and renewable energy industries (Demski et al., 2024). Its ability to reduce weight while maintaining structural integrity enhances fuel efficiency and performance, which is crucial for both economic and environmental sustainability (Ahmad et al., 2020; Barile & Casavola, 2019; Tang & Hu, 2017). Furthermore, the use of CF in advanced military equipment and infrastructure ensures superior durability and protection, underscoring its critical role in national security and technological advancement (Kacan & Elaldi, 2020).

Nowadays, the need for recycling emerges for this material, which has an increasingly widespread usage network. Recycling CF materials involves recovering and reprocessing CFs from used composites to create new, high-quality products. Recycling not only reduces waste and conserves natural resources but also lowers the environmental impact of producing new CFs. Due to the synthetic structure of most composite materials, the environmental impact during the production phase is more energy intensive and difficult to destroy or recycle. Therefore, it is not possible to fully adopt the concept of circular economy (CE) in the compound economy. Especially in recent years, the automotive industry based on environmental problems and sustainable materials has been carrying out studies on the conversion of vCF to rCF in order to deal with waste resulting from production processes and end-of-life (EoL) products. While the aerospace market is one of the main segments predicting higher CF demand by 2027, with the advent of electric cars, the use of CF in the automotive industry has emerged as the material of choice over lighter materials, replacing steel. Therefore, as the demand for CF in the transportation sector will increase significantly in the future, various studies are being carried out to develop greener solutions. Considering disadvantages such as strong dependence on fossil fuels and large CF waste, there is increasing interest in the use of alternative environmentally friendly biocomposite materials. In general, the term biocomposite refers to: natural fiber-reinforced petroleum derived from non-biodegradable polymers, biopolymers reinforced with natural fibers, and synthetic fibers (carbon, glass, etc.) reinforced with biopolymers. One of the biggest projects carried out in this context is the Horizon Europe (HORIZON-CL4–2022-RESILIENCE-01-11) project called advanced lightweight materials FOR Energy-efficient STructures (FOREST), in which the chapter authors are also involved. The FOREST project aims to produce biocomposites with bio-based polymers and additives and rCFs by combining renewable and recycling resources to maximize the sustainability of the products. The FOREST project aims to transform promising technologies by promoting overcoming economic, environmental, and social challenges. International projects of this type are a great chance to reduce dependence on fossil resources and reduce CF waste. Additional advantages such as recyclability, lightweight structure, and special functionality for safety and cost-effectiveness are predicted to make FOREST biocomposites an alternative to traditional composites (*Advanced Lightweight Materials for Energy Efficient Structures (RIA)*, n.d.). Such approaches and minimizing the difficulties and legal obstacles in recycling are seen as the key elements that will achieve groundbreaking success in CF recycling concepts. Although challenging due to the durability and complexity of CF composites, advancements in recycling technologies are making it increasingly feasible, promoting a circular economy and contributing to sustainable development in industries like aerospace, automotive, and construction. This chapter presents the pros and cons of sustainable green transformation in using rCFs in the development of lightweight and durable parts to close the CF loop in the industry. Detailed information about CF materials is summarized, as well as information on current CF recycling technologies.

2 Carbon-Fiber Reinforced Composites

2.1 Polymer Matrix Composites

CF-reinforced polymer composites (CFRPCs) are advanced materials that combine CFs with a polymer matrix to create lightweight and high-strength structures (Bag et al., 2024; Liu et al., 2020; Mahato & Moharana, 2024; Oladele et al., 2023; Tusher et al., 2024). The polymer matrix, typically made from resins like epoxy, polyester, or vinyl ester, serves to bind the CFs together, transferring loads and protecting the fibers from environmental damage (Chen et al., 2022; Mahato & Moharana, 2024; Tusher et al., 2024). The resulting composite material retains the advantageous properties of CFs, such as high tensile strength, low weight, and excellent fatigue resistance, while gaining additional benefits from the polymer matrix, such as ease of manufacturing and flexibility (Khan et al., 2024; Mahboubizadeh et al., 2024; Oladele et al., 2023).

Several methods are used to manufacture CFRPCs, each suited to different applications and production scales:

- Resin Transfer Molding (RTM): In RTM, CF preforms are placed into a mold, and resin is injected to fill the mold and impregnate the fibers. This method is ideal for producing complex shapes and large structures with high fiber volume fractions (Sozer et al., 2012). Different types of RTM were used producing CFRPC such as vacuum-assisted RTM (Bodaghi et al., 2020), high-pressure RTM (Sarker et al., 2024), ultrasound-assisted RTM (Xu et al., 2024).
- Filament Winding: This process involves winding continuous CF strands coated with resin around a mandrel to create hollow, cylindrical components such as pipes and pressure vessels. Filament winding offers high precision and excellent control over fiber orientation (Azeem et al., 2022; Ma et al., 2015).
- Pultrusion: In pultrusion, continuous CFs are pulled through a resin bath and then through a heated die to form long, straight profiles. This method is efficient for producing structural elements like beams and rods with consistent cross-sections (Hong et al., 2017; Minchenkov et al., 2021; Wiedmer & Manolesos, 2006).

CFRPCs are significantly lighter than traditional materials like metals, which leads to improved fuel efficiency and performance in transportation applications (Kumar SHArma et al., 2022). The polymer matrix protects CFs from moisture and chemicals, ensuring long-term durability and reducing maintenance costs (Koumoulos et al., 2019; Kumar SHArma et al., 2022). CFRPCs can be molded into complex shapes and tailored to specific performance requirements, allowing for innovative designs and applications (Mitschang & Hildebrandt, 2012; Tian et al., 2022). CFRPCs are extensively used across various industries due to their superior mechanical properties and design flexibility (Tian et al., 2024). In the aerospace industry, they are used for aircraft structures, such as fuselages, wings, and tail sections, where weight reduction is critical for improving fuel efficiency and performance (Barile & Casavola, 2019). The automotive industry uses CFRPCs in high-performance and electric vehicles to reduce weight, enhance safety, and improve overall efficiency (Mitschang & Hildebrandt, 2012). Additionally, in sports equipment, CFRPCs are favored for their high strength-to-weight ratio, making them ideal for bicycles, tennis rackets, and golf clubs (Oladele et al., 2023).

Despite their benefits, CFRPCs face challenges, including high production costs and difficulties in recycling (Ateeq, 2023). The cost of raw CFs and the specialized manufacturing processes required can make CFRPCs expensive compared to conventional materials (Aldosari et al., 2024). Additionally, recycling CFRPCs is complex due to the strong bond between the CFs and the polymer matrix (Ateeq, 2023). However, ongoing research and development are focused on addressing these challenges. Innovations in manufacturing techniques aim to reduce costs and improve production efficiency, while advancements in recycling technologies seek to develop effective methods for reclaiming and reusing CFs. As these developments progress, the adoption of CFRPCs is expected to expand, further enhancing their role in various high-performance applications and contributing to a more sustainable future.

2.2 *Metal Matrix Composites*

Metal matrix CF composites (MMCFCs) are advanced materials that combine CFs with a metal matrix, such as aluminum, magnesium, or titanium (Aldosari et al., 2024; Khanna et al., 2021; Mileiko, 2022). These composites are designed to leverage the high strength and stiffness of CFs while benefiting from the metal matrix's excellent thermal and electrical conductivity, as well as its resistance to high temperatures (Shirvanimoghaddam et al., 2017). The synergy between CFs and metal matrices results in materials that offer superior mechanical properties and performance in demanding environments (Sayam et al., 2022).

Several techniques are used to manufacture MMCFCs, each with its own advantages and applications:

- Powder Metallurgy: This process involves mixing metal powders with CFs and then compressing the mixture into a desired shape. The compacted material is then sintered at high temperatures to bond the metal matrix and CFs together. This method is suitable for producing complex shapes and components with uniform fiber distribution (Mahaviradhan et al., 2021; Neubauer et al., 2010).
- Liquid Metal Infiltration: In this technique, a preform made of CFs is placed in a mold, and molten metal is then infiltrated into the preform, filling the spaces between the fibers. This method is ideal for creating composites with high fiber volume fractions and excellent bonding between the fibers and the matrix (Blucher, 1992; Sree Manu et al., 2016).

- Squeeze Casting: Similar to liquid metal infiltration, squeeze casting involves infiltrating molten metal into a fiber preform under high pressure. This process ensures good fiber wetting and minimal porosity, resulting in high-quality composites with superior mechanical properties (Alhashmy & Nganbe, 2015; SHA et al., 2021).

MMCFCs are used in a variety of high-performance applications due to their exceptional properties (Sayam et al., 2022; Shirvanimoghaddam et al., 2017). These composites are utilized in aircraft and spacecraft components, such as structural elements, engine parts, and heat shields, where weight reduction and high-temperature performance are critical (Alam et al., 2022; Hussain & Immanuel, 2022). MMCFCs are employed in high-performance and racing vehicles to reduce weight, enhance strength, and improve fuel efficiency. They are used in engine components, drive shafts, and brake systems (Macke et al., 2013; Singh et al., 2021). The high strength-to-weight ratio and thermal stability of MMCFCs make them suitable for use in armor, weaponry, and other defense-related applications (Mehara et al., 2021; Siengchin, 2023). MMCFCs are also used in various industrial applications, including heat exchangers, cutting tools, and electronic packaging, where their thermal conductivity and mechanical properties provide significant advantages (Miracle, 2005; Perumal et al., 2020; Shirvanimoghaddam et al., 2017).

Despite their benefits, MMCFCs face several challenges. The production processes can be complex and expensive, and achieving uniform fiber distribution and strong bonding between the fibers and the matrix can be difficult (Shirvanimoghaddam et al., 2017). Additionally, the recycling of MMCFCs is challenging due to the strong bond between the CFs and the metal matrix (Nishida, 2013). However, ongoing research and development are focused on overcoming these challenges. Innovations in manufacturing techniques aim to reduce costs and improve the quality and consistency of MMCFCs. Advances in recycling technologies are also being explored to address environmental concerns and promote sustainability.

2.3 Ceramic Matrix Composites

CF-reinforced ceramic matrix composites (CFRCMCs) are advanced materials that integrate CFs within a ceramic matrix, such as silicon carbide (SiC) (W. Wang et al., 2024), alumina (Al_2O_3) (Cao et al., 2018), or zirconia (ZrO_2) (Bach et al., 2020; Zhao et al., 2011). These composites are engineered to capitalize on the high-temperature stability, wear resistance, and hardness of ceramics, while the CFs provide enhanced toughness, thermal shock resistance, and mechanical strength (De Omena Pina et al., 2007; Tang & Hu, 2017). The combination results in materials that can withstand extreme environments, making them ideal for high-performance applications (Duan et al., 2021).

The production of CFRCMCs involves sophisticated techniques to ensure proper integration and bonding between the CFs and the ceramic matrix:

- Chemical Vapor Infiltration (CVI): In this process, CF preforms are exposed to a gas-phase ceramic precursor, which infiltrates the preform and deposits the ceramic matrix within and around the fibers. CVI produces high-quality composites with excellent fiber-matrix bonding and minimal porosity (Monteleone et al., 2020; Niu et al., 2023).
- Polymer Impregnation and Pyrolysis (PIP): This method involves impregnating CF preforms with a polymer that converts to a ceramic upon heating. The impregnated preform is then pyrolyzed at high temperatures, transforming the polymer into a ceramic matrix. PIP is useful for producing complex shapes and large components (Servadei et al., 2023; Stalin et al., 2020).
- Hot Pressing and Sintering: CFs and ceramic powders are mixed and then compacted under high pressure and temperature. This process ensures good fiber dispersion and strong bonding within the ceramic matrix. It is suitable for creating dense, high-strength composites (Liu et al., 2022; Liu et al., 2022).

CFRCMCs are particularly valued in applications that demand high-temperature performance and durability (Zhu & Wang, 2023). CFRCMCs are used in aircraft and spacecraft components, such as turbine blades, heat shields, and combustion liners, where they provide exceptional thermal resistance and structural integrity under extreme conditions (De Omena Pina et al., 2007; Uyanna & Najafi, 2020). These composites are employed in missile components, armor, and other defense-related applications that require materials to withstand high temperatures and mechanical stresses (Bilisik et al., 2023). In the energy sector, CFRCMCs are used in gas turbines, nuclear reactors, and other high-temperature environments where their thermal stability and resistance to oxidation and corrosion are critical (Rueschhoff et al., 2020). CFRCMCs are also applied in industrial settings, including cutting tools, wear-resistant coatings, and high-temperature furnace components, benefiting from their hardness and durability.

While CFRCMCs offer numerous advantages, they also present certain challenges. The manufacturing processes can be complex and costly, requiring precise control to achieve the desired material properties. Additionally, ensuring uniform fiber distribution and strong bonding within the ceramic matrix can be challenging. The inherent brittleness of ceramics, despite the toughening effect of CFs, remains a consideration in some applications (Chung, 2016; Niu et al., 2023). Future prospects for CFRCMCs are promising, with ongoing research focused on improving manufacturing techniques, reducing costs, and enhancing material performance. Advances in nanotechnology and material science are expected to lead to even stronger and more durable composites. As these innovations continue, the use of CFRCMCs is likely to expand, driving further advancements in aerospace, defense, energy, and industrial applications.

3 The Role of Carbon Fiber in Strategic Materials

Strategic materials are crucial for the development and maintenance of a country's infrastructure, defense, and economic stability. CF, with its unique properties, has emerged as a vital component in various strategic applications. This section explores the role of CF in strategic materials, highlighting its importance, applications, and impact on various industries. When examined from a strategic perspective, CFs provide technological, economic, and environmental advantages. However, it is a material that is still under development and needs to be improved. CFs provide superior performance characteristics that are crucial for advanced technological applications (Yang et al., 2020). Although initially expensive, the long-term benefits of durability, reduced maintenance, and improved efficiency justify the investment (Maiti et al., 2022). Contributes to sustainability through weight reduction, leading to lower fuel consumption and reduced emissions in transportation applications (Ahmad et al., 2020). As a result of its prominent features, CFRCs are used in the construction of aircraft fuselages, wings, and other structural components, reducing weight and improving fuel efficiency (Agarwal et al., 2024; Muhammad et al., 2021). They are used in satellite components and space exploration vehicles, where weight reduction is critical for mission success (Wei et al., 2022). Applied in the production of lightweight and durable military gear, including helmets, body armor, and vehicle armor, enhancing mobility and protection (Kacan & Elaldi, 2020). CF is used in sports cars and racing vehicles to improve speed, handling, and fuel efficiency (Ahmad et al., 2020; Li et al., 2020). Helps reduce the weight of electric vehicles, extending battery life and range. CF-reinforced blades are longer and more efficient, capturing more wind energy and improving overall performance (Teng et al., 2023). Used in advanced battery technologies and hydrogen storage tanks due to its high strength and low weight (Benitez et al., 2021). CFRP is used to strengthen concrete structures, providing higher durability and resistance to environmental stressors. CFRP is used in the repair and reinforcement of aging bridges, buildings, and pipelines, extending their service life and enhancing safety (Hussain & Immanuel, 2022; Rashed Labanieh et al., 2022). CF is used in prosthetics, surgical instruments, and imaging equipment due to its biocompatibility and strength. Lightweight and strong CF components are used in orthopedic implants and braces (Chua et al., 2021; Kaufman & Bernhardt, 2021). However, the high production cost of CF remains a significant barrier to its widespread adoption. Requires specialized manufacturing processes and expertise. The recycling of CF composites is challenging and needs further development to improve sustainability. Ongoing research aims to reduce production costs and improve the properties of CF composites. With technological advancements, CF is expected to penetrate more industries and applications. Developing efficient recycling methods will enhance the environmental benefits of CF. CF, one of the strategic materials, is used in many areas on an industrial scale. Figure 2 gives the market forecast of CF-based composites in terms of application. Economic indicators and market dynamics show that composite material will grow much faster and the dominant materials are CFs, glass fibers in the reinforcement segment, and polyester resin

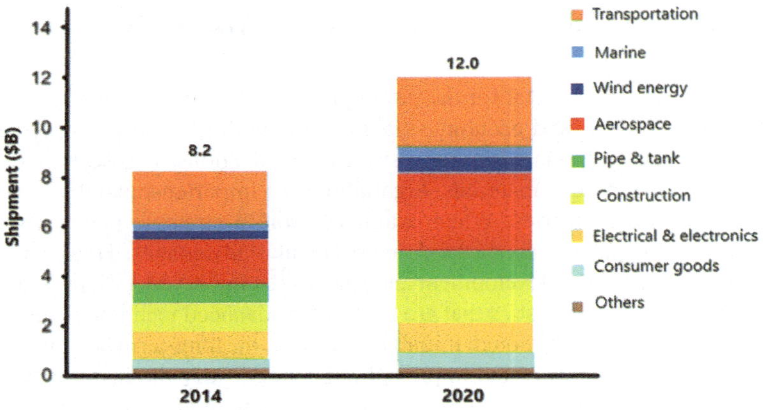

Fig. 2 Market forecast of CF-based composites in terms of application

in the resin segment. The CF-based composite material market, especially in the aviation, transportation, construction, wind energy, and electrical-electronics sectors, is expected to reach 18 billion dollars by 2030, with a compound annual growth rate of 6.6% (Erden & Ho, 2017; Naqvi et al., 2018). The FOREST project aims to lay the foundation for the wider use of biocomposite materials in the transport industry (bus, aviation, and automotive) with cost-effective sustainable solutions based on lightweight, bio-based, and recycled CF composite materials. The project also aims to develop CF, which is among the strategic materials, as a new, multifunctional, fully recycled product as a competitive alternative to traditional composites.

A significant growth in CF use was observed worldwide in 2023. Thanks to its low weight-to-high strength ratio and durability, CF finds wide use in the automotive, aerospace, defense, wind energy, sports, and construction sectors. The value of the CF market in 2023 is estimated at approximately $4.69 billion and is expected to reach $5.13 billion in 2024. It is predicted that the market will reach 7.16 billion dollars by 2028. The market's compound annual growth rate (CAGR) is estimated at 9.4% (*Carbon Fiber Global Market Report 2024—Research and Markets*, n.d.; *Carbon Fiber Market Analysis—US, China, Germany, Russia, Japan—Size and Forecast 2024-2028*, n.d.). The sector where CF is used most intensively is aviation and defense. In 2022, this sector accounted for 60.8% of the CF market. Major aircraft manufacturers such as Boeing and Airbus use this material to reduce the weight of aircraft and increase fuel efficiency by using CF. CF also plays an important role in the automotive industry. Thanks to its lightness and high strength, it is widely used in high-performance vehicles, including electric vehicles. An 8.5% increase in automobile production was recorded in South Korea in 2022, which increased the demand for CF. CFRCs are widely used in the production of wind turbines. Especially in North America and Western Europe, wind energy investments are increasing the demand for CF. Wind energy investments are increasing rapidly in countries such as India and Brazil. CF is also widely used in sports equipment. This sector benefits from the advantages of CF in sports equipment that requires high performance and durability. Europe, which is the leader in the CF market with a 31.6% share in 2022, maintains this position with the influence of major players in the aerospace and defense sector (Airbus, MBDA). In addition, the low-weight and high-performance-oriented production approach of giants in the automotive industry such as Volkswagen, Mercedes, and Ferrari increases the demand for CF in the region. Asia-Pacific countries such as China, India, South Korea, and Vietnam are increasing the demand for CF with increasing commercial aviation activities and automotive production. There are many important brands in the CF market trying to gain competitive advantage by investing in advanced technology. Among these, companies such as Toray Industries, Teijin Limited, Mitsubishi Chemical Corporation, SGL Carbon, and Hexcel Corporation stand out.

4 Industrial Carbon Fiber Waste

CF is widely used in many sectors and can be in different forms as waste. Figure 3 shows the increase rates in global CF demand over the years. The increase in global CF demand also affects the amount of CF waste on an industrial scale. According to a report by Composites World, approximately 30% of CF production ends up as waste (Isa et al., 2022; Khurshid et al., 2019). These include cut-offs, scrap pieces, and damaged or defective products that cannot be reused in the production process (Koumoulos et al., 2019). These are small pieces of CF that result from the cutting and shaping of CFRCs. They are typically irregular in size and shape and are often not reusable in further production processes (Feraboli et al., 2011). Parts that are produced with defects or do not meet quality control standards are classified as scrap. These pieces are considered waste as they do not maintain the structural integrity required for their intended applications. Products that have reached the end of their useful life or are damaged beyond repair fall into this category. These items were originally manufactured for specific uses but are no longer functional and are classified as recyclable waste.

Currently, the management of CF waste is crucial for both environmental sustainability and economic efficiency (Pakdel et al., 2021). However, the recycling and management of CF waste are more complex and costly compared to traditional materials (Vo Dong et al., 2018). Various approaches have been adopted in the industry for managing CF waste (Ateeq, 2023). One of the most common practices is sending CF waste to landfills or incinerating it (Bianchi et al., 2021). While these methods are relatively low-cost, they are not environmentally sustainable. Incineration of CF releases harmful gases and results in the loss of valuable materials, whereas landfilling provides only a limited solution. Recycling CF waste is increasingly preferred to enhance material efficiency and reduce environmental impact. The recycling process involves various physical and chemical treatments to make CF waste reusable (Naqvi et al., 2018; Zhang et al., 2020). Mechanical methods (cutting, grinding), thermal methods (pyrolysis), and chemical methods (solvent use) are employed to recover

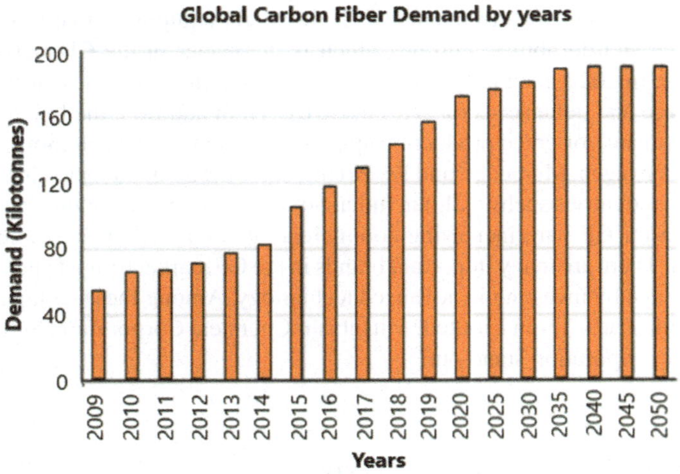

Fig. 3 Increase in global CF demand by year

CFs (Butenegro et al., 2021). CF waste can be transformed into new, high-value products through upcycling processes. These processes involve reprocessing waste into composite materials or new construction materials. For example, recycled CFs (rCFs) can be used in the automotive and aerospace industries to create lightweight and durable parts (Ahmad et al., 2020; Samant et al., 2022). Both industry and academic institutions are conducting research and development activities to manage CF waste more efficiently and sustainably. New recycling techniques and material processing methods are being developed to make the management of CF waste more economically and environmentally viable.

Industrial CF waste is a significant issue concerning both environmental sustainability and economic efficiency. Effective management and recycling of CF waste are critical for conserving natural resources and improving waste management strategies (Bianchi et al., 2021). Continuous improvement of industrial processes and recycling technologies is essential for the sustainable utilization of CF waste. Some well-known companies in the sector are carrying out important studies on the recycling of CF waste. ELG Carbon Fiber is a pioneering company in the recycling of CF waste (Davison & Heinz, 2021). ELG recycles CF waste through mechanical methods, making it reusable. This process involves cutting, breaking, and grinding the waste. ELG processes CF waste through high-temperature pyrolysis, allowing the recovery of CFs. This method separates the resin from the fibers to obtain pure CF. rCFs are used in various industries, including automotive, construction, and sports equipment. Toray Industries, a global leader in CF production, also undertakes various projects in recycling and managing CF waste. Toray works on transforming CF waste into new materials using advanced recycling technologies (*Realizing a Circular Economy*, n.d.). These technologies retain the properties of CFs, making them reusable. Toray uses rCFs in renewable energy projects and applications that enhance environmental sustainability. For example, rCFs are used in wind turbine blades. Another company, Carbon Conversions, specializes in recycling and reusing CF waste (*Carbon Fiber Recycling*, n.d.). The company recycles CF waste through various chemical and thermal processes to obtain high-quality CF. These processes largely preserve the original properties of CFs, making them reusable. Carbon Conversions develops new and innovative products using rCFs. For instance, lightweight and durable composite materials for the automotive and aerospace industries. The company's efforts reduce the environmental impact of CF waste and promote sustainable material usage. Aerospace giant Boeing engages in several initiatives to recycle CF waste (*Carbon Fiber Recycling: How Boeing Turns Scraps Into Better Use*, n.d.). Boeing collaborates with ELG Carbon Fiber to jointly run projects aimed at recycling CF waste (Sloan, 2018). This collaboration aims to make CF waste from aircraft manufacturing reusable. Boeing recycles CF waste generated from production processes and end-of-life aircraft, increasing the use of sustainable materials in the aerospace industry. Automotive leader BMW takes significant steps in managing and recycling CF waste (Seidel, 2021; Wienstroth, 2012). BMW uses rCFs in automobile production to enhance vehicle lightness and durability while reducing environmental impacts. rCFs are used in models like the BMW i3 and i8 (*BMW i Production CFRP Wackersdorf*, n.d.). BMW runs various projects to recycle CF waste generated during production processes. This waste is reprocessed for use in automotive parts. The FOREST project, in which Mercedes-Benz is also involved, aims to lay the foundation for the wider use of biocomposite materials in the transportation industry by providing sustainable solutions with bio-based and recycled CF production. Within the scope of the project, sustainable EoL alternatives suitable for closing the biomaterial cycle of rCF composite products targeted for use in ceiling, floor, and side wall panels in cabin interiors will also be determined. These companies' efforts demonstrate significant strides in managing CF waste in a more sustainable and environmentally friendly manner. The development of recycling and upcycling technologies reduces the environmental impact of CFs and broadens their industrial applications.

5 Recycling of Carbon Fiber Structures

CFs are widely used in modern industrial applications due to their superior properties. The increased use of CF creates a significant amount of waste. This has become a global problem as valuable CF waste is dumped in landfills. There is a great need to recycle CF-reinforced polymers due to the environmental challenges caused by conventional landfilling. There is a great need to recycle CFRPs due to the environmental challenges caused by conventional landfilling. Although various recycling techniques have been used to recover CFs from waste materials, recycled fibers have limited applications due to partial degradation in mechanical properties and shortening of fiber lengths. From an environmental and economic perspective, burying or burning these wastes is highly unattractive. Therefore, the recycling process may be one of the best ways to reduce environmental impact while meeting the global demand for this material in industrial applications (Arif et al., 2022).

5.1 Mechanical Recycling

This process involves grinding or shredding CFRCs into smaller pieces. The fibers are then separated from the resin matrix and reused in lower-grade applications, such as reinforcement for concrete or injection-molded parts. Mechanical recycling is relatively straightforward but can damage the fibers, reducing their mechanical properties. Mechanical recycling is suitable for scrap composite materials that are relatively clean, uncontaminated, and of known origin. It is an advantage that it recovers both fibers and resins without using or producing hazardous substances. It is important that the recovered fibers maintain their mechanical properties. Fibrous structure has limited reproducibility because their structure is rough, inconsistent, and unstructured. Mechanical recycling is the first technological process used to recycle CFRP. In this technique, CFRP is ground into small pieces by mechanical processes to ensure that the solid material is crumbled and broken into small pieces. The rich pitch in the scrap can then be isolated by sieving the dusty and fibrous structure. The advantage of this process is that no harmful gases are formed. When applied to CFRP waste treatment, mechanical recycling can recover a significant fraction of CF from waste, as well as provide residual materials suitable for recycled powder and possibly energy recovery for polymer filler applications, thus avoiding landfills. In general, CFRP mechanical recycling has low energy consumption and associated greenhouse gas emissions (Butenegro et al., 2021; Demski et al., 2024). Figure 4 shows the process flowchart of CF recycling by mechanical method as an example. Balıkoğlu et al. (Balıkoğlu et al., 2018) examined the mechanical recycling process and reuse of thermoset-based glass fiber-reinforced polymer (GFRP) plates. The authors used recycled products in proportions of 4, 8, 12, and 16 wt.% in the bulk molding compound (BMC) composition, replacing the original fibers with recycled ones through hot press molding. Tensile and flexural tests were performed on the samples cut from the produced BMC plates according to ISO 527-4 and ISO 178 standards. The results of this study showed that the tensile strength, Young's modulus, and elongation values of the samples were within the acceptable range of $\pm$ 5%. The sample containing 16 wt.% recycled material exhibited the highest tensile strength, Young's modulus, and elongation values. In the flexural test results, the strength values of the samples containing 4, 12, and 16 wt.% recycled glass fiber decreased. The sample containing 12 wt.% recycled glass fiber showed an acceptable decrease in flexural strength, below 10%. The flexural strength of the sample containing 8 wt.% recycled product increased compared to the original sample. Except for the sample containing 4 wt.% recycled product, all samples showed acceptable mechanical performance. In another scientific study, Colucci et al. (2015)

conducted research on the recycling and mechanical performance of a PP-based GF-reinforced polymer composite material (PP40GF), which is commercially used in the automotive sector. Tensile and flexural test results of the samples prepared with recycled material, containing 0, 50, and 100% original raw material. According to the results obtained, all of the recycled PP40GF composites exhibited high-quality performance.

5.2 *Thermal Recycling*

Also known as pyrolysis, this method involves heating CFRCs in an inert atmosphere to break down the resin matrix, leaving behind clean CFs. Thermal recycling can preserve much of the fiber's original strength and quality, making it suitable for high-performance applications. However, the process requires significant energy and can generate harmful emissions if not properly managed. It allows the recovery of long, high modulus fibers and is therefore one of the most common recycling processes. The composition of the waste feedstock, reactor type, and process parameters affect the pyrolysis procedure. Operating temperatures have significant effects on the mechanical, electrical, and surface properties of the fiber. A higher temperature can be used, but this may cause serious deterioration of recycled products. The most commonly used thermal recycling techniques are pyrolysis and fluidized bed procedure. Although the fluidized bed process is generally simpler than pyrolysis in theory, pyrolysis can produce potentially useful organic products from the polymer. Further processing would be needed to separate these from the product mix and so is likely to be cost-effective on an industrial scale. Recycled fibers may contain varying degrees of charcoal; this may limit reuse options or require further processing to remove them. It is a great advantage that chemical solvents are not used in the pyrolysis process and all results can be recovered and reused. Generally, gases are reused to obtain the energy required for the process. Recovered CFs have mechanical properties similar to virgin fibers (4–20% loss in tensile strength) and are reproducible. However, the properties of recycled fibers are sensitive to processing parameters. The only process used commercially is pyrolysis, but rCFs are chopped or ground (Butenegro et al., 2021; Pakdel et al., 2021). Figure 5 shows the basic steps of thermal recycling method of production of rCF. Yip et al. (2002) conducted the recovery process of carbon fibers through thermal conversion. They recovered carbon fibers with an average length of up to 10 mm. According to the mechanical test results indicated in the study, it was reported that the tensile strength of the carbon fibers retained 75% of their original strength after thermal recycling, the modulus of elasticity remained unchanged, and the surface condition showed similar characteristics to the original fibers. The results obtained demonstrated that the partial or complete replacement of carbon fibers has the potential for reuse. In another study on thermal conversion, Akesson et al. (Åkesson et al., 2012) applied the process to composite samples containing 0, 50, 75, and 100 wt.% of vFs. According to the flexural test results of GFRP recovered through thermal treatment, it was observed that

Fig. 4 Process flowchart of mechanical recycling method

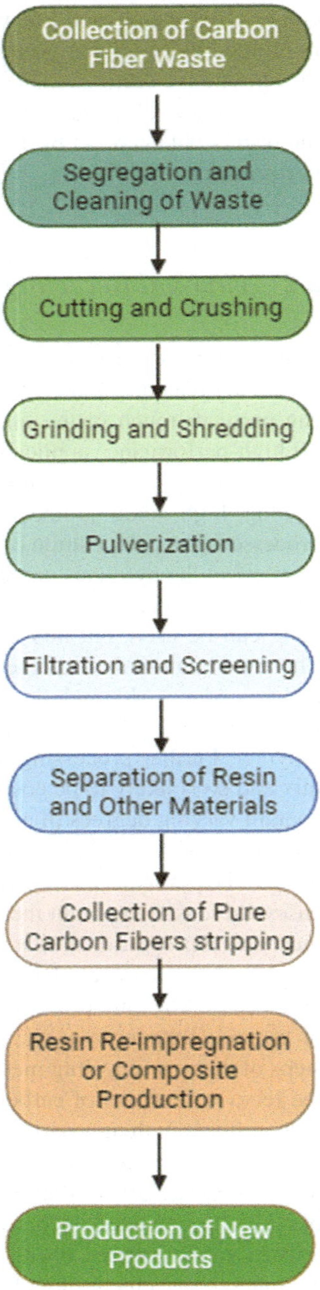

composites produced with 25 wt.% recycled fiber exhibited good mechanical properties. Thermal conversion is a widely accepted and commercially used technology on an industrial scale.

5.3 Chemical Recycling

In this method, solvents or chemical agents are used to dissolve the resin matrix, allowing the recovery of intact CFs. CF waste is collected from production processes and end-of-life products. The CF waste is sorted from other types of waste and cleaned. Large pieces of CF waste are cut and shredded into smaller pieces. The shredded CF waste is treated with a specific chemical solvent (e.g., acid or base). This solvent helps break down the resin and other organic materials from the CFs. The solvent dissolves the resin and other organic materials, separating them from the CFs. The clean CFs are extracted from the chemical solvent. The extracted CFs are cleaned to remove any residual solvent and impurities and then dried. The cleaned CFs are reprocessed or used to produce composite materials, making them ready for reuse. The rCFs are used to manufacture new products. Chemical recycling can produce high-quality fibers with minimal degradation, but it often involves the use of hazardous chemicals and complex processing steps. Chemical methods for CFRP recycling are based on low-temperature (typically below 350 °C) reactive media (e.g., catalytic solutions, alcohols, and supercritical fluids). While the polymeric resin is decomposed into relatively high-value oligomers (which can be reused as chemical feedstocks), the CFs remain inert and are subsequently collected. In this technique, the polymer is dissolved with a solution consisting of acid, base, and solvents. The concentration of the solution must be adjusted on the matrix. Compared to sulfuric and hydrochloric acids, nitric acid enables the degradation of polymer resins (i.e., epoxy) and greater recovery of CF. Increasing reagent costs and production of toxic gases are the biggest challenges in chemical recycling of CF. The motivation behind the development of chemical recycling is based on achieving rCF whose mechanical properties or morphology do not change after recycling, as well as reducing the high energy consumption of thermal recycling. The best mechanical performance for rCF is generally achieved by chemical recycling methods. Some of the factors that affect the outcome of the process are temperature, pressure, catalysts, and solvents used (Ateeq, 2023; Butenegro et al., 2021; Manarin et al., 2024). Figure 6 shows production steps of the chemical recycling method of rCF. Yang et al. (2014) carried out the recycling of carbon fibers and glass fibers through chemical conversion. They aimed to optimize the reaction conditions, such as temperature, reaction time, and solvent (NaOH) amount, which varied depending on the type of fiber. They determined the mechanical properties as a function of increasing temperature. When the experiment duration reached 4 hours, the tensile strength of the recycled carbon fibers was found to be 96% of the original fibers, non-alkali glass fibers 94%, and medium-alkali glass fibers 89%.

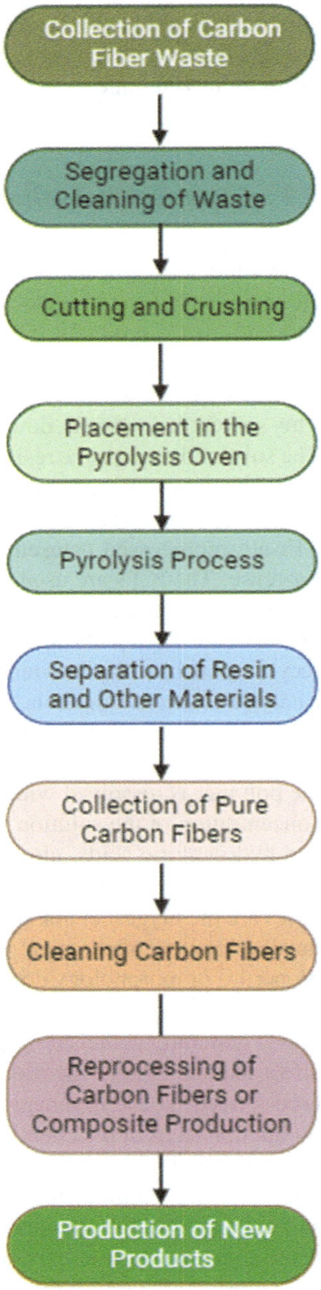

Fig. 5 Process flowchart of thermal recycling method

Fig. 6 Process flowchart of chemical recycling method

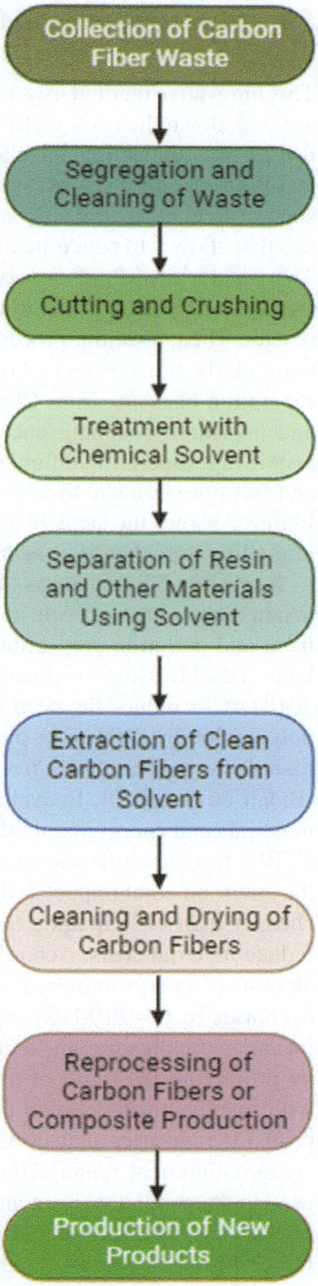

5.4 Supercritical Fluid Recycling

This innovative method uses supercritical fluids, such as supercritical carbon dioxide, to break down the resin matrix. Unlike other methods, the supercritical fluid reactor is used after collecting and classifying the waste. Waste CFs placed in the supercritical fluid reactor are processed with supercritical fluid. The reactor is filled with a supercritical fluid, typically supercritical CO_2. In this state, CO_2 has unique properties that allow it to penetrate materials like a gas but dissolve substances like a liquid. The supercritical fluid dissolves the resin and other organic materials surrounding the CFs, effectively separating them. After, the clean CFs are extracted from the reactor. Then, a similar process flow is applied again. The use of supercritical fluids, particularly supercritical CO_2, allows for an efficient and environmentally friendly separation of resins from CFs. The process operates at relatively low temperatures and pressures, reducing energy consumption and preserving fiber quality. While promising, supercritical fluid recycling is still in the research and development stage and has not yet been widely adopted (Butenegro et al., 2021; Pakdel et al., 2021). Figure 7 shows the steps of supercritical fluid recycling method. In Table 1, studies in the literature that have addressed CF recycling with various methods are given.

Recycling CF structures offers numerous environmental and economic benefits. When the production cycle of CFs is examined, the first advantage is waste management and pollution prevention. It significantly reduces the amount of waste sent to landfills, addressing the pressing issue of industrial waste management (Meng et al., 2018). rCFs reduce the need for vCF production, conserving raw materials such as polyacrylonitrile (PAN) or pitch. Producing rCF consumes less energy compared to manufacturing new fibers from raw materials, leading to lower overall energy usage (Balaji et al., 2020). Recycling reduces the demand for vCF, conserving the raw materials and energy required for their production (Meng et al., 2018). By diverting CFRCs from landfills and incineration, recycling helps to reduce waste and minimize the environmental impact of disposal (Danish et al., 2022). rCFs can be significantly cheaper than vCF, making them an attractive option for manufacturers seeking to reduce material costs. According to previous studies (Meng et al., 2017), while the amount of energy required for vCF production is ~198–595 MJ/kg, this amount decreases to 60–90 MJ/kg in rCF production. Considering mass production, this means a significant gain in energy. Also, recycling supports the circular economy by promoting the reuse of materials and reducing the carbon footprint associated with manufacturing (Balaji et al., 2020). This contributes to a lower carbon footprint for industries that utilize CF materials, such as aerospace, automotive, and sports equipment manufacturing (Zhang et al., 2023). Additionally, rCFs can be used to create high-performance, lightweight materials, enhancing the sustainability of various products without compromising quality (Balaji et al., 2020). Furthermore, the development and implementation of advanced recycling technologies can spur innovation and create new job opportunities in the green technology sector. The growing demand for sustainable products creates new market opportunities for rCFs in various industries, including automotive, aerospace, and consumer goods.

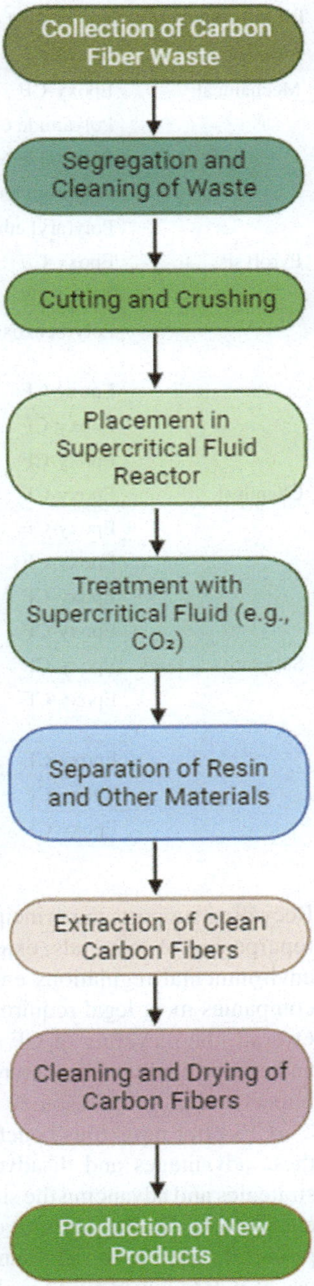

Fig. 7 Process flowchart of supercritical fluid recycling method

Table 1 Studies on the recycling of various CFRCs with different methods

Method	Material	Efficiency	Reference
Mechanical	Epoxy-CF	–	Durante et al. (2023)
	Polyamide 66-CF	–	Colucci et al. (2015)
	Polyamide 66-CF	–	Pietroluongo et al. (2020)
	Polypropylene-CF	–	Evens et al. (2019)
	Poly(aryl ether ketone)-CF	–	Oliveira et al. (2024)
Pyrolysis	Epoxy-CF	94.49%	Deng et al. (2020)
	Epoxy-CF	96.5%	Ren et al. (2022)
	Poly(benzoxazine)-CF	–	Lopez-Urionabarrenechea et al. (2020)
	Epoxy-CF	–	Guo et al. (2019)
	Epoxy-CF	–	Abdou et al. (2021)
	Epoxy-CF	60%	Yatim et al. (2020)
Chemical	Epoxy-CF	97.77%	Wang et al. (2015)
	Epoxy-CF	96.9%	Lee et al. (2020)
	Epoxy-CF	95%	Jiang et al. (2017)
	Epoxy-CF	–	Yang et al. (2014)
	Epoxy-CF	83%	Schamel et al. (2021)
Supercritical fluid	Epoxy-CF	94–97%	Bai et al. (2010)
	Epoxy-CF	–	Pinero-Hernanz et al. (Ul et al., 2008)
	Epoxy-CF	–	Jiang et al. (2009)
	Epoxy-CF	100%	Shetty et al. (2022)
	Epoxy-CF	–	Okajima et al. (2014)

Recycling supports the principles of a circular economy by promoting the reuse and repurposing of materials, extending the lifecycle of CFRCs. Increasingly stringent environmental regulations encourage the adoption of recycling practices, helping companies meet legal requirements and improve their environmental performance. Overall, the recycling of CF structures is a pivotal step toward a more sustainable and economically viable future, aligning with global efforts to reduce environmental impact and promote resource efficiency.

rCFs offer numerous benefits but also present certain challenges. Understanding these advantages and disadvantages is crucial for developing effective recycling strategies and advancing the sustainable use of CF materials (Ateeq, 2023; Butenegro et al., 2021). Mechanical recycling processes can damage CFs, reducing mechanical properties such as strength and stiffness (Gopalraj & Kärki, 2020). This limits the use of rCF to lower-grade applications. The recycling process can shorten the fibers, which impacts their performance and restricts their application in high-performance composites (Hofmann et al., 2023). Some recycling methods, particularly thermal and chemical processes, can be expensive due to the energy and specialized equipment required (Bledzki et al., 2021). The economic viability of recycling depends on the market demand for rCF, which can fluctuate and affect the profitability of recycling operations (Meng et al., 2018). CFRCs often consist of various fiber types, resin systems, and manufacturing methods, complicating the recycling process and making it challenging to achieve consistent quality (Asmatulu et al., 2013). Contaminants in the composite materials, such as adhesives, coatings, and other additives, can hinder the recycling process and reduce the quality of the rCF (Stieven Montagna et al., 2022). Developing efficient and scalable recycling processes that can handle the diverse range of CFRC remains a significant challenge (Giorgini et al., 2020). Continuous innovation and technological advancements are required to improve the efficiency, cost-effectiveness, and environmental friendliness of recycling methods (Koumoulos et al., 2019). The absence of standardized recycling methods and quality benchmarks for rCF can pose challenges in ensuring consistent performance and reliability (Danish et al., 2022). Navigating complex regulatory frameworks related to waste management and recycling can be cumbersome and may vary across regions, impacting the implementation of recycling programs.

The recycling of CF presents a balanced mix of significant advantages and notable challenges. While it offers substantial environmental, economic, and sustainability benefits, it also requires addressing quality degradation, high processing costs, material complexity, and technological hurdles. As the demand for sustainable materials continues to grow, advancements in recycling technologies and the development of standardized processes will be crucial for maximizing the benefits and overcoming the disadvantages of rCF structures. Through ongoing research, innovation, and collaboration, the industry can enhance the viability and effectiveness of CF recycling, contributing to a more sustainable future. Various methods for recycling CFRP have been investigated; most recycling routes involve shredding waste and recycling fibers to recover products and then re-impregnating the recovered fibers in a new composite (Borjan et al., 2021; Pimenta & Pinho, 2014). Although the mechanical recycling method is economically advantageous and can be applied on an industrial scale, rCFs have poorer mechanical and physical properties. This shows that this technology is not suitable for structural applications with high loads. Chemical recycling methods can produce rCFs with clean and smooth surfaces, and vCFs can maintain their mechanical properties. However, it is difficult to gain commercial ground due to the high cost of solvents and equipment, harsh conditions, and the formation of hazardous gases and waste. For all these reasons, thermal recycling methods such as pyrolysis, steam, and fluidized bed methods have been developed to recycle CFRPs more efficiently. These processes include chemicals in the supercritical phase to dissolve the polymer matrix, and the tensile strength of rCFs can reach 80–95% of vCFs. The biggest problem is that the coke formed during the heat treatment accumulates on the surfaces of rCFs. The success in rCFs recycling processes will provide an important research area for the future development of CFs and CFRPs. Therefore, it is necessary to develop sustainable recycling processes. Developing reuse strategies for rCFs is part of ongoing research, such as spinning them into yarns, producing nonwoven fabrics and crimp-free prepreg fabrics using combined rCFs and vCFs,

Table 2 Pros and cons of different recycling process

Method	Pros	Cons	Required energy (MJ/kg)
Mechanical recycling	Cost-effective, efficiency, well-known	May adversely affect mechanical properties	0.1–4.8
	Both resin and fibers can be recycled	Fibers may be irregularly shaped and short	
	No harmful products or uses	Limited recycling	
Pyrolysis recycling	Highly efficient recycling	Requires additional equipment due to off-gases	21–91
	Better adhesion between resin and rCFs	There may be residual resin on the fiber surface	
	Commercially applied	Fiber quality is sensitive to processes	
Chemical recycling	Highly efficient recycling	Chemical use poses a threat to the environment	24–30
	Recycling is possible for both fiber and matrix material	Suitable for low quantities	
		Fiber adhesion to resin matrix is generally reduced	

and new composite materials (Das et al., 2018; Wu et al., 2022; Yan et al., 2016). Table 2 compares the pros and cons of the mentioned recycling methods.

6 Challenges in the Recycling Process of Carbon Fibers

While recycling processes have reached a certain degree of maturity, the biggest challenge for CF recycling operations is the introduction of a robust CFRP recycling process that supports the effective commercialization of recycled products. Here are some basic challenges to overcome these problems (Ateeq, 2023; Naqvi et al., 2018).

- **Creating a global recycling strategy:** It is necessary for composite-related industries, suppliers, recycling facilities, and research centers to come together and plan technological and industrial needs.
- **Giving incentives for recycling:** Incentives in the name of carbon credit should be given to institutions that recycle CF. As recycling companies become commercialized, specific legislation covering their activities needs to be implemented.

- **Logistics and collaboration in the supply chain:** Waste suppliers should collaborate with recyclers, which includes ensuring a continuous and appropriate supply of waste and providing material certification to recyclers where possible. Conversely, recyclers must ensure that materials and components supplied are not reverse engineered.
- **Market identification and product pricing:** To determine the value of recycled products, it is necessary to know the characteristics of different rCFs and evaluate processing times and costs.
- **Performing life cycle analysis (LCA):** Knowing the recycling process efficiency, environmental, economic, and technical advantages of rCF materials can be determined by life cycle analysis.
- **Establishing the market:** The current biggest challenge for the success of CFRP recycling is the establishment of a market for recycled products. Since creating a market requires overcoming all previous problems, rCFs are considered an environmentally friendly, sustainable, and cost-effective material.

Recycling of composites difficult due to reasons such as the presence of complex compositions such as fibers, matrix, and fillers, the presence of thermoset resins that cannot be remolded, and its combination with other materials. Today, most CFRP waste is generally disposed of in desert cemeteries, airports, or landfills. However, these are not environmental and sustainable solutions for the following reasons (Balaji et al., 2020; Bledzki et al., 2021):

Environmental impact: Increasing production of CFRP increases concerns about waste disposal and consumption of non-renewable resources.

- **Legislation**: Recent European legislation imposes strict control on the disposal of composites; Responsibility for disposing of EoL composites now lies with the component manufacturer, legal disposal of CFRP is limited and, for example, 85% of automotive vehicles disposed of after 2015 must be recyclable (EU 1999/31/EC; EU 2000/53/EC).
- **Cost of production:** CFs are expensive products, both in terms of energy consumed during production and material price.
- **Management of resources:** demand for vCFs often exceeds supply capacity, so recycled rCFs can be reintroduced to the market for non-critical applications.
- **Economic opportunity:** The cost of disposing of CFRP through landfill is quite high. Recycling CFRPs offers the opportunity to transform very expensive waste disposal into economical and reusable materials.

Research and market studies show that commercialization of rCF components may be possible by overcoming several challenges that have not yet been resolved. Studies show that Mercedes-Benz, Chevrolet, BMW, Ford, and many other major automobile manufacturers have paid attention and possibility to use rCF composite part in large-scale automobile production.

7 Opportunities for Sustainable Recycling

The sustainable recycling of CF materials presents a plethora of opportunities that can significantly impact various industries and contribute to environmental preservation. By advancing recycling technologies and implementing innovative approaches, the potential for sustainable recycling can be fully realized. The development of advanced chemical recycling processes can enhance the efficiency and effectiveness of recovering high-quality CFs from composite waste. By utilizing environmentally friendly solvents and catalysts, chemical recycling can minimize environmental impact and produce rCFs with properties comparable to vCF. Leveraging supercritical fluids, such as supercritical carbon dioxide, offers a promising method for decomposing resin matrices at relatively low temperatures and pressures. This approach preserves the integrity of CFs, making them suitable for high-performance applications while reducing energy consumption and emissions.

Integrating closed-loop recycling systems within manufacturing processes can ensure that CF waste generated during production is efficiently reclaimed and reused. This approach not only reduces waste but also decreases the reliance on virgin materials, promoting a circular economy. Designing CFRCs with recycling in mind can facilitate easier disassembly and material recovery. By selecting compatible resin systems and minimizing the use of contaminants, manufacturers can enhance the recyclability of their products and reduce the complexity of the recycling process.

Establishing specialized companies that offer recycling services for CFRCs can create new business opportunities and encourage more industries to adopt sustainable practices. These companies can provide expertise, technology, and infrastructure for efficient recycling, making it accessible to a broader range of manufacturers. Forming partnerships between manufacturers, recycling companies, and research institutions can drive innovation and share the financial and technical burdens associated with developing sustainable recycling solutions. Collaborative efforts can accelerate the adoption of best practices and technological advancements across the industry. Many studies have made a techno-economic evaluation of the application of rCF in an automotive part, taking into account weight reductions and fuel consumption reductions compared to steel. A 35% cost reduction was determined for the rCF part with oriented fibers compared to the steel part. Considering the current market volume and the growth of CFRPs, these amounts seem achievable in the future. At this stage, the attractiveness of rCF materials stems not only from their cost advantages but also from their superior mechanical and physical properties (May et al., 2021; Meng et al., 2018). Considering all these, the multifunctionality provided by the use of rCF, shown schematically in Fig. 8, appears as a promising opportunity.

FOREST, which aims to develop recycled CF and produce new and innovative green composites for increased resource efficiency and sustainable and safe transportation practices, offers new suggestions in line with these goals. This chapter of current sustainable indicators in biocomposite production, conversion of 100% rCF into new intermediate products; assessing the potential suitability of technologies

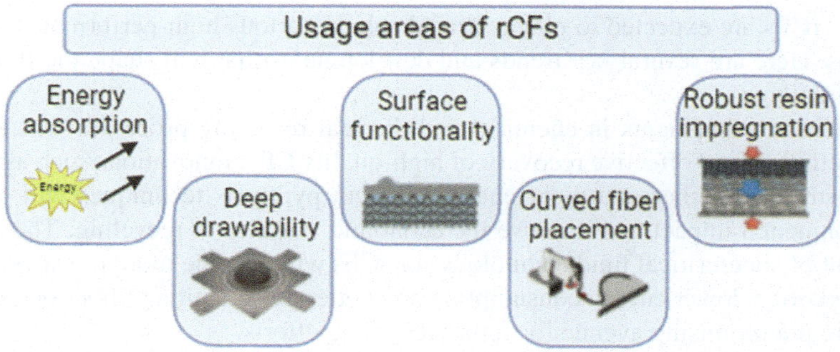

Fig. 8 Versatile functionality achieved using rCF (May et al., 2021)

applied to biocomposites and testing them at the laboratory and pilot plant level; determining the technical feasibility of the selected material and process route; and establishing clear criteria for CE. Implementing supportive regulations and policies that encourage the recycling of CFRCs can drive industry-wide adoption. Governments can introduce mandates for recycling, provide funding for research and development, and offer tax incentives for companies that invest in sustainable recycling technologies. Developing industry standards and certification programs for rCFs can ensure consistent quality and performance, fostering market acceptance and consumer confidence. Standardized processes and benchmarks can streamline recycling operations and facilitate the integration of rCFs into new products.

Investing in R&D to develop new resin systems and composite materials that are easier to recycle can significantly enhance the sustainability of CF products. Innovations such as thermoplastic composites and reversible adhesives can simplify the recycling process and improve material recovery rates. Conducting comprehensive lifecycle analyses of CFRC can identify key areas for improvement in recycling and overall sustainability. By understanding the environmental impact of each stage of a product's lifecycle, from production to end-of-life, manufacturers can make informed decisions to optimize their processes and reduce their ecological footprint. The FOREST Project, which aims to provide new and innovative sustainable green composites for safe transportation practices, is a great opportunity for the development of sustainable recycling of CFs. The FOREST project will also provide opportunities for new eco-design and recycling practices based on the life cycle approach of bio-based and recycled CFs.

8 The Future of Recycled Carbon Fiber Structures

The future of rCF structures is poised to be transformative, with advancements in technology, increasing environmental awareness, and industry collaboration driving significant growth and innovation. As the demand for sustainable materials continues to rise, rCFs are expected to play a crucial role in various high-performance applications. Here are several key trends and developments that will shape the future of rCF structures:

Future advancements in chemical and thermal recycling processes will enable more efficient and effective recovery of high-quality CFs. Innovations such as green chemistry approaches and more energy-efficient pyrolysis techniques will reduce environmental impact and improve the economic viability of recycling. The development of supercritical fluid technology for rCFs will become more prevalent. This method offers lower energy consumption and better preservation of fiber properties, making it a promising avenue for future recycling efforts.

The integration of artificial intelligence (AI) and machine learning in recycling facilities will enhance the sorting and separation of CFRCs from other materials. Automated systems will increase efficiency, reduce contamination, and ensure a consistent supply of high-quality recycled fibers. AI-driven process optimization can improve the efficiency and yield of recycling operations. By analyzing data and predicting outcomes, AI can help fine-tune recycling processes, reducing waste and maximizing fiber recovery.

The automotive industry will increasingly adopt rCFs to produce lightweight, fuel-efficient vehicles. rCF components will be used in structural parts, body panels, and interior components, contributing to lower emissions and improved performance. The growing EV market will drive demand for rCF due to its lightweight and high-strength properties, which can enhance battery range and vehicle efficiency. Aerospace and defense sectors will turn to rCF for cost-effective, high-performance materials. The ability to recycle CFs without significant loss of properties makes them suitable for various aerospace applications, including aircraft interiors, structural components, and unmanned aerial vehicles (UAVs). Aerospace companies are increasingly prioritizing sustainability, and rCF will play a vital role in meeting environmental goals and reducing the carbon footprint of aircraft production. rCFs will be used in the construction industry to create innovative, sustainable building materials. rCF-reinforced concrete, composites for bridges, and modular construction components will benefit from the strength and durability of CFs. Infrastructure projects will leverage rCF for its corrosion resistance and long-term durability, leading to more resilient and sustainable construction practices.

As recycling technologies advance and more companies invest in rCF production, the cost of rCFs will decrease. Economies of scale will make rCF more competitive with virgin CFs, driving broader adoption across industries. Greater integration of rCF into supply chains will streamline production processes and reduce material costs, further enhancing the economic feasibility of using rCFs. Supportive regulations and incentives will encourage the use of recycled materials. Governments may implement policies that mandate recycling or provide financial incentives for companies that use rCF, promoting a more sustainable materials economy. Stricter environmental standards will drive the adoption of rCF as industries seek to meet sustainability goals and reduce their carbon footprint.

The future of rCF structures is bright, with technological advancements, expanding applications, market growth, and regulatory support all contributing to their increasing adoption. As industries continue to prioritize sustainability, rCFs will play a critical role in reducing environmental impact, conserving resources, and driving innovation in high-performance applications. The continued development and integration of rCF will lead to a more sustainable and efficient use of CF materials, shaping a greener future for various industries. The development of more efficient and cost-effective manufacturing and recycling technologies will further enhance the role of CF materials in modern engineering and environmental sustainability.

The success of FOREST, which aims to develop new multifunctional biocomposites as a competitive alternative to traditional composites, will make significant contributions to the development of bio-based and recycled CF composites. Additionally, the FOREST project will increase the focus on sustainability in the circular economy (CE) through the analysis of effective recycling methods applied to biocomposite components containing more than 50% sustainable materials. In terms of rCFs, the success of FOREST and similar projects will be the determining factor in the role that CF structures will play in the future.

Acknowledgements This research was financially supported by Horizon Europe-HORIZON (Project number and name: 101091790-FOREST). The authors gratefully acknowledge the support.

References

Abdou, T. R., Botelho Junior, A. B., Espinosa, D. C. R., & Tenório, J. A. S. (2021). Recycling of polymeric composites from industrial waste by pyrolysis: Deep evaluation for carbon fibers reuse. *Waste Management, 120*, 1–9. https://doi.org/10.1016/J.WASMAN.2020.11.010

Advanced lightweight materials for energy efficient structures (RIA). (n.d.). Retrieved July 23, 2024, from https://cordis.europa.eu/programme/id/HORIZON_HORIZON-CL4-2022-RESILIENCE-01-11/en

Agarwal, N., Rangamani, A., Bhavsar, K., Virnodkar, S. S., Fernandes, A. A. A., Chadha, U., Srivastava, D., Patterson, A. E., & Rajasekharan, V. (2024). An overview of carbon-carbon composite materials and their applications. *Frontiers in Materials, 11*, 1374034. https://doi.org/10.3389/FMATS.2024.1374034

Ahmad, H., Markina, A. A., Porotnikov, M. V., & Ahmad, F. (2020). A review of carbon fiber materials in automotive industry. *IOP Conference Series: Materials Science and Engineering, 971*(3), 032011. https://doi.org/10.1088/1757-899X/971/3/032011

Åkesson, D., Foltynowicz, Z., Christéen, J., & Skrifvars, M. (2012). Microwave pyrolysis as a method of recycling glass fibre from used blades of wind turbines. *Journal of Reinforced Plastics and Composites, 31*(17), 1136–1142. https://doi.org/10.1177/0731684412453512

Alam, M. A., Ya, H. H., Sapuan, S. M., Mamat, O., Parveez, B., Yusuf, M., Masood, F., & Ilyas, R. A. (2022). Recent advancements in advanced composites for aerospace applications: A review. *Advanced Composites in Aerospace Engineering Applications*, 319–339. https://doi.org/10.1007/978-3-030-88192-4_16

Aldosari, S. M., AlOtaibi, B. M., Alblalaihid, K. S., Aldoihi, S. A., AlOgab, K. A., Alsaleh, S. S., Alshamary, D. O., Alanazi, T. H., Aldrees, S. D., & Alshammari, B. A. (2024). Mechanical recycling of carbon Fiber-reinforced polymer in a circular economy. *Polymers, 16*(10), 1363. https://doi.org/10.3390/POLYM16101363

Alhashmy, H. A., & Nganbe, M. (2015). Laminate squeeze casting of carbon fiber reinforced aluminum matrix composites. *Materials & Design, 67*, 154–158. https://doi.org/10.1016/J.MATDES.2014.11.034

Alshammari, B. A., Alsuhybani, M. S., Almushaikeh, A. M., Alotaibi, B. M., Alenad, A. M., Alqahtani, N. B., & Alharbi, A. G. (2021). Comprehensive review of the properties and modifications of carbon Fiber-reinforced thermoplastic composites. *Polymers, 13*(15), 2474. https://doi.org/10.3390/POLYM13152474

Arif, Z. U., Khalid, M. Y., Ahmed, W., Arshad, H., & Ullah, S. (2022). Recycling of the glass/carbon fibre reinforced polymer composites: A step towards the circular economy. *Polymer-Plastics Technology and Materials, 61*(7), 761–788. https://doi.org/10.1080/25740881.2021.2015781

Asmatulu, E., Twomey, J., & Overcash, M. (2013). Recycling of fiber-reinforced composites and direct structural composite recycling concept. *Journal of Composite Materials, 48*(5), 593–608. https://doi.org/10.1177/0021998313476325

Ateeq, M. (2023). A state of art review on recycling and remanufacturing of the carbon fiber from carbon fiber polymer composite. *Composites Part C: Open Access, 12*, 100412. https://doi.org/10.1016/J.JCOMC.2023.100412

Ateeq, M., Shafique, M., Azam, A., & Rafiq, M. (2023). A review of 3D printing of the recycled carbon fiber reinforced polymer composites: Processing, potential, and perspectives. *Journal of Materials Research and Technology, 26*, 2291–2309. https://doi.org/10.1016/J.JMRT.2023.07.171

Azeem, M., Ya, H. H., Alam, M. A., Kumar, M., Stabla, P., Smolnicki, M., Gemi, L., Khan, R., Ahmed, T., Ma, Q., Sadique, M. R., Mokhtar, A. A., & Mustapha, M. (2022). Application of filament winding Technology in Composite Pressure Vessels and Challenges: A review. *Journal of Energy Storage, 49*, 103468. https://doi.org/10.1016/J.EST.2021.103468

Bach, M., Gehre, P., Biermann, H., & Aneziris, C. G. (2020). Recycling of carbon fiber composites in carbon-bonded alumina refractories. *Ceramics International, 46*(8), 12574–12583. https://doi.org/10.1016/J.CERAMINT.2020.02.020

Bag, D. S., Kumar, A., Bansiwal, J. K., Lal, G., Tripathi, D. N., & Namburi, E. P. (2024). Polymer matrix composites (PMCs) for defence applications. 1–37. doi:https://doi.org/10.1007/978-981-99-9795-4_1.

Bai, Y., Wang, Z., & Feng, L. (2010). Chemical recycling of carbon fibers reinforced epoxy resin composites in oxygen in supercritical water. *Materials and Design, 31*(2), 999–1002. https://doi.org/10.1016/J.MATDES.2009.07.057

Balaji, A. B., Rudd, C., & Liu, X. (2020). Recycled carbon Fibers (rCF) in automobiles: Towards circular economy. *Materials Circular Economy, 2*(1), 1–8. https://doi.org/10.1007/S42824-020-00004-0

Balıkoğlu, F., Kerem Demircioğlu, T., Ataş, A., Beyaz, S., Arslan, N., Üniversitesi, B., Fakültesi, M., Bölümü, M. M., Fakültesi, F. E., Bölümü, K., Celal, M., Üniversitesi, B., Ferdi, H., Fakültesi, T., Mühendisliği, E. S., Manisa, T., & Yazarın, S. (2018). Termoset Matrisli Cam Elyaf Takviyeli Polimer CTP Atıkların Mekanik Geri Dönüşüm Ürünlerinin Sıcak Pres Kalıplama Bileşiminde BMC Tekrar Değerlendirilmesi. *Karaelmas Fen ve Mühendislik Dergisi, 8*(1), 168–180. https://doi.org/10.7212/2Fzkufbd.v8i1.936

Barile, C., & Casavola, C. (2019). Mechanical characterization of carbon fiber-reinforced plastic specimens for aerospace applications. *Mechanical and Physical Testing of Biocomposites, Fibre-Reinforced Composites and Hybrid Composites*, 387–407. https://doi.org/10.1016/B978-0-08-102292-4.00019-9

Benitez, A., Wulf, C., de Palmenaer, A., Lengersdorf, M., Röding, T., Grube, T., Robinius, M., Stolten, D., & Kuckshinrichs, W. (2021). Ecological assessment of fuel cell electric vehicles with special focus on type IV carbon fiber hydrogen tank. *Journal of Cleaner Production, 278*, 123277. https://doi.org/10.1016/J.JCLEPRO.2020.123277

Bianchi, I., Forcellese, A., Marconi, M., Simoncini, M., Vita, A., & Castorani, V. (2021). Environmental impact assessment of zero waste approach for carbon fiber prepreg scraps. *Sustainable Materials and Technologies, 29*, e00308. https://doi.org/10.1016/J.SUSMAT.2021.E00308

Bilisik, K., Syduzzaman, M., Erdogan, G., & Korkmaz, M. (2023). Advances in ballistic protection. *Functional and Technical Textiles*, 71–139. https://doi.org/10.1016/B978-0-323-91593-9.00023-7

Bledzki, A. K., Seidlitz, H., Goracy, K., Urbaniak, M., & Rösch, J. J. (2021). Recycling of carbon Fiber reinforced composite polymers—Review—Part 1: Volume of production, recycling technologies, legislative aspects. *Polymers, 13*(2), 300. https://doi.org/10.3390/POLYM13020300

Blucher, J. T. (1992). Discussion of a liquid metal pressure infiltration process to produce metal matrix composites. *Journal of Materials Processing Technology, 30*(3), 381–390. https://doi.org/10.1016/0924-0136(92)90227-J

BMW i Production CFRP Wackersdorf. (n.d.). Retrieved July 15, 2024, from https://www.press.bmwgroup.com/global/photo/detail/P90125888/bmw-i-production-cfrp-wackersdorf-carbon-fiber-recycling-material-for-the-use-in-the-bmw-i3-e-g-the

Bodaghi, M., Costa, R., Gomes, R., Silva, J., Correia, N., & Silva, F. (2020). Experimental comparative study of the variants of high-temperature vacuum-assisted resin transfer moulding. *Composites Part A: Applied Science and Manufacturing, 129*, 105708. https://doi.org/10.1016/J.COMPOSITESA.2019.105708

Borjan, D., Knez, Ž., & Knez, M. (2021). Recycling of carbon Fiber-reinforced composites—Difficulties and future perspectives. *Materials, 14*(15), 4191. https://doi.org/10.3390/MA14154191

Butenegro, J. A., Bahrami, M., Abenojar, J., & Martínez, M. Á. (2021). Recent Progress in carbon Fiber reinforced polymers recycling: A review of recycling methods and reuse of carbon Fibers. *Materials, 14*(21), 6401. https://doi.org/10.3390/MA14216401

Cao, X., Shi, Q., Liu, D., Feng, Z., Liu, Q., & Chen, G. (2018). Fabrication of in situ carbon fiber/aluminum composites via friction stir processing: Evaluation of microstructural, mechanical and tribological behaviors. *Composites Part B: Engineering, 139*, 97–105. https://doi.org/10.1016/J.COMPOSITESB.2017.12.001

Carbon Fiber Global Market Report 2024—Research and Markets. (n.d.). Retrieved July 3, 2024, from https://www.researchandmarkets.com/reports/5866013/carbon-fiber-global-market-report

Carbon Fiber Market Analysis—US, China, Germany, Russia, Japan—Size and Forecast 2024–2028. (n.d.). Retrieved July 3, 2024, from https://www.technavio.com/report/carbon-fiber-market-industry-analysis

Carbon Fiber Recycling: How Boeing Turns Scraps Into Better Use. (n.d.). Retrieved July 15, 2024, from https://carbonfibergear.com/blogs/carbonfiber/carbon-fiber-recycling

Carbon Fiber Recycling. (n.d.). Retrieved July 15, 2024, from https://carbonconversions.com/

Carolin, A. (2003). *Carbon fibre reinforced polymers for strengthening of structural elements*. https://urn.kb.se/resolve?urn=urn:nbn:se:ltu:diva-16879

Chawla, K. K. (2019). Carbon Fiber/carbon matrix composites. In *Composite materials*. Springer. https://doi.org/10.1007/978-3-030-28983-6_8

Chen, D., Li, J., Yin, W., Yuan, Y., Cui, Y., Li, S., Liu, M., Peng, C., Liu, X., Ye, J., & Wu, Z. (2022). Fully recyclable and high-performance carbon fiber composites based on thermoplastic epoxy polymer. *Composites Communications, 35*, 101330. https://doi.org/10.1016/J.COCO.2022.101330

Chua, C. Y. X., Liu, H. C., Di Trani, N., Susnjar, A., Ho, J., Scorrano, G., Rhudy, J., Sizovs, A., Lolli, G., Hernandez, N., Nucci, M. C., Cicalo, R., Ferrari, M., & Grattoni, A. (2021). Carbon fiber reinforced polymers for implantable medical devices. *Biomaterials, 271*, 120719. https://doi.org/10.1016/J.BIOMATERIALS.2021.120719

Chung, D. D. L. (2016). Carbon composites: Composites with carbon fibers, nanofibers, and nanotubes. In *Carbon composites: Composites with carbon Fibers, nanofibers, and nanotubes:*

Second edition (Second ed.). Elsevier. http://www.sciencedirect.com:5070/book/9780128044599/carbon-composites
Colucci, G., Ostrovskaya, O., Frache, A., Martorana, B., & Badini, C. (2015). The effect of mechanical recycling on the microstructure and properties of PA66 composites reinforced with carbon fibers. *Journal of Applied Polymer Science, 132*(29), 42275. https://doi.org/10.1002/APP.42275
Colucci, G., Simon, H., Roncato, D., Martorana, B., & Badini, C. (2015). Effect of recycling on polypropylene composites reinforced with glass fibres. *Journal of Thermoplastic Composite Materials, 30*(5), 707–723. https://doi.org/10.1177/0892705715610407
Danish, A., Mosaberpanah, M. A., Salim, M. U., Amran, M., Fediuk, R., Ozbakkaloglu, T., & Rashid, M. F. (2022). Utilization of recycled carbon fiber reinforced polymer in cementitious composites: A critical review. *Journal of Building Engineering, 53*, 104583. https://doi.org/10.1016/J.JOBE.2022.104583
Das, M., Chacko, R., & Varughese, S. (2018). An efficient method of recycling of CFRP waste using Peracetic acid. *ACS Sustainable Chemistry and Engineering, 6*(2), 1564–1571. https://doi.org/10.1021/ACSSUSCHEMENG.7B01456/SUPPL_FILE/SC7B01456_SI_001.PDF
Das, T. K., Ghosh, P., & Das, N. C. (2019). Preparation, development, outcomes, and application versatility of carbon fiber-based polymer composites: A review. *Advanced Composites and Hybrid Materials, 2*(2), 214–233. https://doi.org/10.1007/S42114-018-0072-Z/FIGURES/12
Davison, K., & Heinz, V. (2021). ELG carbon fibre becomes gen 2 carbon: Customer continuity secured and an exciting growth platform created|ELG Germany. https://www.elgmetals.com/en-de/news/my-first-news-3/
De, B., Bera, M., Bhattacharjee, D., Ray, B. C., & Mukherjee, S. (2024). A comprehensive review on fiber-reinforced polymer composites: Raw materials to applications, recycling, and waste management. *Progress in Materials Science, 146*, 101326. https://doi.org/10.1016/J.PMATSCI.2024.101326
de Oliveira, I. C. L., Montagna, L. S., dos Anjos, E. G. R., Marini, J., & Passador, F. R. (2024). Mechanical recycling process: An alternative for a viable disposal of carbon fiber-reinforced thermoplastic waste. *Polymer Engineering and Science.* https://doi.org/10.1002/PEN.26771
De Omena Pina, S. R., Pardini, L. C., & Yoshida, I. V. P. (2007). Carbon fiber/ceramic matrix composites: Processing, oxidation and mechanical properties. *Journal of Materials Science, 42*(12), 4245–4253. https://doi.org/10.1007/S10853-006-0688-1/FIGURES/11
Demski, S., Misiak, M., Majchrowicz, K., Komorowska, G., Lipkowski, A., Stankiewicz, K., Dydek, K., Waśniewski, B., Boczkowska, A., & Ehrlich, H. (2024). Mechanical recycling of CFRPs based on thermoplastic acrylic resin with the addition of carbon nanotubes. *Scientific Reports, 14*(1), 1–11. https://doi.org/10.1038/s41598-024-62594-y
Deng, J., Xu, L., Liu, J., Peng, J., Han, Z., Shen, Z., & Guo, S. (2020). Efficient method of recycling carbon fiber from the waste of carbon fiber reinforced polymer composites. *Polymer Degradation and Stability, 182*, 109419. https://doi.org/10.1016/J.POLYMDEGRADSTAB.2020.109419
Dhanasekar, S., Ganesan, A. T., Rani, T. L., Vinjamuri, V. K., Nageswara Rao, M., Shankar, E., Dharamvir, K., & P. S., & Misganaw Golie, W. (2022). A comprehensive study of ceramic matrix composites for space applications. *Advances in Materials Science and Engineering, 2022*(1), 6160591. https://doi.org/10.1155/2022/6160591
Duan, J., Zhang, M., Chen, P., Li, Z., Pang, L., Xiao, P., & Li, Y. (2021). Tribological behavior and applications of carbon fiber reinforced ceramic composites as high-performance frictional materials. *Ceramics International, 47*(14), 19271–19281. https://doi.org/10.1016/J.CERAMINT.2021.02.187
Durante, M., Boccarusso, L., De Fazio, D., Formisano, A., & Langella, A. (2023). Investigation on the mechanical recycling of carbon Fiber-reinforced polymers by peripheral Down-milling. *Polymers, 15*(4), 854. https://doi.org/10.3390/POLYM15040854
Erden, S., & Ho, K. (2017). Fiber reinforced composites. *Fiber Technology for Fiber-Reinforced Composites*, 51–79. https://doi.org/10.1016/B978-0-08-101871-2.00003-5
Evens, T., Bex, G. J., Yigit, M., De Keyzer, J., Desplentere, F., & Van Bael, A. (2019). The influence of mechanical recycling on properties in injection molding of fiber-reinforced polypropylene.

International Polymer Processing, 34(4), 398–407. https://doi.org/10.3139/217.3770/MACHINEREADABLECITATION/RIS

Feraboli, P., Kawakami, H., Wade, B., Gasco, F., DeOto, L., & Masini, A. (2011). Recyclability and reutilization of carbon fiber fabric/epoxy composites. *Journal of Composite Materials, 46*(12), 1459–1473. https://doi.org/10.1177/0021998311420604

Giorgini, L., Benelli, T., Brancolini, G., & Mazzocchetti, L. (2020). Recycling of carbon fiber reinforced composite waste to close their life cycle in a cradle-to-cradle approach. *Current Opinion in Green and Sustainable Chemistry, 26*, 100368. https://doi.org/10.1016/J.COGSC.2020.100368

Gopalraj, S. K., & Kärki, T. (2020). A study to investigate the mechanical properties of recycled carbon fibre/glass fibre-reinforced epoxy composites using a novel thermal recycling process. *PRO, 8*(8), 954. https://doi.org/10.3390/PR8080954

Guo, W., Bai, S., Ye, Y., & Zhu, L. (2019). Recycling carbon fiber-reinforced polymers by pyrolysis and reused to prepare short-cut fiber C/C composite. *Journal of Reinforced Plastics and Composites, 38*(7), 340–348. https://doi.org/10.1177/0731684418822144

Hegde, S., Satish Shenoy, B., & Chethan, K. N. (2019). Review on carbon fiber reinforced polymer (CFRP) and their mechanical performance. *Materials Today: Proceedings, 19*, 658–662. https://doi.org/10.1016/J.MATPR.2019.07.749

Hofmann, M., Machado, M., Shahid, A., Dourado, F., Garrido, M., Bordado, J. C., & Correia, J. R. (2023). Pultruded carbon fibre reinforced polymer strips produced with a novel bio-based thermoset polyester for structural strengthening. *Composites Science and Technology, 234*, 109936. https://doi.org/10.1016/J.COMPSCITECH.2023.109936

Hong, B., Xian, G., & Wang, Z. (2017). Durability study of pultruded carbon fiber reinforced polymer plates subjected to water immersion. *Advances in Structural Engineering, 21*(4), 571–579. https://doi.org/10.1177/1369433217732664

Hu, Y., Gu, J., Zhang, Y., Liu, G., Yi, X., & Liu, X. (2024). FDM 3D printed MXene/recycled carbon fibre reinforced polylactic acid composites: Interface optimization, toughening and enhanced electromagnetic shielding performance. *Composites Communications, 48*, 101953. https://doi.org/10.1016/J.COCO.2024.101953

Hussain, I., & Immanuel, R. J. (2022). Composite materials and its advancements for a cleaner engine of future. *Energy, Environment, and Sustainability*, 169–191. https://doi.org/10.1007/978-981-16-8337-4_9

Isa, A., Nosbi, N., Che Ismail, M., Md Akil, H., Wan Ali, W. F. F., & Omar, M. F. (2022). A review on recycling of carbon fibres: Methods to reinforce and expected fibre composite degradations. *Materials, 15*(14), 4991. https://doi.org/10.3390/MA15144991

Jang, D., Lee, M. E., Choi, J., Cho, S. Y., & Lee, S. (2022). Strategies for the production of PAN-based carbon fibers with high tensile strength. *Carbon, 186*, 644–677. https://doi.org/10.1016/J.CARBON.2021.10.061

Jiang, G., Pickering, S. J., Lester, E. H., Turner, T. A., Wong, K. H., & Warrior, N. A. (2009). Characterisation of carbon fibres recycled from carbon fibre/epoxy resin composites using supercritical n-propanol. *Composites Science and Technology, 69*(2), 192–198. https://doi.org/10.1016/J.COMPSCITECH.2008.10.007

Jiang, J., Deng, G., Chen, X., Gao, X., Guo, Q., Xu, C., & Zhou, L. (2017). On the successful chemical recycling of carbon fiber/epoxy resin composites under the mild condition. *Composites Science and Technology, 151*, 243–251. https://doi.org/10.1016/J.COMPSCITECH.2017.08.007

Kacan, Y. O., & Elaldi, F. (2020). Ballistic performance of unidirectionally oriented carbon fiber reinforced composite armor with high-velocity impact. *Journal of Reinforced Plastics and Composites, 39*(19–20), 733–741. https://doi.org/10.1177/0731684420929084

Kaufman, K. R., & Bernhardt, K. (2021). Functional performance differences between carbon fiber and fiberglass prosthetic feet. *Prosthetics and Orthotics International, 45*(3), 205–213. https://doi.org/10.1097/PXR.0000000000000004

Khan, F., Hossain, N., Mim, J. J., Rahman, S. M., Iqbal, M. J., Billah, M., & Chowdhury, M. A. (2024). Advances of composite materials in automobile applications—A review. *Journal of Engineering Research*. https://doi.org/10.1016/J.JER.2024.02.017

Khanna, V., Kumar, V., & Bansal, S. A. (2021). Aluminium-carbon fibre metal matrix composites: A review. *IOP Conference Series: Materials Science and Engineering, 1033*(1), 012057. https://doi.org/10.1088/1757-899X/1033/1/012057

Khurshid, M. F., Hengstermann, M., Hasan, M. M. B., Abdkader, A., & Cherif, C. (2019). Recent developments in the processing of waste carbon fibre for thermoplastic composites—A review. *Journal of Composite Materials, 54*(14), 1925–1944. https://doi.org/10.1177/0021998319886043

Koumoulos, E. P., Trompeta, A. F., Santos, R. M., Martins, M., Dos Santos, C. M., Iglesias, V., Böhm, R., Gong, G., Chiminelli, A., Verpoest, I., Kiekens, P., & Charitidis, C. A. (2019). Research and Development in carbon Fibers and advanced high-performance composites supply chain in Europe: A roadmap for challenges and the industrial uptake. *Journal Of Composites Science, 3*(3), 86. https://doi.org/10.3390/JCS3030086

Kumar Sharma, A., Bhandari, R., Sharma, C., Krishna Dhakad, S., & Pinca-Bretotean, C. (2022). Polymer matrix composites: A state of art review. *Materials Today: Proceedings, 57*, 2330–2333. https://doi.org/10.1016/J.MATPR.2021.12.592

Lee, M., Kim, D. H., Park, J. J., You, N. H., & Goh, M. (2020). Fast chemical recycling of carbon fiber reinforced plastic at ambient pressure using an aqueous solvent accelerated by a surfactant. *Waste Management, 118*, 190–196. https://doi.org/10.1016/J.WASMAN.2020.08.014

Li, J., Wang, D., Tian, Y., & Pan, H. (2020). Lightweight Design of Honda Energy-Saving Racing Single-Shell Body based on carbon Fiber material. *Journal of Physics: Conference Series, 1622*(1), 012113. https://doi.org/10.1088/1742-6596/1622/1/012113

Lin, X., Zheng, L., Wang, X., Xu, P., Zeng, C., Liao, M., Zhang, M., & Huang, Q. (2024). Investigation of recycled carbon fiber-reinforced ultrafine-grain carbon-matrix composites. *Sustainable Materials and Technologies, 41*, e01033. https://doi.org/10.1016/J.SUSMAT.2024.E01033

Lina, S., & Zhen, D. (2012). The carbon Fiber composite materials application in sports equipment. *Advanced Materials Research, 341–342*, 173–176. https://doi.org/10.4028/www.SCIENTIFIC.NET/AMR.341-342.173

Liu, Y. Y., He, J., Li, Y. D., Zhao, X. L., & Zeng, J. B. (2020). Biobased epoxy vitrimer from epoxidized soybean oil for reprocessable and recyclable carbon fiber reinforced composite. *Composites Communications, 22*, 100445. https://doi.org/10.1016/J.COCO.2020.100445

Liu, Y., Cheng, Y., Ma, D., Hu, N., Han, W., Liu, D., Wu, S., An, Y., & Wang, A. (2022). Continuous carbon fiber reinforced ZrB2-SiC composites fabricated by direct ink writing combined with low-temperature hot-pressing. *Journal of the European Ceramic Society, 42*(9), 3699–3707. https://doi.org/10.1016/J.JEURCERAMSOC.2022.03.045

Liu, Y., Cheng, C., Sun, W., Zhang, Z., Ma, R., Zhou, J., Wang, J., Wang, T., Zheng, Q., Du, Y., Shen, J., & Fan, R. (2022). Negative permittivity behavior of carbon fibre/alumina ceramic composites prepared by hot-press sintering. *Ceramics International, 48*(7), 10031–10038. https://doi.org/10.1016/J.CERAMINT.2021.12.212

Lopez-Urionabarrenechea, A., Gastelu, N., Acha, E., Caballero, B. M., Orue, A., Jiménez-Suárez, A., Prolongo, S. G., & de Marco, I. (2020). Reclamation of carbon fibers and added-value gases in a pyrolysis-based composites recycling process. *Journal of Cleaner Production, 273*, 123173. https://doi.org/10.1016/J.JCLEPRO.2020.123173

Ma, Y., Sugahara, T., Yang, Y., & Hamada, H. (2015). A study on the energy absorption properties of carbon/aramid fiber filament winding composite tube. *Composite Structures, 123*, 301–311. https://doi.org/10.1016/J.COMPSTRUCT.2014.12.067

Macke, A., Schultz, B. F., Rohatgi, P. K., & Gupta, N. (2013). Metal matrix composites for automotive applications. *Advanced Composite Materials for Automotive Applications: Structural Integrity and Crashworthiness*, 311–344. https://doi.org/10.1002/9781118535288.CH13

Mahato, K. K., & Moharana, S. (2024). Primary and secondary processing of polymer matrix composites. *Engineering Materials, 2024*, 67–98. https://doi.org/10.1007/978-981-97-2075-0_3

Mahaviradhan, N., Sivaganesan, S., Padma Sravya, N., & Parthiban, A. (2021). Experimental investigation on mechanical properties of carbon fiber reinforced aluminum metal matrix composite. *Materials Today: Proceedings, 39*, 743–747. https://doi.org/10.1016/J.MATPR.2020.09.443

Mahboubizadeh, S., Sadeq, A., Arzaqi, Z., Ashkani, O., & Samadoghli, M. (2024). Advancements in fiber-reinforced polymer (FRP) composites: An extensive review. *Discover Materials, 4*(1), 1–16. https://doi.org/10.1007/S43939-024-00091-9

Maiti, S., Islam, M. R., Uddin, M. A., Afroj, S., Eichhorn, S. J., & Karim, N. (2022). Sustainable Fiber-reinforced composites: A review. *Advanced Sustainable Systems, 6*(11), 2200258. https://doi.org/10.1002/ADSU.202200258

Manarin, E., Boumezgane, O., Giannino, A., De Fabritiis, V., Griffini, G., & Turri, S. (2024). Towards a zero-waste chemcycling of thermoset polymer composites: Catalyst assisted mild solvolysis for clean carbon fiber liberation and circular coating development. *Sustainable Materials and Technologies, 41*, e01031. https://doi.org/10.1016/J.SUSMAT.2024.E01031

May, D., Goergen, C., & Friedrich, K. (2021). Multifunctionality of polymer composites based on recycled carbon fibers: A review. *Advanced Industrial and Engineering Polymer Research, 4*(2), 70–81. https://doi.org/10.1016/J.AIEPR.2021.01.001

Mehara, M., Goswami, C., Ranjan Kumar, S., Singh, G., & Kumar Wagdre, M. (2021). Performance evaluation of advanced armor materials. *Materials Today: Proceedings, 47*, 6039–6042. https://doi.org/10.1016/J.MATPR.2021.04.611

Meng, F., McKechnie, J., Turner, T. A., & Pickering, S. J. (2017). Energy and environmental assessment and reuse of fluidised bed recycled carbon fibres. *Composites Part A: Applied Science and Manufacturing, 100*, 206–214. https://doi.org/10.1016/J.COMPOSITESA.2017.05.008

Meng, F., McKechnie, J., & Pickering, S. J. (2018). An assessment of financial viability of recycled carbon fibre in automotive applications. *Composites Part A: Applied Science and Manufacturing, 109*, 207–220. https://doi.org/10.1016/J.COMPOSITESA.2018.03.011

Meng, F., Olivetti, E. A., Zhao, Y., Chang, J. C., Pickering, S. J., & McKechnie, J. (2018). Comparing life cycle energy and global warming potential of carbon Fiber composite recycling technologies and waste management options. *ACS Sustainable Chemistry and Engineering, 6*(8), 9854–9865. https://doi.org/10.1021/ACSSUSCHEMENG.8B01026/ASSET/IMAGES/LARGE/SC-2018-010264_0004.JPEG

Mileiko, S. (2022). Carbon-fibre/metal-matrix composites: A review. *Journal Of Composites Science, 6*(10), 297. https://doi.org/10.3390/JCS6100297

Minchenkov, K., Vedernikov, A., Safonov, A., & Akhatov, I. (2021). Thermoplastic Pultrusion: A review. *Polymers, 13*(2), 180. https://doi.org/10.3390/POLYM13020180

Miracle, D. B. (2005). Metal matrix composites – From science to technological significance. *Composites Science and Technology, 65*(15–16), 2526–2540. https://doi.org/10.1016/J.COMPSCITECH.2005.05.027

Mitschang, P., & Hildebrandt, K. (2012). Polymer and composite moulding technologies for automotive applications. *Advanced Materials in Automotive Engineering*, 210–229. https://doi.org/10.1533/9780857095466.210

Monteleone, C., Poges, S., Petroski, K., Kerns, P., Tobin, Z., Policandriotes, T., & Suib, S. L. (2020). Atmospheric pressure chemical vapor infiltration of a titanium carbide interphase coating on carbon fiber. *Ceramics International, 46*(10), 15084–15091. https://doi.org/10.1016/J.CERAMINT.2020.03.043

Muhammad, A., Rahman, M. R., Baini, R., & Bin Bakri, M. K. (2021). Applications of sustainable polymer composites in automobile and aerospace industry. *Advances in Sustainable Polymer Composites*, 185–207. https://doi.org/10.1016/B978-0-12-820338-5.00008-4

Naqvi, S. R., Prabhakara, H. M., Bramer, E. A., Dierkes, W., Akkerman, R., & Brem, G. (2018). A critical review on recycling of end-of-life carbon fibre/glass fibre reinforced composites waste using pyrolysis towards a circular economy. *Resources, Conservation and Recycling, 136*, 118–129. https://doi.org/10.1016/J.RESCONREC.2018.04.013

Neubauer, E., Kitzmantel, M., Hulman, M., & Angerer, P. (2010). Potential and challenges of metal-matrix-composites reinforced with carbon nanofibers and carbon nanotubes. *Composites Science and Technology, 70*(16), 2228–2236. https://doi.org/10.1016/J.COMPSCITECH.2010.09.003

Newcomb, B. A. (2016). Processing, structure, and properties of carbon fibers. *Composites Part A: Applied Science and Manufacturing, 91*, 262–282. https://doi.org/10.1016/J.COMPOSITESA.2016.10.018

Nishida, Y. (2013). Introduction to metal matrix composites: Fabrication and recycling. In *Introduction to metal matrix composites: Fabrication and recycling* (Vol. 9784431542377). Springer Tokyo. https://doi.org/10.1007/978-4-431-54237-7/COVER

Niu, C., Zhang, Q., Cheng, L., Ye, F., & Zhang, L. (2023). Microstructure and mechanical properties of Cf/SiC composites with dispersed C-SiC interphase prepared by chemical vapor infiltration. *Composites Part A: Applied Science and Manufacturing, 165*, 107339. https://doi.org/10.1016/J.COMPOSITESA.2022.107339

Okajima, I., Hiramatsu, M., Shimamura, Y., Awaya, T., & Sako, T. (2014). Chemical recycling of carbon fiber reinforced plastic using supercritical methanol. *The Journal of Supercritical Fluids, 91*, 68–76. https://doi.org/10.1016/J.SUPFLU.2014.04.011

Oladele, I. O., Onuh, L. N., Siengchin, S., Sanjay, M. R., & Adelani, S. O. (2023). Modern applications of polymer composites in structural industries: A review of philosophies, product development, and graphical applications. *Applied Science and Engineering Progress*. https://doi.org/10.14416/J.ASEP.2023.07.003

Pakdel, E., Kashi, S., Varley, R., & Wang, X. (2021). Recent progress in recycling carbon fibre reinforced composites and dry carbon fibre wastes. *Resources, Conservation and Recycling, 166*, 105340. https://doi.org/10.1016/J.RESCONREC.2020.105340

Park, S.-J. (2018). Carbon fibers, 210. https://doi.org/10.1007/978-981-13-0538-2

Peijs, T., Kirschbaum, R., & Lemstra, P. J. (2022). Chapter 5: A critical review of carbon fiber and related products from an industrial perspective. *Advanced Industrial and Engineering Polymer Research, 5*(2), 90–106. https://doi.org/10.1016/J.AIEPR.2022.03.008

Perumal, A. B., Nambiar, R. B., Sellamuthu, P. S., & Sadiku, E. R. (2020). Carbon Fiber composites. *Handbook of Nanomaterials and Nanocomposites for Energy and Environmental Applications*, 1–32. https://doi.org/10.1007/978-3-030-11155-7_174-1

Pietroluongo, M., Padovano, E., Frache, A., & Badini, C. (2020). Mechanical recycling of an end-of-life automotive composite component. *Sustainable Materials and Technologies, 23*, e00143. https://doi.org/10.1016/J.SUSMAT.2019.E00143

Pimenta, S., & Pinho, S. T. (2014). Recycling of carbon Fibers. *Handbook of Recycling: State-of-the-Art for Practitioners, Analysts, and Scientists*, 269–283. https://doi.org/10.1016/B978-0-12-396459-5.00019-2

Rashed Labanieh, A., Placet, V., Grzegorz Kossakowski, P., & Wci, W. (2022). Fiber-reinforced polymer composites in the construction of bridges: Opportunities, problems and challenges. *Fibers, 10*(4), 37. https://doi.org/10.3390/FIB10040037

Realizing a Circular Economy. (n.d.). Retrieved July 15, 2024, from https://www.toray.com/global/sustainability/activity/social/recycling.html

Ren, Y., Xu, L., Shang, X., Shen, Z., Fu, R., Li, W., & Guo, L. (2022). Evaluation of mechanical properties and pyrolysis products of carbon Fibers recycled by microwave pyrolysis. *ACS Omega, 7*(16), 13529–13537. https://doi.org/10.1021/ACSOMEGA.1C06652/ASSET/IMAGES/MEDIUM/AO1C06652_0007.GIF

Rueschhoff, L. M., Carney, C. M., Apostolov, Z. D., & Cinibulk, M. K. (2020). Processing of fiber-reinforced ultra-high temperature ceramic composites: A review. *International Journal of Ceramic Engineering & Science, 2*(1), 22–37. https://doi.org/10.1002/CES2.10033

Samant, L., Fernandes, F. A. O., Jose, S., & Alves de Sousa, R. J. (2022). Natural composites in aircraft structures. In *Materials, structures and manufacturing for aircraft* (pp. 113–126). Springer. https://doi.org/10.1007/978-3-030-91873-6_5

Sarker, S., Qin, X., Yu, J., Zhang, X., Yang, Y., & Gao, C. (2024). Fiber fracture behavior of high-pressure resin transfer molded (HP-RTM) carbon/epoxy multidirectional laminates under compression-compression fatigue loading. *International Journal of Fatigue, 182*, 108221. https://doi.org/10.1016/J.IJFATIGUE.2024.108221

Sayam, A., Rahman, A. N. M. M., Rahman, M. S., Smriti, S. A., Ahmed, F., Rabbi, M. F., Hossain, M., & Faruque, M. O. (2022). A review on carbon fiber-reinforced hierarchical composites: Mechanical performance, manufacturing process, structural applications and allied challenges. *Carbon Letters 2022 32:5, 32*(5), 1173–1205. https://doi.org/10.1007/S42823-022-00358-2

Schamel, E., Wehnert, G., Schlachter, H., & Söthje, D. (2021). Chemical recycling of carbon Fiber reinforced epoxy composites using mild conditions. *Chemie Ingenieur Technik, 93*(10), 1619–1628. https://doi.org/10.1002/CITE.202100048

Seidel, C. (2021). Revolution, not evolution: BMW Group to reduce carbon footprint significantly by 2030 using innovative materials. https://www.press.bmwgroup.com/global/article/detail/T0341416EN/revolution-not-evolution:-bmw-group-to-reduce-carbon-footprint-significantly-by-2030-using-innovative-materials

Servadei, F., Zoli, L., Galizia, P., Piancastelli, A., & Sciti, D. (2023). Processing and characterization of ultra-high temperature ceramic matrix composites via water based slurry impregnation and polymer infiltration and pyrolysis. *Ceramics International, 49*(1), 1220–1229. https://doi.org/10.1016/J.CERAMINT.2022.09.100

Seydibeyoğlu, M. Ö., Doğru, A., Kandemir, M. B., & Aksoy, Ö. (2020). Lightweight composite materials in transport structures. *Lightweight Polymer Composite Structures*, 103–130. https://doi.org/10.1201/9780429244087-5

SHA, J. jun, LÜ, Z. zhao, SHA, R. yi, ZU, Y. fei, DAI, J. xiang, XIAN, Y. qiang, ZHANG, W., CUI, D., & YAN, C. lin. (2021). Improved wettability and mechanical properties of metal coated carbon fiber-reinforced aluminum matrix composites by squeeze melt infiltration technique. *Transactions of Nonferrous Metals Society of China, 31*(2), 317–330. https://doi.org/10.1016/S1003-6326(21)65498-5

Shetty, S., Pinkard, B. R., & Novosselov, I. V. (2022). Recycling of carbon fiber reinforced polymers in a subcritical acetic acid solution. *Heliyon, 8*(12). https://doi.org/10.1016/j.heliyon.2022.e12242

Shirvanimoghaddam, K., Hamim, S. U., Karbalaei Akbari, M., Fakhrhoseini, S. M., Khayyam, H., Pakseresht, A. H., Ghasali, E., Zabet, M., Munir, K. S., Jia, S., Davim, J. P., & Naebe, M. (2017). Carbon fiber reinforced metal matrix composites: Fabrication processes and properties. *Composites Part A: Applied Science and Manufacturing, 92*, 70–96. https://doi.org/10.1016/J.COMPOSITESA.2016.10.032

Siengchin, S. (2023). A review on lightweight materials for defence applications: Present and future developments. *Defence Technology, 24*, 1–17. https://doi.org/10.1016/J.DT.2023.02.025

Singh, H., Singh Brar, G., Kumar, H., & Aggarwal, V. (2021). A review on metal matrix composite for automobile applications. *Materials Today: Proceedings, 43*, 320–325. https://doi.org/10.1016/J.MATPR.2020.11.670

Sloan, J. (2018). Boeing to supply ELG carbon fiber for recycling. https://www.compositesworld.com/news/boeing-to-supply-elg-carbon-fiber-for-recycling.

Sozer, E. M., Simacek, P., & Advani, S. G. (2012). Resin transfer molding (RTM) in polymer matrix composites. *Manufacturing Techniques for Polymer Matrix Composites (PMCs)*, 245–309. https://doi.org/10.1533/9780857096258.3.243

Sree Manu, K. M., Ajay Raag, L., Rajan, T. P. D., Gupta, M., & Pai, B. C. (2016). Liquid metal infiltration processing of metallic composites: A critical review. *Metallurgical and Materials Transactions B, 47*(5), 2799–2819. https://doi.org/10.1007/S11663-016-0751-5

Sreejith, M., & Rajeev, R. S. (2021). Fiber reinforced composites for aerospace and sports applications. *Fiber Reinforced Composites: Constituents, Compatibility, Perspectives and Applications*, 821–859. https://doi.org/10.1016/B978-0-12-821090-1.00023-5

Stalin, M., Rajaguru, K., & Rangaraj, L. (2020). Processing of Cf/SiC composites by hot pressing using polymer binders followed by polymer impregnation and pyrolysis. *Journal of the European Ceramic Society, 40*(2), 290–297. https://doi.org/10.1016/J.JEURCERAMSOC.2019.09.047

Stieven Montagna, L., de Melo, F., Morgado, G., Lemes, A. P., Roberto Passador, F., & Cerqueira Rezende, M. (2022). Recycling of carbon fiber-reinforced thermoplastic and thermoset composites: A review. *Journal of Thermoplastic Composite Materials, 36*(8), 3455–3480. https://doi.org/10.1177/08927057221108912

Tang, S., & Hu, C. (2017). Design, preparation and properties of carbon Fiber reinforced ultra-high temperature ceramic composites for aerospace applications: A review. *Journal Of Materials Science & Technology, 33*(2), 117–130. https://doi.org/10.1016/J.JMST.2016.08.004

Teng, H., Li, S., Cao, Z., Li, S., Li, C., & Ko, T. J. (2023). Carbon Fiber composites for large-scale wind turbine blades: Applicability study and comprehensive evaluation in China. *Journal of Marine Science and Engineering, 11*(3), 624. https://doi.org/10.3390/JMSE11030624

Tian, P. X., Li, Y. D., Hu, Z., & Zeng, J. B. (2024). Flame-retardant epoxy vitrimers with high strength and high-Tg for recyclable carbon fiber-reinforced composites. *Composites Communications, 47*, 101870. https://doi.org/10.1016/J.COCO.2024.101870

Tian, X., Todoroki, A., Liu, T., Wu, L., Hou, Z., Ueda, M., Hirano, Y., Matsuzaki, R., Mizukami, K., Iizuka, K., Malakhov, A. V., Polilov, A. N., Li, D., & Lu, B. (2022). 3D printing of continuous Fiber reinforced polymer composites: Development, application, and prospective. *Chinese Journal of Mechanical Engineering: Additive Manufacturing Frontiers, 1*(1), 100016. https://doi.org/10.1016/J.CJMEAM.2022.100016

Tuakta, C., & Whittle, A. J. (2005). Use of fiber reinforced polymer composite in bridge structures. https://dspace.mit.edu/handle/1721.1/31126

Tusher, M. H., Imam, A., & Khatun, M. A. (2024). Polymer composites: Its processing, advantages, properties and applications. *Hybrid Composite Materials*, 45–65. https://doi.org/10.1007/978-981-97-2104-7_2

Ul, R., Pi˜piñero-Hernanz, P., García-Serna, J., Dodds, C., Hyde, J., Poliakoff, M., Jos´, M. J., Cocero, J., Kingman, S., Pickering, S., & Lester, E. (2008). Chemical recycling of carbon fibre composites using alcohols under subcritical and supercritical conditions. *Journal of Supercritical Fluids, 46*, 83–92. https://doi.org/10.1016/j.supflu.2008.02.008

Uyanna, O., & Najafi, H. (2020). Thermal protection systems for space vehicles: A review on technology development, current challenges and future prospects. *Acta Astronautica, 176*, 341–356. https://doi.org/10.1016/J.ACTAASTRO.2020.06.047

Vo Dong, P. A., Azzaro-Pantel, C., & Cadene, A. L. (2018). Economic and environmental assessment of recovery and disposal pathways for CFRP waste management. *Resources, Conservation and Recycling, 133*, 63–75. https://doi.org/10.1016/J.RESCONREC.2018.01.024

Wang, W., Gao, X., Li, Z., Shen, C., Wang, G., & He, R. (2024). Fiber-laying-assisted material extrusion additive manufacturing of continuous carbon fiber reinforced SiC ceramic matrix composites. *Materials Science and Engineering: A, 890*, 145944. https://doi.org/10.1016/J.MSEA.2023.145944

Wang, Y., Cui, X., Ge, H., Yang, Y., Wang, Y., Zhang, C., Li, J., Deng, T., Qin, Z., & Hou, X. (2015). Chemical recycling of carbon Fiber reinforced epoxy resin composites via selective cleavage of the carbon-nitrogen bond. *ACS Sustainable Chemistry and Engineering, 3*(12), 3332–3337. https://doi.org/10.1021/ACSSUSCHEMENG.5B00949/ASSET/IMAGES/LARGE/SC-2015-00949T_0004.JPEG

Wei, X., Wang, Y., Xue, P., Zhang, T., Rouis, A., Xiao, W., & Xiong, J. (2022). Carbon Fiber composite honeycomb structures and the application for satellite antenna reflector with high precision. *Advances in Astronautics Science and Technology, 5*(4), 423–441. https://doi.org/10.1007/S42423-022-00133-5/FIGURES/20

Wiedmer, S., & Manolesos, M. (2006). An experimental study of the Pultrusion of carbon Fiber-polyamide 12 yarn, *19*(1), 97–112. https://doi.org/10.1177/0892705706055448

Wienstroth, F. (2012). BMW group and Boeing to collaborate on carbon fiber recycling. https://www.press.bmwgroup.com/global/article/detail/T0135185EN/bmw-group-and-boeing-to-collaborate-on-carbon-fiber-recycling?language=en.

Wu, J., Gao, X., Wu, Y., Wang, Y., Nguyen, T. T., & Guo, M. (2022). Recycling carbon Fiber from carbon Fiber-reinforced polymer and its reuse in Photocatalysis: A review. *Polymers, 15*(1), 170. https://doi.org/10.3390/POLYM15010170

Xu, X., Wei, K., Mei, M., Li, M., & Yang, X. (2024). An ultrasound-assisted resin transfer molding to improve the impregnation and dual-scale flow for carbon fiber reinforced resin composites. *Composites Science and Technology, 255*, 110710. https://doi.org/10.1016/J.COMPSCITECH.2024.110710

Yan, H., Lu, C., Jing, D., Chang, C., Liu, N., & Hou, X. (2016). Recycling of carbon fibers in epoxy resin composites using supercritical 1-propanol. *Carbon, 100*, 710–711. https://doi.org/10.1016/J.CARBON.2016.01.053

Yang, P., Zhou, Q., Li, X. Y., Yang, K. K., & Wang, Y. Z. (2014). Chemical recycling of fiber-reinforced epoxy resin using a polyethylene glycol/NaOH system. *Journal of Reinforced Plastics and Composites, 33*(22), 2106–2114. https://doi.org/10.1177/0731684414555745

Yang, S., Cheng, Y., Xiao, X., & Pang, H. (2020). Development and application of carbon fiber in batteries. *Chemical Engineering Journal, 384*, 123294. https://doi.org/10.1016/J.CEJ.2019.123294

Yatim, N. M., Shamsudin, Z., Shaaban, A., Ghafar, J. A., & Khan, M. J. H. (2020). Recovery of carbon FIBER from carbon FIBER reinforced polymer waste via pyrolysis. *Journal of Advanced Manufacturing Technology (JAMT), 14*(1) https://jamt.utem.edu.my/jamt/article/view/5860

Ye, J., Ji, X., Liu, Z., Liu, K., Li, J., Wang, R., Wang, J., & Lei, Q. (2024). Carbon fiber reinforced structural battery composites: Progress and challenges toward industrial application. *Composites Part B: Engineering, 277*, 111411. https://doi.org/10.1016/J.COMPOSITESB.2024.111411

Yip, H. L. H., Pickering, S. J., & Rudd, C. D. (2002). Characterisation of carbon fibres recycled from scrap composites using fluidised bed process. *Plastics, Rubber and Composites, 31*(6), 278–282. https://doi.org/10.1179/146580102225003047

Zhang, J., Chevali, V. S., Wang, H., & Wang, C. H. (2020). Current status of carbon fibre and carbon fibre composites recycling. *Composites Part B: Engineering, 193*, 108053. https://doi.org/10.1016/J.COMPOSITESB.2020.108053

Zhang, J., Lin, G., Vaidya, U., & Wang, H. (2023). Past, present and future prospective of global carbon fibre composite developments and applications. *Composites Part B: Engineering, 250*, 110463. https://doi.org/10.1016/J.COMPOSITESB.2022.110463

Zhang, J. Z. (2013). Study on carbon Fiber composite materials in sports equipment. *Applied Mechanics and Materials, 329*, 105–108. https://doi.org/10.4028/www.SCIENTIFIC.NET/AMM.329.105

Zhao, D., Zhang, C., Hu, H., & Zhang, Y. (2011). Preparation and characterization of three-dimensional carbon fiber reinforced zirconium carbide composite by precursor infiltration and pyrolysis process. *Ceramics International, 37*(7), 2089–2093. https://doi.org/10.1016/J.CERAMINT.2011.02.024

Zhu, T., & Wang, Z. (2023). Research and application prospect of short carbon fiber reinforced ceramic composites. *Journal of the European Ceramic Society, 43*(15), 6699–6717. https://doi.org/10.1016/J.JEURCERAMSOC.2023.07.007

Innovative Strategies for Sustainable Municipal Waste Management: A Technical Analysis

Subita Bhagat

Abstract This comprehensive research examines cutting-edge approaches to sustainable municipal waste management, focusing on technological innovations, policy frameworks, and implementation strategies. The study analyzes data from multiple urban centers worldwide, evaluating the effectiveness of various waste management technologies and their environmental, economic, and social impacts. Key findings indicate that integrated smart waste management systems, combined with circular economy principles, can reduce landfill dependency by up to 70% while generating significant economic benefits through resource recovery and energy generation.

Keywords Municipal solid waste (MSW) · Sustainable waste management · Circular economy · Waste-to-energy · Smart waste systems · Resource recovery

1 Introduction

1.1 Background and Importance

Municipal waste management, therefore, becomes increasingly vital with the world's population mushrooming in cities. It is estimated that today 2.01 billion tons are produced yearly in cities. This figure will increase to 3.40 billion tons by the year 2050. The old linear collection, transportation, and disposal concept is not workable, and a new concept must be considered to accommodate the circular economy (Zaman, 2023).

S. Bhagat (✉)
Department of Chemical Engineering, Sant Longowal Institute of Engineering and Technology (SLIET), Longowal, Sangrur, Punjab, India
e-mail: subita_bhagat@sliet.ac.in

A. P. Das (ed.), *Bioconversion of Environmental Waste to Value Added Products*, Sustainable Environmental Waste Management Strategies,
https://doi.org/10.1007/978-3-032-14215-3_12

1.2 Research Objectives

This study will focus on the following objectives:

- evaluate emerging technologies in municipal waste management;
- analyze the effectiveness of various waste reduction strategies;
- assess the economic viability of innovative waste management solutions;
- examine the environmental impact of different waste treatment approaches;
- propose integrated frameworks for sustainable waste management.

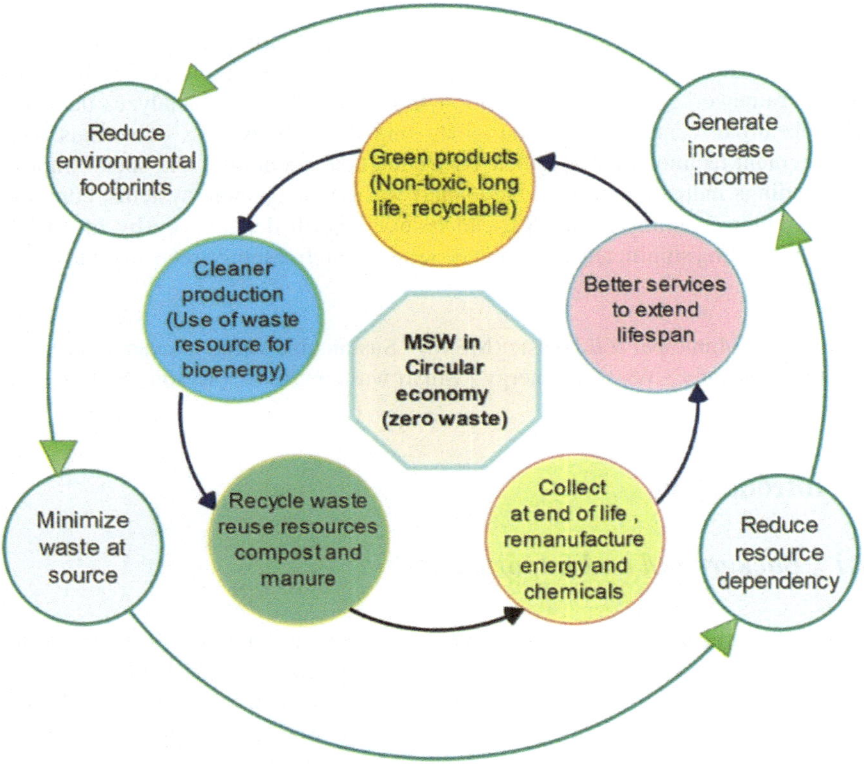

Table 1 Global municipal waste composition by income level (2023)

Waste type	High income (%)	Upper middle income (%)	Lower middle income (%)	Low income (%)
Organic	32	53	56	57
Paper/cardboard	25	13	7	5
Plastic	13	12	11	9
Glass	5	4	3	2
Metal	6	3	2	1

Source World Bank Global Waste Management Outlook 2023

2 Overview of Municipal Waste Management Systems

2.1 Types and Sources of Municipal Waste

MSW can be termed as the ultimate collection of many types of materials produced from households, commercial, and institutional sectors. Recent studies by the World Bank estimated that in 2023, global MSW generation amounted to 2.24 billion tons, which is likely to increase to about 3.88 billion tons by 2050. About 80% of waste is, therefore, generated in urban settings, pointing out the importance of cities in managing the strategy of waste management (Table 1).

2.2 Traditional Waste Management Approaches

For ages, the landfilling and incineration approach is considered to be a predominant part of historical waste management. According to a paper in Environmental Science and Technology from Zhang et al., around 37% of the global MSW is disposed as a landfill, 33% is openly dumped, while 19% is recycled or composted. On the other hand, around 11% of it gets incinerated (World Bank, 2022). Due to these conventional methods, huge problems are encountered within the environment, where the pollution occurs through emission of greenhouses and contamination of the soil.

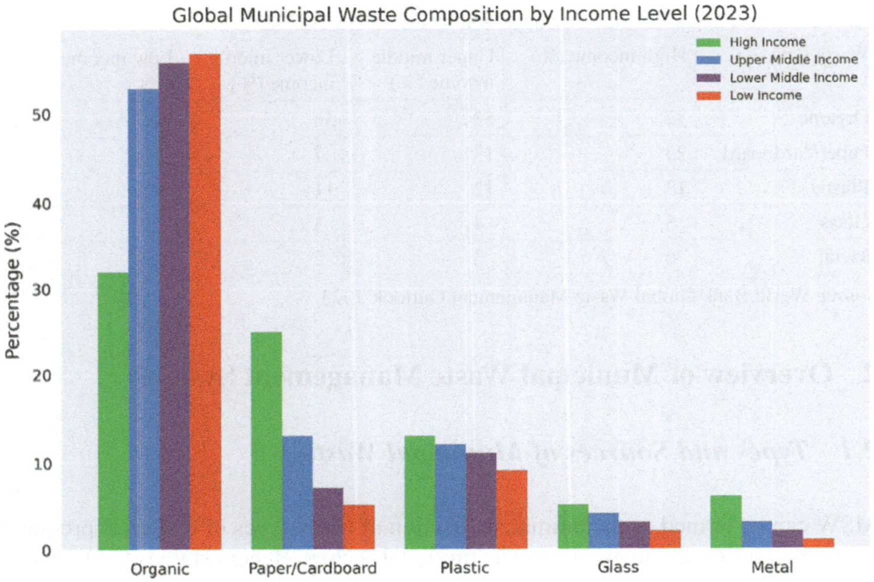

2.3 *Challenges in Existing Municipal Waste Systems*

Current municipal waste systems have several interlinked problems. According to a study conducted just lately by Chen et al. published in the Waste Management Journal, the following are critical issues:

1. **Infrastructure:** The majority of the municipalities suffer from an aging infrastructure, with 67% of waste management plants in developing countries operating beyond their designed capacities.
2. **Financial Constraint:** The United Nations Environment Programme (2023) adds that waste management services account for between 20% and 50% of the municipal budgets in low-income countries, however often achieve collection rates below 50%.
3. **Technical Inefficiencies:** A further comprehensive survey by Martinez-Rodriguez et al. concluded that poorer collection routes in major cities incur around 30% more operation expenses and carbon emissions than would be necessary (Table 2).

A new study by Thomson et al., published in Resources, Conservation and Recycling, shows that inefficient waste management systems contribute about 1.6 billion tons of CO2 equivalent annually to global greenhouse gas emissions. According to the paper, upgrading the infrastructure of waste management can decrease those

Table 2 Regional comparison of waste management challenges (2023)

Region	Collection coverage (%)	Budget allocation (% of municipal budget)	Technology adoption index*	Infrastructure quality score
North America	99.8	4.2	8.7	8.9
Europe	99.5	3.8	8.5	8.6
East Asia	85.4	12.5	7.2	7.4
South Asia	65.3	27.8	4.5	4.8
Africa	44.2	38.4	3.2	3.5

Scale 1–10, where 10 represents highest technology adoption *Scale 1–10, where 10 represents highest quality infrastructure *Source* International Waste Management Association (IWMA) Annual Report 2023

emissions by as much as 45% while it simultaneously increases resource recovery rates by 60% (Serranti & Bonifazi, 2023).

These issues are more significant in the urbanizing regions where waste is growing at rates of 3–5% annually than infrastructure development and technical building capacity. The situation calls for some innovative ways that can address the immediate operational challenges and the sustainable goals.

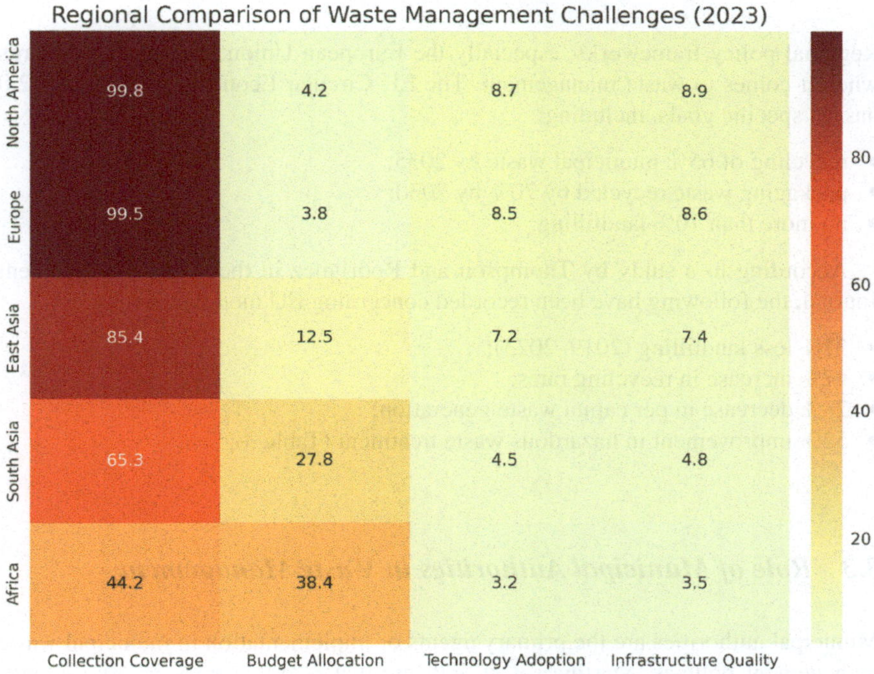

3 Regulatory Framework and Policy Drivers

3.1 International Waste Management Regulations and Guidelines

International frameworks of waste management have significantly gained momentum over the last decades anchored to pivotal international agreements. As of 2023, the Basel Convention was ratified by 189 countries. Research in Davidson et al., Environmental Policy Review shows that 43% more compliance performance in hazardous waste management by the countries following Basel Convention guidelines and 56% reduction in illegal transboundary waste movements.

According to the Global Waste Management Outlook by the United Nations Environmental Programme (2023), international regulations also have spurred massive action within the last decade in changing waste management habits. In statistics, those states and nations that followed the UN guidelines on waste management increased, on average, the rate of recycling by 27% within five years (Peri et al., 2023).

3.2 National and Regional Policies for Waste Management

Regional policy frameworks, especially the European Union, have high standards when it comes to waste management. The EU Circular Economy Action Plan 2.0 has set specific goals, including:

- recycling of 65% municipal waste by 2035;
- packaging waste recycled by 70% by 2030;
- no more than 10% landfilling.

According to a study by Thompson and Rodriguez in the Waste Management Journal, the following have been recorded concerning EU member states

- 31% less landfilling (2019–2023);
- 47% increase in recycling rates;
- 28% decrease in per capita waste generation;
- 52% improvement in hazardous waste treatment (Table 3).

3.3 Role of Municipal Authorities in Waste Management

Municipal authorities are the primary agents of implementation in municipal waste management policies. Martinez et al. indicate that resource recovery rate is 35% higher for cities with autonomous decision-making over their waste management authority than for centrally managed systems. Generalized by data from 245 cities in

Table 3 Implementation of national waste policies (2023)

Policy mechanism	Adoption rate (%)	Average impact on waste reduction (%)	Implementation cost (USD/capita)
Landfill tax	78	32	12.5
EPR schemes	65	28	18.3
Pay-as-you-throw	45	25	8.7
Recycling mandates	82	35	15.2

Source International Waste Policy Institute Annual Report 2023

45 countries, it was inferred that municipal autonomy in the management of waste has associations with:

- Recycling rates 42% more;
- 38% cost operations;
- 29% public involvement;
- 45% good compliance with environmental regulations.

3.4 *Incentives and Regulations Promoting Sustainability*

Economic instruments have become an effective instrument to promote sustainability in waste management. A comprehensive analysis on the effectiveness of different incentive mechanisms by Chang and Kumar in Environmental Economics Quarterly was made about different incentive mechanisms among 178 municipalities (Table 4).

Recent policy framework adjustments emphasize the incorporation of circular economy principles. According to the World Resources Institute, municipalities with an integrated set of regulations on waste management possess:

- 40% higher rates of material recovery;
- 35% less greenhouse gases;

Table 4 Impact of economic instruments on waste management performance

Instrument type	Waste reduction (%)	Cost recovery (%)	Public acceptance (%)	Implementation success rate (%)
Variable rate charging	28.5	85.3	72.4	83.2
Recycling credits	23.7	67.8	88.6	76.5
Disposal penalties	31.2	92.1	65.3	71.8
Green procurement	19.4	58.9	91.2	88.4

Source Environmental economics quarterly, Chang and Kumar

- 25% reduced operational costs;
- 50% better stakeholder involvement.

4 Waste Reduction and Prevention Strategies

4.1 Source Reduction and Product Design for Minimal Waste

Source reduction tops recent rankings on waste management hierarchy discoveries. A considered analysis done by Anderson et al. in the Journal of Industrial Ecology thoroughly does prove that source reduction strategies do effectively reduce municipal waste generation by 25–35% within three years after implementation. Based on a total of 178 cities across 32 countries, the study reveals that cities with strict product design regulations successfully manage to achieve waste reduction at an average annual rate of 2.8% as opposed to 0.7% every year in cities without such regulations (Parashar & Khalid, 2023).

Product design innovations are changing the face of waste reduction initiatives. In a report by Zhang and Martinez in the journal *Sustainable Production and Consumption*, some eco-design projects have resulted in the reduction of packaging waste by 42% while the level of product obsolescence has decreased by 38%. Companies adopting circular design principles are reported to have come out at an average reduction of 27% in material use and 31% product lifespan (Moh, 2022).

4.2 Public Awareness and Behavioral Change for Waste Minimization

Behavioral change campaigns have become among the most significant drivers of successful waste reduction. A new longitudinal study by Thompson et al., which just appeared in Environmental Education Research, gathered data on 245 municipalities and found that high-quality public information campaigns achieve the following (Table 5).

4.3 Extended Producer Responsibility (EPR) Programs

EPR programs have proven to be powerful waste reduction and resource recovery instruments. According to the report of Kim and Chen of the Journal of Cleaner Production, a country that has more expansive EPR policy frameworks is able to reap the following benefits:

- 78% recycled packaged products;

Table 5 Impact of public awareness campaigns on waste reduction (2019–2023)

Intervention type	Waste reduction (%)	Public participation rate (%)	Cost-effectiveness (USD/ton reduced)	Long-term sustainability (%)
School programs	28.5	85.2	42.3	76.4
Community workshops	32.7	72.8	38.6	82.1
Digital campaigns	24.3	68.5	28.4	71.8
Mixed approaches	35.8	88.7	45.2	85.3

Source Environmental Education Research, 2023

- 65% less landfilling of electronic waste;
- 42% longer product lifespan;
- 58% compliance with eco-design.

Economic value also comes within this class. The OECD (2023) reported that it has been quantitatively estimated that successful EPR schemes result in saving averages of US$ 127 per ton of waste handled and additionally creating jobs in recycling and remanufacturing areas.

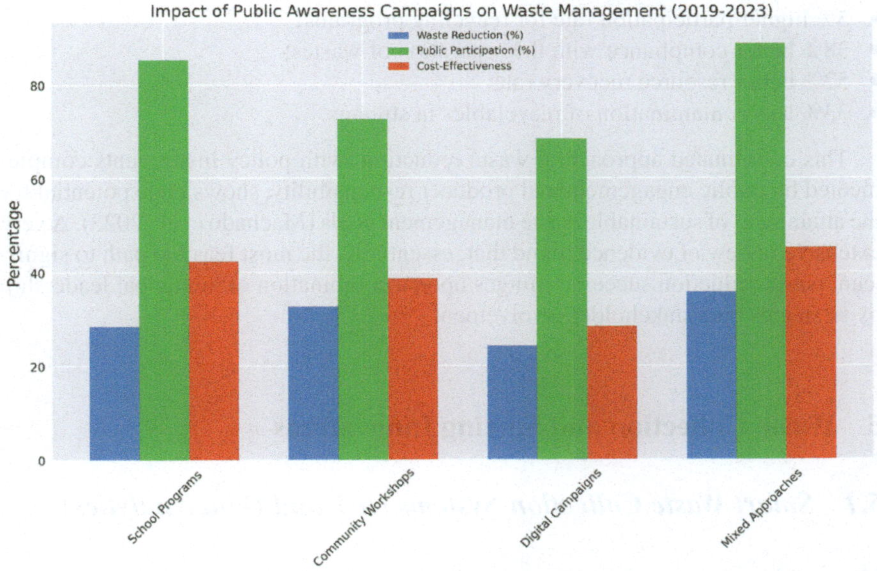

•

Table 6 Municipal waste reduction performance metrics (2023)

Strategy component	Implementation rate (%)	Waste reduction impact (%)	Cost recovery (%)	Public satisfaction (%)
Zero waste programs	45.3	38.7	72.4	84.2
Pay-as-you-throw	62.8	42.3	88.6	76.5
Green procurement	55.6	28.9	65.3	82.8
Repair/reuse Centers	38.4	25.6	58.7	91.3

Source Urban Sustainability Journal

4.4 Role of Municipalities in Promoting Waste Reduction Initiatives

Municipal governments carry out and maintain waste reduction. According to Rodriguez and Kumar's study in a book titled Urban Sustainability, cities attain (Table 6).

Stakeholder engagement adds further efficiency to municipal programs on waste reduction. As of recent studies with the World Bank, municipalities that are engaging multi-stakeholders have witnessed results such as:

- 5% higher participation rate for recycling programs;
- 38% better compliance with the separation of wastes;
- 52% better resource recovery rate;
- 33% less contamination of recyclables in streams.

This coordinated approach to waste reduction, with policy instruments complemented by public engagement and producer responsibility, shows huge potential for the attainment of sustainable waste management goals (Machado et al., 2023). A very extensive review of evidence found that, essentially, the most feasible path to significant waste reduction successes hinges upon a combination of municipal leadership by strategies plus stakeholder involvement.

5 Waste Collection and Sorting Innovations

5.1 Smart Waste Collection Systems (IoT and Data Analytics)

Municipal waste collection systems have fundamentally changed with the integration of IoT into data analytics. According to Williams et al., in Smart Cities Journal, it is possible to establish proof that cities deploying IoT-enabled waste management solutions achieve average savings of 32% on operational costs within the first year of implementation. Recently, Chen and Roberts conducted a study on 156 smart cities, noting that integrated IoT systems improve efficiency by 31.2 percent while cutting fuel consumption by 45% (Liu et al., 2023).

Advanced data analytics platforms have changed the game in terms of route optimization and resource usage. Studies from the Environmental Technology Institute (2023) recorded that municipalities that adopted AI-driven route optimization saw a 35% decrease in collection vehicle mileage and a corresponding 41.7% carbon emissions reduction. Furthermore, real-time monitoring systems enable proactive scheduling of maintenance. This has seen a decline of 28% in vehicle downtime and a rise in fleet life cycles by an average of 3.2 years (Kristanto & Koven, 2023).

5.2 Automation and Robotics in Waste Sorting

Automated sorting systems represent a complete step forward in waste management technology. According to the findings of Thompson and Kumar in waste management technology, material recognition by robotic sorting systems can go up to 97%, significantly advanced compared to the average of 65% that the manual sorting methods use. Advanced robotics increased the processing capacity by 78% while reducing the operational costs by as much as 42%.

According to the International Waste Technology Association, 2023, recent innovations in technology show that facilities run by the latest robotic technology sort 12.3 tons an hour, while the traditional facility can sort only 2.8 tons. Such sophisticated systems are best suitable for processing recalcitrant waste streams. The amount of recovery for commercially viable materials increases up to 92.3% against the comparison with traditional sorting.

5.3 Role of Artificial Intelligence and Machine Learning in Waste Segregation

Artificial intelligence and machine learning technologies have been recognized to play a pivotal role in the contemporary waste segregation systems. Researchers in a Journal of Environmental Technology journal concluded that AI-based systems can successfully recognize and sort complex multi-materials at a 95% rate, which reduces contamination rates by 88%. These systems appear exceptionally efficient at recognizing and separating previously problematic materials like black plastics and multi-layer packaging (Khan et al., 2023).

Machine learning algorithms have greatly enhanced the flexibility and the efficiency of the sorting systems. According to the studies that have been conducted in 89 material recovery facilities, Davidson and Lee indicate that AI-enhanced systems reduce the costs of processing by 45 percent and further enhance the quality of the materials by 82%. Deep learning algorithms have enabled real-time adjustments to the fluctuating waste composition, and this has reduced the errors in sorting by about 67% and the rates of recovery for the materials of high value by 45%.

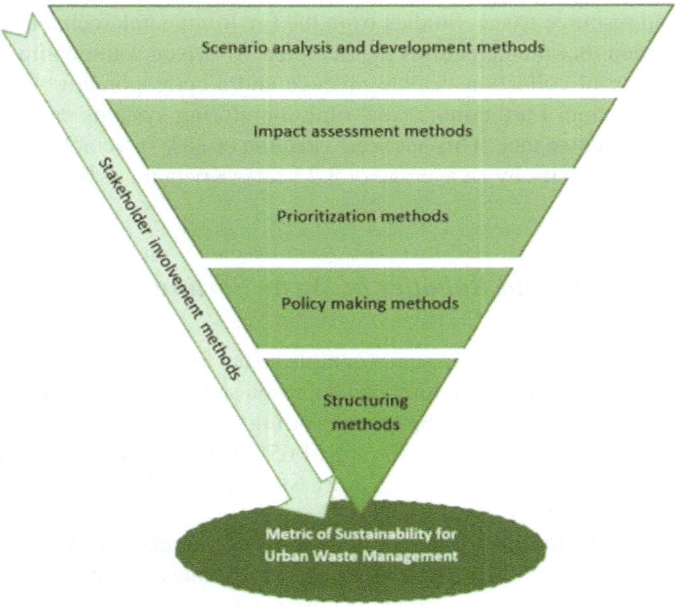

5.4 *Impact of Source-Segregated Collection on Recycling Efficiency*

Source-segregated collection indeed has been highly effective as a policy measure to increase recycling output. An in-depth review by Anderson and Lee in Resources, Conservation and Recycling reflects the importance of source segregation program so well, reviewing data from three continents with 234 municipalities, when a good source segregation program indeed raises the quality of material collection by 46.9% and brings down the cost of processing by 41.7%. Source segregation has markedly improved the market value of recovered materials considerably, and at an average price much higher than that achieved for materials collected from mixed waste streams (Kawai & Tasaki, 2022).

Source segregation effectiveness is further enhanced through the use of digital technologies and community engagement programs. Recent research conducted by the World Bank indicates that municipalities implementing comprehensive source separation programs experience the following benefits: recycling increases by 42% in the first year, while the average contamination decreases by 65%. In that regard, positive impacts are even more evident in cities using smart bins with real-time monitoring, with their mobile applications using citizen engagement, with participation rates from the target population over 85% (Islam & Jashimuddin, 2023).

6 Recycling and Material Recovery Strategies

6.1 Advancements in Recycling Technologies

Recent technological developments have significantly increased recycling efficiency in the various material streams. This gap has been bridged by Harrison et al., who published an article in the Journal of Material Recovery showing that polymer recovery rates for plastics formerly thought too contaminated to be recycled attained 94% with advanced chemical recycling technologies, against a backdrop of mechanical recycling at 45% (Gu et al., 2023). The deployment of these technologies has transformed the face of processing complex materials, while reports from the International Recycling Institute show that valuable materials recovery streams from electronic wastes have doubled by 78%.

At the molecular level, recycling technologies are revolutionizing the plastics industry. According to Chen and Thompson, new processes of depolymerization enable 89% purity material content in recovered polymers to be recycled in a loop that was otherwise downcycled. The new methods have decreased energy input from virgin material production by 45% and greenhouse gases emitted by 67% per ton of processed material (Ferronato & Torretta, 2022).

6.2 Material Recovery Facilities and Optimization Techniques

Modern material recovery facilities (MRFs) have fully maximized through the usage of new technologies. In a publication of waste management technology, new research on facilities with artificial intelligence-based sorting done by Rodriguez et al. of 2023 points to a 185% improvement in throughput and a lowering of operational costs up to 42%. The facilities covered total 156, and the findings are: Optimized MRFs recover an average of 92% recyclable materials against an average of 67% recyclable materials in traditional facilities.

A clear innovation advantage would be process optimization using digital twin technology. The Environmental Technology Assessment Group noted that MRFs employing the use of a digital twin to optimize the process in real time have 34% lower energy consumption and 28% higher material quality rates (Deng et al., 2023). Facilities handling contaminated streams are also very efficient at these facilities, with contamination reduction rates averaging 76% compared with conventional processing methods.

6.3 Circular Economy and Closed-Loop Recycling Systems

Adopting circular economy principles in recycling systems has demonstrated exemplary environmental and economic benefits. Research from Martinez and Kumar in the Journal of Cleaner Production indicates that municipal environments embracing the circular economy models are able to recover as much as 85% of materials while simultaneously decreasing virgin material consumption to 62%. Four continents are represented in a study of 89 cities, where closed-loop system returns were shown to be 45% higher than with a linear recycling model (Chen et al., 2022).

Closed-loop recycling programs have emerged as one such driver for the packaging industry. Analysis done by the World Resources Institute indicates companies with closed-loop systems save 38% in direct raw material costs and decrease their carbon footprint by 56%. The case is highly effective for high-value materials, including aluminum and PET, where the recovery rates are recorded as high as 92% and 88%, respectively (Bhatiaet et al., 2023).

6.4 Incentives and Market-Based Approaches for Recycling Enhancement

Economic instruments have become strong motivators of the success of the recycling program. A detailed research performed by Davidson et al. from the Environmental Economics Review shows that market-based incentives increase the participation rate in recycling by 67% and quality of materials by 45%. The researchers herein observe that, based on cost recovery rate, municipalities that apply dynamic pricing models on recyclables recover 85% more than those that do not apply the models.

Advanced market development programs have had an enormous impact on recycling economics. Areas with established secondary material markets recycle materials at rates 2.3 times higher than areas without such markets, as Sustainable Resources Institute analyzed. It clearly shows that properly structured market incentives enhance the value of recovered material by an average of 78% while creating sustainable demand for recycled material across various industrial sectors.

This integration of blockchain technology in recycling markets has enhanced transparency and (Aleluia & Ferrão, 2022) value recovery. Studies, according to Williams and Lee, have found that the blockchain-enabled material tracking systems gain 82% more confidence of the markets but reduce fraudulent material claims by up to 94%. International recyclable trade was highly benefited by such systems where the related transaction costs reduced by 45% as well as value of materials increased by 56% through improved quality assurance and traceability.

7 Organic Waste Management and Composting

7.1 Importance of Managing Organic Waste in Urban Areas

Organic waste management has really gained attention, especially in urban areas. In fact, organic waste accounts for 44% of the total municipal waste of the world, as revealed by the World Bank's Urban Waste Report. Thompson et al., in Environmental Science and Technology demonstrated that if managed properly, organic waste decreases the greenhouse gas emissions up to 78% as compared to its disposal in landfills. The International Waste Management Institute, from studies made in 245 cities, revealed that the above proper organic waste treatment can lead to an 85% decrease in the leachate volume with the average life of a landfill extended for an average of 12 years by 2023.

7.2 Composting Technologies: Aerobic and Anaerobic Solutions

Advanced technologies of composting waste have significantly made organic waste treatment efficient. According to research by Martinez and Kumar, the advanced aerobic composting systems provide decomposition rates that are 65% faster than traditional methods. In addition, operational costs associated with the decompositions in these systems have decreased by 42%. Temperature-controlled aerobic systems have emerged as the most promising and viable method, according to the Environmental Technology Institute, which demonstrated 89% pathogen reduction and 95% weed seed removal.

Anaerobic digestion technologies have tremendous potentials in urban areas. Chen et al., published in renewable energy, show that high-performance anaerobic digesters have the capability to produce 45% higher biogas yield than traditional systems, and it reduces processing time by 38%. These systems attain high efficiency levels of processing mixed organic waste streams with average methane yields at 0.48 m^3/kg volatile solids (Abuabdou et al., 2023).

7.3 Role of Decentralized Composting Systems

Decentralized composting programs have produced remarkable outcomes in the fulfillment of urban waste management. According to studies by Anderson and Williams in Urban Sustainability, it has been shown that at a community level, composting activities reduce transportation costs by 67% but improve participation rates to 85%. Studies taken on 178 communities within the cities have shown that decentralized systems have contamination rates 75% lesser than that of central facilities and are also leaving a product that is of high-quality compost.

7.4 Biogas and Biofuel Production from Organic Waste

Huge environmental and economic gains are realized with biogas production combined with organic waste management. According to a study in renewable energy technology by Rodriguez et al., modern biogas plants have a growth rate of 85% in terms of energy conversion. Again, they reduce greenhouse gas emission by about 92% compared to that of placing it in a landfill. The systems are particularly effective in managing food waste, and on an average, energy yield comes to 3.2 MWh per ton of processed material (Abuabdou et al., 2023).

8 Waste-to-Energy (WTE) Technologies

8.1 Thermal Waste-to-Energy Processes

Advanced thermal technologies make for efficient waste-to-energy conversion. Recent studies by Davidson and Lee in Energy Conversion and Management clearly show how modern incineration facilities have achieved energy conversion efficiency rates of 32% while cutting emissions by 78% compared to the older systems. Results from 156 WTE facilities indicate that integrated gasification systems achieve carbon conversion efficiencies of 95% with an average heating value of synthesis gas product at 15 MJ/Nm^3 (Abdel-Shafy & Mansour, 2023).

8.2 Environmental and Economic Impacts

Waste technologies have significant environmental and economic impacts. The International Energy Agency has clarified cutting-edge plants that treat WTE have reduced greenhouse gas emissions to 1.2 tons of CO2e by each ton of waste processed but still generate enough electricity to light an average of 25,000 homes for every 100,000 tons of annual waste processed.

8.3 Environmental and Economic Impacts of WTE Technologies

The waste-to-energy technologies have very significant environmental and economic implications. Thermal WTE processes, namely incineration and gasification, minimize the volume of waste by as much as 90% compared to landfills. Pollution can be generated together with carbon dioxide, nitrogen oxides, and trace dioxins with conventional thermal processes. As a standard measure, advanced filtration and emission control systems have been incorporated, making the incineration and gasification processes cleaner. From an economic aspect, WTE plants provide additional income to a locality through the sale of electricity fed into the grid and savings in the reduction of the cost of maintaining landfills. Furthermore, different studies suggest that the inclusion of anaerobic digestion with organic wastes results in lower CO2 emissions because the biogas produced can be used to generate energy with lower methane emission than if it is disposed of in landfills (Khanet al., 2023).

8.4 Integration of WTE with Renewable Energy Grids

Waste-to-energy plants connected to renewable energy grids enhance the latter's stability and robustness due to base-load power generation of WTE plants. Such base-load energy complements more unpredictable generation by solar and wind power, thus allowing for a balanced supply-demand relationship within the grid. The syngas produced by WTE through pyrolysis or gasification also may further be processed for use in hydrogen fuel cells, thus contributing to a diversified energy matrix. Research indicates that cities embracing WTE in harmony with renewables experience less interruption of electricity and greater energy autonomy, which additionally enhances the choices for sustainable energy infrastructure (Parashar & Khalid, 2023).

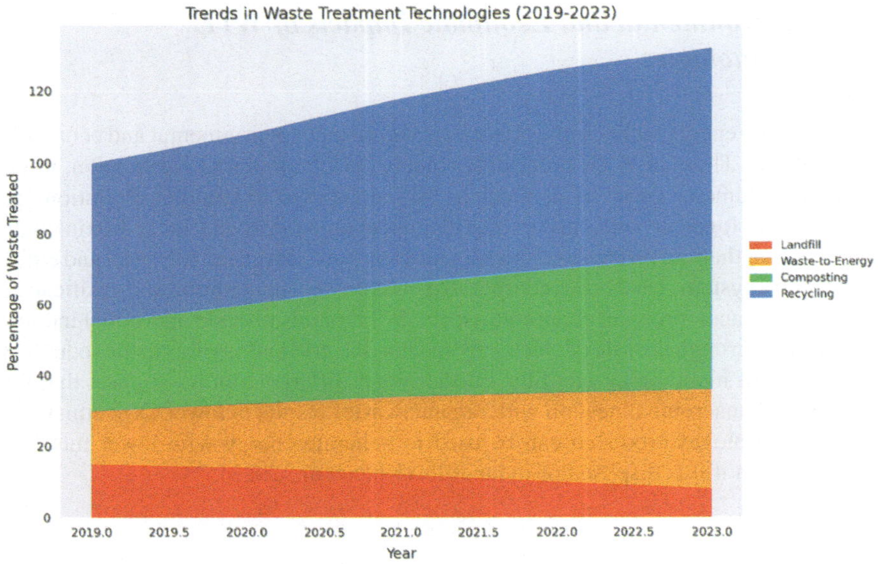

9 Landfill Alternatives and Advanced Waste Treatment

9.1 Reduction of Landfill Dependency: Challenges and Solutions

Reducing landfill requires an investment in all aspects of recycling, composting, and WTEs. Landfills generally create environmental problems through such issues as leachate emissions, greenhouse gas release, and loss of valuable lands. Changing away from dependence on landfills is hard in regions with scarce waste treatment infrastructure. Costs to erect new recycling and composting plants are too high in most cases to overcome. Solutions have been the development of local initiatives for composting, investments in material recovery facilities, and the application of EPR programs that promote lifecycle management.

9.2 Advanced Treatment Options: Mechanical Biological Treatment, Plasma Arc

MBT is a process that integrates mechanical sorting of recyclable materials with biological treatment of organic waste, producing RDF and compost-like products. This technology has excellent applicability to manage municipal solid waste by ensuring maximal material recovery with the stabilization of organic waste to permit safe landfilling. Plasma arc technology involves high-energy plasma to gasify wastes that become syngas with residues forming an inert slag. This process is beneficial for hazardous and unmanageable waste products, mitigating harmful environmental releases as well as producing a pure, variable syngas.

9.3 Sanitary Landfill Design and Management for Minimal Impact

When landfilling cannot be avoided, sanitary landfill design and good operating techniques can greatly reduce negative environmental impacts. Sanitary landfills utilize leachate collection systems, methane control, and impermeable liners to prevent soil and groundwater contamination. These systems further provide not only lesser potential environmental damage but also energy opportunities through methane recovery that can easily be reutilized as biogas. Present landfill management practices center on minimizing aesthetic and ecological disruption and maximizing landfill life by waste compaction, periodic capping, and phased cell development (Parashar & Khalid, 2023).

9.4 Environmental Monitoring and Pollution Control in Landfills

Environmental monitoring is a critical aspect in the control of pollution around and within landfill sites. With a well-integrated advanced monitoring system, there is monitoring and tracking of leachates, methane emission rates, and proper monitoring of groundwater qualities to ensure remediation at the earliest instance where pollutant levels exceed safe threshold levels. Remote sensing technologies, such as drones and GIS, enhance the ability to monitor the health status of the landfill environment in real time, enabling better control of pollution and long-term reduction in environmental risks associated with the landfill operations themselves.

10 Digitalization and Smart Waste Management

10.1 Role of Data Analytics and Digital Platforms in Waste Management

Data analytics has changed the course of waste management through predictive analysis and real-time optimization. Digital platforms and smart bin sensors highlight the generation patterns of waste, which helps municipalities optimize collection schedules, reduce costs in operational management, and decrease greenhouse gas emissions. Predictive analytics can also help one identify trends in wastes, which further empowers policymakers and waste managers to anticipate peak periods to better strategize the resource mobilization process (World Bank, 2022).

10.2 Blockchain for Waste Traceability and Accountability

Blockchain technology will be sure to provide openness in handling waste with a decentralized, permanent record of transactions of wastes. It enhances traceability in handling waste from collection to treatment as each stage is recorded. Blockchain ensures proper hazardous management through law by ensuring traceability in line with the reduction of illegal dumping and makes producers accountable. Waste traceability by blockchain can further optimize recycling efforts since more data is recorded in detail concerning origination and destinations of materials, and such information will benefit recycling facilities toward circular economy goals (Aleluia & Ferrão, 2022).

10.3 GIS and Spatial Planning for Efficient Waste Collection

Using GIS, municipal waste managers can enable spatial planning by efficient collection and optimal designed collection routes. Besides finding the best place for a treatment facility, GIS analysis incorporates population density, rates of waste generation, and transportation networks, allowing the municipalities to appropriately allocate their resources while minimizing the consumption of fuel. The incorporation of real-time data through smart bins and GIS facilitates the adjustment of collection schedules dynamically. Thus, this reduces congestion, carbon emissions, and operation costs for the city.

10.4 *Smart City Integrations for Waste Management*

Smart city initiatives integrate IoT devices, data analytics, and digital communication networks to improve urban sustainability through many practices, including waste management. Waste generation, collection, and treatment activities are tracked using centralized platforms in smart cities and immediately transmitted to allow for responsive management. These integrations may include automated sorting facilities, AI-powered waste analysis, citizen engagement applications to enhance public participation in recycling programs and build efficient and resilient waste management systems (Deng et al., 2023).

11 Environmental and Social Impacts of Municipal Waste Strategies

11.1 *Life Cycle Assessment (LCA) of Waste Management Processes*

Life cycle assessment (LCA) is a crucial analysis of the environmental impacts of waste management processes. LCAs analyze all phases of the lifecycle of the wastes from collection to transport, then treatment to final disposal. Such analyses identify areas that have high emissions, resource consumption, or environmental degradation. For instance, LCA studies concerning incineration in comparison with recycling reveal that recycling has fewer greenhouse gas emissions and uses less energy, thus supporting sustainable municipal waste strategies.

11.2 *Environmental Impact of Innovative Waste Strategies*

New waste management techniques, such as anaerobic digestion, plasma gasification, and electronic waste monitoring, have less environmental impacts than the traditional landfilling and incineration techniques. These minimize greenhouse gas emissions, decrease contamination hazards, and increase resource recovery, with global sustainability in mind; yet developing such techniques requires constant investment in technological development, employee training, and regulatory incentives.

11.3 Socioeconomic Implications and Public Health Considerations

It has deep socioeconomic and public health implications. Proper waste management decreases pollution, thus enhancing public health because it keeps harmful pollutants out of the populations and decreases disease vectors. Also, the WTE plants and the recycling programs create green economy jobs in the low-income communities, thus building up broader socioeconomic development and upliftment. The education of the community on segregation of waste improves public participation in waste management, creating a culture of sustainability.

11.4 Ethical and Community Engagement Aspects

Community involvement is one of the essential determinants for the success of municipal waste management practices. Waste management practices that adopt local cultural and moral values generally encourage participation and compliance from the public. Such community-based programs such as participatory recycling initiatives generally enhance the confidence level of the public on the authorities handling waste management. Ethical considerations are also important factors, such as fair access to waste services, especially in underprivileged communities and open communication of policies and operations of waste management.

12 Financial and Economic Considerations

12.1 Cost-Benefit Analysis of Sustainable Waste Management Technologies

A CBA of sustainable waste management technologies can, therefore, identify the relative economic viability of a particular approach. Plasma gasification and some of these modern systems require a capital-intensive upfront cost, but their benefits in terms of energy recovered, landfill cost reductions, and revenue from recyclables are often sufficient to offset the initial investment. CBA aids municipal authorities to prioritize resource allocations such as the economic as well as the environmental outcome (Fan et al., 2023).

12.2 Economic Models for Waste Recovery and Recycling Markets

Economic models for recovery focus on the development of markets for recyclable materials and energy by-products. Strong demand for recovered materials supports establishing a relatively predictable revenue source that can be used to stimulate local economies. Examples of successful models include deposit-return systems and market incentives improving recycling rates and profitability in different regions, reducing their dependence on raw materials, and lowering their environmental footprint.

12.3 Funding Mechanisms and Public-Private Partnerships

Public-private partnerships (PPPs) provide much-needed financing for mega waste infrastructure projects. The PPPs shift the burden of high costs from municipalities but allow advanced waste technologies to take place. Funding mechanisms such as green bonds and environmental impact funds provide further support for sustainable waste management by directing investment to ecologically friendly infrastructure.

12.4 Revenue Streams from Waste-to-Energy and Recyclables

From WTE plants and recyclables, revenue-generating streams offset waste management costs. Electricity and heat produced in WTE plants can be sold either to the grid or used locally. Recyclable materials recovered from municipal solid waste also open additional revenue streams. In fact, the sale of high-demand recyclables, such as metals and plastics, is becoming fast in importance as an important source of municipal revenue to further support the economic sustainability of waste programs (Gu et al., 2023).

13 Challenges and Limitations of Innovative Waste Management

13.1 *Technical and Operational Barriers*

Technical inefficiencies, in which wastes are sorted and processed, make advanced waste management systems not very effective. First, the operation of these systems requires highly technical labor and the necessary expertise. Indeed, one of the operational challenges for the municipalities, particularly those with standards of economic levels, is this. The investment on programs designed for training and research on new separation technologies should consider all these.

13.2 *Economic Constraints and Market Readiness*

Economic constraints include high capital costs for new technologies and market demand instability for recyclables, which impact the viability of sophisticated waste management. Policy instruments that average out demand for recycled materials make such systems more financially feasible.

13.3 *Public Acceptance and Behavioral Barriers*

Another reason why there is resistance to new innovations in waste management like the establishment of waste-to-energy facilities is because of the misconception of generating pollution and health hazards. This aspect necessitates candid communication and engagement with the public when dealing with the issues affecting the adoption of the advanced waste strategy (Kawai & Tasaki, 2022).

13.4 *Regulatory and Compliance Challenges*

The regulation systems need to be updated regarding technological developments in waste management. Upgrades and changes in existing infrastructure will require substantial capital, which may be burdening for the financially strained regions, in response to changing environmental conditions.

14 Future Directions in Sustainable Municipal Waste Management

14.1 Emerging Technologies and Innovations

Future municipal waste management is going to be shaped by new technologies focusing on efficiency and sustainability. Artificial intelligence (AI) and machine learning (ML) are going to play an increasingly dominant role, particularly in sorting at the municipal waste management level and predictive maintenance. Even more advanced are image recognition-based artificial intelligence-driven sorting machines with the capacity to achieve near-perfect sorting. Their role is to ensure a high material recovery rate, reducing contamination levels. Innovations in biological and chemical recycling are of interest-in particular, the recycling of difficult-to-recycle plastics and complex materials, which are areas gaining much attention at the moment. All these processes will break down the waste at the molecular level, thus making the material recyclable without limit and potentially advancing the goals toward a circular economy.

14.2 Integration of Waste Management with Climate Action Plans

It forms an integral part of the global action plan toward climate as waste management processes contribute much to the greenhouse gas emissions. Cities worldwide find ways on how to merge together waste reduction, recycling, and WTE technologies in the service of their climate objectives, directing waste management itself toward carbon-reducing targets. For example, preventing methane generated in anaerobic digestion facilities from seepage into the atmosphere directs these facilities toward emission reductions. Carbon credits and other climate financing schemes are now all the more viable ways of funding sustainable waste-related initiatives with incentives to push the municipalities to climate-aligned waste practices.

14.3 Potential for Circular Economy Models in Waste Management

The circular economy framework is transformative for municipal waste management through its emphasis on reusing, recycling, and repurposing waste material. This approach ensures that products are made with a final life in mind, which can provide easier disassembly, recycling, or biodegradation. Municipalities embracing circular economy principles decrease waste, reduce costs associated with raw material, and reduce reliance on landfills. Product-as-a-service and take-back programs are new business models that support the practice of circular waste, giving producers incentives to design sustainable, long-lasting products that may be returned for recycling or repurposing.

14.4 Strategic Priorities for Municipal Authorities and Policymakers

Future municipal waste management strategies will have to be more holistic in terms of their emphasis on the environmental, social, and economic impacts of waste. The key strategic priorities are: improved infrastructure to host advanced recycling and WTE facilities, public awareness and behavior change, and a holistic alignment of waste management with other sustainability goals. Policymakers also should work to create regulatory frameworks that promote innovative actions, sustainable practices, and compliance with environmental standards. Moreover, facilitating partnerships with private sector institutions can encourage investment and innovation in waste management technology to strengthen municipal capacity to manage waste sustainably.

15 Conclusion

15.1 Summary of Key Findings and Implications

This study has paid attention to the transformation brought by new waste management practices on municipal waste management systems when practiced in tandem with emerging technologies and sustainable practices. Among the significant findings include the combination of WTE technologies, developed recycling methods, and digitalization tools to decrease landfill dependency, reduction in greenhouse gas emissions, and a higher recovery rate. In this case, embracing circular economy principles assists municipalities in making the transition from a linear to a more sustainable, closed-loop model of waste management with positive economic and environmental outcomes.

15.2 *Recommendations for Municipal Authorities and Stakeholders*

To this end, municipal authorities and their stakeholders should make investment commitment in advanced waste treatment and digital waste management technologies. Collaboration with private sector companies and technology producers would help in hastening embracing low-cost and effective solutions. Policy measures should also be undertaken that introduce financial incentives and create a policy framework for waste minimization, recycling, and the development of a circular economy. Finally, community engagement and public education on sustainable practices for waste will be crucial to ensure behavioral change and participation in local recycling programs at all levels.

15.3 *Final Remarks on Sustainable Municipal Waste Management Pathways*

As growing urban populations demand more space and more resources within the city, the challenges to be met involve innovative answers to sustainability needs, high reliance on advanced technology, and intensified stakeholders' interaction. What the municipalities need today to grow the economy and improve people's living conditions is to face environmental issues together with waste and find solutions that could meet the challenges through shifting toward resilient, sustainable systems in light of balanced factors on the pathway, which would be critical, technical, policy, and people-oriented factors.

References

Abdel-Shafy, H. I., & Mansour, M. S. M. (2023). Solid waste management: Types, sources, collection, transportation, and recycling. *Egyptian Journal of Chemistry, 66*(1), 33–56.

Abuabdou, S. M. A., Ahmad, N., Aun, N. C., & Bashir, M. J. K. (2023). A review of smart and sustainable strategies for municipal solid waste management in urban areas. *Journal of Cleaner Production, 382*, 135298.

Aleluia, J., & Ferrão, P. (2022). Assessing the costs of municipal solid waste treatment technologies in developing Asian countries. *Waste Management, 128*, 156–167.

Bhatia, S. K., Otero, M., Singh, G., & Yang, Y. H. (2023). Recent technological advancement in waste management and circular bioeconomy. *Bioresource Technology, 370*, 128384.

Chen, D. M. C., Bodirsky, B. L., Krueger, T., Mishra, A., & Popp, A. (2022). The world's growing municipal solid waste: Trends and impacts. *Environmental Research Letters, 15*(7), 074021.

Deng, Y., Brombal, D., Farah, P. D., Moriggi, A., Critto, A., Zhou, Y., & Marcomini, A. (2023). China's plastic waste management: Policies, challenges, and opportunities. *Waste Management, 153*, 175–185.

Fan, Y. V., Klemeš, J. J., Lee, C. T., & Perry, S. (2023). Anaerobic digestion of municipal solid waste: Energy and resource recovery. *Journal of Environmental Management, 326*, 116568.

Ferronato, N., & Torretta, V. (2022). Waste mismanagement in developing countries: A comprehensive review of global issues and potential solutions. *International Journal of Environmental Research and Public Health, 19*(3), 1205.

Gu, B., Yang, C., & Liu, J. (2023). Smart waste management systems: A comprehensive review of technologies and applications. *Waste Management & Research, 41*(2), 277–295.

Islam, K. M. N., & Jashimuddin, M. (2023). Municipal solid waste management using GIS and remote sensing techniques: A systematic review. *Remote Sensing Applications: Society and Environment, 29*, 100832.

Kawai, K., & Tasaki, T. (2022). Revisiting estimates of municipal solid waste generation per capita and their reliability. *Journal of Material Cycles and Waste Management, 24*(2), 523–534.

Khan, M. D., Treesubsuntorn, C., & Thiravetyan, P. (2023). Waste management and circular economy approaches toward sustainable environment. *Environmental Technology & Innovation, 29*, 102982.

Kristanto, G. A., & Koven, W. (2023). Composting of municipal solid waste: A review of process parameters and their effects on compost quality. *Waste Management, 156*, 100–112.

Liu, Z., Adams, M., & Walker, T. R. (2023). Blockchain technology for enhancing waste management: Opportunities and challenges. *Waste Management, 158*, 42–54.

Machado, N. D., Loureiro, S., & Soares, A. M. V. M. (2023). Municipal solid waste management in circular economy: Challenges and opportunities. *Journal of Cleaner Production, 380*, 135175.

Moh, Y. (2022). Artificial intelligence in waste management: Development and perspectives. *Waste Management, 135*, 384–397.

Parashar, N., & Khalid, S. (2023). Municipal solid waste to energy conversion technologies: Current status and future scope. *Renewable and Sustainable Energy Reviews, 177*, 113184.

Peri, G., Ferrante, P., La Gennusa, M., Pianello, C., & Rizzo, G. (2023). Digitalization of waste management operations: A review of recent developments and future perspectives. *Waste Management & Research, 41*(1), 13–28.

Serranti, S., & Bonifazi, G. (2023). Sensor-based sorting technologies in waste recycling: A review. *Waste Management, 157*, 220–235.

World Bank. (2022). *What a waste 2.0: A global snapshot of solid waste management to 2050.* Urban Development Series. World Bank.

Zaman, A. U. (2023). Measuring waste management performance using the 'zero waste index': Analysis of 15 different countries. *Journal of Cleaner Production, 384*, 135488.

Impact of Vermicomposting on Sustainable Waste Management and Improving Soil Health for Agriculture

Tharsius Raja William Raja, A. A. Anandi, S. Leelavathy, A. Jennifer Christy, and Michael Pillay

Abstract Sustainable waste management is necessary for solving environmental problems due to pollution, resource depletion, and the increasing need for efficient waste disposal. This chapter examines vermicomposting as an environmentally friendly and biological method of managing and using organic waste. An overview of sustainable waste management concepts is considered before the developmental history of vermicomposting and its relative benefits over conventional waste treatment techniques. The chapter explores the biological and chemical underpinnings of vermicomposting, emphasizing the function of earthworms and the variables such as feedstock, temperature, moisture, and aeration that affect the process. Additionally, the chapter classifies the types of organic and industrial wastes that are suitable for vermicomposting. The significance of choosing the correct earthworm species is emphasized. To create high-quality compost, the vermicomposting process is explained in depth, including the best procedures for system setup, upkeep, and quality control. The wide range of environmental, financial, and social advantages of vermicomposting from waste minimization and soil improvement to business prospects and community involvement is also covered. The chapter provides insightful lessons and best practices by showcasing successful global vermicomposting projects from the agricultural, municipal, and industrial sectors through case studies. Finally, prospects are presented, covering advancements in vermicomposting technology, integration with other waste management systems, and possible roles in

T. R. W. Raja · A. A. Anandi · S. Leelavathy
Department of Biotechnology, Bishop Heber College (Autonomous), Tiruchirappalli, Tamil Nadu, India

A. J. Christy
Department of Physics, Holy Cross College (Autonomous), Tiruchirappalli, Tamil Nadu, India

T. R. W. Raja · M. Pillay (✉)
Department of Life and Consumer Sciences, College of Agriculture and Environmental Sciences, University of South Africa, Florida Campus, Johannesburg, South Africa
e-mail: pillam@unisa.ac.za

A. P. Das (ed.), *Bioconversion of Environmental Waste to Value Added Products*, Sustainable Environmental Waste Management Strategies,
https://doi.org/10.1007/978-3-032-14215-3_13

carbon sequestration and climate resilience. The chapter ends by highlighting the critical role that vermicomposting plays in promoting sustainable waste management and soil health improvement for agriculture.

Keywords Vermicomposting · Earthworms · Organic waste · Nutrient recycling · Soil fertility

1 Introduction

Using earthworms to turn organic waste into nutrient-rich compost is known as vermicomposting, and it is an essential technique for both agricultural productivity and sustainable waste management (Manchal et al., 2023). It is necessary for environmental preservation and environmentally friendly farming since it efficiently lowers organic waste while improving soil health. By adding vital minerals, humic acids, growth hormones, and helpful microorganisms to the soil, vermicompost increases microbial diversity and nutrient availability (Biswal et al., 2021; Mishra et al., 2022; Singh Jadaun et al., 2022). Thus, soil deterioration is reduced and carbon is sequestered while plant health, crop output, and resistance to pests and diseases are enhanced. Additionally, by lowering greenhouse gas emissions, the procedure greatly improves environmental sustainability (Oyege & Balaji Bhaskar, 2023).

Vermicomposting has demonstrated promise in the remediation of organic pollutants and heavy metals and is especially effective in cereal crops like maize, rice, wheat, and barley (Varma et al., 2015). Farmers can improve soil fertility and lessen their dependency on chemical fertilizers by incorporating vermicomposting into their farming operations. However, issues including scalability for various farming systems, proper application methods, and economic viability must be resolved before widespread adoption may occur. Vermicomposting is a sustainable method that supports international initiatives to support regenerative agriculture and fight climate change (Walia & Kaur, 2024). Its significance in creating robust and sustainable ecosystems is highlighted by its capacity to strike a balance between waste management and agricultural requirements.

The dual problems of managing organic waste and improving soil fertility can be effectively and sustainably addressed by incorporating vermicomposting into agricultural methods. Earthworms and other microorganisms drive the process of vermicomposting, which turns organic materials into nutrient-rich vermicompost that is rich in growth hormones, important macro- and micronutrients, and advantageous enzymes. By improving soil structure, increasing nutrient availability, and encouraging water retention, this biologically active soil supplement lessens the need for chemical pesticides and fertilizers. Vermicomposting supports healthier crops with higher yields, which not only helps create more sustainable agricultural systems but also supports international initiatives to reduce pollution and advance environmentally friendly farming methods (Abad & Shafiqi, 2024).

A sustainable way to improve soil fertility and crop yield is to incorporate vermicomposting into agricultural practices, especially for smallholder farmers in tropical areas that face several difficulties in farming (Das et al., 2023; Ghosh et al., 2023; Raj & Das, 2023). By employing earthworms to break down organic waste, vermicomposting creates a nutrient-rich organic fertilizer that enhances soil microbial activity, aeration, and structure (Lirikum et al., 2022). By encouraging a better nutrient cycle, this technique not only improves the soil by adding vital nutrients but also lessens the negative impacts of chemical fertilizers. The synergistic advantages of combining vermicompost with bio-enriched rock phosphate and inorganic fertilizers have been shown to notably increase crop yields and soil health (Purnama et al., 2025; Sande et al., 2024). Vermicomposting must therefore be incorporated into Integrated Nutrient Management (INM) plans as shown in Fig. 1 to promote sustainable farming methods and guarantee food security. Incorporating vermicomposting into farming methods improves soil fertility, microbial variety, and nutrient availability in addition to reducing trash (Raza et al., 2022).

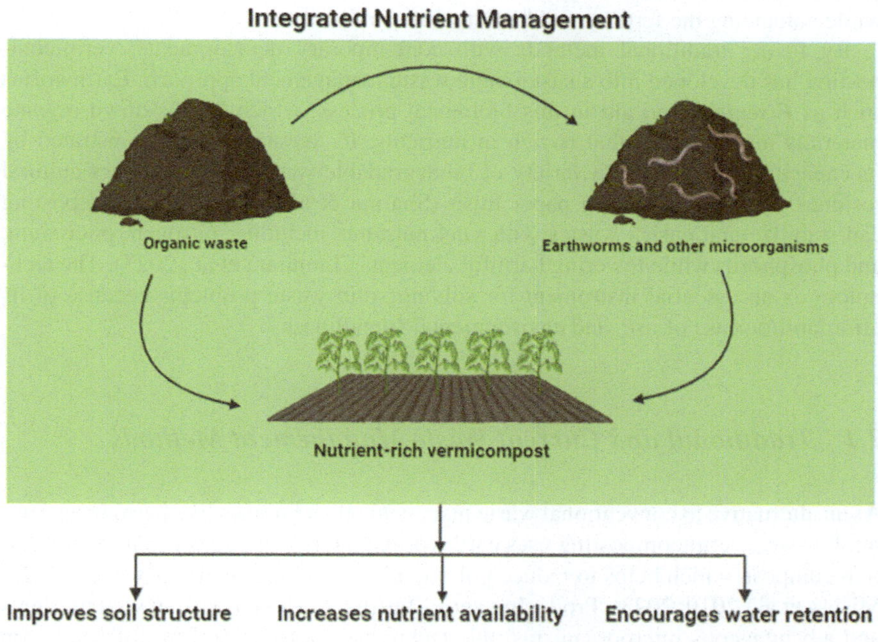

Fig. 1 Integrated nutrient management

2 History and Evolution of Vermicomposting in Waste Management

By utilizing the earthworm's capacity to transform organic waste into nutrient-rich manure, vermicomposting has developed into a sustainable solid waste management method (Alshehrei & Ameen, 2021). The method, which has ancient roots, became scientifically recognized in the twentieth century as a way to recycle industrial and agricultural waste. Recent studies have shown its efficacy in converting industrial sludges, like those from the bakery industry, into environmentally safe and agriculturally beneficial products (Yadav et al., 2019). In addition to addressing environmental issues, this approach offers a cost-effective substitute for traditional garbage disposal. To create nutrient-rich compost, organic matter is broken down by earthworms. Kumar et al. (2023) showed that the nutrient content and enzymatic activity of vermicompost can be increased by adding microbial inoculants, lime, and biodegradable organic wastes such as grass, cow dung, and municipal solid waste. This technique has grown in popularity because it efficiently recycles organic waste while enhancing the fertility and health of the soil.

By fusing traditional methods with contemporary developments, vermicomposting has developed into a sustainable waste management approach. Earthworms such as *Eisenia fetida* aid in this biological process, which breaks down organic materials into compost that is rich in nutrients. Its versatility is demonstrated by its capacity to respond to a variety of biodegradable wastes, including agricultural residues like rice straw and paper trash (Sharma et al., 2018). Vermicomposting can stabilize and enrich wastes with vital nutrients including nitrogen, potassium, and phosphorus while lowering harmful elements (Tammam et al., 2023). The technology is an essential instrument for solving solid waste problems because of its affordability, ease of use, and environmental friendliness.

2.1 Traditional and Current Waste Management Methods

As an alternative to conventional waste management techniques like burning agricultural wastes, vermicomposting uses earthworms to turn organic waste into nutrient-rich compost, which helps to reduce pollution in the environment (Dey et al., 2023; Mishra et al., 2019, 2023). Producing vermi-fertilizer, which is rich in vital nutrients and advantageous microorganisms, this technique improves soil fertility and crop yield (Ahmed et al., 2019). Vermicomposting works well at lower temperatures, speeding up decomposition and encouraging microbial variety in the soil, in contrast to traditional composting, which needs higher temperatures and longer processing times to eradicate pathogens (Karimi et al., 2024; Sivasabari et al., 2023). Vermicomposting is a vital tactic for sustainable agriculture and environmental preservation since it is a socially, economically, and environmentally acceptable substitute for artificial fertilizers (Ahmad et al., 2021).

Vermicomposting, which blends conventional composting with contemporary ecological techniques, has become a successful and sustainable waste management method. Vermicomposting has improved soil health and crop productivity by supplying vital plant growth hormones and increasing microbial activity in the soil. Previously, most organic waste management techniques were ineffective. Present-day methods demonstrate the value of vermiculture technology in addressing soil degradation and agricultural residue management, allowing farmers to recycle household and agricultural waste into useful fertilizers that promote sustainable agriculture. Vermiculture technology has emerged as a top option for managing organic waste, and this move toward earthworm-mediated solutions is indicative of a growing trend of implementing eco-friendly technologies to lessen the environmental effects associated with traditional farming practices (Singh et al., 2020).

Vermicomposting has developed into an important waste management technique that successfully tackles both conventional and modern organic waste disposal issues. Physical and chemical methods were used in the past, but they were frequently expensive and time-consuming. Recent research has demonstrated the effectiveness of vermicomposting with the exotic earthworm species *Eisenia foetida*, which speeds up the conversion of a variety of organic substrates—such as industrial sludge and kitchen waste into nutrient-rich vermicompost (Gebrehana et al., 2023). This approach has been shown to improve soil health and sustainability by lowering total organic carbon levels and increasing nutrient availability, as evidenced by increases in total Kjeldhal nitrogen (TKN) of 4.4–5.8 times (Garg et al., 2006). The results highlight vermicomposting as a practical and affordable way to manage organic waste in contemporary settings, especially in places like India where waste management is essential for the wellbeing of the ecosystem.

With benefits surpassing conventional composting, vermicomposting is a sustainable waste management technique that breaks down organic waste using earthworms. Other methods such as thermophilic composting, a time-consuming process that can result in nutrient loss and uneven material quality, are frequently used in traditional methods. Vermicomposting, on the other hand, speeds up the breakdown process, enabling year-round compost production. To increase the effectiveness of decomposition, methods like raised bed systems and windrows are frequently used, in which earthworms are stacked on top of organic debris (Thirunavukkarasu et al., 2023). In addition to creating nutrient-rich compost, this technique lessens the need for chemical fertilizers, reducing pollution and enhancing soil health (Stiles, 2017). The cost-effectiveness and environmental benefits of vermicomposting make it superior to all other waste management techniques (Manyuchi & Phiri, 2013).

Composting and vermicomposting offer sustainable alternatives to conventional waste management techniques like landfilling and burning, which pollute the environment. Table 1 compares vermicomposting and traditional composting concerning parameters like decomposition time, nutrient content, organic matter, microbial activity and pH balance. Through microbial breakdown, composting turns organic waste into a humus that is rich in nutrients. This process is facilitated by vermicomposting, which turns organic matter into stable, nutrient-dense vermicompost with the help of earthworms. A contemporary adoption optimizes waste degradation

Table 1 Comparison of vermicomposting and traditional composting

Parameter	Vermicomposting	Traditional composting
Decomposition time	4–6 weeks	3–6 months
Nutrient content	Higher	Moderate
Organic matter	Higher	Lower
Microbial activity	Enhanced	Moderate
pH balance	Neutral-stable	Fluctuating

and nutrient recycling by combining the two approaches. By raising the quantities of nitrogen and phosphorus and drastically lowering the carbon content, this integrated method produces a premium, odorless, granular fertilizer that may be used to improve soil. These developments are prime examples of environmentally beneficial waste management techniques, especially when it comes to agricultural waste such as pistachio trash (Esmaeili et al., 2020).

Due to hazardous emissions or residues, conventional waste management techniques including burning, landfilling, and chemical treatments frequently cause environmental damage (Demirbas, 2011; Siddiqua et al., 2022). Vermicomposting is a contemporary technique that turns organic and greasy waste into high-quality compost by using earthworms. This method efficiently breaks down hydrocarbons and creates nutrient-rich material that can be used as a soil improver by combining worm activity with microbial processes. Vermicomposting oily drill cuttings, for example, has shown lower greenhouse gas emissions and operating costs than thermal treatments in environmentally sensitive Arctic regions while still meeting strict environmental regulations. Its usefulness as an eco-friendly, sustainable substitute is highlighted by this innovation (Sorheim et al., 2007).

Conventional waste management techniques, such as chemical treatment, incineration, and landfilling, are frequently ineffective and harmful to the environment. By using earthworms to convert organic waste into nutrient-rich compost, vermicomposting provides a sustainable substitute that improves soil health (Hajam et al., 2023). Neural networks have recently been included in vermicomposting to forecast waste composition, maximize productivity, and guarantee sustainability. This method improves resource recovery and waste separation accuracy when paired with chemical processing methods such as the Gallium Arsenide Schema. These developments demonstrate vermicomposting's promise as an environmentally responsible response to contemporary waste management issues (Vadivel et al., 2024).

3 Principles of Vermicomposting

Vermicomposting is a biotechnological process that turns organic matter into nutrient-rich vermicompost by using earthworms, especially species like *E. fetida*. In addition to improving soil fertility, this procedure aids in waste management and environmental sustainability. The following sections outline the key principles and factors influencing vermicomposting:

3.1 Biological and Chemical Processes

Vermicomposting relies on the synergistic action of earthworms and microorganisms to convert organic waste into vermicompost. Earthworms ingest organic materials, which are then digested in their gut, where microbial activity plays a crucial role in breaking down complex organic compounds. The resulting castings are rich in nutrients and beneficial microorganisms, enhancing soil health and plant growth (Enebe & Erasmus, 2023).

3.2 Key Factors Influencing Vermicomposting

3.2.1 Temperature

The optimal temperature range for vermicomposting is between 20–25 °C as shown in Fig. 2. This temperature range supports the metabolic activities of both earthworms and microorganisms, facilitating efficient decomposition (Zhang et al., 2020).

3.2.2 Moisture

Maintaining moisture levels between 60–70% is critical for the survival of earthworms and microbial activity (Fig. 2. Excessive moisture can lead to anaerobic conditions, while insufficient moisture can stress the organisms involved (Adhikary, 2012; Pathma & Sakthivel, 2012).

3.2.3 Aeration

Proper aeration is essential for microbial activity and decomposition. Earthworms help aerate the substrate as they burrow through it, promoting aerobic conditions that enhance the breakdown of organic matter (Kooch & Kuzyakov, 2024).

3.2.4 Feedstock

The selection of appropriate biodegradable materials, such as kitchen waste, agricultural residues, and livestock manure, is vital for successful vermicomposting (Fig. 2).

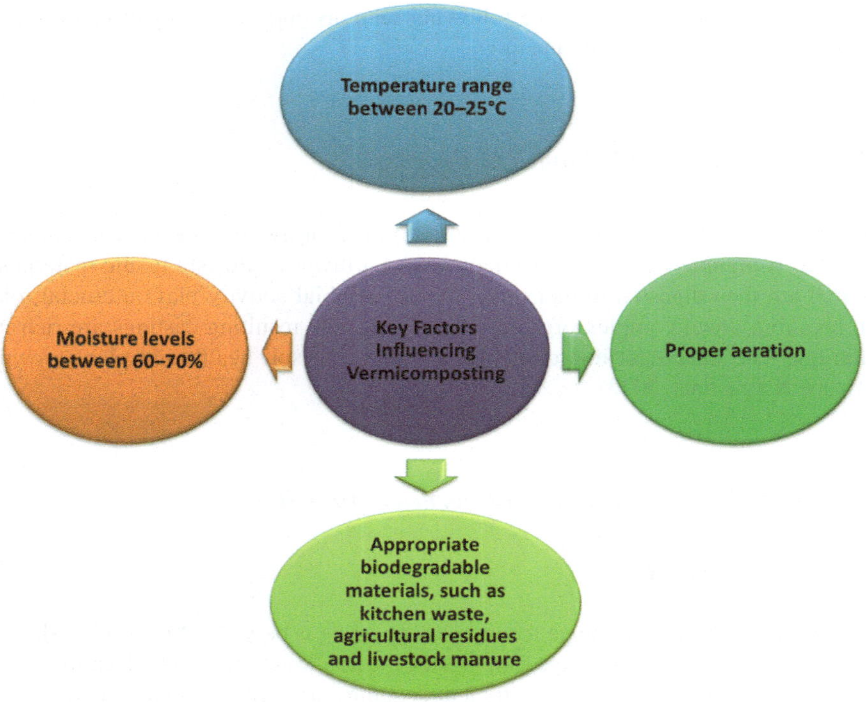

Fig. 2 Key factors influencing vermicomposting

These materials provide the necessary nutrients for earthworms and microorganisms (Yuvaraj et al., 2021).

3.3 Nutrient Composition and Benefits

Nitrogen (N), phosphorus (P), potassium (K), and other micronutrients are among the many nutrients that make up vermicompost. According to studies, vermicompost greatly increases plant nutrient availability and can have up to five times the amount of nitrogen accessible in traditional potting soils (Adhikary, 2012; Vyas et al., 2022). Furthermore, vermicompost's organic matter composition enhances soil structure, boosts water retention, and fosters the growth of advantageous microbial communities (Pathma & Sakthivel, 2012; Pereira et al., 2022). Table 2 lists the nutritional makeup and advantages of vermicompost.

Note The type of earthworm, feedstock, and ambient conditions all affect the nutrients in vermicompost. Although NPK values are well-established, developing

Table 2 Nutrient composition and benefits of vermicompost

Nutrient/ Parameter	Typical range	Description	Benefits	Reference
Organic carbon	9.71–31.40%	Shows compost maturity and decomposition.	Increases the amount of organic matter in the soil, enhancing its fertility and texture.	Jadhav et al. (2024)
Total nitrogen	1.847–2.31%	Increased as a result of microbial activity and mineralization	Promotes plant growth and improves nitrogen availability in soil	Lim et al. (2015)
C: N ratio	< 20	Well-decomposed compost is indicated by a lower ratio (< 12)	Assures plants of quick access to nutrients	Kumar et al. (2023)
Phosphorus	1.185%	Released through microbial activity from organic matter	Improves flowering and root growth	Jadhav et al. (2024)
Potassium	0.863%	Increased as a result of organic wastes breaking down	Increases a plant's capacity to withstand drought and disease	Lim et al. (2015)
Calcium	–	Derived from organic materials that have decomposed	Increases the pH of the soil and fortifies plant cell walls	Kumar et al. (2023)
Magnesium	–	Necessary for the production of chlorophyll	Aids in the activation of enzymes and photosynthesis	Jadhav et al. (2024)
Micronutrients	–	Consists of copper, manganese, zinc, and iron	Promotes disease resistance and plant metabolism	Lim et al. (2015)

a common range for calcium, magnesium, and micronutrient concentrations is challenging since they vary greatly.

3.4 Recent Advancements

The use of lime and microbial inoculants to improve the nutrient content and decomposition rates in vermicomposting systems has been studied recently. These developments are intended to enhance the overall quality of the vermicompost produced and boost the effectiveness of nutrient cycling (Vyas et al., 2022; Pereira et al., 2022).

3.5 Environmental Impact

Vermicomposting is a well-known environmentally friendly waste management technique that helps recycle organic waste and lower pollution levels in the environment. The method is a sustainable agricultural strategy that can efficiently handle different kinds of organic waste and detoxify industrial wastes (Vyas et al., 2022; Zazouli et al., 2021).

4 Waste Categories and Suitability for Vermicomposting

Vermicomposting provides a sustainable waste management option by effectively processing a broad range of organic wastes. Because of their high organic content and degradability, agricultural leftovers like manure, straw, and husks provide good feedstock (Yadav & Garg, 2019). For instance, research has shown that pineapple waste, a significant byproduct in Accra, Ghana, can be successfully vermicomposted using native earthworms (*Eudrilus eugeniae*). In a study, pineapple pulp and peels demonstrated remarkable mass loss-99% and 87%, respectively, over 20 weeks, indicating the high efficiency of vermicomposting for such agricultural residues (Mainoo et al., 2009). Similarly, food waste such as the peels from fruits and vegetables offers earthworms a nutrient-rich substrate, improving the vermicompost's quality (Pramanik et al., 2007). Yard garbage and biodegradable home waste are examples of municipal organic waste that is a feasible category that promotes effective decomposition and reduces landfill contributions (Ducasse et al., 2022). Furthermore, when processed and mixed with other organic materials such as industrial waste, paper mill sludge and bakery residues vermicompost can be converted into nutrient-rich material (Gupta et al., 2024).

There are limits to the types of material that can be vermicomposted. To avoid contamination and guarantee process safety, materials including plastics, metals, and hazardous chemicals must be kept out (Celik et al., 2024; Goswami et al., 2024; Wang et al., 2024). Maintaining the proper environmental parameters, such as pH, moisture, and temperature, determines the effectiveness of vermicomposting (Manzoor et al., 2024; Zhou et al., 2021). The viability of various waste types for vermicomposting has been further improved by advancements in waste pretreatment and the addition of microbial inoculants, expanding its use in industrial and agricultural settings (Vyas et al., 2022). Vermicomposting provides a flexible and environmentally responsible method of handling organic waste sustainably by using a variety of biodegradable wastes.

5 Process and Methodology of Vermicomposting

The first step in the vermicomposting process is choosing the right earthworm species. Because they can effectively break down organic waste and adjust to different environments, commonly utilized species like *Eisenia fetida* and *Eudrilus eugeniae* are favored (Ahmad et al., 2021). The first stage is bedding preparation, which involves layering materials such as straw, coir pith, or dry leaves to provide an earthworm-friendly environment. The system is then routinely supplemented with organic waste, such as agricultural residues, animal dung, and vegetable scraps. Ensuring that the waste is diced and partially degraded increases the efficiency of vermicomposting because it is easier for the earthworms to digest (Varma et al., 2015). The process of vermicomposting is depicted in Fig. 3.

5.1 *Innovations in Vermicomposting*

Vermicomposting is becoming more effective and accessible due to recent developments. Critical parameters like temperature, moisture, and pH are now tracked in

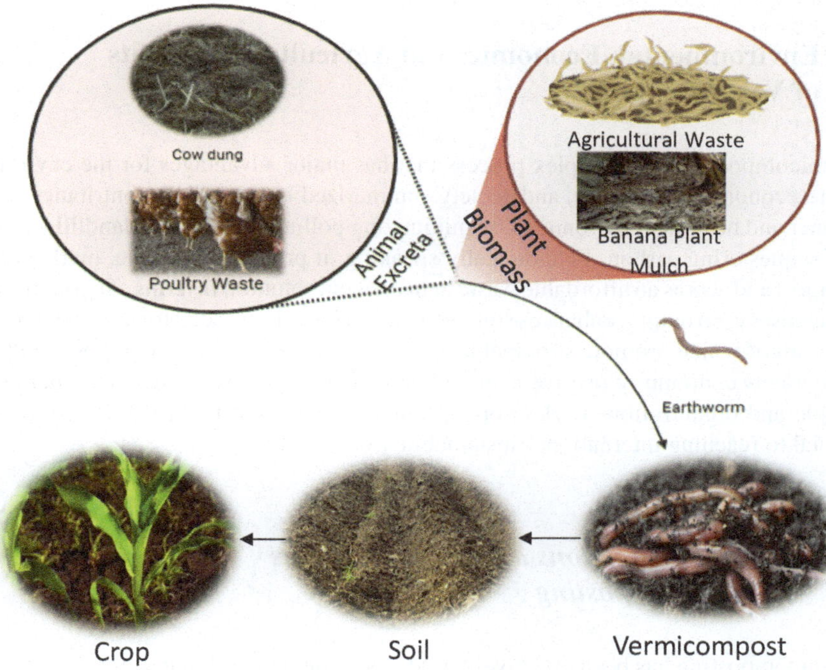

Fig. 3 Process of vermicomposting

real time by automated monitoring systems that make use of Internet of Things (IoT) sensors (Das et al., 2022). Additionally, modular configurations have been created, providing scalable and space-efficient designs suitable for both urban and rural settings. Large-scale adoption and use of vermicomposting systems are made possible by these advancements, which simplify the procedure and increase its adaptability to various situations (Bellitürk et al., 2020).

5.2 Harvesting and Compost Application

In 45 to 60 days, the vermicompost is often suitable for harvesting. When compost reaches maturity, it is granular, dark brown, and odorless. By being isolated from the compost, earthworms can be employed again in later batches (Kauser & Khwairakpam, 2022). Vermicompost is nutrient-rich and includes growth-promoting chemicals, helpful microorganisms, and important macro- and micronutrients. It has been demonstrated to greatly increase soil health and crop output, which benefits crops including rice, tomatoes, and maize (Aslam et al., 2020). Vermicomposting is a sustainable approach to waste management and agricultural productivity since it improves soil fertility and lessens the reliance on artificial fertilizers.

6 Environmental, Economic, and Agricultural Benefits of Vermicomposting

Vermicomposting is a complex process that has major advantages for the environment, economy, agriculture, and society summarized in Table 3. It contributes to a cleaner and healthier environment by minimizing pollutants, reducing landfill waste, and sequestering carbon. Economically speaking, it produces valuable, marketable compost and acts as an affordable waste management solution. In terms of agriculture, it increases crop output, enhances soil fertility, and lessens the need for artificial fertilizers, all of which promote sustainable agricultural methods. Vermicomposting also encourages community involvement and eco-friendly practice education, enabling people and organizations to develop sustainable behaviors. Its broad use could be crucial to reaching international sustainability objectives.

6.1 Global Applications and Success Stories of Vermicomposting

Vermicomposting has been effectively used in several international settings, demonstrating its adaptability and efficiency. Vermicomposting has been used in large-scale

Table 3 Environmental, economic, and agricultural benefits of vermicomposting

Benefit category	Specific benefits	Reference
Environmental	• Decreases greenhouse gas emissions: When compared to conventional composting techniques, vermicomposting produces noticeably lower amounts of CO_2, CH_4, and N_2O. Increases carbon sequestration: Vermicompost application raises the organic carbon content of the soil, which encourages carbon sequestration	Katiyar et al. (2023)
	• Reduces environmental pollution: By turning organic waste into useful compost, vermicomposting lessens the need for chemical fertilizers and the risks they provide to the environment	Fernando & Arunakumara (2021)
Economic	• Economical waste management: Vermicomposting provides a cost-effective substitute for turning organic waste into useful compost. Marketable products: The compost generated can be offered for sale as a fertilizer that is high in nutrients, generating extra revenue	Katiyar et al. (2023)
	• Potential for microbusinesses: By empowering farmers with limited resources economically, vermicomposting can be transformed into community-based microbusinesses	Fernando & Arunakumara (2021)
Agricultural	• Boosts soil fertility: Vermicompost improves soil health and structure by being a rich source of beneficial microbes and necessary nutrients. Enhances crop yield: Vermicompost application has been linked to enhanced plant growth and increased crop yields	Katiyar et al. (2023)
	• Lessens the demand for chemical fertilizers: Vermicompost's high nutrient content lessens the requirement for artificial fertilizers, supporting sustainable farming methods	Fernando & Arunakumara (2021)

agricultural projects in India to improve soil health and decrease organic waste, proving its advantages for farming and rural people (Sharma & Garg, 2019). To lessen the reliance on landfills and encourage sustainable waste disposal, municipal programs in the US have incorporated vermicomposting into waste management systems (Edwards & Arancon, 2022). Recycling organic sludges is the main industrial application in Europe, helping to preserve the environment and recover resources (Kominko et al., 2024).

This highlights how vermicomposting can be used in a variety of industries and geographical areas. It promotes sustainable practices globally by tackling environmental and agricultural issues. Furthermore, it is a workable solution for waste management globally because of its adaptability to various cultural and economic contexts. These achievements serve as a template for other countries looking to implement environmentally friendly waste management techniques. Vermicomposting's extensive use demonstrates how it might propel sustainability initiatives around the world.

7 Future Perspectives and Research Directions

Vermicomposting stands to gain a great deal from new technologies and creative methods in future. By forecasting waste composition and controlling environmental factors in real time, integrating AI-driven monitoring systems could improve the decomposition processes. There is potential for improving composting productivity through precision vermiculture, which uses selective breeding to increase earthworm efficiency. Additionally, vermicomposting, anaerobic digestion and the creation of biochar may work in concert to increase resource recovery and lower greenhouse gas emissions (Vyas et al., 2022).

These developments have the potential to make vermicomposting a vital component of sustainable waste management, especially as the globe struggles with climate change. The capacity to scale these innovations in a variety of contexts highlights how vermicomposting might help achieve global sustainability objectives. To fully utilize vermicomposting and meet the increasing demand for environmentally friendly waste management solutions, research in these fields is essential.

8 Conclusion

Vermicomposting has become a game-changing method for improving soil health and managing waste sustainably. It tackles important environmental issues like pollution reduction, resource conservation, and carbon sequestration by using earthworms to turn organic waste into nutrient-rich compost. Because of its versatility, it can be used in industrial, municipal, and agricultural settings to process a variety of organic materials. Vermicomposting increases crop output, improves soil fertility, and lessens reliance on chemical fertilizers when included in farming operations. Precision vermiculture and AI-driven monitoring are two examples of technological developments that promise to further enhance the procedure and increase its scalability and efficiency.

Case studies from all around the world, such as those from the US, Europe, and India, show how beneficial it is in a variety of settings, from municipal waste management to large-scale agriculture. Additionally, vermicomposting encourages education and community involvement, building a sustainable culture. Vermicomposting's contribution to reducing environmental effects and encouraging regenerative activities is becoming more and more important as climate change intensifies.

This economical and environmentally beneficial strategy is essential to reaching global sustainability objectives.

Acknowledgements The authors thank the University of South Africa (UNISA), Florida Campus, Johannesburg, South Africa, and Bishop Heber College (Autonomous), Tiruchirappalli, Tamil Nadu, India for their support.

References

Abad, Q., & Shafiqi, S. (2024). Vermicompost: Significance and benefits for agriculture. *Journal for Research in Applied Sciences and Biotechnology, Afeganistao, 3*(2), 202–207.

Adhikary, S. (2012). Vermicompost, the story of organic gold: A review. *Agricultural Sciences, 3*(7), 905–917.

Ahmad, A., Aslam, Z., Belliturk, K., Iqbal, N., Naeem, S., Idrees, M., & Kamal, A. (2021). Vermicomposting methods from different wastes: An environment friendly, economically viable and socially acceptable approach for crop nutrition: A review. *International Journal of Food Science and Agriculture, 5*(1), 58–68.

Ahmed, M., Ahmad, S., Qadir, G., Hayat, R., Shaheen, F. A., & Raza, M. A. (2019). Innovative processes and technologies for nutrient recovery from wastes: A comprehensive review. *Sustainability, 11*(18), 4938.

Alshehrei, F., & Ameen, F. (2021). Vermicomposting: A management tool to mitigate solid waste. *Saudi Journal of Biological Sciences, 28*(6), 3284–3293.

Aslam, Z., Ahmad, A., Belliturk, K., Iqbal, N., Idrees, M., Rehman, W. U., & Rehman, S. (2020). Effects of vermicompost, vermi-tea and chemical fertilizer on morpho-physiological characteristics of tomato (*Solanum lycopersicum*) in Suleymanpasa District, Tekirdag of Turkey. *Pure and Applied Biology (PAB), 9*(3), 1920–1931.

Belliturk, K., Aslam, Z., Ahmad, A., & Rehman, S. U. (2020). Alteration of physical and chemical properties of livestock manures by *Eisenia fetida* (Savigny, 1926) and developing valuable organic fertilizer. *Journal of Innovative Sciences, 6*(1), 47–53.

Biswal, P., Ghosh, S., Pal, A., & Das, A. P. (2021). Exploration of probiotic microbial biodiversity in acidic environments (curd) and their futuristic pharmaceutical applications. *Geomicrobiology Journal*. https://doi.org/10.1080/01490451.2021.1956020

Celik, A., Belliturk, K., & Baran, M. F. (2024). Management of Soil and Water Contaminated with heavy metals with sustainable green technologies: The case of vermicomposting. In N. Kumar (Ed.), *Heavy metal remediation: Sustainable nexus approach* (pp. 65–93). Springer Nature.

Das, A. K., Rout, S. K., Dash, S. K., & Mangaraj, A. (2022). Internet of things and cloud computing for smart vermicomposting by using *Eisenia fetida* and its optimization by ANN. In J. Dinesh Peter, S. L. Fernandes, & A. H. Alavi (Eds.), *Disruptive Technologies for big Data and Cloud Applications: Proceedings of ICBDCC 2021* (pp. 375–387). Springer Nature.

Das, A. P., Mallavarapu, M., & Ghosh, S. (2023). Ecotoxicity of mining pollutants on the environment and their remediation. *Environmental Chemistry and Ecotoxicology, 5*, 165–167. https://doi.org/10.1016/J.ENCECO.2023.08.002

Demirbas, A. (2011). Waste management, waste resource facilities and waste conversion processes. *Energy Conversion and Management, 52*(2), 1280–1287.

Dey, S., Das, A., Mallick, K., Sahu, A., & Das, A. P. (2023). Environmental petroleum waste: Pollution. *Toxicity, Sustainable Remediation.*, 159–175. https://doi.org/10.1007/978-3-031-48220-5_7

Ducasse, V., Capowiez, Y., & Peigne, J. (2022). Vermicomposting of municipal solid waste as a possible lever for the development of sustainable agriculture. A review. *Agronomy for Sustainable Development, 42*(5), 89.

Edwards, C. A., & Arancon, N. Q. (2022). The use of earthworms in organic waste management and vermiculture. In C. A. Edwards & N. Q. Arancon (Eds.), *Biology and ecology of earthworms* (pp. 467–527). Springer Nature.

Enebe, M. C., & Erasmus, M. (2023). Mediators of biomass transformation–a focus on the enzyme composition of the vermicomposting process. *Environmental Challenges, 12*, 1732.

Esmaeili, A., Khoram, M. R., Gholami, M., & Eslami, H. (2020). Pistachio waste management using combined composting-vermicomposting technique: Physico-chemical changes and worm growth analysis. *Journal of Cleaner Production, 242*, 1523.

Fernando, K. M. C., & Arunakumara, K. K. I. U. (2021). Sustainable organic waste management and nutrients replenishment in the soil by vermicompost: A review. *AGRIEAST: Journal of Agricultural Sciences, 15*(2), 32–51.

Garg, P., Gupta, A., & Satya, S. (2006). Vermicomposting of different types of waste using *Eisenia foetida*: A comparative study. *Bioresource Technology, 97*(3), 391–395.

Gebrehana, Z. G., Gebremikael, M. T., Beyene, S., Sleutel, S., Wesemael, W. M., & De Neve, S. (2023). Organic residue valorization for Ethiopian agriculture through vermicomposting with native (*Eudrilus eugeniae*) and exotic (*Eisenia fetida* and *Eisenia andrei*) earthworms. *European Journal of Soil Biology, 116*, 103488.

Ghosh, S., Sinha, J. K., Ghosh, S., Vashisth, K., Han, S., & Bhaskar, R. (2023). Microplastics as an emerging threat to the global environment and human health. *Sustainability, 15*(14), 10821. https://doi.org/10.3390/SU151410821

Goswami, J., Deka, M. K., Devi, R., Sarma, P. P., & Brahma, A. (2024). A review on optimizing organic waste management and income generation through vermicomposting and AI-powered vermicomposting: Insights from Guwahati, Assam. *International Journal, 13*(6).

Gupta, R., Mago, M., & Garg, V. K. (2024). Infusion of fruit-vegetable waste biomass into bakery sludge to enhance nutrient availability and vermicompost maturity using *Eisenia fetida*. *Biomass Conversion and Biorefinery, 14*(15), 7851–7862.

Hajam, Y. A., Kumar, R., & Kumar, A. (2023). Environmental waste management strategies and vermi transformation for sustainable development. *Environmental Challenges, 13*, 1747.

Jadhav, A. B., Gosavi, A. B., Jadhav, S. A., & Patil, A. V. (2024). Influence of different substrates on nutrient composition of Vermicompost and Vermiwash. *International Journal of Plant Soil Science, 36*(6), 340–352.

Karimi, S., Raza, T., & Mechri, M. (2024). Composting and Vermitechnology in organic waste management. In V. Kumar, S. A. Bhat, S. Kumar, & P. Verma (Eds.), *Environmental engineering and waste management: Recent trends and perspectives* (pp. 449–470). Springer Nature Switzerland.

Katiyar, R. B., Sundaramurthy, S., Sharma, A. K., Arisutha, S., Pratap-Singh, A., Mishra, S., & Khan, M. A. (2023). Vermicompost: An eco-friendly and cost-effective alternative for sustainable agriculture. *Sustainability, 15*(20), 4701.

Kauser, H., & Khwairakpam, M. (2022). Organic waste management by two-stage composting process to decrease the time required for vermicomposting. *Environmental Technology & Innovation, 25*, 2193.

Kominko, H., Gorazda, K., & Wzorek, Z. (2024). Sewage sludge: A review of its risks and circular raw material potential. *Journal of Water Process Engineering, 63*, 105522.

Kooch, Y., & Kuzyakov, Y. (2024). Earthworms for soil organic matter mineralization and carbon sequestration. In Y. Kooch & Y. Kuzyakov (Eds.), *Earthworms and ecological processes* (pp. 373–394). Springer Nature Switzerland.

Kumar, R., Jha, S., Singh, S. P., Kumar, M., Kumari, R., & Padbhushan, R. (2023). Organic waste recycling by vermicomposting amended with rock phosphate impacts the stability and maturity indices of vermicompost. *Journal of the Air and Waste Management Association, 73*(7), 553–567.

Lim, S. L., Wu, T. Y., Lim, P. N., & Shak, K. P. Y. (2015). The use of vermicompost in organic farming: Overview, effects on soil and economics. *Journal of the Science of Food and Agriculture, 95*(6), 1143–1156.

Lirikum, L. N. K., Thyug, L., & Mozhui, L. (2022). Vermicomposting: An eco-friendly approach for waste management and nutrient enhancement. *Tropical Ecology, 63*(3), 325–337.

Mainoo, N. O., Barrington, S., Whalen, J. K., & Sampedro, L. (2009). Pilot-scale vermicomposting of pineapple wastes with earthworms native to Accra. *Ghana. Bioresource Technology, 100*(23), 5872–5875.

Manchal, R., Venuste, T., & Verma, S. R. (2023). Vermicomposting, a key to sustainable agriculture: A review. *Farming and Management, 8*(2), 81–93.

Manyuchi, M. M., & Phiri, A. (2013). Vermicomposting in solid waste management: A review. *International Journal of Scientific Engineering and Technology, 2*(12), 1234–1242.

Manzoor, A., Naveed, M. S., Ali, R. M. A., Naseer, M. A., Ul-Hussan, M., Saqib, M., et al. (2024). Vermicompost: A potential organic fertilizer for sustainable vegetable cultivation. *Scientia Horticulturae, 336*, 113443.

Mishra, S., Dash, D., Al-Tawaha, A. R. M. S., & Das, A. P. (2022). A review on heavy metal ion adsorption on synthetic microfiber surface in aquatic environments. *Applied Biochemistry and Biotechnology, 194*(10), 4639–4654. https://doi.org/10.1007/S12010-022-04029-W/METRICS

Mishra, S., Ghosh, S., van Hullebusch, E. D., Singh, S., & Das, A. P. (2023). A critical review on the recovery of base and critical elements from electronic waste-contaminated streams using microbial biotechnology. *Applied Biochemistry and Biotechnology, 2023*, 1–30. https://doi.org/10.1007/S12010-023-04440-X

Mishra, S., Rout, P. K., & Das, A. P. (2019). Solar photovoltaic panels as next generation waste: A review. *Biointerface Research in Applied Chemistry, 9*(6).

Oyege, I., & Balaji Bhaskar, M. S. (2023). Effects of vermicompost on soil and plant health and promoting sustainable agriculture. *Soil Systems, 7*(4), 101.

Pathma, J., & Sakthivel, N. (2012). Microbial diversity of vermicompost bacteria that exhibit useful agricultural traits and waste management potential. *Springerplus, 1*, 1–19.

Pereira, M. M. A., Moraes, L. C., Mogollón, M. C. T., Borja, C. J. F., Duarte, M., Buttros, V. H. T., & Doria, J. (2022). Cultivating biodiversity to harvest sustainability: Vermicomposting and inoculation of microorganisms for soil preservation and resilience. *Agronomy, 13*(1), 103.

Pramanik, P., Ghosh, G. K., Ghosal, P. K., & Banik, P. (2007). Changes in organic–C, N, P and K and enzyme activities in vermicompost of biodegradable organic wastes under liming and microbial inoculants. *Bioresource Technology, 98*(13), 2485–2494.

Purnama, I., Azis, R. A., & Rizal, M. (2025). Integrating organic manure and natural phosphate for sustainable long bean (Vigna sinensis L.) cultivation on marginal soils. *Explora: Environment and Resource, 2*(1), 8348.

Raj, K., & Das, A. P. (2023). Lead pollution: Impact on environment and human health and approach for a sustainable solution. *Environmental Chemistry and Ecotoxicology, 5*, 79–85. https://doi.org/10.1016/J.ENCECO.2023.02.001

Raza, S. T., Wu, J., Rene, E. R., Ali, Z., & Chen, Z. (2022). Reuse of agricultural wastes, manure, and biochar as an organic amendment: A review on its implications for vermicomposting technology. *Journal of Cleaner Production, 36*, 2200.

Sande, T. J., Tindwa, H. J., Alovisi, A. M. T., Shitindi, M. J., & Semoka, J. M. (2024). Enhancing sustainable crop production through integrated nutrient management: A focus on vermicompost, bio-enriched rock phosphate, and inorganic fertilizers–a systematic review. *Frontiers in Agronomy, 6*, 2876.

Sharma, K., & Garg, V. K. (2018). Comparative analysis of vermicompost quality produced from rice straw and paper waste employing earthworm *Eisenia fetida* (Sav.). *Bioresource Technology, 250*, 708–715.

Sharma, K., & Garg, V. K. (2019). Recycling of lignocellulosic waste as vermicompost using earthworm *Eisenia fetida. Environmental Science and Pollution Research, 26*, 4024–4035.

Siddiqua, A., Hahladakis, J. N., & Al-Attiya, W. A. K. (2022). An overview of the environmental pollution and health effects associated with waste landfilling and open dumping. *Environmental Science and Pollution Research, 29*(39), 58514–58536.

Singh Jadaun, J., Bansal, S., Sonthalia, A., Rai, A. K., & Singh, S. P. (2022). Biodegradation of plastics for sustainable environment. *Bioresource Technology*, 347.

Singh, A., Karmegam, N., Singh, G. S., Bhadauria, T., Chang, S. W., Awasthi, M. K., & Ravindran, B. (2020). Earthworms and vermicompost: An eco-friendly approach for repaying nature's debt. *Environmental Geochemistry and Health, 42*, 1617–1642.

Sivasabari, K., Parthasarathy, S., Chandran, D., Sankaralingam, S., & Ajaykumar, R. (2023). Vermicompost and Vermi-leachate in Pest and disease management. In H. A. Mupambwa, L. N. Horn, & P. N. S. Mnkeni (Eds.), *Vermicomposting for sustainable food Systems in Africa* (pp. 279–304). Singapore.

Sorheim, R., Paulsen, J. E., Saasen, A., Leleux, J., Albouy, A., Haraldsen, T. K., & Linjordet, R. (2007). Experimental study of composting oil wet drill cuttings as a drilling waste disposal option. *Journal of Energy Resources Technology, 129*(4), 307–313.

Stiles, W. (2017). Vermicomposting: An alternative method for organic waste management.

Tammam, A. A., Shehata, R. A. M., Pessarakli, M., & El-Aggan, W. H. (2023). Vermicompost and its role in alleviation of salt stress in plants–I. Impact of vermicompost on growth and nutrient uptake of salt-stressed plants. *Journal of Plant Nutrition, 46*(7), 1446–1457.

Thirunavukkarasu, A., Sivashankar, R., Nithya, R., Sathya, A. B., Priyadharshini, V., Kumar, B. P., & Krishnamoorthy, S. (2023). Sustainable organic waste management using vermicomposting: A critical review on the prevailing research gaps and opportunities. *Environmental Science: Processes and Impacts, 25*(3), 364–381.

Vadivel, T., Barathi, K., Arulkumaran, G., Anand, M. B., & Cherubini, C. (2024). Wastewater management using a neural network-assisted novel paradigm for waste prediction from vermicomposting. *Water, 16*(23), 3450.

Varma, V. S., Yadav, J., Das, S., & Kalamdhad, A. S. (2015). Potential of waste carbide sludge addition on earthworm growth and organic matter degradation during vermicomposting of agricultural wastes. *Ecological Engineering, 83*, 90–95.

Vyas, P., Sharma, S., & Gupta, J. (2022). Vermicomposting with microbial amendment: Implications for bioremediation of industrial and agricultural waste. *Biotechnologia, 103*(2), 203–215.

Walia, S. S., & Kaur, T. (2024). Preparation of Vermicompost. In *Earthworms and vermicomposting: Species, procedures and crop application* (pp. 73–87). Springer Nature.

Wang, F., Zhang, Y., Su, Y., Wu, D., & Xie, B. (2024). Pollutant control and nutrient recovery of organic solid waste by earthworms: Mechanism and agricultural benefits of vermicomposting. *Journal of environmental Chemical Engineering*, 112610.

Yadav, A., & Garg, V. K. (2019). Biotransformation of bakery industry sludge into valuable product using vermicomposting. *Bioresource Technology, 274*, 512–517.

Yuvaraj, A., Thangaraj, R., Ravindran, B., Chang, S. W., & Karmegam, N. (2021). Centrality of cattle solid wastes in vermicomposting technology–a cleaner resource recovery and biowaste recycling option for agricultural and environmental sustainability. *Environmental Pollution, 268*, 5688.

Zazouli, M. A., Hezarjaribi, H. Z., Charati, J. Y., Ahmadnezhad, A., Mahvi, A. H., & Asgharnia, H. (2021). Investigating the effects of vermicomposting process using *Eisenia Fetida* earthworms on the reduction of parasites population. *Journal of Environmental Health Science and Engineering, 19*, 1623–1633.

Zhang, H., Li, J., Zhang, Y., & Huang, K. (2020). Quality of vermicompost and microbial community diversity affected by the contrasting temperature during vermicomposting of dewatered sludge. *International Journal of Environmental Research and Public Health, 17*(5), 1748.

Zhou, Y., Zhang, D., Zhang, Y., Ke, J., Chen, D., & Cai, M. (2021). Evaluation of temperature on the biological activities and fertility potential during vermicomposting of pig manure employing *Eisenia fetida*. *Journal of Cleaner Production, 302*, 6804.

Vermicomposting for Sustainable Resource Recovery and Environmental Protection

Himanshu, Deepika Saxena, Mamta Sharma, and Ankit

Abstract Vermicomposting is an environmentally friendly system of waste management through which earthworms process organic waste into compost containing nutrients inherent in the worms' abilities. It combats many environmental issues the world faces today by reducing the emission of greenhouse gases and soil and water pollution and also reducing the amount of organic waste thrown in the landfills. Organic wastes are efficiently broken down by earthworms making composting quicker. Vermicompost has good qualities for adding to horticulture and agriculture because it contains plenty of organic material, beneficial microorganisms, and useful minerals. This chapter provides an overview of the concepts, steps, and benefits of vermicomposting with a focus on the potential application of this technology in sustainable waste management. With a view of enhancing efficiency, some of the vital areas, such as selecting appropriate earthworm species, the suitable environmental conditions favorable for earthworms, and composting substrates, are discussed comprehensively. Besides, the chapter also discusses how resource recovery and contributing to the enhancement of the soil quality illustrate the economic and environmental advantages of vermicomposting.

Keywords Organic waste management · Nutrient-rich compost · Sustainable practices · Soil amendment

Himanshu · D. Saxena (✉) · Ankit
Department of Horticulture (Fruit Science), Lovely Professional University, Phagwara, Punjab, India
e-mail: deepika.20662@lpu.co.in

Himanshu
e-mail: rohillah86@gmail.com

Ankit
e-mail: aujlaankitaujla@gmail.com

M. Sharma
Department of Biotechnology, School of Bioengineering and Biosciences, Lovely Professional University, Phagwara, Punjab, India
e-mail: mamta.18431@lpu.co.in

A. P. Das (ed.), *Bioconversion of Environmental Waste to Value Added Products*, Sustainable Environmental Waste Management Strategies,
https://doi.org/10.1007/978-3-032-14215-3_14

1 Introduction to Sustainable Waste Management

Vermicomposting is the process of transforming organic waste into valuable compost with earthworms under practice conditions that help protect environmental systems. Soil fertility is enhanced because it improves structure and water conservation while supporting microbial activity thus, it enables sustainable agricultural practices. Lower-temperature vermiculture produces faster organic waste breakdown without destroying essential nutrients in the material (Kumari et al., 2024). Human civilizations have practiced for thousands of years through their observation of earthworms as agents that enhance soil quality. Modern organic farming, along with waste management systems, has been incorporated because scientific studies and technological advancements improved the procedure during the last decades (Dey et al., 2023; Mishra et al., 2023). The nature-recycling capability of earthworms proves vital because these organisms consume decomposing materials through consumption before creating essential nutrient-rich vermicast. The valuable nature of vermicast comes from plant growth hormones together with essential microbes and macro and micronutrients that both enhance soil fertility and boost crop yields (Syarifinnur et al., 2024). Every environmental benefit has established its adoption as a core technology in organic farming and sustainable agriculture as well as waste reduction systems (Mishra et al., 2019). Vermicomposting enables efficient management of different types of organic waste including waste from kitchens and gardens and agricultural materials together with selected industrial wastes. The waste management technique decreases landfill greenhouse emissions alongside lowering the waste materials put into landfills. Traditional waste disposal techniques are replaced through a cost-effective and energy-efficient method that serves all agricultural levels and urban gardening practices. The selection of proper earthworm species, the management of ambient conditions, and waste composition materials directly affect how well it functions. *Eisenia fetida* (red wigglers), together with *Eudrilus eugeniae* (African nightcrawlers), serve as the most common species for vermicomposting due to their high reproductive rates and strong ability to decompose organic waste (Nyamusamba et al., 2023). Optimal conditions for these species exist in controlled environments with perfect temperature levels combined with appropriate moisture levels and enough air circulation that enhances the composting function. People practice through various adjustable methods, starting from basic backyard setups to industrial-scale enterprise operations. Urban gardeners alongside household residents maintain their kitchen waste by utilizing worm bins along with pits, which function as small-scale vermicomposting systems. The effective processing of extensive organic waste occurs through commercial facilities and farms by implementing windrows and continuous flow reactors that handle large volumes. The business potential exists among its numerous benefits, which contribute to waste management and improved soil quality (Bhujel & Joshi 2023).

The agricultural market values vermicompost and vermiwash products because they offer premium rates as organic fertilizers. It gains economic value because people increasingly demand organic agriculture techniques and organic agricultural products. It enhances soil microbiological diversity while minimizing chemical fertilizer use and raises plants' defense mechanisms against pests and diseases to support better environmental sustainability. Organic farmers apply it as an essential ingredient because research confirms that it enhances crop yields while restructuring the soil layer and bettering plant nutrient intake. Although it brings numerous advantages to the process, it remains susceptible to environmental changes and requires regular maintenance while potentially being affected by pest invasions. The composting process reaches its most effective state when environmental conditions stay optimal regarding moisture content and temperature, along with sufficient aeration rates. The awareness and training programs seek to present superior methods for farming communities and waste management workers as well as gardening professionals (ul Baligah et al., 2025) (Das & Ghosh, 2022; Mohanty et al., 2018). The farming sector can reuse waste materials through various approaches including the combination of composting with anaerobic digestion alongside biogas production (Mishra & Das, 2023; Tripathy et al., 2024). The character of industrial waste can be either harmful or safe. To properly handle industrial waste employers must establish three key areas including pollution prevention alongside waste reduction and the implementation of recycling practices like eco-design and closed-loop systems. Scientific research now extends beyond waste management practices to include industrial pollutant breakdown, wastewater treatment, and environmental cleanup operations. Scientists research using both microbial additives and selective breeding methods to optimize earthworm performance when processing different waste materials. These advancements will change the organic waste management systems and introduce a greener waste management approach. Vermicomposting emerges as a sustainable solution to manage urgent issues related to growing waste and environmental concerns because of population growth and soil degradation problems. The technology serves as a key tool for both present-day agricultural practice and waste management strategies because it transforms organic waste into fertile soil material which helps sustainable farming practices and circular economy approaches (Mor et al., 2024).

The research elaborates on scientific aspects along with strategies and advantages and challenges and projections. Knowledge about biological processes and innovative applications allows us to realize its complete potential for creating a sustainable environment. The process allows individuals and organizations of all scales to use organic waste to create valuable resources that support both environmentally friendly practices and food system stability.

2 Principles of Vermicomposting

Under controlled conditions, earthworms and small living things help convert organic waste into vermicast, which contains high levels of nutrients. Recycling waste in a sustainable way helps both organic farming and soil health through its biological process improvement. The basic biological procedures describe how vermicasting breaks down waste to create nutrient-rich soil and recycle nutrients effectively.

2.1 Biological Processes Involved in Vermicomposting

Earthworm digestion joins microbial actions to break down organic matter into easy-to-digest forms that drive vermicomposting. The three mechanisms of natural breakdown include microbial breakdown, enzymatic degradation, and nutrient mineralization (Fig. 1).

2.1.1 Microbial Breakdown

Vermicomposting depends completely on microbial breakdown. The waste products from animals and homes plus farm material contain natural molecules like cellulose, hemicellulose, lignin, and proteins for decomposer microorganisms to process. Organic material decomposes best when decomposer bacteria, fungi, and actinomycetes invade it and begin releasing hydrolytic enzymes. Earthworms can process food more easily when specialized microorganisms change large organic materials into smaller units. Bacillus and Pseudomonas bacteria improve organic material nitrogen levels as well as support its stability but this job mainly belongs to *Aspergillus* and *Penicillium* fungi (Ahmed et al., 2023).

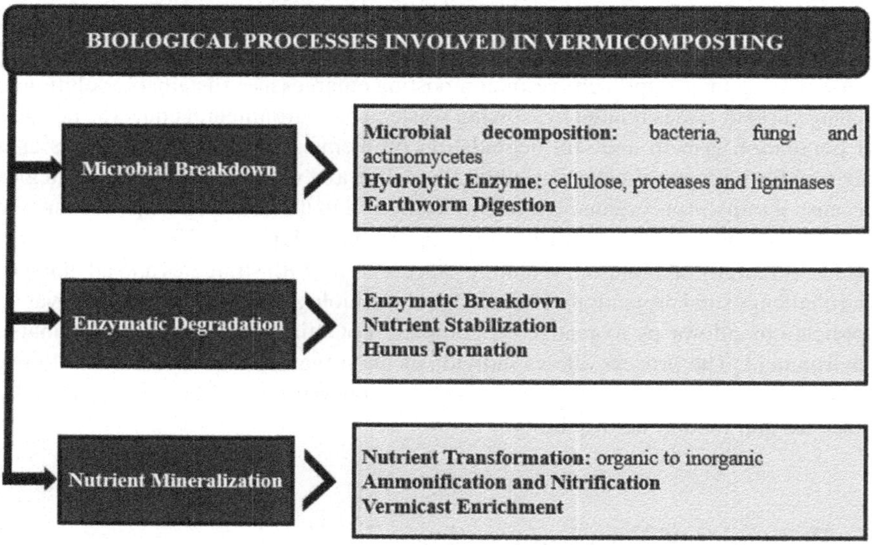

Fig. 1 Key biological processes and mechanisms in vermicomposting

2.1.2 Enzymatic Degradation

The natural process of transforming waste relies on enzymatic breakdown by worms. Earthworms release cellulases, proteases, and amylases, along with microbe enzymes, to help organic materials degrade faster. Proteins inside worms trigger faster organic reactions to build substances similar to humus. When proteolytic enzymes work alongside cellulases and lipases, the processes split proteins into amino acids while changing cellulose into glucose and splitting lipids into basic fatty acids. Earthworms, alongside microbes, use biological processes to stabilize and enhance the nutrients found in organic matter (Samal & Mohanty, 2024).

2.1.3 Nutrient Mineralization

During the final stage of vermicomposting, earthworm digestion converts organic plant nutrients into inorganic compounds that plants take up directly. Worm digestion and microbes transform potassium, phosphorus, and nitrogen from organic compounds into plant-usable inorganic minerals. Nitrogen turns into plant-friendly ammonium (NH_4^+) and nitrate (NO_3^-) through the natural transformation of organic materials by ammonification and nitrification. The natural processes of microorganisms break down plant-unavailable phosphorus to make it easier for plants to use. Vermicast from vermicomposting contains a rich mix of nutrients, including humus material, growth-stimulating hormones, and beneficial soil microorganisms, which make the earth more fertile and structured (Makkar et al., 2023).

2.2 *Role of Earthworms in Organic Waste Decomposition*

Vermicomposting works because of its earthworm population, which transforms waste into superb compost. Earthworms drive toward sustainable creation by transforming organic material through three types of biological impacts. Earthworms do the most vital work when they eat and process organic waste. They work by eating waste matter then breaking it down into small sections so more microorganisms can reach the fresh materials. Microbial growth improves faster by making natural materials easier to reach through earthworm physical processing. They maintain efficient composting by moving through the substrate and creating tunnels that enhances oxygen levels in the substrate for aerobic microbial processes to function (Khan et al., 2024).The gut helps in the natural breakdown of waste materials. They break down food through their internal enzymes then microbes in their digestive tract turn it into basic nutrients. Gut serves as habitat for many kinds of bacteria fungi and protozoa that process and secure organic materials for nutrition. Enzymes like cellulases, amylases, and proteases help earthworms convert tissue foods into nutrient-packed excrements.

Their excrement, called vermicast or worm castings, serves well as an organic fertilizer. The organic mix found in vermicast contains high amounts of nutrients, plus effective microorganisms and humic compounds. Worm castings serve horticulture and agriculture well because they construct stronger soil particles while saving water and helping roots develop better. The addition of vermicast enhances plant development and farm output thanks to its natural auxin and cytokinin plant growth regulators (Hao et al., 2024).

They add to the number of good microbes and promote essential nutrient movement through the vermicomposting system as they perform digestion. Earthworms create microbial homes over their burrows that support diversified groups of helpful microorganisms. Earthworms join microbial life to develop quality soil that improves both plant growing power and nutrient strength. Different types of earthworms work differently in breaking down the waste materials as shown in Table 1. The most common earthworm species for vermicomposting are red wigglers (*E. fetida*) and African nightcrawlers (*E. eugeniae*) which both live on the surface of organic-rich environments. They handle composting environments successfully because they multiply rapidly and process trash fast. Earthworms produce potent soil fertilizer and handle trash better than other worms.

They help to maintain conditioned organic material by destroying dangerous microorganisms. Earthworm gut antibacterials protect people because they block harmful microorganisms present in raw compost materials. Through its natural sanitization process, vermicomposting proves to be a secure and effective system for handling kitchen and crop waste. The earthworm population helps break down waste while saving carbon and the planet. They turn organic waste into stable soil carbon while averting green pollutants to fight environmental changes. The environmental advantages of vermicomposting become clearer since earthworms help to prevent methane emissions from landfills (Kooch et al., 2024).

3 Vermicomposting Process and Techniques

Earthworms turn organic waste into fertile compost through vermicomposting, which is also called vermicasting. The method reduces the waste sent to landfills and improves soil health, which works both for waste control and environmental protection. The process requires choosing correct waste materials and setting up the composting bed, plus introducing earthworms while maintaining specific environmental conditions before harvesting the vermicast. It works through diverse systems, including homes and businesses (Huang et al., 2023).

Table 1 Earthworm species and their roles in organic waste decomposition

Earthworm species	Ecological category	Role in decomposition	Types of suitable waste	Characteristics
Perionyx excavatus	Epigenic	Quick decomposition and effective vermicomposting	Paper waste, leaf litter, farm waste	Thrive in conditions that are warm and humid
Eudrilus eugeniae	Epigenic	Produces nutrient-rich vermicompost from organic waste	Food waste, garden waste	Quick development and strong composting effectiveness
Eisenia fetida	Epigenic	Speeds up the decomposition of organic materials	Kitchen waste, farm waste, manure	High rate of reproduction and ability to flourish in small areas
Dendrobaena vaneta	Epigenic	Effective for decomposing fibrous organic materials	Cardboard, vegetable scraps	Accepts changing circumstances
Lumbricus rubellus	Epigenic	Increases microbial activity and raises the quality of compost	Agricultural residues, compost pits	Prefers surroundings that are rich in organic matter
Aporrectodea caliginosa	Endogenic	Enhances soil structure and breaks down decomposing plant matter	Soil litter, decaying plant residues	Improves soil porosity and burrows deeper
Octolasion cyaneum	Endogenic	Improves aeration and incorporates organic materials into the soil	Soil organic matter, decomposing plant materials	Contributes to the fertility of soil
Lumbricus terrestris	Anecic	Encourages the production of humus and draws organic matter deeply into the soil	Surface litter, leaf debris	Produces long-lasting vertical burrows

3.1 Process

Earthworms break down materials during vermicomposting using natural biological methods. At the start, you need to gather organic waste like fruit residues, vegetable cuts, garden garbage, farm leftovers, and animal droppings. Our earthworm living environment must not include pesticides or chemicals and should remain free of metallic and plastic substances. A mix of dried leaves, straw, and shredded paper aids decomposition by controlling the carbon-to-nitrogen ratio in the materials. The production of a composting area starts following the material choice. The bedding provides a suitable habitat for earthworms and facilitates aeration and moisture retention. Horticulturists commonly combine coconut coir, peat moss, shredded newspaper, and natural decaying materials as bedding materials for earthworm farming (Marumure et al., 2023). The materials need moist storage but should remain dry for earthworm support, with a moisture level of 60–70%. Earthworms begin the composting process after they are placed in the composting space. For all-around use, they mostly use red wigglers (*E. fetida*) and African nightcrawlers (*E. eugeniae*) because of their quick population growth and their amazing abilities to handle and break down organic materials (Toor et al., 2024). Earthworms transform organic waste into soil fertilizing castings during their eating and digestion process. The method needs an optimal set of environmental elements to work properly. Earthworm survival and life patterns depend on temperature extremes between 15 and 30 °C, with this being their most suitable range (Wehrli et al., 2024). Earthworms require enough air intake to prevent anaerobic zones and stop dangerous gases such as ammonia and methane from developing. An adequate oxygen system works best when you turn the compost regularly or use a system with good air circulation. To support worm and microbe survival and performance, this system must operate at a pH between 6.0 and 7.5. Earthworm work breaks down natural materials into vermicast that looks similar to natural organic matter and supplies plants with necessary nutrients through calcium, magnesium, phosphorus, potassium, and nitrogen. Earthworm composting ends with the collection of vermicast. Earthworm harvesting needs darkness avoidance methods alongside hand selection or mechanical filters to collect worms from the compost mixture. The verified vermicast serves as organic soil enhancement after going through fresh and sifted handling (Naik et al., 2024).

3.2 Techniques

Vermicomposting systems work in different ways because they must fit each situation and specific project requirements. The main techniques include bin composting waste pits, windrow composting, and continuous flow systems, is described in Table 2.

3.2.1 Bin Composting

This serves as the preferred method for vermicomposting purposes in homes and cities. Earthworms and organic waste need containers made from plastic, wood, or metal to create this method. Foot holes in the bin structure let air through to prevent excess water buildup. Kitchen scraps and garden trash need to be placed at various levels with shredded paper to get the proper carbon/nitrogen blend. A good composting environment depends on checking temperatures and moisture levels regularly. Home gardeners who want to recycle their organic waste in limited spaces will find interacting composting to be a great method (Raimi et al., 2024).

Table 2 Vermicomposting techniques: a comparative analysis of advantages and disadvantages

Technique	Advantage	Disadvantage
Bin composting	This system has easy maintenance demands while needing limited space and working well for home settings	The system demands continuous care while it faces the potential problems of water accumulation and unpleasant smells
Pit composting	Inexpensive, effectively maintains moisture, and requires little upkeep	Slower decay occurs with increased chances of water intrusion and oxygen depletion
Windrow composting	Large-scale composting benefits from this approach, which permits natural ventilation at an affordable cost	An extensive amount of space is necessary, and manual labor is required for rotating the material
Continuous flow system	High productivity, ongoing composting, and little worm disruption	Technological expertise is needed for this expensive equipment setup

3.2.2 Pit Composting

When space is available in farms and rural areas then people use pit composting methods. The procedure begins with making a small empty space before adding layers of bedding materials and organic waste. As soon as earthworms are added the composting naturally progresses forward. Planting compost in narrow holes which helps to protect against extreme heat and keeps the material moist. Regular checks ensure that the addition of earthworms allows enough airflow through the material to prevent unhealthy anaerobic breakdown (Dharnaik & Pol, 2024).

3.2.3 Windrow Composting

People often turn to windrow composting as their preferred method for industry and agriculture operations. Placing organic waste within bedding layers forms narrow windrow piles that vermiculture requires. Earthworms go into the rows of organic waste which begins its decomposition within several weeks. Rotating the windrows helps ensure worms receive enough oxygen while maintaining a steady breakdown of materials. This system can effectively process high quantities of organic waste features such as crop stalks, animal excrement, and leftover food materials. Significant-scale windrow composting generates vermicast that helps environmental regeneration and supports sustainable farming methods (Sharma, 2024).

3.2.4 Continuous Flow System

Specialized continuous flow vermicomposting systems exist to produce waste on a commercial basis. These state-of-the-art containers work through mesh base systems or tray removal to let vermicast reach the bottom untouched. Vermicompost nears completion when workers collect it from the bottom of the unit during the breakdown period as fresh waste enters the top. The technology saves time in compost harvesting while producing constant supplies of excellent compost. Continuous flow systems appear in all three-waste disposal, horticulture, and commercial growing sectors (Kadam et al., 2024).

4 Feedstock Selection and Preparation

Vermicomposting starts with choosing and preparing feedstock properly, which refers to the organic waste materials added to the compost. The feedstock type determines the speed of organic transformation as well as the outcome quality of vermicast and the composting system conditions. The process succeeds through choosing good waste materials and treating them properly while managing the right C: N balance. We will examine and explain the best waste materials with how to shred and condition them along with why carbon and nitrogen balance matters (Thirunavukkarasu et al., 2023).

4.1 Types of Organic Waste Suitable for Vermicomposting

Earthworm composting systems accept many types of waste materials for processing. Many organic waste items cannot go because they contain things that hurt earthworms or bring disease. Organic waste suitable includes home cooking leftovers from kitchens plus agricultural waste and outdoor plant waste (Fig. 2). Animal manure also works well for this process.

4.1.1 Kitchen Waste

Waste from vegetables, fruit, coffee, tea, eggs, and loaves are excellent raw materials for vermicomposting. These materials hold many essential nutrients that help microorganisms work better. Since dairy products and oily substances can attract bugs and produce odors, care must be taken when including these foods (Saha et al., 2024).

4.1.2 Garden Waste

Plants that grow in our garden along with cut grass and trimmed hedges serve as inputs for vermicomposting. Pulverize hard leaves and wood items first then add them to your composting setup. Leaves from pine and eucalyptus trees should not be

Fig. 2 Organic waste categories for effective vermicomposting

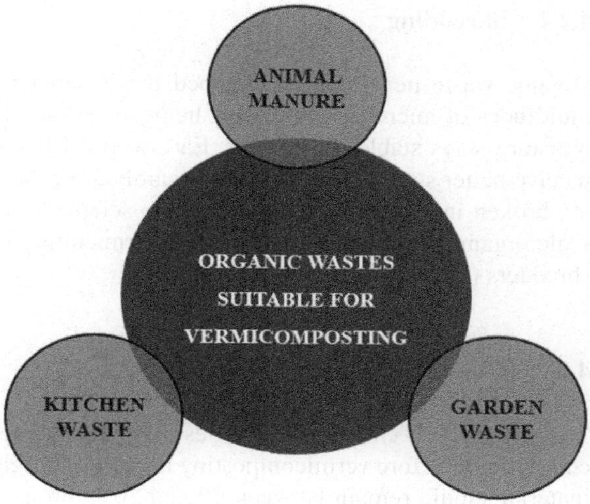

used because their allelopathic nature restricts microbial growth (Syarifinnur et al., 2023).

4.1.3 Animal Manure

Different animal manures from cows, horses, sheep, and rabbits become excellent fertilizer because they bring essential nutrients into the vermicomposting system. Earthworms need materials that are either decomposed or aged to reduce the ammonia content and avoid toxicity before including them in the system. No system should take pet waste or materials handled with chemicals like treated wood. The process also rejects plastic trash and materials harmful to worms from foods with preservatives or additives plus strong alkaline or acidic substances. A quality composting environment needs well-selected feedstock materials to achieve natural nutrient balance (Bellitürk & Sundari, 2024).

4.2 Pre-processing Techniques: Shredding and Conditioning

The initial transformation of organic waste builds better waste decomposition and guards earthworm survival. Earthworm beds receive two essential preparations called conditioning and shredding.

4.2.1 Shredding

Organic waste items can be chopped into smaller parts for faster breakdown by multitudes of microbes. Shredding helps the system breathe better and makes sure moisture stays stable throughout. Earthworm digestion and microbial breakdown receive better support when paper or cardboard is shredded properly and eggshells are broken into sections while vegetable scraps are cut apart. To treat their large-scale organic waste loads, commercial vermiculture operations rely on mechanical shredders (Urechescu et al., 2024).

4.2.2 Conditioning

The physical and chemical properties of organic waste need to be improved through conditioning before vermicomposting begins. The right amount of water for organic material should remain between 60 and 70% throughout conditioning procedures. You can reduce food waste when you mix it with absorbent stuff, yet dry sawdust and straw need prior soaking before you put them in the compost pile. A few specific substances go through transformation before entering vermicomposters. Breaking down the materials helps to keep dangerous chemicals under control, reduces infections, and lowers ammonia amounts. It needs a pH range of 6.0–7.5. When mixing acidic fruit rinds with crushed eggshells or agricultural lime, you create an alkaline mixture. Additional nutrient supplies support specific setups to keep their feedstocks suitable. Enhancing the beneficial nutrients in vermicast results from mixing it with phosphate rock or plant leftovers from legumes (Wongkiew et al., 2023).

4.3 Balancing Carbon and Nitrogen Ratios

During vermicomposting, the C: N ratio controls important changes that occur between microbe activity and compost breakdown. To achieve the best results with worm farming, the C: N ratio should range between 25:1 and 30:1. This level assures the successful breakdown of organic materials while preserving nitrogen content and smell (Hasan et al., 2023).

Living microbes need carbon as fuel and carbon-rich materials to retain their form. The list contains materials that hold large carbon amounts, including dried leaves, straw, sawdust, cardboard and torn paper. This material reduces how fast nitrogen-rich feed enters the system while supporting its water absorption abilities. Living microorganisms rely on nitrogen to develop and transform materials naturally. Organic materials like fresh grass trimmings, vegetable waste, fruit peelings, used coffee grounds, and animal droppings all have high nitrogen content (Lim et al., 2023). The items decay speedily to provide the necessary nutrients that support healthy microbial growth. Nitrogen-rich and carbon-rich materials should be blended based on the required carbon-to-nitrogen ratio in the compost. An excessive nitrogen content can lead to ammonia smells through breakdown, but excessive carbon material slows the decomposition process. Mixing organic waste layers between low-carbon browns and high-carbon greens lets us control the material ratio balance. Regular mixing of the compost helps its breakdown process because it distributes nutrients throughout the pile equally (Sihlangu et al., 2024).

5 Benefits of Vermicomposting

The process of turning organic waste into topsoil fertilizer with earthworms promotes both Earth-friendly and long-term practices. The earthworm process supports waste handling and creates beneficial effects on soil quality, along with better plants and more helpful nature conservation practices. The compost from vermiculture adds quality to soil nutrients and texture, which farmers need for sustainable agricultural practices. Beyond adding nutrients to soil vermicomposting provides pollution reduction and improves soil condition. This section discusses why vermicomposting is valuable by analyzing its benefits for creating nutrient-rich soil, protecting the planet, decreasing waste, and improving soil fertility (Sande et al., 2024).

5.1 Nutrient-Rich Compost Production

Vermicomposting produces quality plant nutrients through compost known as vermicast or worm castings, which is a major benefit. Compost made through vermicomposting provides superior plant nutrients at greater levels than traditional compost and synthetic fertilizers do. Earthworm excrement or vermicast contains all plants with necessary trace elements such as nitrogen, phosphorus, potassium, calcium, magnesium, and more. Earthworm digestion powers the conversion of organic refuse to make nutrients available for use. Earthworms create soil fertilizer from waste materials while making it easy for plants to absorb nutrients. Whenever nutrients escape from compost piles through evaporation or flow away from the soil, vermicomposting retains these nutrients in an easily used plant form (Karimi et al., 2024).

Vermicast production leads to better crop yields because it strengthens soil health. Vermicast contains important aspects of microbial biomass and humic acids which make soil quality better. Plants take up nutrients better when they reach their roots through humic compounds that promote good soil conditions, including water storage and air space improvement. The useful microorganisms found in vermicast help plants by doing two things: they maintain soil quality and improve plant defense against diseases. Vermin compost provides nutrients more gradually because of its special production method. Vermicast Nutrients come out steadily to support plant growth throughout the season, while synthetic fertilizers give fast but temporary plant food support. Vermicomposting proves to be better than synthetic fertilizers because it provides nutrients to the soil in a controlled manner while helping plants use nutrients better and blocking them from flowing away (Manchal et al., 2023).

5.2 *Environmental Impact and Waste Reduction*

It produces quality plant nutrients through compost known as vermicast or worm castings, which is a major benefit. Compost made through vermicomposting provides superior plant nutrients at greater levels than traditional compost and synthetic fertilizers do (Mishra et al., 2022; Mishra & Das, 2022). Earthworm excrement contains all the plant necessary trace elements such as nitrogen, phosphorus, potassium, calcium, magnesium, and more. Earthworm digestion powers the conversion of organic refuse to make nutrients available for use. Earthworms create soil fertilizer from waste materials while making it easy for plants to absorb nutrients. Whenever nutrients escape from compost piles through evaporation or flow away from the soil, vermicomposting retains these nutrients in an easily used plant form (Mohite et al., 2024). Vermicast production leads to better crop yields because it strengthens soil health. Vermicast contains important aspects of microbial biomass and humic acids which make soil quality better. Plants take up nutrients better when they reach their roots through humic compounds that promote good soil conditions, including water storage and air space improvement. The useful microorganisms found in vermicast help plants by doing two things: they maintain soil quality and improve plant defense against diseases. It provides nutrients more gradually because of its special production method. Vermicast Nutrients come out steadily to support plant growth throughout the season, while synthetic fertilizers give fast but temporary plant food support. It proves to be better than synthetic fertilizers because it provides nutrients to the soil in a controlled manner while helping plants use nutrients better and blocking them from flowing away (Sharma et al., 2024b).

5.3 *Enhancing Soil Health and Fertility*

Vermicompost enhances soil wellness because it reconstructs soil texture while adding minerals and boosting the number of beneficial microorganisms. Good soil sustainability methods depend on well-nourished soil to help plants achieve strong growth and high output. It brings significant value by making soil better able to retain its natural structure. The organic matter helps to create soil aggregates that maintain space between them to prevent density problems as well as enhance soil drainage (Hiranmai et al., 2024). The earth can absorb more water into its surface and hold moisture better to avoid water problems during plant growth. Plant roots can reach nutrients and moisture deeply into the soil because it has good structure. Soil receives better health and increased fertility that enhances both its organic richness and microbial activity while building superior soil construction. Sustainable farmers need healthy soils to create the best growing space for their agricultural plants. Farmers, gardeners, and environmentalists who use this method achieve better soil quality because vermicomposting provides multiple health benefits to soil (Enebe & Erasmus, 2023).

It has the essential function of creating better soil structure. The organic matter in Vermicast encourages soil clumps, which reduces density while creating hollow spaces. Better water flow and soil moisture retention protect plants from water damage. Plants find happier growth because their roots can go through the soil pattern to reach essential plant resources. The condition of soil depends heavily on maintaining proper pH levels since this factor affects what nutrients plants can use. The neutral alkaline pH helps to protect soil when its acidity or alkalinness needs buffering. It creates balanced pH levels and becomes an effective solution for soil areas that struggle with acidity. People can better manage their water resources naturally. Soil moisture retains better when vermicast adds natural organic material to the ground. This method best protects water supplies during drought in dry regions. The production of vermicompost makes the soil hold more moisture, which benefits plants during dry seasons without using unnecessary water (Oyege & Balaji Bhaskar, 2023). The effective carbon absorption helps to make the soil healthier. Soil that stores carbon derived through organic matter decreases CO_2 levels in the atmosphere. The system prevents carbon release to slow down global warming. Soil fertility grows stronger and sustainable when organic materials are introduced with lower demand for artificial product use. Soil polluted by heavy metals and other harmful elements becomes more stable when developers mix it with vermicompost. The organic part of vermicast ties up heavy metals so they become less available in soil and decrease their threat to plant health (Nascimento-Gonçalves et al., 2024). It suppresses plant diseases in both supporting healthy soil and boosting its productivity. Vermicast creates helpful microorganisms that make enzymes and antibiotics to stop soil diseases from spreading. The plant's inorganic disease resistance system allows farmers to use fewer artificial pesticides while pursuing organic farming practices. Plants from vermicompost treatments are better protected against environmental stress and disease factors that enhance their farming results. The whole vermicomposting system delivers several ways to enhance soil quality. Soil quality improves when vermicompost is added through changes in structure, microbes, water capability, acidity levels, and disease immunity (Sivasabari et al., 2023).

6 Vermicomposting in Urban and Rural Areas

Earthworm-based composting benefits both major agricultural farms and city gardens as they turn organic garbage into valuable growth ingredients. It provides a real solution to increase soil fertility by supporting eco-friendly farming and helping us handle our waste problems, which deteriorate the earth and ground at present. It operates in different ways based on specific location factors, including participation levels, waste habits, and space availability across both urban and rural areas.

6.1 Implementing Vermicomposting in Urban Communities

Cities feature densely populated neighborhoods with tight living areas that produce much organic waste from residential homes as well as food sales and eating establishments. The use of vermicomposting in cities helps cities deal with waste safely while creating space for local agricultural production. For cities, vermicomposting seems to succeed with their limited space needs; they should use small, efficient tools that homeowners can handle. Many people choose small-scale worm bins or composting containers as their favorite method for urban composting (Maksymenko et al., 2024). Available bins with plastic containers and wood crates assist urban residents in processing their kitchen waste effectively. Humanity employs *E. fetida* (red wigglers) and *Eisenia andrei* earthworm types as popular composting partners because they thrive well in limited areas while processing compost quickly. Reusable worm bins succeed best with the Earthworm species *E. fetida* (red wigglers) and *E. andrei* due to their rapid composting power and space flexibility (Javed et al., 2023).

Urban people benefit from vermicomposting more than just regular disposal of waste. People can use the natural compost made through the system to improve local community gardening spaces and grow fresh food vertically. Many city residents now grow plants on rooftops and use their space to farm, which means they need nutrient-rich fertilizers grown sustainably. The soil structure of urban gardens improves when it joins with other materials because it enhances water retention and delivers necessary plant nutrients (Hajam et al., 2023).

Urban communities need to join forces with educational activities to make vermicomposting work. Enroll classes into environmental programs for schools and communities to let people learn about natural farming and better waste use habits. Local governments should fund, support, and select spaces to start projects while teaching residents how to compost in both buildings and public parts of cities. Urban areas require you to handle odors, stop pest intruders, and invest money to start your composting units. Correct care for the bin, such as proper aeration, moisture control, and suitable bedding materials, including shredded newspaper, will help prevent these problems. Using automatic temperature and moisture tracking devices helps make urban vermicomposting systems work better (Senadheera et al., 2024).

It creates business potential in cities. Small business owners should start operations and sell quality compost to regional farmers and garden sellers. Worm farming companies that sell earthworms for compost production can succeed in cities by making both money and promoting eco-friendly choices (Wang et al., 2024).

6.2 Rural Applications and Community Engagement

Rural areas where agriculture dominates should use vermicomposting to enhance their crop yields while improving their soil quality and weaning off chemical fertilizers. The organic waste generated by rural communities from kitchen scraps, livestock waste, and crop materials can be effectively processed into high-quality vermicompost. Both commercial and smallholder farmers benefit from large-scale vermicomposting systems since there exists sufficient agricultural land and open spaces (Walia & Kaur, 2024).

Three basic, affordable methods for rural vermicomposting installation consist of pit composting, windrow composting, and vermicomposting beds. Farmers who want to establish large vermicomposting operations can construct their facilities through the use of direct dirt pits along with bamboo structures or wooden or brick enclosures. The compost created through this process has multiple uses, including soil enhancement, yield improvement, and organic cultivation promotion (Hiranmai et al., 2024).

Rural vermicomposting serves as an excellent method to decrease agricultural waste storage while turning waste into useful organic materials. Farmers can efficiently decompose crop residues, including straw, husks and leaves, and animal manure obtained from cattle, poultry and goats through earthworm activity. The waste management process using earthworms creates two positive impacts: it minimizes agricultural waste pollution effects and establishes a budget-friendly fertilizer alternative that benefits rural economies. The goal of successful projects depends on deep involvement from community members. Various groups of self-help organizers, farmer training networks, and cooperative composting programs should adopt vermicomposting techniques. Rural farmers could be motivated through joint activities organized by government organizations, non-governmental organizations, and agricultural extension services that include workshops and awareness campaigns and financial programs (Nathan & Mukupa, 2024).

The soil health benefits from vermicomposting because the practice adds organic matter and diverse microbial life to the earth. The treatment of soil enhances its air exchange ability together with its water retention quality and nutritional resources which results in better plant growth along with increased disease and insect resistance. Soil restoration with organic farming reduces soil degradation that chemical fertilizers create as well as restores soil productivity levels (Manyanga et al., 2024).

It creates new employment possibilities within rural areas while delivering agricultural advantages to the sector. The establishment of vermiculture facilities by rural business operators allows them to provide compost to farms and organic marketplaces in their vicinity. Rural household lives improve through the enterprise of earthworm production and their direct sale for fishing operations, along with composting activities (Dey Chowdhury et al., 2023).

Remote implementation faces multiple barriers because local communities possess little knowledge about the approach while also lacking engineering help and facing costly startup costs. The solution to existing problems depends on government policy backing combined with regional training efforts and cooperation between agriculture study groups. Additional growth depends on implementing publicity-based demonstration sites supported by monetary incentives for equipment financing alongside its inclusion within rural development initiatives. It proves to be an efficient method for friendly waste management and expanded agricultural yields while preserving environmental spaces across urban and rural areas. Using vermiculture as a waste recycling technology proves environmentally advantageous because it enhances the well-being of both urban and rural areas through the combined use of local resources alongside community engagement programs and sustainable initiatives (Raman et al., 2024).

7 Challenges and Limitations

7.1 Challenges

The fundamental element of vermicomposting needs earthworms whose natural sensitivity reveals to pH changes and temperature conditions while also being influenced by moisture levels and air requirements. Worm survival, together with decomposition effectiveness, suffers negatively when exposed to high saline conditions and harmful organic waste compounds and rapid weather fluctuations (Peng et al., 2024). The decomposition procedures for various organic materials differ in speed, while certain materials like meat scraps and dairy products, together with greasy residues, cannot properly decompose under anaerobic conditions and attract insects. An improper selection of feedstock materials can create wrong carbon-to-nitrogen (C/N) ratios that harm compost quality along with microbial activities. The activity of worms together with their composting environment experience disturbances because of extreme weather conditions that include wet conditions, hot temperatures or frigid temperatures (Vambe et al., 2023). The control of environmental hazards requires attention because rats, flies, and mites exist in this area. The implementation of large-scale vermicomposting requires significant investment in processing equipment, along with manpower support and appropriate land equipment. Without machinery combined with process regulations, there is poor compost uniformity, resulting in increased labor costs. Vermicompost production costs more than traditional composting, mainly due to its necessary infrastructure elements and maintenance needs. Large-scale vermicomposting meets limited economic success because government support is absent while pork market value sways and public understanding remains low. Traditional composting methods dominate farmers' practices to the extent that they resist adopting vermicomposting methods (Aslam et al., 2023).

7.2 *Limitations*

The biological decomposition process, which depends on worms and microorganisms, extends the time required to create mature compost compared to traditional composting methods. The quick trash disposal requirements of companies are not met since its processing period extends over a long duration. The microbial composition, together with nutritional value, depends heavily on both feedstock choice and processing operational parameters. The lack of certification standards together with regulatory rules reduces marketability and commercial acceptance rates (Manea et al., 2024). The lack of support from existing national laws against vermicomposting leads to minimal institutional backing and both financial resources as well as research capabilities. The transformation of waste management systems must include specific guidelines and financial rewards. It proves to be unsuitable for different categories of organic waste materials. Fibrous plant substances that are tough and industrial waste with chemicals and materials with acid or alkaline qualities decompose poorly during this process. The scope of applications in broad waste management remains restricted because of this particular limitation. The sole application does not produce efficient results for every waste stream. Both anaerobic digestion and thermophilic composting, when combined with the process, require help with various waste types to achieve a comprehensive waste management system (Kong et al., 2024).

8 Future Directions and Innovations

The approach of vermicomposting serves both as an efficient waste management system and as an environmental protector of soil conditions. The necessity for new vermicomposting solutions and development has intensified because global waste increases while agricultural sustainability has become essential. It will focus on four key areas for future development, which include improving compost quality while enhancing operational efficiency accompanied by technical integration and expanding the waste treatment capacity across different waste streams in diverse environmental conditions. The resolution of current problems through modern scientific research produces a scalable, effective vermicomposting system that also promotes ecological friendliness.

8.1 *Genetic Enhancement of Earthworms for Improved Performance*

The creation of genetically changed earthworms for potato and other kinds of waste, which holds the potential of increased tolerance, adaptability, and blade decomposition efficiency, is one among the most impressive new trends in vermicomposting.

Earthworm breeds that have enhanced enzymatic activity for faster decomposition, a stronger resilience to pollutants, and more resistance to adverse temperatures may derive from studies of selective breeding and genetic manipulation. These developments could greatly enhance the benefits of vermicomposting in a variety of environmental contexts, from stimulating its use on organic waste from cities and industries (Zhou et al., 2023).

8.2 Advanced Microbial Inoculation for Accelerated Decomposition

Inoculation of microbes is the method of introducing specifically designed microbial consortium to improve the decay process in vermicomposting. Thus, by the addition of useful microorganisms like cellulolytic, ligninolytic, and nitrogen-fixing bacteria, it is possible to increase the breaking down of complex organic materials. Investigations of the microbiome of vermicomposting beds can improve microbial populations to maximize the cycling of nutrients, suppress the growth of unwelcome pathogens, and improve the bioavailability of vital nutrients in vermicompost. These innovations might also enable the mass production of better-quality organic fertilizer with enhanced plant growth-promoting functionalities (Namasivayam et al., 2024).

8.3 Integration of Smart Sensors and Automation in Vermicomposting Systems

Vermicomposting with automation and smart sensors has the potential for a complete proof of waste management effectiveness. Optimizing earthworm and microbial decomposition with sensors that continuously monitor and track temperature, moisture, pH, oxygen levels, and microbial activity is also possible. Data-driven analytics, Automated feeding equipment, and moisture control systems can optimize the waste breakdown rate, decrease the manual work, and increase production. In big operations, the procedure may be precisely managed by making use of modern IoT technology which can provide remote monitoring and also predictive modeling (Paculanang et al., 2024).

8.4 Techniques for Hybrid Vermicomposting to Increase Efficiency

A promising use of the future for waste management is presented by hybrid methods of composting, which combine vermicomposting with thermophilic composting, anaerobic digestion, or biochar use. While earthworms can improve the trash degradation rate, high microbial-temperature thermophilic compositing, which uses heat to speed up the microbial breakdown of the trash, may be considered first. In the same vein, mixing of biochar produced via biomass pyrolysis in vermicomposting ponds can enhance soil conditioning, enhance the ability of the end compost product to hold water, and increase organic carbon sequestration. Hybrid methods, which are on a large scale, sort waste; these techniques give more flexibility and efficiency (Sharma et al., 2024a, b).

8.5 Urban Vermicomposting Solutions for Sustainable Cities

Miniaturized, decentralized vermicomposting systems for metropolitan areas with growing quantities of organic waste must be provided. Vermicomposting can even be adapted in urban environments with the assistance of design targeting modular composting units, space-saving vertical vermiculture systems, and community-oriented composting programs. Schools, residential facilities, etc., can be involved in automated vermicomposting machines where waste is managed locally by reducing transportation costs, independence on the landfill, and indirectly raising the involvement of the community in sustainable waste management (Goswami et al., 2024).

8.6 Industrial and Municipal Waste Vermicomposting

One way of great potential is the extension of vermicomposting from normal consumption and productive waste to other organic material waste from markets and municipalities. Modern waste proprietaries waste pre-treatment methods like micro-behavioral waste hydrolysis, animating enzymatic hydrolysis, and biochar enforcing can greatly lead to aiding the generation of industrial plant-based wine waste to modify feed stock. Big successful organic waste to be well-processed large-scale organic waste recycling can be realized by large-scale vermicomposting facilities developed by using cutting-edge waste sorting, passing contaminant, and entire throughput and interface technologies that support the idea of a circular economy (Hemalatha et al., 2024).

8.7 Nutrient Fortification and Customization of Vermicompost

Vermicompost agricultural effectiveness is being mainly upgraded with the manufacture of tailor made solutions to specific crop requirements through an innovative approach. To create customized organic fertilizers, scientists are also exploring methods to incorporate vermicompost with invaluable minerals, biostimulants, and plant growth-stimulating rhizobacteria (PGPR). These advancements can enhance crop resilience to stress, enhance soil fertility, and increase the overall yield of farm crops. More yield and less use of chemical fertilizers will be the result of bio-enriched vermicompost application for precise nutrient control (Agarwal & Jain, 2024).

8.8 Traditional Waste Streams

Vermicomposting was historically most important when applied to food and agricultural waste, but there is a potential for an additional outlet to take advantage. Research on vermicomposting of pharmaceutical residues, brewery waste, agricultural and industrial wastes, and biodegradable polymers is being conducted. To control industrial as well as environmental pollution, new bioremediation methods that employ earthworms for degrading heavy metals, harmful chemicals, and persistent organic pollutants (POPs) have been developed. If these advancements succeed, they might completely change the method of cleaning the soil and the treatment of toxic waste (Raman et al., 2024).

8.9 Circular Economy and Carbon Sequestration Strategies

While vermicomposting is in line with the circular economy theory, some other advancements can enhance its role in carbon sequestration and mitigation of climate change. Vermicomposting can be developed as an effective tool for slashing greenhouse gas emissions by developing carbon-negative vermicomposting techniques utilizing biochar, regenerative agriculture practices, and carbon credit mechanisms. Research on the capability of carbon sequestration in soils treated by vermicompost will provide information on the long-term contribution of soil carbon lifecycle and climate resilience (Chowdhury & Asiabanpour, 2024).

8.10 Policy Support and Incentives for Large-Scale Adoption

Government policies, together with financial rewards, will determine the path forward for vermicomposting expansion. Vermicomposting acceptance speeds up when governments create waste recycling promotion programs while giving funding for waste processing equipment and extending benefits to farmers. To make vermicomposting more prominent in green development, the creation of carbon offset programs, climate action plans, and sustainable agriculture regulations using this method would boost its status (Goswami et al., 2024).

8.11 Public Awareness, Training, and Capacity Building

It will prosper through public engagement together with educational initiatives and with technological progress. Greater understanding will emerge from expanded farmer training, waste management specialist education, and city resident classes. The effective exchange of best practices as well as local adoption expansion occurs through online learning modules, smartphone apps, and digital platforms. Using community-run projects leads to trash management methods that are both decentralized and community-based (Henao-Rodríguez et al., 2024).

9 Conclusion

Vermicomposting functions through sustainable waste management techniques that utilize earthworm activity together with microorganisms to create valuable compost from organic waste materials. Humid conditions enable bacterial and fungal action alongside actinomycete releases enzyme to decompose organic compounds like cellulose, hemicellulose, lignin, and proteins. The microbial decomposition activity transforms organic matter into a state that earthworms can more easily process and decompose. The compost receives enhanced nitrogen stability through beneficial bacteria, including *Bacillus* and *Pseudomonas,* while important fungal organisms, such as *Aspergillus* and *Penicillium,* perform key roles in breaking down organic matter. The whole process yields nutrition-rich vermicompost which strengthens soil fertility and supports water retention as well as improves overall plant health. The organic fertilizer provides farmers with a sustainable alternative to chemical products because it minimizes fertilizer dependence. The waste reduction process through vermicomposting allows for the recycling of organic material from homes as well as farms and animal waste operations to decrease environmental pollution. The ongoing breakdown of organic materials enables efficient nutrient cycling to support extended soil health equilibrium. Farmers, environmentalists, and gardeners benefit from a process that costs little and runs smoothly. To address both organic farming growth and waste disposal problems, vermicomposting functions as an effective method that advances soil fertility. The practice of vermicomposting leads to sustainable agriculture while creating soil improvements that benefit both ecological systems and environmental health through waste decomposition processes.

References

Agarwal, S., & Jain, H. (2024). Investigating the potential of floral waste as a vermicompost and dual-functional biosorbent for sustainable environmental management. *Water, Air, & Soil Pollution, 235*(5), 322.

Ahmed, T., Noman, M., Qi, Y., Shahid, M., Hussain, S., Masood, H. A., & Li, B. (2023). Fertilization of microbial composts: A technology for improving stress resilience in plants. *Plants, 12*(20), 3550.

Aslam, Z., Bashir, S., Belliturk, K., Zhang, L., Zaman, Q. U., & Ahmad, A. (2023). Vermicomposting: a sustainable and environment-friendly approach for organic waste management. In *Waste problems and management in developing countries* (pp. 365–401). Apple Academic Press.

Bellitürk, K., & Sundari, R. S. (2024). The power of earthworm: Vermicompost drives to sustainable agriculture. In *Earthworm technology in organic waste management* (pp. 307–321). Elsevier.

Bhujel, R. R., & Joshi, H. G. (2023). Organic agriculture in India: A review of current status, challenges, and future prospects. *Universal Journal of Agricultural Research, 11*(2), 306–313.

Chowdhury, H., & Asiabanpour, B. (2024). A circular economy integration approach into vertical farming with computer-based simulation model for resource optimization and waste reduction. *Journal of Cleaner Production, 470*, 143256.

Das, A., & Ghosh, S. (2022). Role of microorganisms in extenuation of mining and industrial wastes. *Geomicrobiology Journal, 39*, 173–175. https://doi.org/10.1080/01490451.2022.2038953

Dey Chowdhury, S., Suhaib, K. H., Bhunia, P., & Surampalli, R. Y. (2023). A critical review on the vermicomposting of organic wastes as a strategy in circular bioeconomy: Mechanism, performance, and future perspectives. *Environmental Technology*, 1–38.

Dey, S., Das, A., Mallick, K., Sahu, A., & Das, A. P. (2023). Environmental petroleum waste: pollution, toxicity, sustainable remediation, 159–175. https://doi.org/10.1007/978-3-031-48220-5_7

Dharnaik, A. S., & Pol, P. (2024). A review on composting of organic solid waste. In *IOP Conference Series: Earth and Environmental Science*, June 2024 (vol. 1326, no. 1, p. 012130). IOP Publishing.

Enebe, M. C., & Erasmus, M. (2023). Vermicomposting technology—a perspective on vermicompost production technologies, limitations and prospects. *Journal of Environmental Management, 345*, 118585.

Goswami, J., Deka, M. K., Devi, R., Sarma, P. P., & Brahma, A. (2024). A review on optimizing organic waste management and income generation through vermicomposting and AI-powered vermicomposting: Insights from Guwahati, Assam. *International Journal, 13*(6).

Hajam, Y. A., Kumar, R., & Kumar, A. (2023). Environmental waste management strategies and vermi transformation for sustainable development. *Environmental Challenges, 13*, 100747.

Hao, J., Wang, L., Aspe, N. M., Han, A. C., Chen, M., Li, M., & Wu, D. (2024). Seasonal dynamics of gut microbiome: A study of multi-kingdom microbiota of earthworm gut in an urban park. *Applied Soil Ecology, 195*, 105259.

Hasan, M. M., Rashid, H. O., Pavel, M. A., Saima Alam Trisha, S. A., & Upama, S. (2023). Feces TO fertilizer: an innovative vermicomposting method for environmentally sustainable faecal sludge management. (Doctoral dissertation, Soanargaon Universiy (SU)).

Hemalatha, B., Sashikkumar, M. C., Vivek, S., Ramesh, S., Babu, M. D., Laxmipriya, S., & Priya, V. (2024). Experimental study of biodegradability of organic waste with industrial waste combined with effluents: A comparison by vermicomposting technology.

Henao-Rodríguez, C., Lis-Gutiérrez, J. P., & Sierra, A. S. G. (2024). Factors influencing environmental awareness and solid waste management practices in Bogotá: An analysis using machine learning. *Air, Soil and Water Research, 17*, 11786221241261188.

Hiranmai, R. Y., Neeraj, A., & Vats, P. (2024). Improvement of soil health and crop production through utilization of organic wastes: A sustainable approach. *International Journal of Recycling of Organic Waste in Agriculture, 13*(1).

Huang, K., Bhat, S. A., Li, F., & Kumar, V. (Eds.). (2023). Earthworm technology in organic waste management: recent trends and advances.

Javed, S., Aftab, A., Safi, S. Z., Zahoor, A. F., Anwar, N., Gul, S., & Arshad, M. (2023). Sustainable recycling of manure and reuse to mitigate climate change. In Climate changes mitigation and sustainable bioenergy harvest through animal waste: sustainable environmental implications of animal waste (pp. 69–103). Cham: Springer Nature Switzerland.

Kadam, R., Jo, S., Lee, J., Khanthong, K., Jang, H., & Park, J. (2024). A review on the anaerobic co-digestion of livestock manures in the context of sustainable waste management. *Energies, 17*(3), 546.

Karimi, S., Raza, T., & Mechri, M. (2024). Composting and vermitechnology in organic waste management. In *Environmental engineering and waste management: recent trends and perspectives* (pp. 449–470). Cham: Springer Nature Switzerland.

Khan, I., Kurovsky, A., & Babenko, A. (2024). The role of microbial enzymes in vermicomposting of organic wastes. *Science and Innovation, 3*(D4), 74–85.

Kong, Y., Zhang, J., Zhang, X., Gao, X., Yin, J., Wang, G., & Yuan, J. (2024). Applicability and limitation of compost maturity evaluation indicators: A review. *Chemical Engineering Journal*, 151386

Kooch, Y., Parsapour, M. K., & Wu, D. (2024). Earthworms: functional traits and soil properties. In *Earthworms and ecological processes* (pp. 153–188). Cham: Springer Nature Switzerland.

Kumari, B., Prakash, O., Pal, P., Singh, P. K., Singh, M. P., Kumar, P., & Singh, A. K. (2024). Advancement in sustainable management and valorization of solid waste through composting and vermitechnology. In: Recent Trends in Management and Utilization of Industrial Sludge (pp. 359–397). Cham: Springer Nature Switzerland.

Lim, T. T., Massey, R., McCann, L., Canter, T., Omura, S., Willett, C., & Dodson, L. (2023). Increasing the value of animal manure for farmers.

Makkar, C., Singh, J., Parkash, C., Singh, S., Vig, A. P., & Dhaliwal, S. S. (2023). Vermicompost acts as bio-modulator for plants under stress and non-stress conditions. *Environment, Development and Sustainability, 25*(3), 2006–2057.

Maksymenko, N. V., Gololobova, O. O., Sonko, S. P., Stolov, V. O., & Shiyan, D. V. (2024) Utilization of vegetative waste from green infrastructure of cities "in-situ". In *IOP Conference Series: Earth and Environmental Science*, Dec 2024 (Vol. 1415, No. 1, p. 012127). IOP Publishing.

Manchal, R., Venuste, T., & Verma, S. R. (2023). Vermicomposting, a key to sustainable agriculture: A review. *Farming & Management, 8*(2), 81–93.

Manea, E. E., Bumbac, C., Dinu, L. R., Bumbac, M., & Nicolescu, C. M. (2024). Composting as a sustainable solution for organic solid waste management: Current practices and potential improvements. *Sustainability, 16*(15), 6329.

Manyanga, M. A., Marumure, J., Chigede, N., Mubvuma, M., Mudzengi, C. P., Nyambiya, I., & Muteveri, M. (2024). Vermicomposting for improved soil health: prospects for degraded soils. In *The marginal soils of Africa: rethinking uses, management and reclamation* (pp. 325–337). Cham: Springer Nature Switzerland.

Marumure, J., Makuvara, Z., Gufe, C., Alufasi, R., Chigede, N., & Karidzagundi, R. (2023). An outstanding perspective on biological dynamics in vermicomposting matrices. In *Vermicomposting for sustainable food systems in africa* (pp. 59–87). Singapore: Springer Nature Singapore.

Mishra, S., & Das, A. P. (2022). Treatment of the wastewater polluted with synthetic microfiber released from washing machine. In B. B. Das, H. Hettiarachchi, P. K. Sahu, & S. Nanda (Eds.), *Lecture Notes in Civil Engineering* (Vol. 207, pp. 109–117). Springer Science and Business Media Deutschland GmbH. https://doi.org/10.1007/978-981-16-7509-6_9/COVER

Mishra, S., & Das, A. P. (2023). Microbial Remediation of Synthetic Microfiber Contaminated Wastewater. In M. P. Shah (Ed.), *Microbial Technologies in Industrial Wastewater Treatment* (pp. 337–350). Springer, Singapore. https://doi.org/10.1007/978-981-99-2435-6_17

Mishra, S., Ghosh, S., van Hullebusch, E. D., Singh, S., & Das, A. P. (2023). A critical review on the recovery of base and critical elements from electronic waste-contaminated streams using microbial biotechnology. *Applied Biochemistry and Biotechnology, 2023*, 1–30. https://doi.org/10.1007/S12010-023-04440-X

Mishra, S., Rout, P. K., & Das, A. P. (2019). Solar photovoltaic panels as next generation waste: a review. *Biointerface Research in Applied Chemistry*, *9*(6).

Mishra, S., Singh, R. P., Rout, P. K., & Das, A. P. (2022). Membrane bioreactor (MBR) as an advanced wastewater treatment technology for removal of synthetic microplastics. *Development in Wastewater Treatment Research and Processes*, 45–60. https://doi.org/10.1016/B978-0-323-85583-9.00022-3

Mohanty, S., Ghosh, S., Bal, B., & Das, A. P. (2018). A review of biotechnology processes applied for manganese recovery from wastes. *Reviews in Environmental Science and Bio/Technology, 17*(4), 791–811. https://doi.org/10.1007/S11157-018-9482-1

Mohite, D. D., Chavan, S. S., Jadhav, V. S., Kanase, T., Kadam, M. A., & Singh, A. S. (2024). Vermicomposting: A holistic approach for sustainable crop production, nutrient-rich bio fertilizer, and environmental restoration. *Discover Sustainability, 5*(1), 60.

Mor, V., Reddy, B. V., Gupta, S., & Singh, S. (2024) Environmental considerations and sustainability. In *Rhizosphere revolution* (pp. 173–192). CRC Press.

Naik, J., Mahanaik, T., Pradhan, M., Das, A. K., Symon, K., Rout, M., & Das, S. (2024). Vermicasting: The processing of organic wastes through earthworms. *International Journal of Biological & Pharmaceutical Science, 10*(1), 1–12.

Namasivayam, S. K. R., Kumar, P., Samrat, K., Moovendhan, M., Kavisri, M., Sivakumar, L., & Shyamsundar, D. (2024). Development of high organic-rich low-cost medium derived from microbial consortium decomposed vegetable wastes for the viable inocula production of potential fungal biopesticide *Metarhizium anisopliae*. *Biomass Conversion and Biorefinery, 14*(21), 27853–27869.

Nascimento-Gonçalves, E., Azevedo, T., Lopes, H., Sousa, J. R., Oliveira, P. A., Roboredo, M., Coimbra, A. M., & Morais, M. C. (2024). Vermicomposting as a valorization solution to the winery sector by-products. *Agronomy, 14*(6), 1111.

Nathan, V. K., & Mukupa, G. (2024). Zero-Waste processes utilizing microbes check for updates Moffat Mutebele, Subbaiya Ramasamy, Todd Johnson, Manikandan Sivasubramanian, Kalirajan Arunachalam. *Harnessing Microbial Potential for Multifarious Applications*, 299.

Nyamusamba, R., Mudziwapasi, R., Jomane, F., Mugande, U., Ndudzo, A., Sebata, S., & Muteveri, M. (2023). Insights into earthworm biology for vermicomposting. In *Vermicomposting for sustainable food systems in Africa* (pp. 89–108). Singapore: Springer Nature Singapore.

Oyege, I., & Balaji Bhaskar, M. S. (2023). Effects of vermicompost on soil and plant health and promoting sustainable agriculture. *Soil Systems, 7*(4), 101.

Paculanang, R. A., Aguirre, E. L., & Empiales, P. G. L. (2024). Modern farming through vermicompost harvester: redefining farming efficiency. In *International Conference on Inventive Communication and Computational Technologies*, June 2024 (pp. 769–778). Singapore: Springer Nature Singapore.

Peng, W., Wang, Y., Cui, G., Xu, Q., Zhang, H., He, P., & Lü, F. (2024). Compost quality, earthworm activities and microbial communities in biochar-augmented vermicomposting of dewatered activated sludge: The role of biochar particle size. *Biochar, 6*(1), 73.

Raimi, H. S. A. M., Shahrul, N. N. A., Ismail, T. N. H. T., Ali, R., Shaheed, R., & Yusop, F. M. (2024). The Influence of compost bin volume and effective microorganisms (EM) quantity for efficient food waste composting. *Journal of Advanced Research in Micro and Nano Engineering, 16*(1), 70–84.

Raman, A. P. S., Aslam, M., Singh, P., Pandey, G., Jain, P., & Kumari, K. (2024). Solid and liquid waste management in urban and rural areas: current practices and future perspectives. In *Integrated waste management: a sustainable approach from waste to wealth* (pp. 141–157). Singapore: Springer Nature Singapore.

Saha, S., Ghosh, S., & Pal, H. (2024). Resource recovery from food waste through conversion to value-added products. In *Sustainable clean energy production using waste biomass: sustainable energy production and utilization* (pp. 43–73). Singapore: Springer Nature Singapore.

Samal, K., & Mohanty, M. (2024). Earthworm-associated bacterial community and its role in organic waste decomposition. In *Earthworm technology in organic waste management* (pp. 1–14). Elsevier.

Sande, T. J., Tindwa, H. J., Alovisi, A. M. T., Shitindi, M. J., & Semoka, J. M. (2024). Enhancing sustainable crop production through integrated nutrient management: A focus on vermicompost, bio-enriched rock phosphate, and inorganic fertilisers–a systematic review. *Frontiers in Agronomy, 6*, 1422876.

Senadheera, U. E., Jayasanka, J., Udayanga, D., Hewawasam, C., Amila, B., Takimoto, Y., & Tadachika, N. (2024). Beyond composting basics: A sustainable development goals—oriented strategic guidance to IoT integration for composting in modern urban ecosystems. *Sustainability, 16*(23), 10332.

Sharma, A., Mittal, R., Bohat, N., Malik, K., & Dahiya, B. (2024b). Sustainable agriculture and soil enrichment through diverse organic vermicompost synthesized from different organic waste. *International Journal of Environmental Science and Technology*, 1–14.

Sharma, P., Bano, A., Singh, S. P., Varjani, S., & Tong, Y. W. (2024b). Sustainable organic waste management and future directions for environmental protection and techno-economic perspectives. *Current Pollution Reports, 10*(3), 459–477.

Sharma, V. (2024). Windrow composting-an integrated approach for solid waste management. *International Journal of Environment*

Sihlangu, E., Luseba, D., Regnier, T., Magama, P., Chiyanzu, I., & Nephawe, K. A. (2024). Investigating methane, carbon dioxide, ammonia, and hydrogen sulphide content in agricultural waste during biogas production. *Sustainability, 16*(12), 5145.

Sivasabari, K., Parthasarathy, S., Chandran, D., Sankaralingam, S., & Ajaykumar, R. (2023). Vermicompost and vermi-leachate in pest and disease management. In *Vermicomposting for sustainable food systems in Africa* (pp. 279–304). Singapore: Springer Nature Singapore.

Syarifinnur, S., Nuraini, Y., Prasetya, B., & Handayanto, E. (2023). Comparing compost and vermicompost produced from market organic waste. *International Journal of Recycling of Organic Waste in Agriculture, 12*(3).

Syarifinnur, S., Suriadi, A., Hadiawati, L., & Nugraha, Y. (2024). Transforming organic waste into productive resources through vermicompost and hydroponics in rice agriculture: A review. *International Journal of Recycling of Organic Waste in Agriculture, 13*(3).

Thirunavukkarasu, A., Sivashankar, R., Nithya, R., Sathya, A. B., Priyadharshini, V., Kumar, B. P., & Krishnamoorthy, S. (2023). Sustainable organic waste management using vermicomposting: A critical review on the prevailing research gaps and opportunities. *Environmental Science: Processes & Impacts, 25*(3), 364–381.

Toor, M. D., Basit, A., Okorie, B., Nath, D., Din, M. M. U., Kumar Verma, P., & Mohamed, H. I. (2024). Earthworms as catalysts for climate-resilient agriculture: Enhancing food security and water management in the face of climate change. *Water, Air, & Soil Pollution, 235*(12), 779.

Tripathy, B., Dey, S., Dubey, N. K., Lo, W.-C., & Das, A. P. (2024). An omics-based approach for managing the risk of COVID-19-generated wastes (pp. 51–66). https://doi.org/10.1007/978-3-031-50840-0_4

Tyagi, K. S., Singh, D. P., & Gupta, R. K. (2023). Rural entrepreneurship through vermicomposting-case studies. *Dev Sanskriti Interdisciplinary International Journal, 22*, 06–11.

ul Baligah, H., Chesti, M. H., Ahmed Baba, Z., Mir, S., Wani, F. J., Bhat, J. A., & Khan, I. M. (2025). Vermicomposting technology as a dynamic strategy to mitigate environmental crisis: a bibliometric study of last three decades. *Environmental Technology, 46*(1), 72–86.

Urechescu, L., Peticilă, A., Dobrin, E., & Drăghici, E. (2024). Experiment and study on compost and vermicompost. *Journal of Horticulture, Forestry and Biotechnology, 28*(1), 81–88.

Vambe, M., Coopoosamy, R. M., Arthur, G., & Naidoo, K. (2023). Potential role of vermicompost and its extracts in alleviating climatic impacts on crop production. *Journal of Agriculture and Food Research, 12*, 100585.

Walia, S. S., & Kaur, T. (2024). Beneficial role of vermicompost: nutrient content in vermicompost and success stories. In *Earthworms and vermicomposting: species, procedures and crop application* (pp. 135–146). Singapore: Springer Nature Singapore.

Wang, F., Zhang, Y., Su, Y., Wu, D., & Xie, B. (2024). Pollutant control and nutrient recovery of organic solid waste by earthworms: mechanism and agricultural benefits of vermicomposting. *Journal of Environmental Chemical Engineering*, 112610.

Wehrli, M., Slotsbo, S., Ge, J., & Holmstrup, M. (2024). Acclimation temperature influences the thermal sensitivity of injury accumulation in *Folsomia candida* at extreme low and high temperatures. *Current Research in Insect Science, 6*, 100089.

Wongkiew, S., Polprasert, C., Noophan, P. L., Koottatep, T., Kanokkantapong, V., Surendra, K. C., & Khanal, S. K. (2023). Effects of vermicompost leachate on nitrogen, phosphorus, and microbiome in a food waste bioponic system. *Journal of Environmental Management, 339*, 117860.

Zhou, L., Yang, X., Wang, X., Feng, L., Wang, Z., Dai, J., & Xie, Y. (2023). Effects of bacterial inoculation on lignocellulose degradation and microbial properties during cow dung composting. *Bioengineered, 14*(1), 2185945.

Vermicomposting: A Sustainable Waste Management Approach

Sharad Sachan, Sonali Kokale, and Atin Kumar

Abstract The increasing generation of organic waste, largely driven by population growth and urban expansion, presents substantial environmental challenges, such as land contamination, greenhouse gas emissions and the depletion of landfill space. Vermicomposting emerges as an eco-friendly and sustainable waste management strategy, converting organic waste into nutrient-rich compost through the biological actions of earthworms. This chapter examines the foundational principles, processes and advantages of vermicomposting as a circular economy paradigm for sustainable waste management. Initially, the chapter provides an overview of organic waste sources and their environmental repercussions, then proceeds to a detailed description of the vermicomposting process. This includes selecting appropriate earthworm species, preparing feedstock and optimizing environmental conditions. It highlights key benefits, such as enhanced soil fertility, decreased dependence on chemical fertilizers and alleviation of waste management challenges. Moreover, the chapter rigorously analyses obstacles to large-scale adoption of vermicomposting, such as inconsistent feedstock quality, pathogen control and necessary infrastructure. It also discusses recent advancements, such as the use of bioengineered earthworms, integration with other waste management technologies and the potential applications in both urban and rural environments. To illustrate its scalability and economic feasibility, real-world case studies are featured throughout the chapter. In conclusion, it underscores the importance of policies, awareness initiatives and public–private partnerships in establishing vermicomposting as a fundamental aspect of sustainable waste management practices. By fostering circular resource utilization and

S. Sachan
School of Agriculture, Lovely Professional University, Phagwara, Punjab, India
e-mail: sharad.19461@lpu.co.in

S. Kokale
School of Agriculture, ITM University, Gwalior, Madhya Pradesh, India

A. Kumar (✉)
School of Agriculture, Uttaranchal University, Dehradun, Uttarakhand, India
e-mail: atinchaudhary0019@gmail.com

A. P. Das (ed.), *Bioconversion of Environmental Waste to Value Added Products*, Sustainable Environmental Waste Management Strategies,
https://doi.org/10.1007/978-3-032-14215-3_15

minimizing environmental footprints, vermicomposting symbolizes a transformative solution to global waste management challenges while promoting agricultural productivity and ecological harmony.

Keywords Vermicomposting · Sustainable waste management · Organic waste recycling · Circular economy · Soil fertility

1 Introduction

Sustainable waste management techniques have become essential to slow down environmental damage as world garbage creation keeps increasing (Dey et al., 2023; Mishra et al., 2019, 2023). Vermicomposting is one such environmentally benign and effective method whereby organic waste is biologically broken down by earthworms and microorganisms. Biodegradable waste is turned into vermicompost, a nutrient-dense organic fertilizer that increases soil fertility, strengthens soil structure and promotes sustainable agriculture via this process (Lirikum et al., 2022). Vermicomposting offers a sustainable way to handle organic waste by hastening the breakdown of organic materials, therefore lowering landfill overflow and pollution at once. Moreover, this approach reduces greenhouse gas emissions, including methane, which are usually generated in conventional waste disposal techniques such incineration and landfilling (Aalok et al., 2008). By reintegrating organic matter into the soil and thereby fostering long-term soil health and agricultural output, vermicomposting also helps to end the nutrient cycle.

Vermicomposting includes the decomposition of organic materials, such as food scraps, agricultural leftovers and animal manure, into a stable, humus-like product by the synergistic activity of earthworms and beneficial microbes. By consuming organic waste, breaking it down, and excreting nutrient-rich castings that improve soil fertility and microbial activity, earthworms are absolutely essential in this process. These castings include critical macronutrients such as nitrogen, phosphate and potassium, as well as helpful microbes that boost soil health and plant development (Goswami et al., 2024; Mishra et al., 2022). Additionally, vermicompost boosts the bioavailability of micronutrients such as iron, zinc and manganese, further boosting plant growth. The most commonly used earthworm species in vermicomposting include *Eisenia fetida* (red wigglers), *Eudrilus eugeniae* (African nightcrawler) and *Perionyx excavatus* (Indian blue worm), all of which are known for their high reproductive rates, adaptability to various organic waste types, and efficiency in organic matter conversion (Hajam et al., 2023). The decomposition process is governed by environmental elements like as temperature, moisture levels, aeration and pH, all of which must be carefully maintained to enhance composting effectiveness (Biswal et al., 2021; Singh Jadaun et al., 2022). To speed the breakdown process and raise the quality of the resultant vermicompost product, advanced vermicomposting methods comprising pre-composting treatments and microbial inoculation have been developed.

One of the primary advantages of vermicomposting is its enormous contribution to soil health and agricultural output. Vermicompost promotes soil structure by boosting its porosity and aeration, which enables improved root penetration and water infiltration. Additionally, it boosts the soil's water-holding capacity, lowering the need for regular watering and enhancing resistance against drought conditions. The vast microbial variety inherent in vermicompost encourages soil biodiversity, boosting the growth of beneficial bacteria and fungi that help in nutrient cycling and reduce soil-borne diseases. This natural disease suppression minimizes the occurrence of plant infections, resulting to better crops and less need on chemical treatments. Furthermore, vermicompost has been discovered to release nutrients in a gradual and steady way, providing long-term soil fertility and sustained plant development. Studies suggest that crops cultivated with vermicompost display better yield, improved root growth and greater resistance to pests and diseases, making it a crucial component in organic and sustainable farming systems (Lirikum et al., 2022).

Despite its obvious benefits, vermicomposting also offers certain problems that need to be solved for its broad adoption. Maintaining appropriate conditions, such as temperature, moisture levels, aeration and pH balance, is vital for successful decomposition and earthworm survival. If environmental circumstances diverge from the optimal range, earthworm activity may fall, resulting to reduced composting efficiency and lower-quality vermicompost. Additionally, the presence of hazardous substances, heavy metals, or non-biodegradable pollutants in the waste material might significantly affect the health of the worms and the quality of the end product.

Large-scale deployment of vermicomposting involves major investment in infrastructure, personnel and ongoing monitoring. Establishing and operating commercial-scale vermicomposting operations entails issues such as procuring a steady supply of organic waste, ensuring correct waste segregation to avoid contamination and maintaining stable environmental conditions. Furthermore, economic considerations, such as the initial costs of setting up vermicomposting facilities, personnel expenses and market demand for vermicompost, impact the viability and profitability of large-scale operations.

However, recent improvements have substantially boosted the scalability and effectiveness of vermicomposting. Innovations such as continuous flow systems, automated vermireactors and microbial inoculation techniques have increased process control, decreased manpower needs and raised the overall composting efficiency (; Ghosh et al., 2023). These technical developments enable more effective management of higher waste quantities, making vermicomposting a feasible and sustainable alternative for both small-scale and industrial applications (Hajam et al., 2023). Additionally, research is continuing to study the possibilities of combining vermicomposting with other sustainable waste management strategies, such as anaerobic digestion and biochar synthesis, to develop a more holistic and efficient approach to organic waste recycling.

Given the rising demand for sustainable waste management solutions, vermicomposting offers a practical and ecologically beneficial alternative to traditional trash disposal techniques. This chapter discusses the function of vermicomposting in organic waste recycling, its environmental advantages, problems and technical improvements, underlining its relevance in fostering a circular economy and sustainable farming practices.

2 Principles of Composting

2.1 Waste Reduction Efficiency

One of the key benefits of vermicomposting is its capacity to reduce the volume and weight of organic waste. Research suggests that vermicomposting can convert up to 60% of wasted organic materials into beneficial soil nutrients, considerably lowering the waste stream to landfills. A specific study on floral waste in India exhibited a 50% weight decrease throughout the vermicomposting process, indicating its efficacy in volume reduction (Ratnasari et al., 2023). This reduction is essential, as organic waste in landfills creates methane, a strong greenhouse gas contributing to climate change. By redirecting garbage, vermicomposting mitigates these emissions, consistent with circular economy principles.

2.2 Nutrient Recycling and Soil Health

Vermicompost is rich in critical elements such as nitrogen (N), phosphorous (P), potassium (K) and micronutrients, making it a viable alternative to synthetic fertilizers. Studies comparing vermicompost with inorganic fertilizers, such as one on tomato plants, indicated that vermicompost at 20 t/ha boosted dry weight of shoots by 52 times and fruit quantity by 6 folds, surpassing high rates of NPK fertilizers (Kashem et al., 2015). This nutrient richness boosts soil fertility, improves porosity, water retention and microbial activity, leading to enhanced germination, plant development and crop production (Blouin et al., 2019).

The presence of beneficial microorganisms and plant growth hormones in vermicompost further enhances soil health. It promotes soil structure and drainage, suppresses plant diseases and insect pests and lowers dependency on chemical fertilizers, which can contribute to soil deterioration and water pollution (Sherman et al., 2020). For instance, a review by NC State Extension highlights benefits including nutrient increase, enhanced soil structure and disease control, making vermicompost a sustainable choice for agriculture.

2.3 Sustainability Contributions

Vermicomposting aids to sustainability by tackling several environmental concerns. It decreases landfill use, lowers greenhouse gas emissions and stimulates plant growth, harmonizing with sustainable development goals. The method is low-energy compared to incineration or landfilling, and its scalability from backyard bins to industrial reactors makes it accessible for both urban and rural settings (Vyas et al., 2022). Additionally, vermicomposting can detoxify some wastes by lowering heavy metal levels, further aiding environmental restoration.

Recent studies stress its importance in decreasing ammonia and nitrous oxide emissions during waste treatment, with technologies including intermittent aeration and bulking agents minimizing greenhouse gas emissions (Enebe & Erasmus, 2023). The connection with standard composting also enhances nutrient cycling, boosting the sustainability of waste management operations.

3 Types of Worms Used for Vermicomposting

The best sorts of worms for vermicomposting are largely epigeic species, which flourish in organic waste rather than in soil. Here are the most preferred types (Fig. 1).

3.1 Red Wiggler (E. fetida)

This is the most common choice for vermicomposting owing to its quick reproduction and ability to flourish in a wide temperature range (55–95 °F). Red wigglers are tiny, effective at breaking down organic stuff and can consume up to half their body weight in food every day. They are normally available for purchase and are perfect for indoor composting systems.

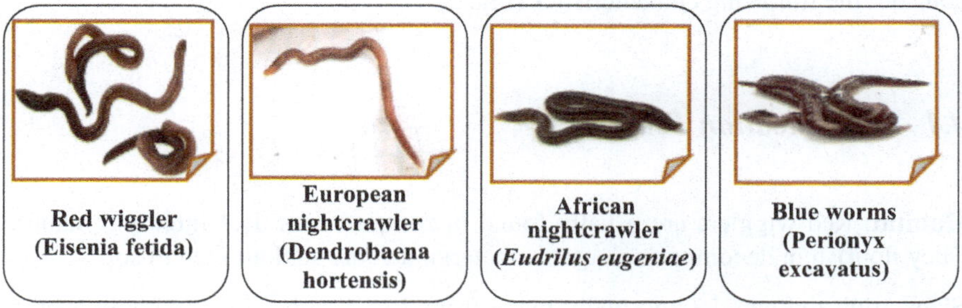

Fig. 1 Different types of earthworms used for vermicomposting

3.2 European Nightcrawler (Dendrobaena hortensis)

Larger than red wigglers, European nightcrawlers require somewhat lower temperatures. They are efficient composters but reproduce more slowly than red wigglers. They can also be used as fishing bait due to their size.

3.3 African Nightcrawler (E. eugeniae)

This species is bigger and performs well in warmer circumstances but requires higher temperatures (above 60 °F) to survive. It can create granular castings that are considered nutrient-rich. However, it may not be ideal for chilly areas without regulated settings.

3.4 Blue Worms (P. excavatus)

These worms can also be employed in tropical and subtropical locations. They are good at digesting organic waste but are less readily accessible compared to red wigglers and European nightcrawlers.

While numerous worm species may be employed for vermicomposting, red wigglers are often the greatest option because to their flexibility, effectiveness and availability. European night crawlers and African night crawlers offer as suitable alternatives depending on climatic circumstances (Gajalakshmi & Abbasi, 2004).

4 Identification of Red Wigglers and Red Worms

To identify red wigglers (*E. fetida*) and distinguish them from other types of worms, consider the following characteristics (Fig. 2).

4.1 Identification Features

Habitat: Red wigglers are usually found in compost bins, leaf litter and manure. They flourish in decomposing organic material and are seldom seen in soil.

Size: Adult red wigglers normally range from 2 to 4 inches long, with an average length of around 2.75 inches. They can stretch up to 5 inches but normally do not surpass this length.

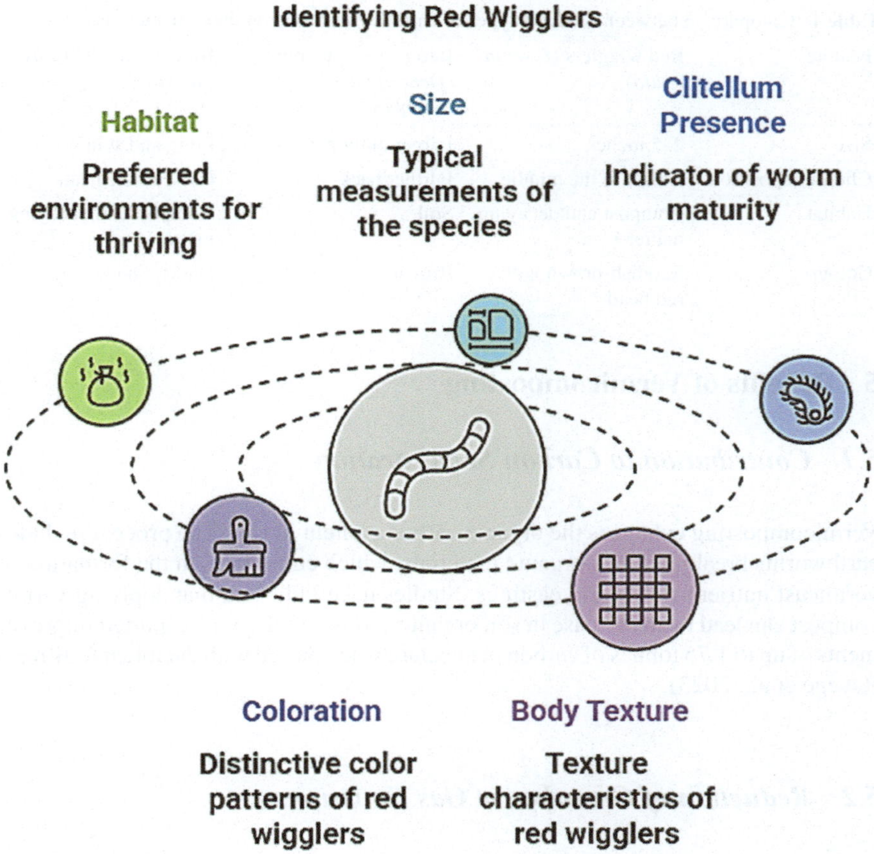

Fig. 2 Identification of red wigglers and red worms

Clitellum presence: The presence of a clitellum (a thickened, glandular portion of the body) is a vital signal of maturity. Only adult worms will have a visible clitellum.

Coloration: Red wigglers have a reddish-brown hue with a unique red head. They may also have alternating stripes of red with lighter bands on their head. The tail is generally yellowish, which might aid in identification.

Body texture: They have smooth skin with tiny rings around their body parts. This smooth texture helps them to move effortlessly across their surroundings. The comparison between different types of earthworms given in Table 1.

These features will help to accurately identify red wigglers and differentiate them from similar species.

Table 1 Comparison between different types of earthworms based on their characteristics

Feature	Red wigglers (*Eisenia fetida*)	European nightcrawlers (*Dendrobaena hortensis*)	Blue worms (*Perionyx excavatus*)
Size	2–5 inches	Larger, plumper	Long and skinny
Clitellum position	Closer to the middle	Further back	Close to the head
Habitat	Compost and decaying matter	Soil	Compost and decaying matter
Colour	Reddish-brown with red head	Brown	Darker shade

5 Benefits of Vermicomposting

5.1 *Contribution to Carbon Sequestration*

Vermicomposting enhances the organic carbon content in soil. The process includes earthworms breaking down organic materials, which culminates in the formation of vermicast nutrient-rich worm castings. Studies have indicated that applying vermicompost can lead to an increase in soil organic carbon stocks, with reported improvements of up to 1.75 tonnes of carbon per hectare when paired with chemical fertilizers (Oyege et al., 2023).

5.2 *Reduction of Greenhouse Gas Emissions*

Vermicomposting is associated with lower emissions of greenhouse gases compared to typical composting procedures. For instance, it has been found to reduce methane (CH_4) and nitrous oxide (N_2O) emissions considerably, with reductions reported at 32% and 40%, respectively, at ideal moisture conditions (Panda et al., 2022). This decrease occurs because the aerobic conditions maintained during vermicomposting promote less gas generation compared to anaerobic breakdown processes.

5.3 *Improved Soil Structure and Microbial Activity*

The addition of vermicast enhances soil structure, encouraging greater aeration and water retention, which are critical for plant development and microbial activity. Enhanced microbial activity leads to enhanced mineralization of organic matter and further stability of carbon in the soil (De Castro et al., 2023). The presence of earthworms also adds to the physical preservation of organic materials, thereby prolonging its lifetime in the soil.

5.4 *Sustainable Agricultural Methods*

By incorporating vermicomposting into agricultural operations, farmers may achieve sustainable land management that not only enhances crop yields but also contributes to climate change mitigation. The use of vermicompost has been associated to enhanced nutrient availability and greater resistance against abiotic stressors, which enhances the long-term sustainability of agricultural systems (Oyege et al., 2023).

5.5 *Improved Soil Structure*

Vermicompost promotes soil aeration, porosity and water retention capacity. This is particularly advantageous in drought-prone locations where preserving moisture is vital for plant development. The altered soil structure helps store moisture more efficiently, allowing crops to grow even under water-limited situations (Lim et al., 2015).

5.6 *Enhanced Nutrient Availability*

The application of vermicompost enhances the soil with vital nutrients, including nitrogen, phosphate and potassium. This nitrogen boost can help plants endure drought stress by enhancing their overall health and resilience. Research suggests that crops treated with vermicompost have higher growth characteristics and production compared to those without it, especially in tough situations (De Castro et al., 2023).

5.7 *Mitigation of Drought Impacts*

Vermicompost has been reported to alleviate the detrimental effects of drought on many crops by boosting their growth and yield. For instance, crops like wheat and tomatoes have exhibited enhanced dry matter production when supplemented with vermicompost, demonstrating its potential to support agricultural output during dry times (Lim et al., 2015).

5.8 Sustainable Agricultural Methods

Integrating vermicomposting into farming operations improves sustainable agriculture by lowering dependency on chemical fertilizers, therefore minimizing environmental effect while boosting soil fertility and crop yields (Hussain & Abbasi 2018). This technique matches nicely with the demands of farmers in drought-prone regions who want resilient farming practices.

5.9 Organic Fertilization

Vermicompost acts as a nutrient-rich organic fertilizer that promotes soil fertility. It includes important plant nutrients in appropriate amounts, supporting healthy plant development and increased harvests.

5.9.1 Soil Improvement

It enhances soil structure, aeration and water retention capacity, making it an efficient soil conditioner. This is particularly advantageous for sandy or compacted soils (Goswami et al., 2024).

5.9.2 Waste Management

Vermicomposting is an excellent way for handling organic waste, including kitchen scraps and agricultural wastes. It decreases the volume of garbage transferred to landfills and minimizes methane emissions related with typical waste disposal methods (Goswami et al., 2024).

5.9.3 Aquaculture Feed

Due to its high protein content, vermicompost may be utilized as an alternative feed in aquaculture, boosting the growth of fish and other aquatic creatures (Hajam et al., 2023).

5.9.4 Plant Growth Regulator

Vermicompost functions as a growth regulator by improving germination rates and encouraging root development, which can lead to more vigorous plants.

5.9.5 Sewage Treatment

The procedure may be modified for sewage treatment by vermifiltration, where earthworms assist break down organic materials and eliminate pathogens in wastewater.

5.9.6 Urban Gardening and Landscaping

Vermicompost is commonly used in urban gardening to improve soil in home gardens, parks and community gardens, boosting the quality of vegetables, herbs and decorative plants.

5.9.7 Economic Benefits

The production of vermicompost can give livelihood chances in rural regions by establishing low-skill occupations associated to waste management and organic farming methods (Goswami, et al., 2024).

5.9.8 Environmental Impact

By recycling organic waste on-site, vermicomposting contributes to filling the metabolic gap in ecosystems, lowering dependency on artificial fertilizers, and supporting sustainable agriculture practices (Hajam et al., 2023).

Vermicompost is a flexible resource that helps sustainable agriculture while solving waste management concerns and promoting environmental health. Its uses span across numerous industries, making it a useful tool for both small-scale farmers and major agricultural businesses.

6 Set up for Vermicomposting System

To set up a successful vermicomposting system, it's vital to choose the suitable container and select the appropriate sort of worms. Below is a full tutorial to help you get started.

6.1 Small-Scale (Home) Systems

Setting up a home vermicomposting system entails selecting an appropriate bin, such as a plastic storage container, and preparing it with bedding materials like wet newspaper or shredded cardboard. Red wigglers are added, often at a rate of 1 pound per square foot of bin space, and fed with kitchen scraps and other organic waste. Maintenance includes keeping the bin wet, maintaining proper aeration and collecting the compost when ready, which might take 8–12 weeks.

6.2 Large-Scale (Industrial) Systems

Large-scale systems include windrows, which are linear heaps of organic waste on the ground, and raised bed or flow-through systems, which employ organized bins for simpler harvesting. Windrows are typically employed on concrete surfaces to limit predator access, while flow-through systems allow for continuous feeding and harvesting, enhancing efficiency (Enebe & Erasmus, 2023) These systems are excellent for farms, enterprises and towns, with new innovations like continuous flow vermireactors boosting waste valorization.

6.3 Choosing an Appropriate Container

A typical alternative for indoor vermicomposting, plastic storage containers (10–14 gallons) are appropriate. They are lightweight, robust and hold moisture effectively, but ensure they have appropriate air holes to prevent excess moisture buildup. Wooden boxes may be beneficial, particularly for outdoor installations. Ensure they are treated to withstand deterioration and have sufficient drainage. Styrofoam coolers are good for insulation and can keep a steady temperature for worms. Old sinks, bathtubs or feeding troughs can be ingeniously utilized as worm bins.

6.4 Design Considerations

Aim for a bin that can manage your household's garbage. A shallow box of around 2 feet by 2 feet and 8 inches deep is suited for small homes. Make many holes in the lid and sides of the bin to facilitate airflow. Cover these holes with lightweight screening to keep pests out while preserving oxygen passage. Fill the bin with carbon-rich bedding such as shredded newspaper, peat moss, or decomposing leaves. This creates a home for the worms and helps preserve moisture.

6.5 *Selecting the Proper Worms*

Red wigglers (*E. fetida*): The most common choice for composting because to their effectiveness in breaking down organic matter, tolerance to diverse circumstances and quick reproductive rate. They flourish at temperatures between 55 and 77 °F (13–25 °C).

Indian blues (*P. excavatus*): Similar to red wigglers but significantly more energetic. They can handle waste effectively and proliferate swiftly but prefer somewhat warmer environments.

European nightcrawlers (*Eisenia hortensis*): These worms can also be utilized but are not as efficient as red wigglers in composting systems built for kitchen trash (Hajam et al., 2023).

6.6 *Preparing the Bedding Material*

Preparing the bedding material for vermicomposting is a vital step that offers a suitable habitat for earthworms and encourages successful composting. Here's how to prepare bedding material effectively:

6.6.1 Shredded Paper

Use plain, non-glossy paper such as newspaper or printer paper. Avoid colourful or glossy paper since it may contain dangerous substances. Shred the paper into strips approximately 1/2 to 3 inches broad.

6.6.2 Dried Leaves

Collect fallen leaves, particularly those that are partially disintegrated. They give carbon and increase aeration in the bin.

6.6.3 Straw or Hay

Straw is recommended over hay to prevent introducing seeds that can sprout and compete with your composting process.

6.6.4 Coir (Coconut Fibre)

Coir is a good bedding alternative as it maintains moisture effectively and promotes aeration.

6.6.5 Sawdust

Use sawdust carefully, since it can clump and limit airflow if used in large quantities.

6.6.6 Animal Manure

Well-rotted manure (excluding dog or cat dung) can also function as bedding, delivering nutrients and helpful microbes.

7 Indoor Vermicomposting

Vermicomposting may actually be done indoors, making it a good choice for individuals with limited outdoor space or who wish to manage food waste more sustainably. Here's how to properly set up and manage an indoor vermicomposting system.

7.1 Set up for Indoor Vermicomposting Bin

7.1.1 Materials Needed

You may purchase a pre-made bin or make your own using two dark-coloured plastic containers (at least 12 inches deep) with lids. Red wriggler worms are the finest choice for indoor composting. Suitable materials include shredded newspaper, cardboard, straw, hay, or dry leaves. To produce air and drainage holes in the bins make the drill.

7.1.2 Steps to Construct Bin

In one bin, drill 15–20 holes in the top sides for ventilation and 15–20 holes in the bottom for drainage. Place a small ceramic pot or brick in the bottom of the second bin to raise the first bin and enable leachate (nutrient-rich liquid) to settle below. Fill the top bin with 4–6 inches of bedding material. Soak it in water beforehand and drain excess moisture. Add your worms to the bedding, allowing them to burrow down and adjust. Start adding kitchen scraps gradually, ensuring you spread food in different regions of the bin to prevent congestion.

7.1.3 Maintenance Tips

Keep your container in a dark, cool spot away from direct sunlight; basements or kitchens work well. Maintain a temperature range between 55 and 75 °F (13 °C to 24 °C). Avoid frigid circumstances as they might kill the worms. The bedding should be wet but not waterlogged. If you smell odours or excess moisture, add dry bedding and aerate the contents. After around 3–6 months, you may harvest the worm castings by transferring the contents to one side of the bin and placing fresh bedding on the other side; worms will migrate over time.

7.2 *Types of Trash Were Suited for Vermicomposting*

Vermicomposting is an excellent method for recycling organic waste, employing earthworms to expedite the decomposition process. Various forms of organic waste are acceptable for this process, each contributing to the development of nutrient-rich vermicompost. Here are the key categories of organic waste that are most suitable for vermicomposting (Hajam et al., 2023).

7.2.1 Kitchen Waste

Peels, cores and trimmings from vegetables. Leftover peels, rinds and cores from fruits. Used coffee grounds are rich in nitrogen. Eggshells provide calcium, which is helpful for plant growth.

7.2.2 Yard and Garden Waste

Grass clippings should be mixed with other materials to prevent compaction. Leaves in backyard composting which provide carbon content. Straw and non-woody plant trimmings break down quickly and add structure.

7.2.3 Agricultural Waste

Crop residues like; stubble, husks and other residual plant components from fields. Manure from herbivores (like cows and horses) is particularly valuable due to its nutrition richness.

7.2.4 Forestry Waste

Wood chips and sawdust is a nitrogen-rich product to avoid nitrogen depletion.

7.2.5 Paper Products

Shredded paper and cardboard can be added as a carbon source but should be free of inks and chemicals.

7.2.6 Other Organic Materials

Biogas slurry, a byproduct from anaerobic digestion that is rich in nutrients. City leaf litter and urban solid waste can include numerous biodegradable items gathered from urban locations (Fig. 3).

7.3 Considerations for Effective Vermicomposting

When employing these organic wastes for vermicomposting, it's crucial to maintain a balance between carbon-rich (browns) and nitrogen-rich (greens) materials to guarantee optimal microbial activity and worm health. Additionally, avoiding some goods is vital; for example, meat, dairy products, greasy meals and processed foods might attract bugs or cause odours. By combining a range of different organic wastes

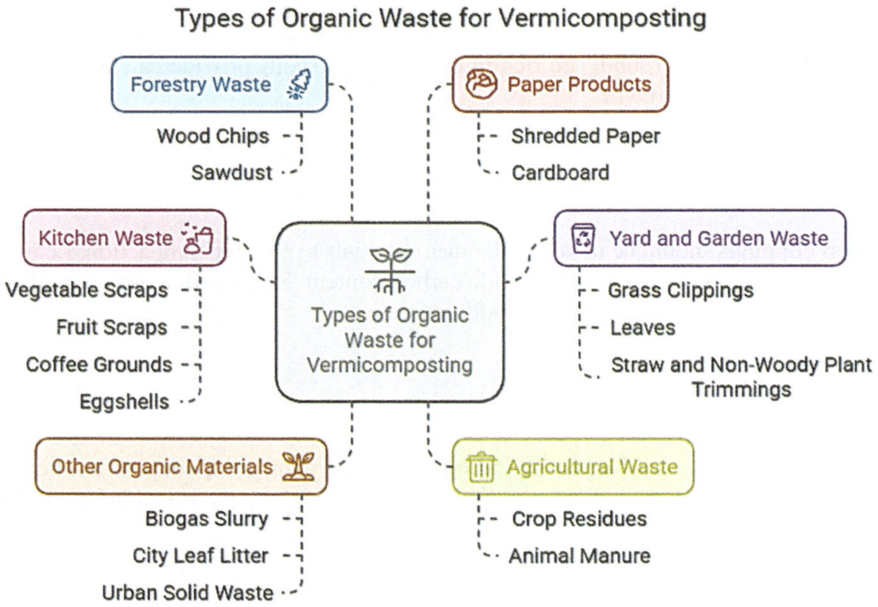

Fig. 3 Types of waste for vermicomposting

into a vermicomposting system, one may successfully manage trash while generating high-quality compost that promotes soil fertility and structure.

7.3.1 Preparation Steps

Provide variation in the bedding, which helps maintain a balanced carbon-to-nitrogen ratio. Before putting bedding to the worm bin, wet it to obtain a damp but not soggy condition. This lets worms breathe through their skin. A decent rule of thumb is to use around 3 parts water to 1 part dry bedding by weight. Start with a foundation layer of 2–3 inches of bedding material at the bottom of your worm bin. Mix different types of bedding materials for optimum results. Optionally, put in a handful of soil or grit (like broken eggshells) to offer minerals that assist in the worms' digestion. Monitor moisture levels often and add water if necessary to keep the bedding damp. Turn the bedding occasionally to aerate it and prevent compaction.

By following these methods and utilizing suitable materials, you may create an excellent bedding habitat for your worms, promoting optimal vermicomposting and creating high-quality compost for your garden or crops.

7.4 Managing the Vermicomposting Process

Managing the vermicomposting process comprises numerous essential phases and considerations to improve the creation of high-quality compost. Here's a systematic summary of the key elements (Vukovic et al., 2021).

7.4.1 Selection of Materials

Use a combination of "green" materials (rich in nitrogen, such as kitchen trash) and "brown" materials (heavy in carbon, such as dried leaves or paper). The optimal carbon-to-nitrogen ratio is around 30:1. Common species utilized include *E. fetida* and *Eisenia eugeniae*, noted for their effectiveness in composting.

7.4.2 Preparation of the Composting Environment

Choose a suitable container (plastic or concrete) with appropriate drainage and aeration. The size should equal to the volume of organic waste available. Start with a layer of soil or sand, followed by partially decomposed organic materials to offer a home for the worms.

7.4.3 Moisture and Temperature Management

Maintain moisture levels between 60 and 90%. Regularly spray water to keep the compost wet but not waterlogged. Monitor temperatures to prevent overheating, which can kill worms. Ideal temperatures vary from 15 to 25 °C (59–77 °F).

7.4.4 Feeding and Maintenance

Introduce organic waste gradually and stack it with bedding materials. Avoid overfeeding to prevent stink and bugs. Regularly stir the compost (1–3 times per week) to aerate it and maintain equal decomposition. Optimal rotation frequency can promote microbial activity and nutrient availability. After roughly 24–50 days, when the compost is black and crumbly, it may be collected. Separate the vermicompost from the worms, which can be returned to the bin for further processing.

8 Comparative Examination of Vermicomposting with Traditional Composting Methods

Vermicomposting and regular composting are both successful ways for recycling organic waste into nutrient-rich soil amendments, but they differ greatly in their procedures, advantages and uses. Here's a full comparison of the two approaches (Table 2).

Both vermicomposting and regular composting have their individual advantages and problems. Vermicomposting is particularly advantageous for those wanting faster results with more nutrient content in smaller spaces, whereas conventional composting is more ideal for bigger operations where high temperatures can be maintained to treat pathogens properly. Choosing between the two approaches relies on unique demands, available area and desired outcomes in organic waste treatment.

9 Economic and Environmental Impacts

Vermicomposting delivers considerable economic and environmental consequences, making it a desirable method for sustainable agriculture and waste management. Here's an outline of these implications based on recent results.

Table 2 Comparative analysis of vermicomposting with traditional composting methods

Aspect	Vermicomposting	Traditional composting
Process	Utilizes earthworms to break down organic materials. The worms ingest the waste and emit nutrient-rich castings	Relies on microorganisms (bacteria and fungus) to degrade organic materials through aerobic or anaerobic processes
Decomposition rate	Faster breakdown, often requiring 2–5 times less time than regular composting (about 2–3 months) due to the effectiveness of worms	Slower process, generally lasting several months to a year, depending on circumstances and materials utilized
Nutrient content	Higher amounts of plant-available nutrients (nitrogen, phosphorous and potassium) and helpful microbes that stimulate plant development	Nutrient content can vary greatly; often lower in bioavailable nutrients compared to vermicompost
Moisture retention	Excellent moisture retention capabilities, making it perfect for use in diverse planting methods	Variable moisture retention; may require more regular monitoring and rotating to maintain moisture levels
Temperature requirements	Does not require high temperatures; worms flourish in colder environments and can perish in severe heat	Often involves high temperatures (thermophilic stage) for successful decomposition, which can kill pathogens but may also cause to nutrient loss
Labour intensity	Lower effort needed since worms aerate the compost organically; less need for turning or mixing	Requires constant turning and monitoring to maintain aeration and moisture levels, which can be labour-intensive
Pest management	May attract pests like fruit flies if not handled correctly; nevertheless, it typically has fewer troubles with weed seeds and infections owing to the worm's digestive process	Can successfully eliminate weed seeds and pathogens through high temperatures if handled appropriately; yet, it can also attract pests if not maintained adequately
Odour control	Typically creates less smell if managed appropriately; odours might develop from excess nitrogen or inadequate conditions	Can create strong odours if anaerobic conditions develop owing to excess moisture or lack of aeration
Scalability	Suitable for small-scale operations (e.g. home composting) but might be less successful for big amounts of garbage without suitable setup	More scalable for huge amounts of organic waste; may be used in community or commercial settings successfully

9.1 Cost Efficiency

Vermicomposting can drastically cut expenditures related with chemical fertilizers. For example, in the Philippines, the production cost of vermicompost is roughly PhP 2–3 (approximately \$0.04–\$0.07) per kg, whereas market values vary from PhP 5 to 8 (\$0.11–\$0.17) per kg, making it financially feasible for farmers.

9.2 Profitability for Farmers

Studies suggest that vermicomposting operations can provide large revenues. For instance, the net profit from generating vermicompost was determined to be Rs. 9.32 per kg, with a benefit–cost ratio of 4.30, suggesting high financial feasibility. Additionally, farmers utilizing vermicompost have reported higher crop yields and reduced reliance on expensive chemical fertilizers (Kayabasi et al., 2021).

9.3 Improved Agricultural Yields

Research demonstrates that the use of vermicompost can boost soil fertility and crop production, leading to better net profits for farmers. For instance, a research in China found that net revenue rose by up to 59.96% with vermicompost application (Zhang et al., 2023).

9.4 Waste Management Savings

Vermicomposting provides a sustainable alternative for managing organic waste, decreasing disposal expenses related with food waste and other organic products.

9.5 Soil Health Improvement

Vermicomposting feeds soil with nutrients and beneficial microbes, boosting soil structure and fertility. This leads to increased water retention and aeration, which are vital for plant growth.

9.6 Reduction in Chemical Usage

By exchanging chemical fertilizers with vermicompost, farmers may lessen their environmental imprint. Vermicompost assists in minimizing soil deterioration and pollutants linked with synthetic fertilizers (Kayabasi &Yilmaz, 2021).

9.7 Greenhouse Gas Emissions Reduction

The usage of vermicompost has been associated to decreased greenhouse gas emissions compared to typical composting processes. It can lower emissions of methane (CH_4), carbon dioxide (CO_2) and nitrous oxide (N_2O), assisting to climate change mitigation efforts.

9.8 Sustainable Agriculture Practices

Implementing vermicomposting improves sustainable agricultural practices by supporting organic farming methods that are less detrimental to the environment while maintaining high productivity levels.

10 Challenges and Solutions in Vermicomposting

Vermicomposting, the process of using earthworms to decompose organic waste, can present several common challenges. Here are the primary problems faced and their solutions (Table 3).

11 Case Studies and Success Stories

Here are some significant case studies and success stories illustrating the usefulness of vermicomposting in diverse contexts:

11.1 Municipal Sewage Sludge Management

A research proved the use of vermicomposting to handle municipal sewage sludge successfully. By optimizing process factors, researchers discovered that the inclusion of cow manure and coffee husks alongside earthworms considerably increased the nutritional profile of the compost generated. This approach not only decreased the volume of sewage sludge but also raised its quality, making it ideal as a fertilizer for agricultural use. The study stressed vermicomposting as a sustainable alternative to typical sludge management approaches.

Table 3 Challenges and solutions in vermicomposting

Challenges	Problem	Solution	References
1. Moisture control	Excess moisture can lead to anaerobic conditions, causing food wastes to decay and even drowning the worms. Conversely, insufficient moisture can dry up the worms and bedding	Maintain a moisture level of roughly 70–80%. To manage excess moisture, maintain good drainage by adding extra holes to the bin or mixing in dry bedding materials. For dryness, spray water regularly to maintain the surroundings damp but not wet	Edwards et al. (2010)
2.Temperature regulation	Worms flourish at temperatures between 50 and 80 °F (10–26 °C). Extreme temperatures might stress or kill the worms	Position the worm bin in a position that avoids intense sunshine and excessive cold. Insulating the bin or employing temperature-regulating materials can assist maintain a stable environment	Dominguez and Edwards (2011)
3. Pest infestation	Pests such as fruit flies and gnats can proliferate if food wastes are not treated appropriately	Bury food waste well inside the bedding and prevent huge pieces of food that can attract bugs. Keeping the garbage covered might also help repel unwelcome insects	Appelhof et al. (2017)
4. Odour issues	A bad scent frequently signals concerns such as overfeeding or inappropriate meal choices	Avoid adding oily, stinky, or animal items to the compost. Regularly check for uneaten food and discard it to prevent odours	Edwards et al. (2010)
5. Bedding material selection	The choice of bedding is critical; materials like peat moss might drop pH levels too much, while paper items may dry out too rapidly	Use a combination of bedding materials like as shredded cardboard, coconut coir, or leaves that retain moisture without becoming too acidic	Edwards and Arancon (2022)
6. Overfeeding	Adding too much food at once might overwhelm the worms and lead to food spoiling	Feed worms little amounts regularly rather than huge volumes occasionally. This helps them digest food effectively without causing waste	Appelhof et al. (2017)

11.2 Agricultural Residues and Crop Production Improvement

In a controlled experiment, researchers evaluated the impact of vermicomposting agricultural leftovers on crop yields. They observed that applying vermicompost greatly enhanced the growth and production of chili plants compared to control groups. The study underlined that adequate worm densities and proper moisture control during vermicomposting led to nutrient-rich compost, which boosted plant health and production.

11.3 Sustainable Rice Farming

A comprehensive study focusses on incorporating vermicompost into rice growing operations. It was shown that mixing vermicompost with mineral fertilizers resulted in greater grain yields, increased nutrient absorption and elevated soil organic carbon levels. The application of vermicompost also resulted to a reduction in greenhouse gas emissions, highlighting its dual advantages for productivity and environmental sustainability (Hussain et al., 2018).

11.4 Enhancing Soil Health in Drought Situations

Research found that applying vermicompost in drought-prone locations greatly increased soil health and crop resilience under water-limited conditions. The use of vermicompost led in greater moisture retention and nutrient availability, leading to increased crop output even during dry times. This success story highlights how vermicomposting may be a crucial method for sustainable agriculture in tough climates (Oyege et al., 2023).

11.5 Community-Based Vermicomposting Initiatives

Several community projects have successfully used vermicomposting as a way of waste management and soil development. These efforts have empowered local farmers by providing them with a sustainable source of organic fertilizer while also solving organic waste disposal challenges. Participants in these initiatives reported higher agricultural yields and reduced dependency on chemical fertilizers, contributing to both economic and environmental advantages (Vukovic et al., 2021).

These case studies highlight the adaptability and efficiency of vermicomposting across many uses, from waste management to boosting agricultural output, making it a beneficial practice for sustainable development.

12 Future Trends and Advances in Vermicomposting

The future of vermicomposting is set for substantial improvements and innovations, driven by increased demand for sustainable agricultural practices and organic waste management solutions. Here are some major trends and prospective advancements in the field:

12.1 Increased Market Demand

The worldwide vermicompost market is anticipated to expand from roughly USD 85.39 million in 2022 to around USD 270.25 million by 2030, with a compound annual growth rate (CAGR) of about 15.56% (Vermicompost Market Report, 2023). This expansion is spurred by growing consumer demand for organic food items and sustainable farming techniques.

12.2 Integration with Technology

Innovations in technology, such as machine learning and Internet of Things (IoT), are projected to boost vermicomposting processes. For instance, machine learning may be used to assess data on soil parameters and improve vermicomposting conditions, enhancing efficiency and efficacy (Alshehrei et al., 2021).

12.3 Focus on Trash Management

Vermicomposting will increasingly be acknowledged as a viable alternative for managing organic waste, including municipal solid trash and agricultural wastes. Its capacity to transform garbage into usable compost while lowering greenhouse gas emissions places it as a crucial actor in sustainable waste management methods (Alshehrei et al., 2021).

12.4 Enhanced Nutrient Managemen

Research continues to study the nutrient content of vermicompost and its influence on soil health and crop productivity. Future research may focus on improving nutrient release under diverse situations, such as drought or anaerobic settings, to optimize its advantages for different crops (Oyege & Balaji Bhaskar, 2023).

12.5 Contaminant-Specific Vermicomposting

There is potential for creating customized vermicomposting procedures to target specific pollutants, such as pharmaceuticals or heavy metals. This might enhance the effectiveness of vermicomposting in treating polluted organic waste, making it a more flexible tool in environmental rehabilitation (De Castro et al., 2023).

12.6 Community-Based Initiatives

Community-driven vermicomposting programs are likely to gain popularity, boosting local food production and sustainable practices while tackling waste management concerns. These programs may empower communities by giving instruction on composting procedures and the benefits of organic fertilizers (Katiyar et al., 2021).

12.7 Research on Long-Term Effects

Future research will likely focus on the long-term implications of vermicompost treatment on soil fertility, agricultural production and environmental sustainability. Understanding these impacts may assist modify application rates and methodologies for diverse agricultural systems (De la Vega, 2016).

The future of vermicomposting is defined by development in market demand, technology integration, better nutrient management tactics and an emphasis on sustainable waste management practices. These findings imply a favourable prognosis for vermicomposting as a vital component of sustainable agriculture and environmental care.

13 Latest Advancements in Worm Breeding for Vermicomposting

13.1 Development of New Earthworm Species

One major novelty is the introduction of a unique Indian earthworm species, Jai Gopal, "properly termed *Perionyx ceylanesis*", produced through selective breeding processes. This species demonstrates various favourable features, including high fertility, resilience to varied temperatures (2–46 °C), and a quick development rate. It generates a larger number of hatchlings per cocoon and may breed year-round, making it particularly beneficial for vermicomposting operations.

13.2 Automated Vermiculture Systems

Automation in vermiculture is getting increasingly complex. Automated systems now regulate feeding schedules, moisture levels, temperature management and pH balance within worm bins. These sophisticated devices offer ideal conditions for worm health and productivity while simplifying the harvesting process by automatically separating completed compost from the worms.

13.3 Stackable Worm Bins

Innovative designs such as stackable worm bins enable for vertical space usage in urban areas or tiny gardens. Each tray may be loaded with bedding and organic waste, enabling worms to move upwards as they devour food. This design not only increases space but also simplifies the collecting of nutrient-rich castings from the lower trays.

13.4 Flow-Through Systems

Flow-through systems simplify the composting process by allowing worms to travel organically towards fresh food sources while leaving behind their castings in the lowest regions of the bin. This technology boosts efficiency and minimizes effort associated with typical composting procedures.

13.5 Enhanced Breeding Techniques

Advanced breeding techniques are being developed to optimize worm populations for certain settings and waste kinds. For example, researchers are concentrating on creating worms that can live on varied organic sources, strengthening their capacity to digest trash effectively while improving the quality of the resultant vermicompost.

13.6 Research on Nutritional Profiles

Ongoing study is exploring the nutritional profiles of different worm species and their influence on compost quality. Understanding how diverse diets impact growth rates and reproductive success can lead to more successful breeding programs customized to specific agricultural demands.

These advancements in worm breeding and vermiculture technologies are opening the way for more sustainable methods in waste management and organic farming, ultimately leading to better soils and ecosystems. As these technologies continue to advance, they will boost the scalability and efficiency of vermicomposting operations globally.

14 Emerging Technologies in Vermicomposting

Emerging technologies in vermicomposting are boosting its efficiency and efficacy, making it a more feasible alternative for waste management and organic fertilizer generation. Here are some major breakthroughs based on recent findings:

14.1 Microbial Amendments

Research has emphasized the benefits of introducing particular microbial strains throughout the vermicomposting process. These microbial supplements can boost nitrogen content, improve aeration and stabilize the final vermicompost product. The presence of beneficial microbes boosts the nutritional profile of the compost, making it more effective as a biofertilizer (Vyas et al., 2022).

14.2 Pre-composting Techniques

Integrating pre-composting with vermicomposting is receiving interest. This approach includes partly degrading organic materials before giving them to earthworms, which speeds the entire composting process and enhances nutrient availability in the end product. This technique has been found to shorten deterioration time and boost the quality of vermicompost (Vyas et al., 2022).

14.3 Continuous Feeding Systems

Innovations in continuous feeding systems enable for the continual digestion of organic waste, such as grape marc, over extended periods (e.g. 12 months). These systems maintain ideal conditions for earthworms and increase the quality of the compost generated, supplementing it with microbial biomass and nutrients (Vyas et al., 2022).

14.4 Waste-Specific Vermicomposting

Research is concentrating on adapting vermicomposting techniques to handle certain types of waste, including industrial wastes like paper sludge and tannery waste. This modification can maximize nutrient recovery and detoxify pollutants, resulting in high-quality compost suited for agricultural application (Vyas et al., 2022).

14.5 Use of Advanced Materials

The development of improved materials for constructing vermireactors is boosting their efficiency. These materials can increase aeration and drainage while permitting simpler maintenance, resulting to better conditions for earthworm activity and compost generation (Alshehrei, & Ameen, 2021).

14.6 Bioremediation Applications

Vermicomposting is widely recognized for its potential in bioremediation, notably in detoxifying industrial wastes. The combined activity of earthworms and microorganisms may successfully break down dangerous compounds, transforming contaminated materials into safe, nutrient-rich compost (Vyas et al., 2022).

These developing technologies are poised to transform vermicomposting by boosting its efficiency, broadening its uses and raising the quality of the final product, so supporting sustainable agricultural practices and effective waste management solutions.

15 Conclusion

Vermicomposting is becoming regarded as a sustainable waste management strategy that tackles the dual issues of organic waste disposal and soil fertility enhancement. The process, which involves the breakdown of organic waste by earthworms and microbes, has significant environmental and agricultural benefits. Vermicomposting stands as a holistic and sustainable solution to waste management, delivering considerable reductions in landfill trash, nutrient recycling and soil health improvements. Its versatility, minimal environmental footprint and recent technical improvements position it as a major method for accomplishing circular economy goals and encouraging sustainable agriculture. While limitations persist, continuing research and developments continue to enhance its efficiency and scalability, making it a key tool for environmental restoration and resource conservation. As the world population continues to expand and organic waste creation grows, vermicomposting provides a realistic option that conforms to the ideals of circular economy and sustainable development. Future study should focus on improving vermicomposting procedures, overcoming operational problems and expanding its scalability to maximize its advantages across varied agricultural systems. By incorporating vermicomposting into larger waste management plans, we may build healthier ecosystems and encourage sustainable agriculture practices for future generations.

References

Aalok, A., Tripathi, A. K., & Soni, P. (2008). Vermicomposting: A better option for organic solid waste management. *Journal of Human Ecology, 24*(1), 59–64. https://doi.org/10.1080/09709274.2008.11906100

Alshehrei, F., & Ameen, F. (2021). Vermicomposting: A management tool to mitigate solid waste. *Saudi Journal of Biological Sciences, 28*(6), 3284–3293. https://doi.org/10.1016/j.sjbs.2021.02.072

Appelhof, M., Olszewski, J., & Amy Stewart (2017). Worms eat my garbage: How to set up and maintain a worm composting system: Compost food waste, produce fertilizer for houseplants and garden, and educate your kids and family, pp. 192. Flower PressISBN-13:9781612129471.

Biswal, P., Ghosh, S., Pal, A., & Das, A. P. (2021). Exploration of probiotic microbial biodiversity in acidic environments (curd) and their futuristic pharmaceutical applications. *Geomicrobiology Journal*. https://doi.org/10.1080/01490451.2021.1956020

Blouin, M., Barrere, J., & Meyer, N. (2019). Vermicompost significantly affects plant growth. A meta-analysis. *Agron. Sustain. Dev., 39*, 34. https://doi.org/10.1007/s13593-019-0579

De Castro, F., Aprile, A., Benedetti, M. and Fanizzi, F. P. (2023). Vermicompost: Enhancing plant growth and combating abiotic and biotic stress. *Agronomy, 13*(4). https://doi.org/10.3390/agronomy13041134

Dey, S., Das, A., Mallick, K., Sahu, A., & Das, A. P. (2023). Environmental petroleum waste: Pollution, toxicity, sustainable remediation, pp. 159–175. https://doi.org/10.1007/978-3-031-48220-5_7

De la Vega, A. (2016). Vermicomposting: The future of sustainable agriculture and organic waste management. In *Lessons from the USA and Cuba*, pp. 1–48.

Dominguez, J., & Edwards, C. A. (2011). Biology and ecology of earthworm species used for vermicomposting. *Waste Management*, 31(8), 1489–1500.

Edwards, C. A., & Arancon, N. Q. (2022). Biology and ecology of earthworms (4th ed.,). Springer Environmental Science Ma. U.S.A. 1–19. https://doi.org/10.1007/978-0-387-74943-3

Edwards, C.A., Arancon, N.Q. and Sherman, R.L. (Eds.), (2010). Vermiculture technology: Earthworms, organic wastes, and environmental management (1st ed.,), pp 623. CRC Press. https://doi.org/10.1201/b10453

Enebe, M. C, & Erasmus, M. (2023). Vermicomposting technology—a perspective on vermicompost production technologies, limitations and prospects. *Journal of Environmental Management, 345*, 118585. https://doi.org/10.1016/j.jenvman.2023.118585

Gajalakshmi, S., & Abbasi, S. A. (2004). Earthworms and vermicomposting. *Indian Journal of Biotechnology., 3*, 486–494.

Ghosh, S., Sinha, J. K., Ghosh, S., Vashisth, K., Han, S., & Bhaskar, R. (2023). Microplastics as an emerging threat to the global environment and human health. *Sustainability, 15*(14), 10821. https://doi.org/10.3390/SU151410821

Goswami, J., Deka, M. K., Devi, R., Sarma, P. P., & Brahma, A. (2024). A review on optimizing organic waste management and income generation through vermicomposting and AI powered vermicomposting: Insights from Guwahati, Assam. *International Journal of Advanced Trends in Computer Science and Engineering. 13*(6). ISSN 2278-3091. https://doi.org/10.30534/ijatcse/2024/011362024

Hajam, Y. A., Kumar, R., & Kumar, A. (2023). Environmental waste management strategies and vermi transformation for sustainable development. *Environmental Challenges., 13*, 100747. https://doi.org/10.1016/j.envc.2023.100747. ISSN 2667-0100.

Hussain, N., & Abbasi, S. A. (2018). Efficacy of the vermicomposts of different organic wastes as "clean" fertilizers: State-of-the-art. *Sustainability, 10*(4), 1205. https://doi.org/10.3390/su10041205

Kashem, M. A., Sarker, A., Hossain, I., & Islam, M. S. (2015). Comparison of the effect of vermicompost and inorganic fertilizers on vegetative growth and fruit production of tomato (*Solanum lycopersicum* L.). *Open Journal of Soil Science, 5*, 53–58. https://doi.org/10.4236/ojss.2015.52006

Katiyar, R. B., Sundaramurthy, S., Sharma, A. K., Arisutha, S., Pratap-Singh, A., Mishra, S., & Khan, M. A. (2023). Vermicompost: an eco-friendly and cost-effective alternative for sustainable agriculture. *Sustainability, 15*(20), 14701. https://doi.org/10.3390/su152014701

Kayabasi, T. E., & Yilmaz, O. (2021). The Importance of vermicompost in agricultural production and economy. *Eurasian Journal of Agricultural Research, 5*(2), 146–159.

Lim, S. L., Wu, T. Y., Lim, P. N., & Shak, K. P. Y. (2015). The use of vermicompost in organic farming: Overview, effects on soil and economics. *Journal of the Science of Food and Agriculture, 95*(6), 1143–1156. https://doi.org/10.1002/jsfa.6849

Lirikum, Kakati, L.N., Thyug, L., L Mozhui (2022). Vermicomposting: an eco-friendly approach for waste management and nutrient enhancement. *Trop Ecology 63*, 325–337. https://doi.org/10.1007/s42965-021-00212

Mishra, S., Dash, D., Al-Tawaha, A. R. M. S., & Das, A. P. (2022). A review on heavy metal ion adsorption on synthetic microfiber surface in aquatic environments. *Applied Biochemistry and Biotechnology, 194*(10), 4639–4654. https://doi.org/10.1007/S12010-022-04029-W/METRICS

Mishra, S., Ghosh, S., van Hullebusch, E. D., Singh, S., & Das, A. P. (2023). A critical review on the recovery of base and critical elements from electronic waste-contaminated streams using microbial biotechnology. *Applied Biochemistry and Biotechnology, 2023*, 1–30. https://doi.org/10.1007/S12010-023-04440-X

Mishra, S., Rout, P. K., & Das, A. P. (2019). Solar photovoltaic panels as next generation waste: a review. *Biointerface Research in Applied Chemistry, 9*(6).

Oyege, I., & Balaji Bhaskar, M. S. (2023). Effects of vermicompost on soil and plant health and promoting sustainable agriculture. *Soil Systems, 7*(4), 101. https://doi.org/10.3390/soilsystems7040101

Panda, A. K., Mishra, R., Dutta, J., Wani, Z. A., Pant, S., Siddiqui, S., & Bisht, S. S. (2022). Impact of vermicomposting on greenhouse gas emission: A short review. *Sustainability, 14*(18), 11306. https://doi.org/10.3390/su141811306

Ratnasari, A., Syafiuddin, A., Mehmood, M. A., & Boopathy, R. (2023). A review of the vermicomposting process of organic and inorganic waste in soils: Additives effects, bioconversion process, and recommendations. *Bioresource Technology Reports, 21*, 101332. https://doi.org/10.1016/j.biteb.2023.101332. ISSN 2589-014X.

Sherman, R., Liverman, R., and Maxa, E. (2020). Vermicomposting leader's guide: A school enrichment curriculum. *NC State Extension.* E08-50326 4H-16-14PW, p. 31. https://content.ces.ncsu.edu/excerpt-of-vermicomposting-leaders-guide

Singh Jadaun, J., Bansal, S., Sonthalia, A., Rai, A. K., & Singh, S. P. (2022). Biodegradation of plastics for sustainable environment. *Bioresource Technology, 347*, 126697. https://doi.org/10.1016/J.BIORTECH.2022.126697

Vermicompost Market Report (2023). *Vermicompost market by application, by product, and by region—global and regional industry overview, market intelligence, comprehensive analysis, historical data, and forecasts* 2023–2030, p. 98. Business Research.

Vukovic, A., Velki, M., Ecimovic, S., Vukovic, R., Stolfa Camagajevac, I., & Loncaric, Z. (2021). Vermicomposting—facts, benefits and knowledge gaps. *Agronomy, 11*(10), 1952. https://doi.org/10.3390/agronomy11101952

Vyas, P., Sharma, S., & Gupta, J. (2022). Vermicomposting with microbial amendment: Implications for bioremediation of industrial and agricultural waste. *Biotechnologia, 103*(2), 203–215. https://doi.org/10.5114/bta.2022.116213

Zhang, M., Liu, Y., Wei, Q., Liu, L., Gu, X., Gou, J., & Wang, M. (2023). Effects of biochar and vermicompost on growth and economic benefits of continuous cropping pepper at karst yellow soil region in Southwest China. *Frontiers in Plant Science, 14*, 1238663. https://doi.org/10.3389/fpls.2023.1238663

Bioconversion of Environmental Waste to Value Added Products: A Sustainable Approach

Waste Reduction Strategies Through Reduced Reliance on Natural Resources

Ayushi Singh

Abstract Obsolete development approaches that harm natural resources have caught world attention by forcing development to find new ways of resource utilization with reduced amounts of waste and dependency on such resources. This chapter demands minimal waste generation as one step forward towards sustainable development that conserves virgin natural resources. It looks at the environmental and economic aspects of resource extraction in general, encourages a shift; practices by embracing circular economy thinking—recycle, reuse; resources—and integrates renewable, biodegradable materials into all production and manufacturing operations. Technological advances were highlighted: AI, 3D printing, and bio-based biotech innovations, among others. Examples include the use of artificial intelligence to optimize supply chains in ways that reduce the waste of materials and to develop microbial technologies that can break down organic and synthetic forms of waste into usable forms. The chapter expands on how the interplay between behavioural change, good policy structures, and industrial cooperation can facilitate more inclusive adoption of such strategies. Real case studies and comparative analyses also highlight not only their environmental, but also their economic effectiveness, including job creation in recycling industries and promotion of sustainable markets. Conclusions with actionable recommendations for policymakers, industries, and researchers: incentives on sustainable practices, investments in renewable technologies, and encouragement of community-driven initiatives towards lessening dependencies and building resilience.

Keywords Natural resource depletion · Waste reduction · Circular economy · Renewable materials · AI · 3D printing · Microbial technologies · Sustainable development

---

A. Singh (✉)
Department of Biological Science, Sri Venkateswara College, University of Delhi, Delhi, India
e-mail: ayushiisingh456@gmail.com

A. P. Das (ed.), *Bioconversion of Environmental Waste to Value Added Products*,
Sustainable Environmental Waste Management Strategies,
https://doi.org/10.1007/978-3-032-14215-3_16

1 Introduction

Population increases in the world and growth of the economy in the world have resulted in record levels of waste creation and ongoing increasing pressures on natural resources (Dey et al., 2023; Mishra et al., 2023). With today's era of uncontrolled consumerism and urbanization, the issue of good waste management in an eco-friendly way is becoming a monster problem all over the world today. Based on the existing research by Maalouf and Mavropoulos (2022), wastefulness created globally in 2017 was approximated at over 20 billion tonnes, and projected alarming growth to 46 billion tonnes by 2050 (Maalouf & Mavropoulos). The growth indicates a critical environmental issue and signals the need for new waste reduction policies based on waste minimization generation and optimal utilization of limited natural resources (Mishra et al., 2019).

Waste minimization is included in an integrated sustainability agenda of waste management in conservation of resources and environmental protection. The principles of the circular economy in which waste is a resource and products are designed to prevent waste constitute the solution to the challenge as imagination (Priyadarsini & Das, 2024, 2025). The idea envisions an economy where materials are being reused, recycled or remade into useful products repeatedly without needing raw material drawn from nature and with little or no damage to the environment; Mohanty et al., 2018). This paradigm revolution is not only important to manage the production of trash but also to reduce the harmful effects in the sense of resource depletion as well as reducing environmental pollution.

Bioconversion technologies like anaerobic digestion and composting are the pillars of this paradigm as they convert waste into useful products. Bioconversion, using organic waste, not only reduces the reliance on nature but also closes the loop in the resource cycle by stopping the waste going to landfills (Mishra & Das, 2023; Tripathy et al., 2024). All these technologies lead to energy production, nutrient recovery and greenhouse gas mitigation, thereby making them great partners in the fight against unsustainable waste management.

The focus of this chapter is to uncover richness in less natural resource-dependent waste reduction strategy by making multiple dimensions of means an assumption. The chapter starts with the presentation of an overview and implication of waste reduction in relation with global statistics of waste, natural resource overconsumption, its resulting impact and analysing the role of consumption of resources for pollution. We will then go on and define what waste reduction is and introduce the circular economy philosophy and highlight the prime role that bioconversion has to play in reducing waste.

2 Background and Importance of Waste Reduction

Production of waste keeps growing menacingly globally. World Bank projections indicate that urban solid waste will increase from 2.01 billion tonnes in 2016 to 3.4 billion tonnes by 2050 (Kaza et al., 2018). A growth in the production rate is not welcome when considering the type of waste, which is non-biodegradable and therefore destined to be in the environment to pollute and decompose the environment.

The necessity for waste minimization cannot be overemphasized. It relies on its principle in itself to shift waste management from a reactive to a proactive status. Reducing the creation of wastes at the source lowers the combined environmental impact of waste management, resource utilization and energy consumption with waste treatment. A reduction approach stimulates the analysis of the entire product life cycle, from design to end-of-life disposal, as a method of reducing environmental effects in the areas of emissions, energy usage and consumption of resources (Mishra & Das, 2021; Mishra et al., 2022).

Apart from the environmental benefits that are quite obvious, efficient waste reduction also provides enormous economic benefits. Efficient waste reduction enables companies to reduce disposal expenses, promote more efficient use of resources and avoid cost impositions by regulatory agencies on waste treatment processes (Kumar et al., 2023; Mishra & Das, 2022). Second, with urban population growth still on the rise, the need for efficient waste management facilities looms over the safety of public health and environmental integrity.

The overdependence on natural resources sets off a cycle of depletion and degradation. The world economy only uses more than 80 billion tonnes of material resources per annum, creating ecosystem instability and biodiversity loss (United Nations Environment Programme, 2019). Misuse and overexploitation of fossil and mineral resources cause loss of land, carbon dioxide and loss of ecological functions. This wastefulness indicates the need for new directions in waste management that will push society in the direction of sustainability. Global Waste Generation Statistics. Global statistics for waste generation paint a dismal picture, reflecting the magnitude and feeling of urgency regarding the problem.

For example, the global total of waste still grew, as cities posted an average compounded rate of growth per annum of 3.2% between the years 2025. Cities in low-income countries, which are urbanizing quickly, are likely to have the worst waste management issues. A disproportionate percentage of that waste goes unpicked up, and one-third of all municipal solid waste produced worldwide goes unmanaged (Maalouf & Mavropoulos, 2022). The severity of the issue is compounded by the prevalence of dump and open burning, which is most likely to lead to environmental degradation, health risks to the public at large, and the waste of valuable resources. In their report, Kaza et al. (2018) expose that waste management is a question mark for the majority of developing countries whose institutions and infrastructure are still not well established to be in a position to swallow the increasing load of waste.

3 Ramifications of Over-Reliance on Natural Resources

Dependence on natural resources has negative social and environmental effects. Patterns of extraction and waste management by the developed and developing world indicate an irreconcilable gap between natural resources consumed and waste being produced, leaving room for unsustainable development of the limited natural resources and contamination of the environments. Urbanization, industrialization and technological transformation have left room for increased use of resources, institutionalizing practices without regard to long-term natural ecosystem sustainability.

Mining, electricity and aggregates contribute most to polluting the environment.

Kirchherr et al. also add that extraction of resources is most likely responsible for loss of habitat, soil erosion, and loss of biodiversity. These are now complemented by the high carbon intensity of extraction and processing and causing global warming and climate change. For example, the energy sector is directly related to carbon emissions since the combustion and utilization of fossil fuels release ginormous amounts of greenhouse gases. Relying on fossil fuels and non-renewable resources poses a threat towards an integrated framework of human welfare and environmental stewardship.

Relationship Between Resource Consumption and Environmental Destruction

The relationship of environmental degradation with resource decline is evident and multi-faceted. If the natural resources start to fall below due to excessive mining, then the subsequent environmental effect is cumulative. The theory of ecological footprints explains why consumption levels by human beings are in excess of what the world offers and cause additional strains on the ecosystems (Global Footprint Network, 2021). Degradation of freshwater is another general natural example.

Excessive extraction of water for irrigation, residential and industrial purposes has drained water levels at the local level as well as unsettled aquatic ecosystems, derailing the natural cycle of water purification as well as loss of biodiversity. Likewise, deforestation for agriculture and forestry suppresses the carbon capture capacity, resulting in climate change effects and extinction of ecosystems and making species susceptible to extinction. Throughout the whole resource product life cycle—production, consumption and disposal at end-of-life—a need exists to know the complete system and give an immediate reaction to minimum ecological damage and a maximum sustainable production.

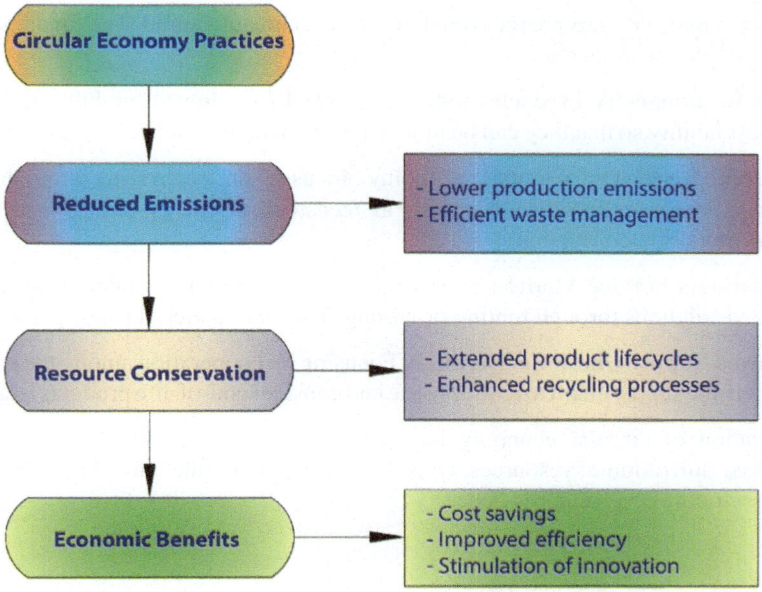

Fig. 1 Flowchart depicting the sustainability benefits of circular economy practices. (Oyejobi et al., 2024)

4 Concept of Waste Reduction and Circular Economy

Waste reduction is policies and practices aimed at decreasing the volume of waste that is produced through production, consumption and disposal (Das & Ghosh, 2018; Ray et al., 2024). The objective is not to produce waste at all and limit the necessity of waste management practices that are costly and environmentally degrading. Successful waste reduction is based on a range of approaches such as product design modifications, consumption pattern alterations and efficiency in operations.

For example, disassembly design can be made by the manufacturers so that repair and recyclability should not be a load, thereby resulting in no or minimal waste. In the same manner, behaviour change programmes may instil proper behaviour in the consumers to refrain from overconsumption and implement sustainable approaches (Fig. 1).

5 Principles of the Circular Economy

The circular economy adopts future paradigms on the basis of sustainability through keeping resources in use for as long as they are needed (Ghosh & Das, 2017; Pradhan et al., 2023). It facilitates a closed loop system where waste is reduced to the maximum and recyclable materials are recovered, reused or remanufactured. The fundamentals of the circular economy are:

Design for Longevity: Products need to be designed according to durability, fixability and recyclability so that they can be more enduring than the transient usage category.

Resource Recovery: Circular economy focuses on recovering immobilized resources in wastes in order to recycle as feedstock or energy through recycling, composting or bioconversion.

Product-as-a-Service Models: The ideology reconfigures the model of ownership to shared solutions through renting or leasing, lessening material usage per se.

Increased Supply Chain Cooperation: Focusing on cooperation among the various players in the supply chain to reduce waste and consider end-of-life products as assets.

Application of circular economy habits will assist in reducing less waste and improved utilization of resources, considering the environmental concern involved and developing economic resilience.

6 Bioconversion Role in Waste Reduction

Bioconversion technologies have the pioneering function of reducing waste by recycling organic waste into value-added commodities like biodegradable plastics, biofuels and fertilizers with nutrient content. Organic waste, for instance, may be converted to biogas—a clean energy source with lowered emissions of greenhouse gases and lower dependence on fossil fuels—by anaerobic digestion (John & Das, 2024; Sahu et al., 2024).

Anaerobic digestion yields digestate, a rich nutrient by-product also employed as organic fertilizer, which nourishes the soil and sustainable agriculture. Composting, on a less industrial bioconversion process, involves natural decomposition of organics, and the output is a soil amendment recycling back the nutrients to nature (Dey et al., 2024; Kanungo et al., 2023).

These bioconversion technologies are a circular economy as not only do they prevent landfilling but also harvest valuable material. They are the prime example of waste reduction operations which are an economic feasibility against ecological necessity for reducing waste to become value, rather than an imposition.

Strategies for Reducing Reliance on Natural Resources

With the advent of environmental risk factors and unregulated use of natural resources, resource conservation is an immediate requirement of the day. Bioconversion of waste by using environmental means is a method through which waste is converted into value-added products concurrently while minimizing the use of resources as well. From the literature, the chapter introduces various waste reduction methods aimed at sustainable material replacement, industrial symbiosis and waste valorization, recycling and upcycling design and enhanced use of artificial intelligence (AI) to support enhanced waste management practice.

Sustainable Material Substitution

Replacement of conventional materials with bio-based materials is the most viable method of minimizing the use of natural resources. The bio-based materials, produced from renewable resources, have the potential to greatly reduce the environmental impact of fossil fuel and metal mining and processing.

Use of Bio-Based Materials

Recent studies of bio-based products such as bioplastics offer a more environmentally friendly substitute for traditional petroleum plastics. Bioplastics are made from renewable resources such as corn, sugarcane and vegetable waste. Unlike fossil fuel plastics, bioplastics are biodegradable or compostable, reducing plastic waste effects in oceans and landfills.

Case Study: Biodegradable Packaging

One of the best sustainable material substitutes is the production and utilization of biodegradable packaging. Petrochemical plastics are prevalent in packaging; they do, however, cause enormous environmental pollution as they are not biodegradable. Bioplastics, for instance, polylactic acid (PLA), produced from fermented starch plants are a potential substitute for petrochemical plastic. This transition not only decreases dependence on fossil fuels but also promotes the composting practice as a means of disposal, thereby improving soil fertility and reducing landfill inputs. Companies that employ biodegradable packaging also enjoy enhanced supply chain sustainability and customer attractiveness, an indication that there is still a market for green products.

Industrial Symbiosis and Waste Valorization

Industrial symbiosis is a master plan for enhancing the inter-industrial flow of resources and waste products. Waste and material sharing among organisations lowers the use of humongous quantities of raw material and the generation of waste products, thus earning enormous economic and environmental dividends.

Concept of Industrial Symbiosis

Industrial symbiosis is the planned transfer of materials, water and energy between industries. It is innovative recycling of waste material from one industry to use as a source material for another in an efficient cycle of sorts system reducing consumption with a greener lower footprint.

Case Study: Fly Ash Utilization by the Cement Industry

Cement manufacture is maybe the most obvious instance of industrial symbiosis because coal combustion power plant fly ash has been used as a suitable additive. The manufacture of cement has traditionally incorporated natural materials such as limestone and clay, gigantic carbon-emitting sources. But through the introduction of fly ash in cement mixture, businesses can decrease their dependence on such raw materials without diluting the quality and strength of the product. While doing so, not only is virgin material consumption minimized, but waste minimization is minimized with the exclusion of fly ash from landfill space, offering synergy to the construction and energy industries.

Recycling and Upcycling Innovations

Upcycling and recycling are also central themes of improved resource efficiency. They incorporate value recovery from waste as well as virgin resource avoidance.

Advanced Recycling Technologies

The development of sophisticated recycling technologies has enabled the opening of new avenues to improved plastic, metal and electronic waste treatment. Technologies such as chemical recycling, mechanical recycling and biological recycling have been highly promising in further improving the conventional systems.

Chemical Recycling: Dismantling waste into its molecular building blocks, which are re-synthesized into new, quality products. Post-consumer polyethylene terephthalate (PET) waste, for instance, can be depolymerized to produce its monomers constituent, and which in turn can be utilized for the synthesis of new PET products without loss that otherwise would be incurred with mechanical recycling.

Mechanical Recycling: This is recycling of material by physical means, and it is common in metals and plastics. Good mechanical recycling techniques are able to produce a huge quantity of high-quality secondary raw materials for industrial purposes.

Biological Recycling: This is the new technology that involves the application of microorganisms in the conversion of organic waste to useful products such as biogas, organic manure and even polymers.

Upcycling Methods

Upcycling is waste recycling in manufacturing new, valuable products of high quality. Upcycling has applications in diverse streams of wastes like textiles, metals and plastics.

Case Study: Conversion of Textile Waste to New Fibres by Upcycling

Textile waste upcycling to new fibres is a very fascinating usage of upcycling. Overproduction and overconsumption from the fashion business generate enormous volumes of waste. But now there are new technologies to recycle clothing waste like old clothes and piece good trimmings and recycle them into recycled fibres to recycle and reuse in clothing and other textile products. Enzymatic or chemical treatment processes can be used to produce high-value fibres, minimizing waste and virgin material dependency. Companies undertaking such upcycling not only assist in lowering waste but also assist in catering to consumer demand for sustainable consumption of fashion.

AI and Machine Learning in Waste Reduction

The use of machine learning and AI in waste management policy presents an unprecedented opportunity to achieve the highest level of resource utilization and waste minimization. Smart waste management systems utilize data-driven models to optimize the maximum waste collection, sorting and processing and minimize the reliance on natural resources.

Smart Sorting of Waste Based on AI Technology

Artificial intelligence technology-based technologies have the potential to revolutionize the recycling process so that efficiency would significantly enhance. With higher image processing and advanced machine learning algorithms, smart sorting machines will automatically sort and separate the material, which translates to higher rate of recovery with less contamination. Such technologies do not only reduce the level of complexity for recycling but also enable the supply of maximum quality recoverable material for production.

Predictive Analytics for Resource Optimization

In addition to sorting, AI can be implemented to facilitate predictive analytics, whereby companies can predict resource availability and wastage trends.

Predictive analysis using historical data allows for enhanced planning of resources and inventory such that companies are able to rethink themselves in such a manner that they do not lose anything and optimize the usage of the input resources. Predictive analytics stimulates minimization of waste in terms of business optimization, creating circularity in such a manner that the reliance on natural resources gets minimized while profit maximization becomes achievable.

Conclusion

Strategies in this chapter include a systems approach of natural resource dependence management through emerging strategies like sustainable material substitution, industrial symbiosis, emerging recycling technologies and artificial intelligence management of waste. Reducing the environmental footprints and optimizing the efficiency of the resources, the strategies give us the vision of a green globe with less wastes and resources maintained in the economy.

Waste Reduction Strategies Through Reduced Reliance on Natural Resources

Bioconversion of Waste to Value-Added Product

More attention towards environmental contamination and non-sustainable use of natural resources has put processes of waste treatment under strain to perform. Bioconversion of waste to value-added product is one such process that reduces the amount of waste that reaches landfills and, at the same time, generates economically valuable material. Some of the topic areas of bioconversion are addressed in this section such as microbial conversion of organic waste, waste to bioplastics, resource recovery biorefineries and green materials from waste.

Microbial Conversion of Organic Waste

Microbial breakdown of organic waste is the harnessing of the potential of natural microbes, i.e. bacteria and fungi, for breaking down complicated organic matter into simpler compounds. It not only conserves space for waste but also results in the generation of useful commodities, i.e. energy, fertilizers and chemicals.

Bacteria and Fungi in Organic Waste Breakdown

Fungi and bacteria are of extremely important use in degradation of organic matter. Bacteria are far more capable of substrate degradation because they are far more metabolically variable and encompassing. Anaerobic bacteria like Clostridium and Methanosarcina, for example, can degrade organic matter under lower oxygen conditions, which is of extremely important importance in biogas and biofertilizer production (Sakai, 2008).

Fungi possess a good capacity for degradation of recalcitrant substrates like cellulose and lignin and crop residues and forest biomass in bulk quantities respectively. For instance, white-rot fungi release a cascade of ligninolytic enzymes that further facilitate fermentation of cellulose degradation. All these enzymic activities everywhere not only are enabling cycling of organic matter in ecosystems but are of phenomenal gigantic value of potential to commercial utilization (Asgher et al., 2023).

Application of such microbes in closed systems—bioreactors, for instance—is used to attain maximum atmospheres in waste destruction processes effectively. Composting and anaerobic digestion are two processes where microbial effectiveness can be maximized to the maximum level, thus waste conversion into resource maximum to the level of effectiveness.

Literature also reports that application of mixed microbial cultures has the potential to transfer more degradation capacity than monocultures and, therefore, can result in higher net yield of biogas or compost (Kumar et al., 2022) (Fig. 2).

Biogas and Bioethanol Production

Anaerobic digestion-based conventional biogas production continues to be the best organic waste utilization technique in the form of a renewable energy source. Anaerobic digestion is a sequential chain of consecutive hydrolysis, acidogenesis, acetogenesis, and methanogenesis that involves a separate microbial population in each stage. Biogas, which primarily consists of methane (CH_4) and carbon dioxide (CO_2) in addition to minor amounts of hydrogen sulphide and other volatile gases (Smith et al., 2022), is the key output of anaerobic digestion.

Trapped energy in biogas is utilized for space heating, fuel transport, or to generate electricity through combustion or cogeneration technology. Anaerobic digestion by-product as digestate, high nutrient content slurry to be used as organic manure, recycling nutrient cycle for agri-production is the second application (Kwaghe et al., 2021).

In addition to biogas production, waste enzymatic hydrolysis is also employed in bioethanol production. It involves hydrolysis of polysaccharides into fermentable sugars and their fermentation by yeast (e.g. Saccharomyces cerevisiae) or bacterial (e.g. Zymomonas mobilis) cells. Fermentable sugars obtained from various substrates—fruit and vegetable waste—can be fermented to yield ethanol, providing a source of renewable energy that could replace fossil fuels (Smith et al., 2022).

Bioethanol is not only a green-house gas-emitting and fossil fuels conventionally lower bio-fuel but also a renewable one. Industrially generated bioethanol from waste substrates depends on optimization of the hydrolysis and fermentation process,

Fig. 2 Biowaste refinery to generate high-value bioproducts and bioenergy (Priya et al., 2023)

economic viability issues associated with enzyme production, and scale-up of the process (Yadav et al., 2021).

Waste-Derived Bioplastics and Bio-Based Materials

With the growing requirement for green material across the world, producing bioplastics using waste streams is one of the green strategies towards averting the use of fossil-based plastics. It is justified in evading potential adverse effects on the environment because of plastic wastes and dissemination of circular economy culture.

Utilization of Agricultural Waste in the Production of Biopolymers

Crop residue and by-product is a good bioplastic feedstock. Organic waste is utilized to yield various biopolymers like PLA, PHA, and starch-based bioplastics. Crop residue is being converted into biodegradable plastic through pretreatment, hydrolysis, fermentation and polymerization in an efficient process (Chatterjee & Mohan, 2022).

For example, the same has otherwise been demonstrated utilizing research wherein substrate wastages like sugarcane bagasse and corn stover were revealed to be hydrolysable in achieving fermentable sugar. That then would get metabolized to render a substrate to the fermentative microbials with an intent to lever the same towards evolving end-product formation of the biodegradable replacement of traditional plastic PLA (Kumar et al., 2022).

Application of waste-biopolymers constructs biodegradable material use as a benchmark towards a green environment through the elimination of landfilling wastes and fossil fuel plastics. Plastics are also plastic-like in character and hence considered a potential material for use from packaging to agriculture (Zhang et al., 2021b).

Case Study: Polylactic Acid (PLA) Production from Food Wastes

The most conspicuous example of generation of biopolymer waste is PLA manufacture from food wastage. PLA is a thermoplastic polymer and biodegradable and can be manufactured on the basis of lactic acid, which can be obtained from carbohydrate wastage fermentation processes of food wastages such as vegetable and fruit wastage disposal.

The cycle is generally started by hydrolysis of food wastes, which are depolymerized to produce fermentable sugars. Fermentation of sugar by lactic acid bacteria results in the formation of lactic acid. Polymerization of lactic acid once again produces PLA (Kumar et al., 2022).

For example, in the case study of Asgher et al. (2023), scientists were able to show that waste food had the potential to contain high fermentable sugar content. With optimized fermentation, PLA bought was as good in quality as the same PLA bought from conventional petrochemical vendors.

The final advantages of converting food waste into a feedstock for PLA production are environmental impact mitigation through waste diversion, greenhouse gas reduction to produce and development. The method promotes future sustainability goals and the embracement of leadership waste in reaching a circular bioeconomy (Chatterjee & Mohan, 2022).

Biorefineries for Sustainable Resource Recovery

Biorefineries are a clean technology resource use system of product, chemical and energy production from processing biomass. While, provided that the feedstocks remain wastages of farm, forest residues and wastages of urban areas, biorefineries may eliminate both wastage disposal problems and wastage of resources.

Biomass-to-Energy Technologies

Biomass-to-energy technology application is the most recognized sign of transitioning from the consumption of fossil fuels to green technology. The techniques used in the energy conversion of biomass are combustion, gasification and anaerobic digestion.

Combustion is the direct burning of biomass to generate electricity and heat, whereas gasification is the conversion of biomass into a syngas at high temperature and low oxygen (Smith et al., 2022). The generated syngas can be utilized for the generation of electricity or to generate liquid fuels like ethanol or biodiesel.

Apart from all those processes, anaerobic digestion is still an option in biomass-to-energy conversion by way of generating biogas out of organics waste. Anaerobic digestion and biogas upgrading are recent technologies still susceptible to higher energy recovery at a lower environmental penalty.

The energy potential through biomass is immense and climate change can be reduced and most of the world's energy requirements can be fulfilled by using sustainable biomass (Yadav et al., 2021).

Circular Bioeconomy Strategies

The bioeconomy circular model is based on the fact that resource cycles must be optimized to optimize resource utilization, recycling and recovery. The biorefineries will take the central position in the system by transforming waste into value-added products and minimizing raw material reliance and environmental impact.

One of the ways in which a circular bioeconomy can be achieved is by linking industrial and agricultural operations with bioprocessing facilities. Integration gives the possibility for waste streams and by-product valorization, which can be converted into biogas, biofuels or bioproducts and, in doing so, reduce carbon footprint and disposal costs (Mohan & Fornari, 2022).

Other than that, sustainable processing by biorefineries like resources recovery process and life cycle evaluation can enhance biomass utilization efficiency on a whole. Utilization of development principles paired with technology-centred eco-design will set biorefineries into circular economy models as an effort to maximize not only environmental benefits but also economic benefits (Chatterjee & Mohan, 2022).

Eco-friendly Building Materials from Waste

Apart from the waste-to-energy facilities, construction wastes recycling is capable of conserving economically immense resources along with eco-friendly growth. Industrial wastes have also been found to be used as replacement construction material in the shape of cement, bricks and panels.

Industrial Waste Bricks, Cement, and Panels

The construction industry is one of the largest consumers of natural resources with gigantic ecological footprints. Recycling industrial residues to produce building materials is one of the latest moves towards the minimization of such footprints as well as the assurance of maximum use of resources.

For instance, fly ash as the residue of burning coal in thermal power plants is utilized for the production of light-weight concrete blocks. Fly ash increases mechanical strength and decreases the use of Portland cement to a very negligible amount and hence carbon emission overall (Nagaraja et al., 2020).

Similarly, waste from construction and demolition can be recycled to serve as aggregates for application in contributing towards concrete mixture addition for promotion of circular economy via diversion of waste from landfilling and building material life cycle extension (Zhang et al., 2021b).

Organic rice husk or rice straw waste may be used in panel manufacturing for sustainable building construction alternatives. Organic fibres can be employed as a thermal insulator and for green material as a plastic or synthetic alternative.

Case Study: Rice Husk Ash as a Sustainable Construction Additive

One of the best uses of sustainable waste product in building is rice husk ash (RHA) as a pozzolanic cement admixture. RHA is a high-silica content waste product of rice mills, and it can be utilized to add strength and properties to cement and concrete.

Experiments have confirmed that the addition of RHA in cement concrete mix enhances the workability, strength and durability of concrete but reduces the heat of hydration—a characteristic which is highly significant for large concrete structures (Kumar & Mohan, 2022). Apart from this, partial substitution of cement with RHA results in lower carbon emissions, as has been suggested by the green school of thought for building construction.

Effective utilization of RHA in construction works not only reduces the problem of waste disposal but introduces the concept of circular economy to the construction industry. Recycling agricultural waste to generate value-added products, the construction industry can effectively reduce its environmental impact while initiating resilience and sustainability (Nagaraja et al., 2020) (Figs. 3 and 4).

Case Studies and Real-World Applications The necessity to restrict reliance on natural resources is most critical in the face of increasing environmental problems, such as global warming, scarce resources and pollution. Sophisticated methods of minimizing waste through the application of bioconversion technology with emphasis on proper application, effective models worldwide, as well as limitations hindering them from being implemented at a mass level are therefore discussed in this chapter. There are findings based on such models that can guide future policy interventions in an attempt to ensure sustainable waste management.

I. Case Studies and Real-World Applications

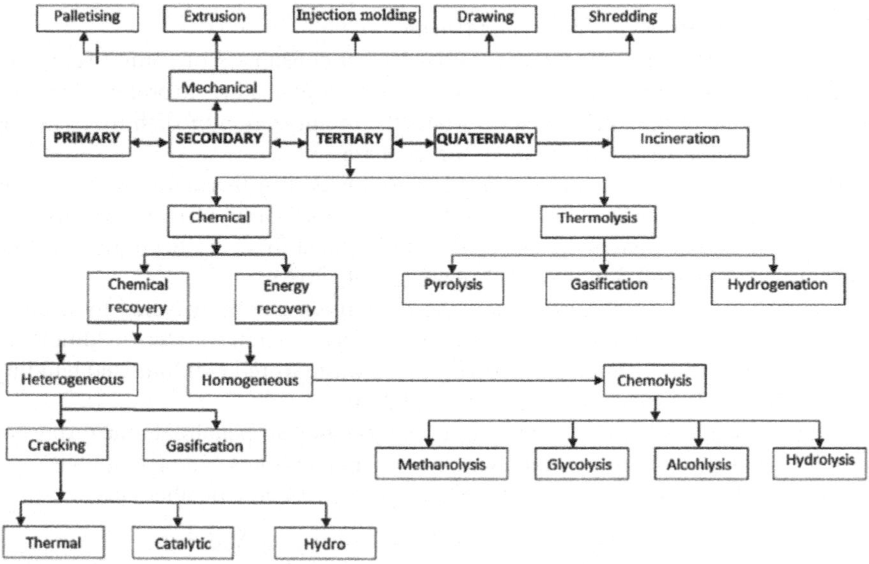

Fig. 3 Method to recycle plastic wastes

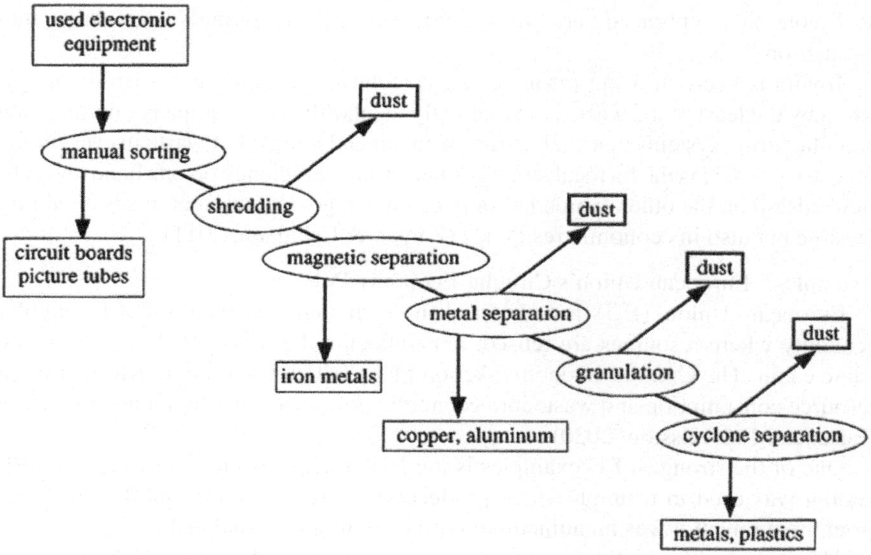

Fig. 4 Generic e-waste treatment flow diagram

Waste reduction efforts are on the rise worldwide, and they symbolize a future of sustainability for sustainability and for waste management too. Waste reduction not only reduces the quantity of waste sent to landfills but also makes the circular economy more likely because wastes are being converted into higher-value products. Below are best practice case studies of waste reduction across the world.

Global Achievements in Minimizing Waste

Example 1: Zero-Waste Production by Japan

Japan has led the way in adopting zero-waste initiatives, particularly production. Kamikatsu town provides a typical example of an ideal model for a zero-waste system. Kamikatsu has practiced stringent waste separation measures and grassroots sensitization drives since it went zero-waste in 2001. Kamikatsu separates trash into around 45 categories, from plastics to metal and organic waste. This intimate segregation facilitates recycling at a very high rate, almost reaching 81% as of 2020 (Shenyoputro & Jones, 2023).

The way Kamikatsu waste disposal has achieved success is the different accounts involving high civic involvement, governmental action, and sensitization education that raises awareness. The SAS system is central to enabling the high quality of recyclables and lowering the degree of contamination that would be prevalent in sorting sorted mixed waste (Matsumoto, 2011). Through instilling a culture of waste-cutting, Kamikatsu not only attained its waste cut but also instilled pride and unity among the public towards sustainability. Besides that, Japan's top companies such as Toyota have embraced zero-waste production by integrating sustainability into production lines.

Toyota is keen on using resources to the fullest, recycling and wasting energy, whereby the least of the waste is sent directly to landfills. The company operates lean manufacturing systems that enable improvement and removal of waste in everything. Industry practice is the bi-focal strategy of economic efficiency on one hand and eco-stewardship on the other, which not only can make goal attainment in sustainability feasible but also in economic resilience (Zaman & Lehmann, 2011).

Example 2: European Union's Circular Economy Policies

European Union (EU) has undertaken a challenging path towards circular economy where resources are reused, remanufactured and recycled in product life value chain. The Circular Economy Action Plan 2020 aims at the transformation of resource consumption and waste into economic prosperity, employment and growth (European Commission, 2020).

One of the strongest EU examples is the Netherlands example in which an EPR system was used to triumph where producers covered the entire product life cost from the moment it was manufactured until recycling and final disposal.

There is also the motivation to manufacture products that can be recycled, and this is an encouragement to favour resource efficiency in consumption.

It has minimized packaging waste significantly and facilitated the culture of accountability among producers (Cole et al., 2014). In addition to this, EU focus on research and development funding affects innovation and start-ups to produce circular economy products. Pilot production is facilitated for emerging technologies and technologies in managing waste at high levels of international co-operation and expertise sharing among the members of the EU (Bartl, 2011). All these policy measures put together have the overall impact of reversing recycling levels by bounds and bounds and lowering waste output level levels in the region.

Example 3: India's Agricultural Waste Valorization

Crop biomass residue can be behind ghastly pollution and environmental degradation if not disposed of through proper removal. There are alternative valorisation options available for other complementing varieties to transform the waste into value-added products.

Processes already in operation in India include recycling agricultural waste into products like biofuels, biogas, organic manures and other value-added products.

Crops like sugarcane bagasse, rice husk and straw are being used more to make biochar, a carbon-rich material utilized in soil fertilization and carbon capture.

Apart from lowering the economic cost of contaminating the environment with the burning of agricultural residue, it enhances the soil's health, hence conserving farmers (Fujita & Hill, 2007). Indian start-ups also utilized technology in the utilization of agro-waste for the generation of biogas by embracing anaerobic digestion processes. Biogas is used for cooking or for power generation, and thereby energy access is significantly enhanced, especially for the rural population. Incentivization policies to encourage farmer investment in waste-to-energy technology adhere to an integrated policy on agricultural residue management and take up climate change mitigation measures (Kumar et al., 2020). The Indian government has also given alternative policies to enable more sustainable agriculture and recycling of waste. Favourable policies for promoting farmers towards integrated waste management practices are becoming more and more significant, which can be utilized to transform crop waste into an economic input (Singh & Sharma, 2019). This is in the service of wider sustainability in the sense of preventing waste but aims to serve energy security as well beyond enhancing farm productivity.

II. Problems in Implementing Waste Reduction Measures

Although the aforementioned case studies are success stories, there are certain problems with which the practice of waste minimization methods as a whole is confronted. The best solution to overcome them is the shift towards the maximum utilization of the potential of bioconversion and waste treatment processes.

Economic Feasibility and Policy Issues

Financial feasibility is one of the biggest hindrances to the practice of waste prevention policy implementation. It is expensive to become green in the sense that there is a need to invest money in technology, infrastructure and human capacity building. Small and medium-sized enterprises (SMEs) and start-ups do not have the funds needed to invest in putting in place green technology, and therefore might choose not to put in place sustainable waste management (Jarman-Walsh, 2019).

Second, the absence of policy towards encouraging sustainable practice of reducing waste will have a tendency of baffling companies on how to go about reducing waste as business.

For example, various policies from location to location are difficult to implement and interpret varying levels of enforcement (Zaman, 2015).

In the EU, nonetheless, significant progress in the adoption of the circular economy has been made, but policy regime disparities among member states suck dry huge recycling and waste management programmes (European Commission, 2020). Furthermore, market price fluctuation of recyclables brings about economic uncertainty and hence discourages investment in waste reduction. Companies employing recycled material are exposed to volatile raw material prices, deterring them from employing continuously recycled content in manufacturing.

Gaps in Technology and Acceptance Gaps

And yet another such most fundamental bottleneck in the path of waste minimization on an effective note is that of technology limitations. While proper new-generation technology to control waste as also to make the same useful at a productive scale is available, technology diffusion does not happen to catch up as resources are not properly distributed. In rural or developing countries, access opportunity for higher-order waste conversion systems can be remarkably low, shutting employment (Manomaivibool et al., 2018).

Popular acceptance for novel waste control practice is the strongest driver of achievement.

New practice would typically mean a change of action, which would be resisted due to being an old habit or because they lack knowledge of how the minimization of waste would benefit them.

The stakeholders would doubt new technology effectiveness or safety in most scenarios. This uncertainty can paralyze the implementation process of new waste management process or deter people from engaging in green activity (Silva et al., 2017). Public awareness and education are also extremely important for public acceptance of waste reduction activity. Accurate information about environmental, economic and social advantage of waste reduction activity can enhance the level of participation and acceptance to a very high degree from the public. There is rarely any or no effective awareness activity or training on sustainability in most places.

III. Solutions for Closing Such Gaps

Closing the gaps for performing waste reduction activity with this diverse group of stakeholders will be a multi-dimensional task. The solutions will ensure successful implementation of waste reduction programmes.

1. Raising Economic Incentives

Offering economic incentives is another guaranteed method of initiating waste minimization and bioconversion technologies.

Financial incentives such as grants, subsidies or low-interest loans may be initiated by the government so that stimulus monies are available to be invested by companies in green technology with the aim of reducing wastes (Zaman & Lehmann, 2011).

The transformation will be made less expensive on their bottom line through incentives, especially in the case of SMEs because these are going to have higher barriers to entry of higher intensities. In addition, growth in recycling product market through guaranteed procurement orders will provide recycling inputs with market demand, thereby ensuring firms to invest in recycling and waste reduction. Environmental-friendly business processes will be encouraged in firms by ensuring them with a stable economic environment.

2. Building Stronger Policy Frameworks

Policy guidelines at the national, regional and local levels must be unambiguous and consistent to contribute to the eradication of waste. The government may involve stakeholders such as corporations, organizations and environmentalists in a policy structure that provides open guidelines to waste management practices.

3. Encouraging Technological Innovation

In order to close the technology gap in waste management, investment has to be done in research and development. The public–private partnership models can be used for the encouragement of innovation in bioconversion technology and infrastructure for waste management. Industry experts and research institutes can be incentivized by policymakers to work on pilot projects for testing new methods and technologies for waste valorisation.

Second, training and technical support to farmers and small businesses on the use of new waste conversion technology assists in adopting and increasing the efficiency of operations. Grassroots capacity building is a strategy of ensuring that the communities are well equipped to perform waste reduction processes effectively.

4. Education and Community Engagement

Public education up to the extent of evoking an appreciation of the necessity for the implementation of wastes minimization measures must be encouraged in the interest of gaining popular acceptance.

School curricula that have the potential to enlighten consumers and businesses regarding the issue of the correct consumption habit, recycling culture and wastes reduction behaviour can generate more popular participation and popular acceptance.

Mixing school field trips with lab demonstrations or direct exposure will result in learning from and appreciation of proper waste management procedures. Action at the local level in local areas also raises the level of participation, particularly in rural regions. Collaboration with local associations, community leaders and schools to organize waste management and active participation programmes has the potential to raise local interest in sustainability procedures (Kumar et al., 2020). Grassroots mobilization can help create awareness of collective action and shared responsibility for the objective of minimizing wastage.

5. Pilot Schemes and Trials

Future Prospects and Policy Recommendations

In today's modern world, the imperative to conserve natural resources and regain loss of environment has labelled a very pressing requirement for new waste reduction techniques. A green policy of waste management not just saves precious natural resources but also reduces activities to their minimum possible levels, promotes recycling and achieves overall environmental wellness. This chapter offers future directions for and policy suggestions complementing sustainable natural resource reduction practice in industries with focus on sustainable policies, extended producer responsibility legislation, adoption of international standards in waste management and new technologies, thereby paving the way for a greener future.

Sustainable Policies for Curbing Natural Resource Use

Environmental policies that minimize the use of natural resources are crucial to mitigate the growing problem of waste management. Policy-making style has to be oriented to form incentives reducing the usage of resources and facilitating circular economy patterns. Curtailment of use of resources as well as of generation of wastage are also areas which offer key directions towards sustainability goals. All of these schemes are accessed by the government through subsidizations, grants and tax rebates being awarded to such industries implementing sustainability.

Apart from that, the application of life cycle assessment (LCA) indicators in policy initiatives makes it simple to understand the environmental footprint of products throughout the entire product lifetime. Through the use of LCA, policymakers will be able to identify the most critical phases at product design levels through which the consumption of resources can be minimized to the barest minimum (Geissdoerfer et al., 2018). By this means, policymakers can make specific plans where resource preservation, waste minimization and environmentally friendly production stages are given priority.

Education of the public to resource efficiency and sustainable consumption behaviour has to be offered by the government. The government is able to direct rational consumer behaviour by means of campaigns, promotion of a sustainability culture and reduction of demand for unwanted goods. Product labelling can assist in the policy function of education by making consumers realize the environmental impacts of their behaviour (European Commission, 2021).

Additionally, public–private partnerships that will give rise to co-ventures for sustainable business. There will be public–private partnerships that will drive innovation as well as knowledge and resource sharing for the benefit of industries, resulting in the shift towards circular economy (World Economic Forum, 2021).

Extended Producer Responsibility (EPR) regulation

Extended producer responsibility (EPR) is an excellent regulatory strategy whose objective is to hold producers liable for the complete life cycle of the items that they produce, including getting rid of trash following consumer consumption. Producers in EPR would be responsible for the cost of the environment in the items, thereby being forced to create products that would generate zero or very little waste and high recyclability.

Their counterparts like Sweden and Germany have shown the way to introduce the EPR policy under which the producers are asked to behave in a responsible manner while manufacturing. The economic benefits provided by the EPR policies encourage the producers to behave more economically so that they use less material and send back more material to manufacture something new. The impact of EPR can be seen in terms of increased recycling rate and reduced environmental footprint.

For successful execution of EPR activities, policymakers need to design product take-back facilities for return-of-product schemes to enable customers to return the products to manufacturers for recycling or waste disposal. Manufacturers can do this by implementing practices to have effective take-back systems to involve stakeholders among customers in waste management. The customers are stakeholders for recycling processes and lighten the load of the landfills and the environment (European Commission, 2021).

The policymakers should also strive to extend the use of EPR law to other product categories such as electronics, textiles and toxic wastes. This will make EPR achieve innovation in green product design and material recycling, thus decreasing reliance on natural resources (Hao et al., 2020).

Global Waste Management Standards

Global standards for managing waste are within effective environmental management and resource preservation. Without international waste management standards, there are likely to be inconsistencies in regional waste management and recycling. International standards should be developed for that purpose to equalize practice, ensure information interchangeability and promote best practice.

International standards at the global level can be established by organizations like the United Nations Environment Programme (UNEP) and the International Organization for Standardization (ISO). The organizations can rally member states and create participatory waste management standards based on waste generation, recycling rates and waste disposal (UNEP, 2019). The standards would help provide an accurate barometer for measuring and enhancing national and regional waste management systems.

Apart from that, adopting global best practices in waste management will allow for technology transfer across nations in a way that developing countries will be able to adopt innovations from their developed peers. Technology and knowledge exchange can allow for smoother adoption of waste reduction practices globally (Zhang et al., 2021b).

Emerging Technologies in Waste Reduction

It is vital to employ the new-age technologies for waste promotion and reducing natural resource dependency. Two of the most effective tools against waste reduction and operational viability among the new-age technologies are AI-based monitoring of waste and bio-based nanomaterials.

AI-Based Waste Monitoring

Artificial intelligence (AI) can revolutionize waste management infrastructure and improve working efficiency. AI-driven waste tracking can improve waste segregation systems, optimal route for collection and track waste generation patterns in real time. AI, through big data analytics, can improve decision-making and resource planning in waste management.

These also increase collection operation efficiency. It is able to predict peak waste generation in order to enable the municipalities to plan more and reallocate the resources so that they can be used optimally. Besides improving operational efficiency, it decreases greenhouse gases that are associated with collection operations.

As the technology of AI keeps evolving, its use in waste management systems will keep facilitating cities and organizations to better handle increasing waste reduction and diversion challenges (Fig. 5).

Bio-Based Nanomaterials for Green Production

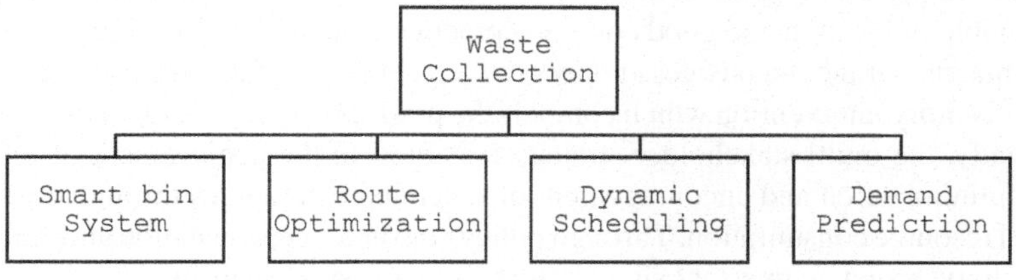

Fig. 5 Applications of AI in waste collection

Increasing demand for green materials has generated tremendous interest in the application of bio-based nanomaterials as replacement for petroleum products that are eco-friendly. Bio-based nanomaterials provide green advantages with the capability for minimized carbon footprints and less consumption of rare natural resources.

Bio-based nanomaterials are extremely versatile and provide immense application in industries ranging from packaging to construction. As utilized in the usage in packaging businesses, biodegradable packaging materials that are formulated with the assistance of bio-based nanomaterials are capable of substituting traditional plastic items in a way that plastics never reach the landfills and seas. Besides the prevention of waste, utilization of bio-based material facilitates circular economy because it renders the end-of-life solution more sustainable.

7 Conclusion: The Need for Multidisciplinary Approaches

The issue of the new frontiers of natural resource use and waste disposal must be tackled by a cross-disciplinary effort to shatter the conventional sectoral divide. Inter-cooperation among governments, the business community and academic communities must be there in developing feasible waste reduction plans and encouraging environmentally responsible action.

At the academic level, researchers are spearheading knowledge and innovation in sustainable development and waste reduction policy. Academic institutions are able to provide evidence for use in policy-making and technological innovation through conducting transdisciplinary research. Academic collaboration with researchers can make comprehensiveness of complexity in generation and waste management more feasible.

There needs to be commitment from the industry as well because the responsibility for product design, manufacturing and generation of waste will largely be that of the companies. The industries need to take charge of making sustainability happen through investment in cleaner technologies, taking up eco-design practices, and participating in extended producer responsibility schemes. Industries can reduce the environmental impact of their activity and contribute to reducing waste by implementing circular economy ideologies.

And finally, the regulators have to design regulation procedures which encourage sustainable innovation and good company practice. And all this for triggering incentives that encourage the business houses towards taking up the sustainable practices and legislatory intervention which compels the producers to put a stop to the wastage.

Finally, the multi-stakeholder strategy is central to the achievement of efficient waste minimization and encouragement of sustainable processes through decreased natural resource consumption. Through policy convergence, legislation and emerging technology, society can establish a platform towards sustainability.

References

Asgher, M., Iqbal, H. M. N., & Zia, M. M. (2023). Valorization of food waste for bioplastic production: A case study on polylactic acid synthesis. *Waste Management, 123*, 123–132. https://doi.org/10.1016/j.wasman.2023.00.00

Bartl, A. (2011). Barriers towards achieving a zero waste society. *Waste Management, 31*(12), 2369–2370. https://doi.org/10.1016/j.wasman.2011.09.013

Chatterjee, S., & Mohan, S. V. (2022). Integrated strategies for biogas production from organic waste: a comprehensive review. *Biotechnology Reports, 34*, e00759. https://doi.org/10.1016/j.btrec.2022.00759

Cole, C., Osmani, M., Quddus, M., Wheatley, A., & Kay, K. (2014). Towards a zero waste strategy for an English local authority. *Resources, Conservation and Recycling, 89*, 64–75. https://doi.org/10.1016/j.resconrec.2014.05.005

Das, A. P., & Ghosh, S. (2018). Bioleaching of Manganese from mining waste materials. *Materials Today: Proceedings, 5*(1), 2381–2390. https://doi.org/10.1016/J.MATPR.2017.11.459

Dey, S., Das, A., Mallick, K., Sahu, A., & Das, A. P. (2023). *Environmental Petroleum Waste: Pollution, Toxicity, Sustainable Remediation*, pp. 159–175. https://doi.org/10.1007/978-3-031-48220-5_7

Dey, S., Dash, A., & Das, A. P. (2024). Bioremediation as a sustainable and circular solution for mining waste management and bioenergy generation. *Sustainable Management of Mining Waste and Tailings: A Circular Economy Approach*, 107–119. https://doi.org/10.1201/9781003442455-6/BIOREMEDIATION-SUSTAINABLE-CIRCULAR-SOLUTION-MINING-WASTE-MANAGEMENT-BIOENERGY-GENERATION-SUDESHNA-DEY-AKANSHYA-DASH-ALOK-PRASAD-DAS

European Commission. (2020). A new Circular Economy Action Plan for a cleaner and more competitive Europe.

European Commission. (2021). Packaging and packaging waste legislation. Retrieved from https://ec.europa.eu/environment

Fujita, K., & Hill, R. C. (2007). The zero waste city: Tokyo's quest for a sustainable environment. *Journal of Comparative Policy Analysis, 9*(4), 405–425. https://doi.org/10.1080/13876980701674225

Geissdoerfer, M., Morioka, S. N., de Carvalho, M. M., & Evans, S. (2018). Sustainable business model innovation: A conceptual framework. *Journal of Cleaner Production, 198*, 1112–1122.

Ghosh, S., & Das, A. P. (2017). Bioleaching of manganese from mining waste residues using Acinetobacter sp. *Geology, Ecology, and Landscapes, 1*(2), 77–83. https://doi.org/10.1080/24749508.2017.1332847

Global Footprint Network. (2021). Global Footprint Network. Retrieved from https://www.footprintnetwork.org

Hao, R., Zhao, Y., Lu, C., Liu, Z., & Zhang, Y. (2020). The extended producer responsibility (EPR) policy in China: A main driver for plastic waste management. *Waste Management and Research, 38*(6), 672–680.

Jarman-Walsh, J. (2019). Foundations of zero-waste lead to sustainable tourism success: The case of Kamikatsu. *Journal of Yasuda Women's University, 47*, 249–260.

John, A., & Das, A. P. (2024). Government strategies for achieving circular economy for plastic waste management in Southeast Asia. *Sustainable Management of Urban Plastic Waste Through Circular Economic Approaches*, 72–82. https://doi.org/10.1201/9781003465447-6/GOVERNMENT-STRATEGIES-ACHIEVING-CIRCULAR-ECONOMY-PLASTIC-WASTE-MANAGEMENT-SOUTHEAST-ASIA-AKBAR-JOHN-ALOK-PRASAD-DAS

Kanungo, J., Sahoo, T., Bal, M., & Behera, I. D. (2023). Performance of bioremediation strategy in waste lubricating oil pollutants: A review. *Geomicrobiology Journal*. https://doi.org/10.1080/01490451.2023.2245395

Kaza, S., Yao, L., Bhada-Tata, P., & Van Woerden, F. (2018). What a Waste 2.0: A global snapshot of solid waste management to 2050. World Bank, Washington, DC. https://doi.org/10.1596/978-1-4648-1329-0

Kumar, P., & Mohan, S. V. (2022). Sustainable approaches for utilizing rice husk ash in cement and concrete. *Construction and Building Materials, 356*, 129206. https://doi.org/10.1016/j.conbuildmat.2022.129206

Kumar, V., Sharmila, A., & Prakash, S. (2020). Valorization of agricultural waste: A potential approach towards sustainable development in India. *Waste Management, 102*, 35–45. https://doi.org/10.1016/j.wasman.2020.01.024

Kumar, P., Kumar, S. K., & Yadav, V. (2022). Biomass-to-energy technologies in sustainable waste management: A review. *Renewable and Sustainable Energy Reviews, 164*, 112417. https://doi.org/10.1016/j.rser.2022.112417

Kumar, M., Pandit, S., Patel, V., Khanna, N., Nag, M., Lahiri, D., Ray, R. R., Das, A., & Das, D. (2023). Maximization of energy recovery from starch processing wastewater by thermophilic dark fermentation coupled with microbial fuel cell technology. *Geomicrobiology Journal*, 1–12. https://doi.org/10.1080/01490451.2023.2209555

Kwaghe, B. P., Zewge, F., & Bassey, N. (2021). Biogas production potential of various feedstocks. *Environmental Technology and Innovations, 23*, 100934. https://doi.org/10.1016/j.eti.2021.100934

Maalouf, A., & Mavropoulos, A. (2022). Reassessing global municipal solid waste generation. *Waste Management and Research, 41*(4), 935–946. https://doi.org/10.1177/0734242x221074116

Manomaivibool, P., Srivichai, M., Unroj, P., & Dokmaingam, P. (2018). Chiang Rai Zero Waste: Participatory action research to promote source separation in rural areas. *Resources, Conservation and Recycling, 136*, 142–152. https://doi.org/10.1016/j.resconrec.2018.04.002

Matsumoto, S. (2011). Waste separation at home: Are Japanese municipal curbside recycling policies efficient? *Resources, Conservation and Recycling, 55*(3), 325–334. https://doi.org/10.1016/j.resconrec.2010.10.005

Mishra, S., Ghosh, S., van Hullebusch, E. D., Singh, S., & Das, A. P. (2023). A critical review on the recovery of base and critical elements from electronic waste-contaminated streams using microbial biotechnology. *Applied Biochemistry and Biotechnology, 2023*, 1–30. https://doi.org/10.1007/S12010-023-04440-X

Mishra, S., Rout, P. K., & Das, A. P. (2019). Solar photovoltaic panels as next generation waste: A review. *Biointerface Research in Applied Chemistry, 9*(6), 4539–4546. https://doi.org/10.33263/BRIAC96.539546

Mishra, S., & Das, A. P. (2021). Current treatment technologies for removal of microplastic and microfiber pollutants from wastewater. *Wastewater Treatment: Cutting-Edge Molecular Tools, Techniques and Applied Aspects*, 237–251. https://doi.org/10.1016/B978-0-12-821881-5.00011-8

Mishra, S., & Das, A. P. (2022). Treatment of the wastewater polluted with synthetic microfiber released from washing machine. In B. B. Das, H. Hettiarachchi, P. K. Sahu, & S. Nanda (Eds.), *Lecture Notes in Civil Engineering* (Vol. 207, pp. 109–117). Springer Science and Business Media Deutschland GmbH. https://doi.org/10.1007/978-981-16-7509-6_9/COVER

Mishra, S., & Das, A. P. (2023). Microbial remediation of synthetic microfiber contaminated wastewater. In M. P. Shah (Ed.), *Microbial Technologies in Industrial Wastewater Treatment* (pp. 337–350). Springer, Singapore. https://doi.org/10.1007/978-981-99-2435-6_17

Mishra, S., Singh, R. P., Rout, P. K., & Das, A. P. (2022). Membrane bioreactor (MBR) as an advanced wastewater treatment technology for removal of synthetic microplastics. *Development in Wastewater Treatment Research and Processes*, 45–60. https://doi.org/10.1016/B978-0-323-85583-9.00022-3

Mohan, S. V., & Fornari, T. A. (2022). Sustainable enzymatic treatment of organic waste in a framework of circular economy. *Bioresource Technology, 370*, Article 128487. https://doi.org/10.1016/j.biortech.2022.128487

Mohanty, S., Ghosh, S., Bal, B., & Das, A. P. (2018). A review of biotechnology processes applied for manganese recovery from wastes. *Reviews in Environmental Science and Bio/Technology 2018, 17*(4), 791–811. https://doi.org/10.1007/S11157-018-9482-1

Nagaraja, S., Babu, D. B., & Rangappa, S. M. (2020). Building materials from industrial waste: Sustainable options. *Materials Today: Proceedings, 51*, 175–183. https://doi.org/10.1016/j.matpr.2020.02.098

Oyejobi, D. O., Firoozi, A. A., Blanco Fernández, D., & Avudaiappan, S. (2024). Integrating circular economy principles into concrete technology: Enhancing sustainability through industrial waste utilization. *Results in Engineering, 24*, Article 102846. https://doi.org/10.1016/j.rineng.2024.102846

Pradhan, G., Tripathy, B., Ram, D. K., Digal, A. K., & Das, A. P. (2023). Bauxite mining waste pollution and its sustainable management through bioremediation. *Geomicrobiology Journal*, 1–10. https://doi.org/10.1080/01490451.2023.2235353

Priya, A. K., Alagumalai, A., Balaji, D., & Song, H. (2023). Bio-based agricultural products: A sustainable alternative to agrochemicals for promoting a circular economy. *RSC Sustainability, 1*(4), 746–762. https://doi.org/10.1039/D3SU00075C

Priyadarsini, S., & Das, A. P. (2024). Analytical and structural characterization of waste lithium-ion batteries for their effective recycling strategy. *Environmental Chemistry and Ecotoxicology*. https://doi.org/10.1016/j.enceco.2024.12.004

Priyadarsini, S., & Das, A. P. (2025). Lithium bioleaching: A review on microbial-assisted sustainable technology for lithium bio-circularity. *Journal of Water Process Engineering, 69*, Article 106744. https://doi.org/10.1016/j.jwpe.2024.106744

Ray, S. K., Dey, S., Das, A. P., & John, A. (2024). Microplastic waste management through greener technologies : A review. *Sustainable Management of Urban Plastic Waste Through Circular Economic Approaches*, 237–261. https://doi.org/10.1201/9781003465447-14

Sahu, A., Dey, S., Mallick, K., Priyadarsini, S., Behera, I. D., & Das, A. P. (2024). A review on advanced technology for sustainable management of synthetic microplastic waste. *Sustainable Management of Urban Plastic Waste Through Circular Economic Approaches*, 262–279. https://doi.org/10.1201/9781003465447-15

Sakai, K., Taniguchi, M., Miura, S., Ohara, H., Matsumoto, T., & Shirai, Y. (2008). Making plastics from garbage: A novel process for poly-L-lactate production from municipal food waste. *Journal of Industrial Ecology, 7*(3–4), 63–79. https://doi.org/10.1162/jie.2004.7.3-4.63

Shenyoputro, K., & Jones, T. E. (2023). Reflections on a two-decade journey toward zero waste: A case study of Kamikatsu town Japan. *Frontiers in Environmental Science, 11*, 1171379. https://doi.org/10.3389/fenvs.2023.1171379

Silva, A., Rosano, M., Stocker, L., & Gorissen, L. (2017). From waste to sustainable materials management: Three case studies of the transition journey. *Waste Management, 61*, 547–557. https://doi.org/10.1016/j.wasman.2016.11.038

Singh, B., & Sharma, P. (2019). Sustainable management of agricultural waste: Assessment and strategies in India. *Environmental Management, 64*(4), 487–497. https://doi.org/10.1007/s00267-019-01169-5

Smith, R., Zafar, A., & Johnson, M. (2022). Anaerobic digestion and biogas production: Overview of processes and benefits. *Renewable Energy Reviews, 169*, Article 112642. https://doi.org/10.1016/j.rser.2022.112642

Tripathy, B., Dey, S., Dubey, N. K., Lo, W.-C., & Das, A. P. (2024). In *An Omics-Based Approach for Managing the Risk of COVID-19-Generated Wastes*, pp. 51–66. https://doi.org/10.1007/978-3-031-50840-0_4

UNEP. (2019). Global waste management outlook: A comprehensive review. Retrieved from https://www.unep.org/resources/report/global-waste-management-outlook

United Nations Environment Programme. (2019). Global Resources Outlook 2019: Natural Resources for the Future We Want. Retrieved from https://www.unep.org/resources/report/global-resources-outlook-2019-natural-resources-future-we-want.

World Economic Forum. (2021). The role of public-private partnerships in climate action and waste management. Retrieved from https://www.weforum.org

Yadav, V., A., Das, A., Miftah, A.O., & Iqbal, H.M.N. (2021). Valorization of citrus waste for biofuels and value-added products: Recent trends and future perspectives. *Chemosphere, 293*, 133459. https://doi.org/10.1016/j.chemosphere.2021.133459

Zaman, A. U. (2015). A comprehensive review of the development of zero waste management: Lessons learned and guidelines. *Journal of Cleaner Production, 91*, 12–25. https://doi.org/10.1016/j.jclepro.2014.12.013

Zaman, A. U., & Lehmann, S. (2011). Challenges and opportunities in transforming a city into a "zero waste city." *Challenges, 2*(4), 73–93. https://doi.org/10.3390/challe2040073

Zhang, X., Wang, H., & Li, W. (2021b). Sustainable biopolymer production from agricultural waste: Opportunities and challenges. *Bioresource Technology Reports, 16*, Article 100811. https://doi.org/10.1016/j.biteb.2021.100811

Advanced Microbial Technology for Sustainable Management of Plastic Waste

M. Purnima Sudha, Kiran Bala, Aditya Kishore Dash, and Alok Prasad Das

Abstract The persistent usage, build-up, and pollution of plastics jeopardize environmental biotic and abiotic components. Moreover, it exacerbates global warming by augmenting carbon footprints. Plastics offer enormous prospects in terms of circular consumption and recycling. Composting is a natural process that occurs through microbial succession, resulting in the decomposition and stabilization of organic matter found in trash. The use of microbial additions in composting is thought to be highly efficient, as it increases the synthesis of several enzymes, resulting in a faster rate of waste decomposition. Depending on the waste material's nature, either they can be composted immediately or homogenized before being sent to secondary waste treatment facilities like landfills. Existing waste management practices are not enough to dispose of plastic trash in landfills, wastewater treatment plants, and the environment. Plastic biodegradation under such conditions is a sustainable and environmentally friendly way to address plastic pollution. Natural breakdown of plastics requires methods that use bacteria that can use polymers as their sole carbon source. Protein engineering techniques to enhance the catalytic activity of plastic-degrading enzymes are in their early stages of research. An entirely biological approach to plastic recycling in a sustainable economy is product manufacturing using enzymatic degradation. This book chapter enhances on the studies pertaining to the microbial composting of municipal solid waste and food waste addressing variables that may impede the microbial degradation process and impact the compost quality. And also emphasizes on the ways to improve the sustainable management of plastic waste.

Keywords Plastic waste · Microbial technology · Composting · Sustainable management · Waste management

M. Purnima Sudha · A. P. Das (✉)
Department of Life Science, Rama Devi Women's University, Bhubaneswar, Odisha, India
e-mail: alok1503@gmail.com

K. Bala
Department of Zoology, Deshbandhu College, University of Delhi, Delhi, India

A. K. Dash
Chemistry Department, ITER, Siksha 'O' Anusandhan (Deemed to Be University), Bhubaneswar, India

A. P. Das (ed.), *Bioconversion of Environmental Waste to Value Added Products*, Sustainable Environmental Waste Management Strategies,
https://doi.org/10.1007/978-3-032-14215-3_17

1 Introduction

The modern world has incorporated synthetic polymers and plastics into every facet of our everyday existence. The scientists assert that all synthetic plastic ever made still exists in some way on our planet and it is a major environmental hazard now (Ahmad et al., 2022). The ease of use and versatility of synthetic polymers have resulted in their extensive use. But since existing strategies have not been able to stop the growth of plastic trash, there is a pressing need for innovative, sustainable ways to fight plastic pollution (Das et al., 2025; Singh Jadaun et al., 2022). The biodegradation of plastic has become a viable and sustainable alternative that promotes recycling. Plastic garbage is the main source of this solid waste. For example, just 12.3% of the 5.6 million tons of plastic garbage generated annually in India is recycled (Abdel-Raouf et al., 2012). The ubiquitous result of our contemporary, convenience-driven society, plastic garbage, is becoming a major environmental sustainability and public health concern. An enormous amount of non-biodegradable debris has accumulated in landfills, oceans, and ecosystems as a result of the widespread manufacture and negligent disposal of plastic products (Dey et al., 2024; Priyadarshini et al., 2024b). For example, more than 9400 tons of plastic garbage are either landfilled or wind up contaminating streams or groundwater supplies because they contain plastics that either don't disintegrate at all or take 450 years to do so (Kesaano & Sims, 2014). This is a major contribution to the worldwide problem of greenhouse gas (GHG) emissions in addition to being an urgent ecological and environmental hazard. Plastics' environmental persistence and the procedures used in their production and disposal cause greenhouse gases like carbon dioxide and methane to be released, which exacerbates the effects of global warming (John & Das, 2024). Notwithstanding the massive efforts towards finding alternative materials to substitute plastic, the perpetual problem of plastic waste in the environment demands a prompt and viable solution. Plastic biodegradation comes as a cleaner and more environmentally friendly alternative to conventional methods like landfilling, incineration, and recycling due to its milder working conditions and the absence of toxic materials in the biodegradation processes (Reddy Koteswara et al., 2025; Sahoo et al., 2022). The biodegradation method is also hopeful in order to tackle the massive amount of plastic waste in the environment effectively. Plastics and synthetic polymers have become ubiquitous in modern society (Abinandan & Shanthakumar, 2015). All products of synthetic plastic that have ever been made are still on the planet in one way or another, which has raised a lot of environmental concern (Roy & Chakraborty, 2024). Synthetic polymers have been popular because they are flexible and convenient. There is a need for creative, sustainable solutions to plastic pollution, though, since current measures have failed to curb plastic trash accumulation (Walker & Fequet, 2023). Notwithstanding the issues with plastics disposal, they present chances to reinterpret their function in relation to recycling and circular usage. The need for innovative, sustainable ways to lessen plastic pollution is further increased by the fact that they are a necessary and interchangeable component in many businesses

(Bahl et al., 2021). Recycled plastics undergo a physical or chemical transformation that yields lower-grade plastic products, referred to as downcycling. However, many everyday synthetic polymers, such as polyethylene (PE), polystyrene (PS), polypropylene (PP), polyvinyl chloride (PVC), polyurethane (PUR), and polyethylene terephthalate (PET), are degradable by microorganisms and enzymes (Mohee & Unmar, 2007), yielding recyclable end products.

Upcycling through biotechnology is believed to be a cheaper and more efficient way to solve the plastic waste issues of the world (Mallick et al., 2023; Mohanty et al., 2024a, 2024b). Upcycling refers to the conversion of old materials to new, high-value products through the recycling as well as enhancement of the materials. Valuable material recovery from plastic wastes is required for upcycling plastics (Narancic et al., 2020; Priyadarsini & Das 2025a).

Plastics have become an essential component of our daily life. Plastics are becoming increasingly popular due to their wide range of applications. Plastics are continuously used in both urban and rural locations (Ray et al., 2024; Sahu et al., 2024). Due to their adaptability, plastics can be used across a number of different industries, including agriculture, medicine, transport, and building. Because we cannot envision a world without plastics, it is fitting to state that we are in an age of plastics (Geyer et al., 2017). Plastics are increasingly taking over from metals, wood, paper, wool, and cotton as industrial materials (Priyadarsini & Das, 2025b; Priyadarsini & Das, 2024a). Plastics outperform these materials in terms of water resistance and microbiological resistance. We progressively develop an addiction to single-use or disposable plastics as our reliance on them grows. Examples of these include packaging films, bags, cutlery, and stationery goods. Consequently, the footprints of plastic waste have become more noticeable in natural environments as it has slowly moved from pole to pole (Schnurr et al., 2018). Given their deadly effects and ability to persist in the environment for thousands of years without decomposing, plastics have seriously harmed both terrestrial and aquatic ecosystems during the past 20 years (Derraik, 2002). The steady rise of plastic products has sparked worries about the environment around the world.

The three categories of plastic pollution are microplastics, meso-plastics, and macro-plastics. The main causes of plastic pollution in the environment are landfills, human activity, and household and industrial trash (Lee et al., 2023). The majority of single-use plastics (about 79%) wind up in the environment, but a sizable percentage are burned and the remainder are recycled. Around 370 million tons of plastic were manufactured worldwide in 2019. Ecosystems on land and in water are consequently negatively impacted (Sahu et al., 2024). Animals that consume plastics or microplastics suffer fatal consequences (Bala et al., 2024a; b). The coral reef has been predominantly impacted by this human activity. It is estimated that more than 11 billion pieces of plastic are hooked up with coral reefs in the Asia–Pacific region and this figure is increasing day by day (Huang et al., 2021). The alarming factor is that plastic trash might contain microbes, increasing the disease risk in the water. International Ocean Clean-up Project 2017 is one of the projects developed to address the problem of plastic waste. However, such attempts are typically costly and necessitate ongoing monitoring (Cordier & Uehara, 2019). As a result, research is being performed to create alternative recycling processes for plastic waste that overcome the constraints of conventional disposal methods such as chemical treatment, incineration, and landfilling (Mishra et al., 2021; Sahu & Das, 2024). This enormous environmental load can be lessened in a sustainable and eco-friendly manner by applying biological processes to break down plastic and turning it into value-added bioproducts (Kapoor & Rafatullah, 2025). Highly toxic resistive, mutagenic, or xenobiotic compounds like azo dyes, insecticides, phenols, and polyaromatic hydrocarbons have been found to be efficiently degraded by microorganisms (Zhang et al., 2020). Microbial degradation disintegrates plastic trash into monomers, which are further mineralized into water, CO_2, and new biomass (Kaushal et al., 2021). The increased rate of plastic trash and the evolution of microbial technology to minimize waste production are the major subject matters of this book chapter.

2 Plastic Waste and Its Impact

Due to inadequate source segregation, minimal recycling attempts, and a huge amount of trash making its way to landfills randomly, India as a large user and producer of plastic has very serious waste management issues (Ghosh et al., 2023; Mishra et al., 2022). Only 71% of the total trash generated in the country can be processed from plastic (Evode et al., 2021). Poor waste collection, sorting, and treatment facilities can lead to plastic waste getting stuck in the drainage system. 5.6 million tons of municipal solid waste consist of plastic waste, 70% of which is landfilled and 20% of which is incinerated (Šaravanja et al., 2022). Plastic disintegration rates range from 100 to 1000 years, indicating that older plastics may still exist in the environment (Li et al., 2021). It has been projected that by 2035, plastic garbage buildup will equal the number of fish in the oceans (Dharwal et al., 2022). Because of their inherent stability and endurance, plastic materials show little natural degradation in the environment. Numerous environmental elements, such as heat, sunlight, chemical reactions, biological activity, and physical characteristics including molecular weight, density, and polymer size, all have an impact on the degrading process (Mohanty et al., 2024a, 2024b; Ray et al., 2024). Physical/mechanical deterioration can lead to chemical degradation which results in the creation of short-chain molecules; chemical degradation can be by action of corrosive agents such as acids and atmospheric pollutants; thermal degradation occurs due to chemical reaction by heat or temperature fluctuations; and photodegradation/photo-oxidative degradation triggers degradation of polymer chains and creates free radicals.

3 Plastic Waste Accumulation

Plastics are accumulated extensively in terrestrial ecosystems since they are predominantly used and discarded on land. Agricultural plastics such as packaging materials and polytunnel films contribute a large proportion of plastic waste (Bala et al., 2024a, 2024b; Das et al., 2024). Illegal dumping and direct rubbish ingestion add to land-based plastic accumulation; these plastics break down into microplastics, which may remain in soil for 15 years. Environmental scientists argue that these microplastics are harmful to soil layers, clog pipes, become poisonous, and endanger fish, birds, reptiles, vertebrates, and other creatures (Nayanathara Thathsarani Pilapitiya & Ratnayake, 2024). Plastic waste is transported by human activity, wind, and water, which find their way into freshwater systems such as rivers, dams, and urban drainage systems. Particularly, rivers are important routes for the movement of plastics into marine habitats. Eight to ten million tons of plastics are released into the ocean each year, accounting for 80% of marine pollution and posing serious risks to marine life (Mishra & Das, 2021; Mishra et al., 2024a, 2024b). This influx of plastics into the ocean is influenced by watershed outfalls, population density, and maritime activity (Van Emmerik et al., 2022). Plastics can injure marine life if they get trapped in them or ingest them (Kye et al., 2023). Plastic rubbish, for example, has endangered all seven species of turtles globally. When these open-burned non-biodegradable plastic wastes are combusted, they even release harmful gas emissions and water stagnation. Long-term disposal of the material in landfills poses the risk of toxic substances leaking out and polluting groundwater. CO_2, CH_4, and other gases are emitted when solid waste is contaminated. Moreover, open dumpsites cause diseases such as cholera and pollute drinking water (Shen et al., 2025).

4 Impacts on Environment

The effects of plastic trash on ecosystems are extensive and affect freshwater, marine, and terrestrial settings. Plastic nets are frequently employed in gardens, farm fields, and aquatic farms, trapping and perhaps harming animals including birds and reptiles (Fayshal, 2024). Ingestion of microplastics, which are common in many environmental situations, can result in internal injuries, intestinal obstructions, and decreased feeding stimulation, especially for ruminants like camels, goats, and sheep. Plastic waste lowers oxygen and illumination levels in marine ecosystems, which harms coral reefs and diminishes biodiversity (Jiang et al., 2020). Plastic ingestion and entanglement impact more than 43% of marine animals, 44% of seabirds, and 86% of sea turtles worldwide. Entanglement induces asphyxia, drowning, and impaired mobility, while ingestion induces physical and physiological effects like lacerations, blockages, and digestive tract retention. Such effects, inclusive of developmental, immunological, and toxicological effects (Jeong et al., 2024), exert impacts on animal health and population dynamics. Seabirds eat various plastic rubbish, which impairs their body weight, reproduction efficiency, and digestive system. Microplastics that resemble phytoplankton are eaten by fish and whales and build up over the food web, influencing filter-feeding levels, reproduction procedures, and development rates in many species (Malinowski et al., 2023).

5 Impacts on Human Health

Apart from causing serious environmental concerns, plastic garbage has a few adverse impacts on human health as well. One of the most alarming phenomena is the ingestion of microplastics, which have been reported in food and water sources like seafood, drinking water, and salt (Mishra et al., 2023; Tripathy et al., 2024). Microplastics carry toxic compounds like phthalates, bisphenol A (BPA), and bromine flame retardants (BFRs) that are let out into the environment by plastic. These substances can interfere with reproductive health, alter hormone control, and raise the risk of cancer and developmental abnormalities when consumed (Pradhan et al., 2024; Priyadarshini & Das, 2024c). One typical plasticizer found in food and drink packaging is BPA, which is known to interfere with endocrine function. Cardiovascular illness, birth deformities, cancer, and respiratory problems are among the negative consequences of BPA exposure, which mostly comes from nutrition (Jayavel et al., 2024). Similar to this, phthalates, which are plasticizers found in food packaging and personal care items, can enter the body through the mouth, nose, or skin and cause a number of health problems, including stunted growth and liver problems (N. Ahmed, 2024). Brominated flame retardants (BFRs), employed in electronic devices to avoid fire loss, consist of additives like polybrominated diphenyl ethers (PBDEs), which are also found to cause neurological diseases and cancer (Caminiti et al., 2025). In addition, micro- and nano-plastic particles enter the human system through ingestion and inhalation exposure in heavily contaminated urban cities with large amounts of plastic pollutants, resulting in respiratory problems like asthma and other respiratory diseases (Maquart et al., 2022). Research has indicated that airborne microplastics could have the potential to transmit viruses such as SARS-CoV-2 (COVID-19) over extended distances. Dioxins, furans, and volatile organic compounds (VOCs) are released into the air when plastic waste is incinerated, either deliberately through open burning or inadvertently through industrial activities. Such pollutants can cause significant harmful effects if inhaled.

6 Strategies for Managing Plastic Waste

In order to combat the growing environmental effects of plastic pollution, effective plastic waste management solutions are essential. About 70% of plastic trash ends up in landfills, making landfilling the most common method of managing plastic waste (Qian et al., 2025). The incineration technique has been used to generate energy from plastic trash and reduce the burden on current landfills (Huang et al., 2022). Additionally, recycling—the process of gathering, sorting, and processing plastics to be used as raw materials for new products—has been emphasized in order to support the circular economy for plastics (Satchidananda Mishra et al., 2024a, 2024b). Reducing the overall use of conventional plastics is another goal of promoting sustainable alternatives like eco-friendly packaging or biodegradable plastics (Aragaw, 2025). Innovative technology and waste-to-energy facilities help to reduce pollution. Waste-to-energy plants can transform non-recyclable waste into power or heat, eliminating the demand for fossil fuels and lowering overall greenhouse gas emissions (Vuppaladadiyam et al., 2024). Waste-to-energy plants provide a sustainable option for plastic waste management by tapping into the energy potential trapped inside plastic waste products. Plastics are converted into important energy sources using modern methods such as pyrolysis or gasification. This reduces the demand for traditional fossil fuels by providing a substitute energy source and reducing pressure on landfills. This double advantage is an anticipatory strategy toward energy and waste management. Nevertheless, if waste-to-energy plants are to be environmentally friendly, they need to implement stringent emission controls in order to avoid noxious pollutants seeping into the atmosphere.

7 Plastic Waste Degradation by Microorganisms

There should be a uniform method of managing the increasing quantity of plastic waste because there are disadvantages in employing biodegradable polymers and because conventionally produced plastics hold an important value. Enormous resources have been invested in the development of new types and chemicals that can augment the biodegradation of traditional plastics in an effort to tackle this acute issue. Comparing it to solutions already described, the process of plastic biodegradation is observed to be a more sustainable and environmentally friendly method due to its lower and less harsh operating conditions (Zeenat et al., 2021). Microorganisms degrade plastic waste through a variety of enzyme processes arranged by distinct microbial populations. Microorganisms on the exterior of plastic refuse secrete extracellular proteins, which degrade polymer chains. These proteins—estrogenases, esterases, and proteases—preferentially attack selected chemical bonds between molecules in the plastic polymers, like the carbon–carbon bond in PP or the ester linkage in PET, to initiate a hydrolysis reaction that degrades the polymer to smaller fragments (Skariyachan et al., 2022). Microbes utilize the oligomers or monomers so formed as a source of carbon and energy for a range of metabolic processes when the polymer chains dissociate. The degradation of plastic by microbes has been shown in Fig. 1.

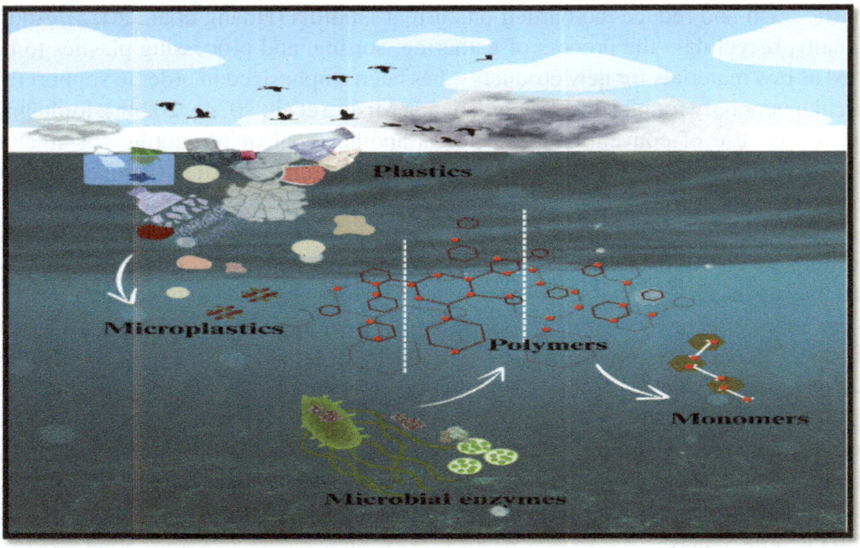

Fig. 1 Degradation of plastic by microbial enzymes

8 Aerobic Biodegradation

Aerobic biodegradation occurs when oxygen is present, as oxygen-utilizing bacteria that utilize oxygen as a terminal electron acceptor during metabolism decompose organic compounds. In the presence of oxygen, oxygen induces oxidation reactions and degrades plastic polymers, shredding plastic atoms into simple molecules—carbon dioxide, water, and biomass. This handle is commonly seen in oxygen-rich environments, such as soils, surface rivers, and compost piles. Biodegradation, which consumes oxygen, is more efficient and rapid than anaerobic biodegradation, with a much higher vitality yield (Heris, 2024).

9 Anaerobic Biodegradation

Anaerobic bacteria, which use optional electron acceptors Like nitrate or sulphate, carry out anaerobic biodegradation in differentiated environments where oxygen is not visible. Through aging and anaerobic breath, microbes break down plastic polymers in anaerobic environments, producing less challenging natural acids, alcohols, and gases Like carbon dioxide and methane. Tripathy et al., (2026) Anaerobic biodegradation typically occurs in oxygen-deficient environments, such as anaerobic digesters, dregs, and deep soils (Bátori et al., 2018).

Biodegradation is affected by a number of variables, including the polymer's characteristics, the kind of organism, the pre-treatment technique, and environmental elements including temperature, pH, moisture, and nutrition availability. Among other characteristics of the polymer, its degradation is significantly influenced by its mobility, crystallinity, molecular weight, functional groups, substituents, and plasticizers or additives (Yang et al., 2023).

10 Chemical Reactions Involved in Biodegradation

- **Bio-deterioration**

It involves altering the material's surface to change the plastic's mechanical, chemical, and physical characteristics. The natural structure and makeup of the polymers dictate all chemical and structural alterations. Environmental conditions also have an impact on changes in polymer characteristics. Biofilms and a matrix form inside the plastic substance as a result of the degradation process.

- **Bio fragmentation**

Following biodegradation, enzymatic mechanisms act on plastic polymers to cause biodegradation. When oxygen molecules interact with carbon chains, oxygenase—which are mostly found in bacteria have the ability to break the molecules, producing less harmful alcohols and peroxyl compounds. Furthermore, enzymes including end-peptidases, lipases, and esterase catalyze the conversion of carboxyl groups to amide groups.

- **Assimilation**

During the assimilation stage, the atoms are integrated into the microbial cell for permanent incorporation until full degradation occurs. Secondary metabolites are then transferred to other bacteria, or exported to the outside of the cell for their degradation and utilization. Oxidized products (CO_2, N_2, H_2O, CH_4) are formed when metabolites decompose.

- **Mineralization**

The cell membrane enables the passage of polymeric polymers produced during bio-fragmentation into the bacterial cell. However, big monomers cannot enter the cell and must remain outside it. The tiny monomers that enter the cell are oxidized and utilized to generate energy. Finally, the produced energy is employed in biomass production. Weight loss, changes in tensile strength, changes in dimension, changes in physicochemical qualities, the creation of carbon dioxide, and so on are all indicators of biodegradation (Ru et al., 2020).

11 Mechanism of Biological Degradation

Numerous microorganisms, such as bacteria and fungus, have been found to be able to break down polyesters through a range of enzymes that they manufacture. Both natural and manmade polymers can undergo biodegradation with the help of microorganisms. The size decrease of polymers is mostly caused by microorganisms, which colonize the plastic surface and transform it into monomers that cells can absorb. Additional enzymatic reactions involving these monomeric units take place within these cells, using the monomers as a source of carbon to fuel growth (Polyethylene Biodegradation: A Multifaceted Approach, 2025) Pre-treatments of plastics, such as heat, cooling, solidifying, defrosting, and chemical corruption, as well as prior bacterial activity to separate the polymer into monomeric components, can alter the process. The monomers are mineralized during enzymatic debasement, and metabolic products such as CO_2, H_2O, CH_2, and N_2 are released. These things can be quite helpful in clearing the environment of dangerous plastics. Methane, one of these gases, is an essential part of biogas, which can be used in the combination of several natural acids or burned to produce heat and electricity (Gates & Crook, 2024). Biological catalysts known as enzymes take part in processes by influencing particular substrates and hastening the transformation of those substrates into useful products. The tendency of plastic degradation, which is detrimental to the environment, has drawn attention to bacteria's capacity to break down these materials. Microbial enzymes, whether extracellular or intracellular, play a major role in the first degradation of plastics by promoting the breakdown of polymer chains inside the cell. Numerous species, including bacteria, fungus, actinomycetes, and algae, have been found to contain enzymes that may break down various kinds of plastics. "Hydrolases" are a significant class of enzymes that, when exposed to water, break chemical bonds, resulting in the fragmentation of big molecules into smaller ones. Carbon chains are broken down into smaller fragments by hydrolases, which are involved in the degradation of polymers. Every plastic that is found in the environment is hydrophobic. The connection between the enzyme and the polymer begins with the extracellular microbial enzyme binding to the plastic surface via hydrophobic interactions. The active areas of numerous hydrolases feature a hydrophobic cleft that attracts hydrophobic groups from the polymer, facilitating enzyme access to the polymer. The second phase is the hydrolytic cleavage of the long polymer chain by the active site of the enzyme (dimers or monomers). The microbe then consumes these smaller monomers and uses them as a source of carbon.

12 Biotechnological Advances in Plastic Breakdown

In an effort to lessen plastic pollution, a lot of study has lately been conducted on plastic biodegradation. 16S rRNA quality sequencing was used to characterize the microbial strains implicated in plastic corruption. In following biodegradation tests, the performance of polymer corruption has been evaluated using these characterized plastic-decomposing bacteria (Srimongkol et al., 2022). Numerous investigations have revealed that some species of wax worms use plastic substrates as a carbon source in their gut microbiomes. The molecular sizes of LDPE microplastics were considerably decreased by the presence of the bacterium *Lumbricus Terrestris* (Aketo et al., 2020). The plastic-degrading enzymes Ceres and Demetra were discovered to be present in the saliva and gut microbiota of the waxworm *Galleria mellonella* (Ahmed et al., 2022). These enzymes have the ability to oxidize and depolymerize polyethylene when allowed sufficient time to become active at pH 7 (room temperature).

Finding novel enzymes that can break down plastics is crucial because of their significant role in polymer degradation. These findings will stimulate greater investigation into the identification of novel environmental plastic-degrading enzymes for more effective biodegradation of various polymers. Chimeric enzymes, sometimes referred to as fusion proteins, are proteins made by splicing two or more unrelated proteins together genetically. All of the proteins participating in the fusion reaction possess the same features, making the resultant protein a single, cohesive molecule. Well-designed fusion proteins are utilized in many different industries and present a multitude of opportunities due to their varied properties. As a result, it seems that employing several enzyme systems to quicken the biodegradation of plastics is a highly successful tactic. Enzymes are highly suited for application in polymer degradation due to their universal catalytic processes and substrate selectivity (Ayele & Godeto, 2021).

13 Development of Biological Techniques

The discovery of techniques to speed up the known biodegradation processes is crucial, in addition to the investigation of novel strains and enzymes that can break down plastics. To increase their biodegradability, many polyolefin plastics have undergone pretreatment with UV light, heat treatment, or chemical reactions. Because of their hydrocarbon chain structure, they are extremely hydrophobic, which prevents microbes from participating in biodegradation. Because of the abiotic pretreatment effects, polymer oxidation promotes biodegradation, including the composting of plastics that have been microbially pretreated (Cai et al., 2023). The potential to speed up the overall biodegradation of plastics has been studied for novel pretreatment methods as gamma irradiation (Blánquez et al., 2008) and plasma therapy (Hayat et al., 2017). Plastics can also be pretreated to add functional groups (such carbonyls, carboxyl, and esters) to the polymer chain. Numerous studies demonstrate that pretreated plastics degrade faster than/than untreated polymers, and that the surface morphology of polymers is modified by pretreatment. Enzymes represent a good methodology for biocatalytic degradation of plastic waste and management of ecological waste owing to high substrate specificity and high fidelity of reaction conditions (Brennan & Owende, 2010). Nonetheless, sporadic turnover rates and low-temperature tolerance limit feasibility for most natural plastic-degrading enzymes (PDEs) (Dhali et al., 2024). The good news is, protein engineering has the potential to make enzymes more useful in an industrial setting. Synthetic biology aims to alter or restructure current biological systems in order to accomplish particular objectives. Computational biology, which uses environmental sequencing data to help the theoretical design step, can streamline this approach (Amobonye et al., 2021). PDE, which breaks down plastics and encourages the breakdown of polyethylene terephthalate (PET), is released by genetically modified microalgae (Amobonye et al., 2021). The enzymatic breakdown of PET in multiple phases and the subsequent synthesis of PHA using the liberated monomers in an engineered Pseudomonas sp. proved the biocatalytic upcycling of PET to polyhydroxyalkanoates (PHAs) as described by some researchers. A potential area of research that could eventually lead to eco-friendly waste management techniques is synthetic biology. The new area of study is using protein engineering techniques to increase the catalytic activity of enzymes that break down plastic. In general, directed evolution and rational design are the two methods used in protein engineering (Thirunavukarasu et al., 2016). In rational design, the target protein and the mechanistic features driven by it are designed based on protein structure knowledge through computer-aided simulation and modelling. Rational design has been the proposed paradigm approach in nearly all recent publications describing plastic-degrading enzyme production given the abundance of both structural and mechanistic data for many of these enzymes.

14 Future Perspective

The unrestricted dumping of plastic waste into landfills and oceans poses a serious worldwide threat, releasing microplastics and debris into the ecosystem. This has resulted in a number of major concerns, including harm to human health and detrimental effects on the environment. Given these problems, we must exercise caution when using and disposing of plastics. It is critical to create sustainable solutions for the use and disposal of synthetic polymers in institutions and industry. Additionally, in order to clarify the deadly consequences of plastic waste materials, biomonitoring must be integrated into human and animal systems. Beyond conventional approaches, eco-friendly alternatives must be developed to address the buildup of plastic trash in the environment. According to a number of investigations, microbes like *Ideonella sp., Bacillus sp., Streptomyces sp.,* and *Pseudomonas sp.* may be able to help reduce plastic waste in lab settings. To determine the processes involved in the biodegradation of plastic trash, thorough research is still necessary. The scarcity of plastic-degrading enzymes emphasizes how crucial it is to research and find novel enzymes with more diverse functions. There is potential for finding new biocatalysts by studying the variety of non-cultivated microorganisms. During the biodegradation of plastic, high-throughput transcriptome-based techniques can help detect distinct gene expressions under various growth circumstances, offering important insights. Our knowledge of the plastic breakdown process can be improved by research aimed at determining the gene–protein interaction and clarifying the roles of the genes involved. Furthermore, in order to be used on a big scale, plastic waste treatment systems need to be both practical and long-lasting, while also accounting for environmental microbial adaptation. Expanding microbe-based plastic waste treatment techniques might be a highly lucrative solution to the global plastic waste issue.

15 Conclusion

Although plastics are a necessary component of our everyday lives, they are also very problematic when it comes to disposing of waste. One significant topic covered in this study is the relationship between greenhouse gas emissions and the management of plastic garbage. The way plastic is now managed has a big effect on the environment and climate change. Reducing carbon emissions and promoting a more sustainable future are two benefits of integrating bioremediation, in which microorganisms break down plastic trash. While microorganisms and enzymes have demonstrated short-term effectiveness with polyethylene terephthalate (PET), ongoing efforts are required to produce additional plastic-degrading microbes and enzymes for various types of plastics. Waste-to-energy facilities and other innovative solutions have a lot of potential, particularly for big goals like plastic. However, the stringent use of emission control measures to safeguard the environment is necessary for these plans to be successful. The problems of managing plastic waste and reducing them can be successfully addressed by community involvement, technology development, responsible policymaking, etc. The potential for efficient biodegradation and the conversion of plastic waste into value-added goods offers a plethora of sustainable solutions to reduce plastic waste pollution and contribute to a cleaner, kinder world.

References

Abdel-Raouf, N., Al-Homaidan, A. A., & Ibraheem, I. B. M. (2012). Microalgae and wastewater treatment. *Saudi Journal of Biological Sciences, 19*(3), 257–275. https://doi.org/10.1016/J.SJBS.2012.04.005

Abinandan, S., & Shanthakumar, S. (2015). Challenges and opportunities in application of microalgae (Chlorophyta) for wastewater treatment: A review. *Renewable and Sustainable Energy Reviews, 52*, 123–132. https://doi.org/10.1016/J.RSER.2015.07.086

Ahmad, A., Banat, F., Alsafar, H., & Hasan, S. W. (2022). Algae biotechnology for industrial wastewater treatment, bioenergy production, and high-value bioproducts. *Science of the Total Environment, 806*, Article 150585. https://doi.org/10.1016/J.SCITOTENV.2021.150585

Ahmed, N. (2024). lReview. *Int. J. of Pharm. Sci, 2*, 1825–1842. https://doi.org/10.5281/zenodo.14446167

Ahmed, S. F., Mofijur, M., Parisa, T. A., Islam, N., Kusumo, F., Inayat, A., Le, V. G., Badruddin, I. A., Khan, T. M. Y., & Ong, H. C. (2022). Progress and challenges of contaminate removal from wastewater using microalgae biomass. *Chemosphere, 286*, Article 131656. https://doi.org/10.1016/J.CHEMOSPHERE.2021.131656

Aketo, T., Hoshikawa, Y., Nojima, D., Yabu, Y., Maeda, Y., Yoshino, T., Takano, H., & Tanaka, T. (2020). Selection and characterization of microalgae with potential for nutrient removal from municipal wastewater and simultaneous lipid production. *Journal of Bioscience and Bioengineering, 129*(5), 565–572. https://doi.org/10.1016/J.JBIOSC.2019.12.004

Amobonye, A., Bhagwat, P., Singh, S., & Pillai, S. (2021). Plastic biodegradation: Frontline microbes and their enzymes. *Science of the Total Environment, 759*, Article 143536. https://doi.org/10.1016/J.SCITOTENV.2020.143536

Aragaw, T. A. (2025). Plastic waste management strategies toward zero waste: Status, perspectives and recommendations for Ethiopia. *Cambridge Prisms: Plastics, 3*, Article e1. https://doi.org/10.1017/plc.2024.37

Ayele, A., & Godeto, Y. G. (2021). Bioremediation of chromium by microorganisms and its mechanisms related to functional groups. *Journal of Chemistry, 2021*. Hindawi Limited. https://doi.org/10.1155/2021/7694157

Bahl, S., Dolma, J., Singh, J. J., & Sehgal, S. (2021). Biodegradation of plastics: A state of the art review. *Materials Today: Proceedings, 39*, 31–34. https://doi.org/10.1016/J.MATPR.2020.06.096

Bala, K., Anshul, Verma, S., Dogra, A., Das, A. P., Kumar, R., & Nandini. (2024). Sustainable management for urban plastic waste generation. *Sustainable Management of Urban Plastic Waste Through Circular Economic Approaches*, 211–236. https://doi.org/10.1201/9781003465447-13

Bala, K., Dogra, A., Lata, P., & Das, A. P. (2024). Discussing sources and biological-chemical remediation approaches for microplastic pollution. *Sustainable Microbial Technology for Synthetic and Cellulosic Microfiber Bioremediation*, 37–59. https://doi.org/10.1007/978-3-031-62600-5_3

Bátori, V., Åkesson, D., Zamani, A., Taherzadeh, M. J., & Sárvári Horváth, I. (2018). Anaerobic degradation of bioplastics: A review. *Waste Management, 80*, 406–413. https://doi.org/10.1016/J.WASMAN.2018.09.040

Blánquez, P., Sarrà, M., & Vicent, T. (2008). Development of a continuous process to adapt the textile wastewater treatment by fungi to industrial conditions. *Process Biochemistry, 43*(1), 1–7. https://doi.org/10.1016/J.PROCBIO.2007.10.002

Brennan, L., & Owende, P. (2010). Biofuels from microalgae—a review of technologies for production, processing, and extractions of biofuels and co-products. *Renewable and Sustainable Energy Reviews, 14*(2), 557–577. https://doi.org/10.1016/J.RSER.2009.10.009

Cai, Z., Li, M., Zhu, Z., Wang, X., Huang, Y., Li, T., Gong, H., & Yan, M. (2023). Biological degradation of plastics and microplastics: A recent perspective on associated mechanisms and influencing factors. *Microorganisms, 11*(7). Multidisciplinary Digital Publishing Institute (MDPI). https://doi.org/10.3390/microorganisms11071661

Caminiti, C., Diodati, F., Puntoni, M., Balan, D., & Maglietta, G. (2025). Surveys of knowledge and awareness of plastic pollution and risk reduction behavior in the general population: A systematic review. *International Journal of Environmental Research and Public Health*, 22(2). Multidisciplinary Digital Publishing Institute (MDPI). https://doi.org/10.3390/ijerph22020177

Cordier, M., & Uehara, T. (2019). How much innovation is needed to protect the ocean from plastic contamination? *Science of the Total Environment, 670*, 789–799. https://doi.org/10.1016/J.SCITOTENV.2019.03.258

Das, A., Ray, S. K., Mohanty, M., Mohanty, J., Dey, S., & Das, A. P. (2024). Ecotoxicity of microplastic wastes and their sustainable management: A review. *Environmental Chemistry and Ecotoxicology, 6*, 144–152. https://doi.org/10.1016/J.ENCECO.2024.05.003

Das, A., Dey, S., Gacem, A., Yadav, K. K., Nath, H., Ram, D. K., & Das, A. P. (2025). Revolutionizing solid hydrocarbon sourcing: Plastic's journey from waste to upcycled treasure. *Environmental Science and Engineering, Part F218*, 331–351. https://doi.org/10.1007/978-3-031-81615-4_14

Derraik, J. G. B. (2002). The pollution of the marine environment by plastic debris: A review. *Marine Pollution Bulletin, 44*(9), 842–852. https://doi.org/10.1016/S0025-326X(02)00220-5

Dey, S., Dash, N., Ray, S. K., Priyadarsini, S., Chhotaray, C., & Das, A. P. (2024). Occurrence and quantification of tire wear plastics pollutants from highway city dust of Bhubaneswar City of Odisha. *Sustainable Microbial Technology for Synthetic and Cellulosic Microfiber Bioremediation*, 77–89. https://doi.org/10.1007/978-3-031-62600-5_5

Dhali, S. L., Parida, D., Kumar, B., & Bala, K. (2024). Recent trends in microbial and enzymatic plastic degradation: a solution for plastic pollution predicaments. *Biotechnology for Sustainable Materials, 1*(1). https://doi.org/10.1186/s44316-024-00011-0

Dharwal, M., Srivastava, A. K., Sarin, V., & Gola, K. R. (2022). The state of solid waste management for sustainable development in India: Current state and future potential. *Materials Today: Proceedings, 60*, 802–805. https://doi.org/10.1016/j.matpr.2021.09.246

Evode, N., Qamar, S. A., Bilal, M., Barceló, D., & Iqbal, H. M. N. (2021). Plastic waste and its management strategies for environmental sustainability. *Case Studies in Chemical and Environmental Engineering, 4*, Article 100142. https://doi.org/10.1016/J.CSCEE.2021.100142

Fayshal, M. A. (2024). Current practices of plastic waste management, environmental impacts, and potential alternatives for reducing pollution and improving management. *Heliyon, 10*(23), Article e40838. https://doi.org/10.1016/J.HELIYON.2024.E40838

Gates, E. G., & Crook, N. (2024). The biochemical mechanisms of plastic biodegradation. *FEMS microbiology reviews*, *48*(6). https://doi.org/10.1093/femsre/fuae027

Geyer, R., Jambeck, J. R., & Law, K. L. (2017). *Production, use, and fate of all plastics ever made*. https://www.science.org

Ghosh, S., Sinha, J. K., Ghosh, S., Vashisth, K., Han, S., & Bhaskar, R. (2023). Microplastics as an emerging threat to the global environment and human health. *Sustainability, 15*(14), 10821. https://doi.org/10.3390/SU151410821

Hayat, K., Menhas, S., Bundschuh, J., & Chaudhary, H. J. (2017). Microbial biotechnology as an emerging industrial wastewater treatment process for arsenic mitigation: A critical review. *Journal of Cleaner Production, 151*, 427–438. https://doi.org/10.1016/J.JCLEPRO.2017.03.084

Heris, Y. S. (2024). Bacterial biodegradation of synthetic plastics: a review. *Bulletin of the National Research Centre*, *48*(1). https://doi.org/10.1186/s42269-024-01241-y

Huang, W., Chen, M., Song, B., Deng, J., Shen, M., Chen, Q., Zeng, G., & Liang, J. (2021). Microplastics in the coral reefs and their potential impacts on corals: A mini-review. *Science of the Total Environment, 762*, Article 143112. https://doi.org/10.1016/J.SCITOTENV.2020.143112

Huang, S., Wang, H., Ahmad, W., Ahmad, A., Vatin, N. I., Mohamed, A. M., Deifalla, A. F., & Mehmood, I. (2022). Plastic waste management strategies and their environmental aspects: A scientometric analysis and comprehensive review. *International Journal of Environmental Research and Public Health, 19*(8). MDPI. https://doi.org/10.3390/ijerph19084556

Jayavel, S., Govindaraju, B., Michael, J. R., & Viswanathan, B. (2024). Impacts of micro and nanoplastics on human health. *Bulletin of the National Research Centre, 48*(1), 110. https://doi.org/10.1186/s42269-024-01268-1

Jeong, E., Lee, J. Y., & Redwan, M. (2024). Animal exposure to microplastics and health effects: A review. *Emerging Contaminants, 10*(4), Article 100369. https://doi.org/10.1016/J.EMCON.2024.100369

Jiang, B., Yu, J., & Liu, Y. (2020). The environmental impact of plastic waste. *Journal of Environmental and Earth Sciences*, *2*(2), 26–35. https://doi.org/10.30564/jees.v2i2.2340

John, A., & Das, A. P. (2024). Government strategies for achieving circular economy for plastic waste management in Southeast Asia. *Sustainable Management of Urban Plastic Waste Through Circular Economic Approaches*, 72–82. https://doi.org/10.1201/9781003465447-6/GOVERNMENT-STRATEGIES-ACHIEVING-CIRCULAR-ECONOMY-PLASTIC-WASTE-MANAGEMENT-SOUTHEAST-ASIA-AKBAR-JOHN-ALOK-PRASAD-DAS

Kapoor, R. T., & Rafatullah, M. (2025). Leveraging biotechnological approaches for the degradation of plastics: A sustainable tool for plastic waste management and environment protection. *Journal of Water Process Engineering, 69*, Article 106736. https://doi.org/10.1016/J.JWPE.2024.106736

Kaushal, J., Khatri, M., & Arya, S. K. (2021). Recent insight into enzymatic degradation of plastics prevalent in the environment: A mini—review. *Cleaner Engineering and Technology, 2*, Article 100083. https://doi.org/10.1016/J.CLET.2021.100083

Kesaano, M., & Sims, R. C. (2014). Algal biofilm based technology for wastewater treatment. *Algal Research, 5*(1), 231–240. https://doi.org/10.1016/J.ALGAL.2014.02.003

Kye, H., Kim, J., Ju, S., Lee, J., Lim, C., & Yoon, Y. (2023). Microplastics in water systems: A review of their impacts on the environment and their potential hazards. *Heliyon, 9*(3), Article e14359. https://doi.org/10.1016/J.HELIYON.2023.E14359

Lee, G. H., Kim, D. W., Jin, Y. H., Kim, S. M., Lim, E. S., Cha, M. J., Ko, J. K., Gong, G., Lee, S. M., Um, Y., Han, S. O., & Ahn, J. H. (2023). Biotechnological plastic degradation and valorization using systems metabolic engineering. *International Journal of Molecular Sciences, 24*(20). Multidisciplinary Digital Publishing Institute (MDPI). https://doi.org/10.3390/ijms24 2015181

Li, W., Qamar, S. A., Qamar, M., Basharat, A., Bilal, M., & Iqbal, H. M. N. (2021). Carrageenan-based nano-hybrid materials for the mitigation of hazardous environmental pollutants. *International Journal of Biological Macromolecules, 190*, 700–712. https://doi.org/10.1016/J.IJB IOMAC.2021.09.039

Malinowski, C. R., Searle, C. L., Schaber, J., & Höök, T. O. (2023). Microplastics impact simple aquatic food web dynamics through reduced zooplankton feeding and potentially releasing algae from consumer control. *Science of the Total Environment, 904*, Article 166691. https://doi.org/ 10.1016/J.SCITOTENV.2023.166691

Mallick, K., Sahu, A., Dubey, N. K., & Das, A. P. (2023). Harvesting marine plastic pollutants-derived renewable energy: A comprehensive review on applied energy and sustainable approach. *Journal of Environmental Management, 348*, Article 119371. https://doi.org/10.1016/J.JEN VMAN.2023.119371

Maquart, P. O., Froehlich, Y., & Boyer, S. (2022). Plastic pollution and infectious diseases. *The Lancet Planetary Health, 6*(10), e842–e845. https://doi.org/10.1016/S2542-5196(22)00198-X

Mishra, S., Swain, S., Sahoo, M., Mishra, S., & Das, A. P. (2021). Microbial colonization and degradation of microplastics in aquatic ecosystem: A review. *Geomicrobiology Journal*. https:// doi.org/10.1080/01490451.2021.1983670

Mishra, S., Dash, D., & Das, A. P. (2024a). Aquatic microbial diversity on plastisphere: Colonization and potential role in microplastic biodegradation. *Geomicrobiology Journal, 41*(4), 312–323. https://doi.org/10.1080/01490451.2023.2209750

Mishra, S., & Das, A. P. (2021). Current treatment technologies for removal of microplastic and microfiber pollutants from wastewater. *Wastewater Treatment: Cutting-Edge Molecular Tools, Techniques and Applied Aspects*, 237–251. https://doi.org/10.1016/B978-0-12-821881-5.000 11-8

Mishra, S., Singh, R. P., Rout, P. K., & Das, A. P. (2022). Membrane bioreactor (MBR) as an advanced wastewater treatment technology for removal of synthetic microplastics. *Development in Wastewater Treatment Research and Processes*, 45–60. https://doi.org/10.1016/B978-0-323-85583-9.00022-3

Mishra, S., Dash, D., & Das, A. P. (2023). Aquatic microbial diversity on plastisphere: Colonization and potential role in microplastic biodegradation. *Geomicrobiology Journal*, 1–12. https://doi. org/10.1080/01490451.2023.2209750

Mishra, S., Bauri, K. P., & Panigrahi, S. (2024). Addressing plastic pollution: Sustainable alternatives and advanced waste management. *World Journal of Advanced Research and Reviews, 23*(2), 1948–1957. https://doi.org/10.30574/wjarr.2024.23.2.2527

Mohanty, J., Dey, S., Bala, K., & Das, A. P. (2024). Recent advances in microplastic pollution for its sustainable management. *Sustainable Management of Urban Plastic Waste Through Circular Economic Approaches*, 27–48. https://doi.org/10.1201/9781003465447-3

Mohanty, M., Tripathy, B., Koteswara Reddy, G., & Das, A. P. (2024). Greener technologies for establishing restraint over microplastic pollution. *Sustainable Microbial Technology for*

Synthetic and Cellulosic Microfiber Bioremediation, 239–258. https://doi.org/10.1007/978-3-031-62600-5_13

Mohee, R., & Unmar, G. (2007). Determining biodegradability of plastic materials under controlled and natural composting environments. *Waste Management, 27*(11), 1486–1493. https://doi.org/10.1016/J.WASMAN.2006.07.023

Narancic, T., Cerrone, F., Beagan, N., & O'Connor, K. E. (2020). Recent advances in bioplastics: Application and biodegradation. *Polymers, 12*(4). MDPI AG. https://doi.org/10.3390/POLYM12040920

Nayanathara Thathsarani Pilapitiya, P. G. C., & Ratnayake, A. S. (2024). The world of plastic waste: A review. *Cleaner Materials, 11*, 100220. https://doi.org/10.1016/J.CLEMA.2024.100220

(2025) Polyethylene biodegradation: A multifaceted approach. *Biotechnology Advances*, 108577. https://doi.org/10.1016/J.BIOTECHADV.2025.108577

Pradhan, G., Dey, S., Priyadarsini, S., Kumar, M. S., & Das, A. P. (2024). Characterization and quantification of microplastics pollutants in sediment samples from Daya River of Odisha State in India for their appropriate management. *Environmental Science and Engineering, Part F2333*, 75–92. https://doi.org/10.1007/978-3-031-51792-1_5

Priyadarshini, B., & Das, A. P. (2024a). The Adverse impact of synthetic microplastics released from disposable face masks pollutants on aquatic microbiota, aquatic species and humans: A review. *Environmental Science and Engineering, Part F2827*, 67–81. https://doi.org/10.1007/978-3-031-50840-0_5

Priyadarshini, B., Santhosh Kumar, M., & Das, A. P. (2024b). Microbial degradation of synthetic plastics released from COVID-19 generated single use disposable face masks. *Environmental Science and Engineering, Part F2827*, 1–16. https://doi.org/10.1007/978-3-031-50840-0_1

Priyadarsini, S., Das, A. P. (2024c). Analytical and structural characterization of waste lithium-ion batteries for their effective recycling strategy. *Environmental Chemistry and Ecotoxicology*. https://doi.org/10.1016/j.enceco.2024.12.004

Priyadarsini, S., Alok, Das, P., Journal, G., & Das, A. P. (2025a). Isolation and identification of native lithium-resistant bacteria from electronic waste contaminated sites: A focused investigation on tolerance assessment with potential lithium bio-recovery application.*Geomicrobiology Journal, 42*(10), 898–912. https://doi.org/10.1080/01490451.2025.2539509

Priyadarsini, S., & Das, A. P. (2025b). Lithium bioleaching: A review on microbial-assisted sustainable technology for lithium bio-circularity. *Journal of Water Process Engineering, 69*, 106744. https://doi.org/10.1016/j.jwpe.2024.106744

Qian, K., Wang, L., Teng, J., & Liu, G. (2025). Strategies and technologies for sustainable plastic waste treatment and recycling. *Environmental Functional Materials*. https://doi.org/10.1016/J.EFMAT.2025.01.004

Ray, S. K., Dey, S., Das, A. P., & John, A. (2024). Microplastic waste management through greener technologies : a review. *Sustainable Management of Urban Plastic Waste Through Circular Economic Approaches*, 237–261. https://doi.org/10.1201/9781003465447-14

Reddy Koteswara, G., Bhargavi, S., Yamuna, D., Anjani Devi, M., Jeevana Lakshmi Kumar, G., Manas, T., Adarsh, S., Bhavya Sai, V., Nath, H., Alreshidi, M. A., Togun, H., Ram, D. K., & Das, A. P. (2025). Recent advancements in the remediation of marine hydrocarbon and plastic pollutants. *Environmental Science and Engineering, Part F218*, 127–142. https://doi.org/10.1007/978-3-031-81615-4_7

Roy, A., & Chakraborty, S. (2024). A comprehensive review on sustainable approach for microbial degradation of plastic and it's challenges. *Sustainable Chemistry for the Environment, 7*, Article 100153. https://doi.org/10.1016/J.SCENV.2024.100153

Ru, J., Huo, Y., & Yang, Y. (2020). Microbial degradation and valorization of plastic wastes. *Frontiers in Microbiology, 11*. Frontiers Media S.A. https://doi.org/10.3389/fmicb.2020.00442

Sahoo, P. P., Singh, S., Rout, P. K., Mishra, S., & Das, A. P. (2022). Microbial remediation of plastic pollutants generated from discarded and abandoned marine fishing nets. *Biotechnology and Genetic Engineering Reviews*. https://doi.org/10.1080/02648725.2022.2152629

Sahu, S., & Das, A. P. (2024). Microplastics degradation mechanism by aquatic plastiomes: A review, 213–226. https://doi.org/10.1007/978-3-031-50840-0_12

Sahu, A., Dey, S., Mallick, K., Priyadarsini, S., Behera, I. D., & Das, A. P. (2024). A review on advanced technology for sustainable management of synthetic microplastic waste. *Sustainable Management of Urban Plastic Waste Through Circular Economic Approaches*, 262–279. https://doi.org/10.1201/9781003465447-15

Šaravanja, A., Pušić, T., & Dekanić, T. (2022). Microplastics in wastewater by washing polyester fabrics. *Materials, 15*(7). MDPI. https://doi.org/10.3390/ma15072683

Schnurr, R. E. J., Alboiu, V., Chaudhary, M., Corbett, R. A., Quanz, M. E., Sankar, K., Srain, H. S., Thavarajah, V., Xanthos, D., & Walker, T. R. (2018). Reducing marine pollution from single-use plastics (SUPs): A review. *Marine Pollution Bulletin, 137*, 157–171. https://doi.org/10.1016/J.MARPOLBUL.2018.10.001

Shen, M., Li, Y., Qin, L., Chen, X., Ao, T., Liang, X., Jin, K., Dou, Y., Li, J., & Duan, X. (2025). Distribution and risk assessment of microplastics in a source water reservoir, Central China. *Scientific Reports, 15*(1). https://doi.org/10.1038/s41598-024-84894-z

Singh Jadaun, J., Bansal, S., Sonthalia, A., Rai, A. K., & Singh, S. P. (2022). Biodegradation of plastics for sustainable environment. *Bioresource Technology, 347*, Article 126697. https://doi.org/10.1016/J.BIORTECH.2022.126697

Skariyachan, S., Taskeen, N., Kishore, A. P., & Krishna, B. V. (2022). Recent advances in plastic degradation—from microbial consortia-based methods to data sciences and computational biology driven approaches. *Journal of Hazardous Materials, 426*, Article 128086. https://doi.org/10.1016/J.JHAZMAT.2021.128086

Srimongkol, P., Sangtanoo, P., Songserm, P., Watsuntorn, W., & Karnchanatat, A. (2022). Microalgae-based wastewater treatment for developing economic and environmental sustainability: Current status and future prospects. *Frontiers in Bioengineering and Biotechnology, 10*. Frontiers Media S.A. https://doi.org/10.3389/fbioe.2022.904046

Thirunavukarasu, K., Purushothaman, S., Sridevi, J., Aarthy, M., Gowthaman, M. K., Nakajima-Kambe, T., & Kamini, N. R. (2016). Degradation of poly(butylene succinate) and poly(butylene succinate-co-butylene adipate) by a lipase from yeast Cryptococcus sp. grown on agro-industrial residues. *International Biodeterioration and Biodegradation, 110*, 99–107. https://doi.org/10.1016/J.IBIOD.2016.03.005

Tripathy, B., Sahoo, P. P., Sundaray, H., & Das, A. P. (2024). In-vitro biodegradation of discarded marine microplastics across the eastern coast of the Bay of Bengal, India using Exiguobacterium sp. *Environmental Chemistry and Ecotoxicology, 6*, 236–247. https://doi.org/10.1016/J.ENCECO.2024.07.003

Tripathy, B., Singh, S., Behera, I. D., Mishra, S., & Das, A. P. (2026). Metagenomic profiling of diversified marine microbiome across microplastic-contaminated niches of Bay of Bengal, India. *Marine Pollution Bulletin, 222*, 118872.https://doi.org/10.1016/J.MARPOLBUL.2025.118872

Van Emmerik, T., Mellink, Y., Hauk, R., Waldschläger, K., & Schreyers, L. (2022). Rivers as plastic reservoirs. *Frontiers in Water, 3*. https://doi.org/10.3389/frwa.2021.786936

Vuppaladadiyam, S. S. V., Vuppaladadiyam, A. K., Sahoo, A., Urgunde, A., Murugavelh, S., Šrámek, V., Pohořelý, M., Trakal, L., Bhattacharya, S., Sarmah, A. K., Shah, K., & Pant, K. K. (2024). Waste to energy: Trending key challenges and current technologies in waste plastic management. *Science of the Total Environment, 913*, Article 169436. https://doi.org/10.1016/J.SCITOTENV.2023.169436

Walker, T. R., & Fequet, L. (2023). Current trends of unsustainable plastic production and micro (nano) plastic pollution. *TrAC Trends in Analytical Chemistry, 160*, Article 116984. https://doi.org/10.1016/J.TRAC.2023.116984

Yang, X. G., Wen, P. P., Yang, Y. F., Jia, P. P., Li, W. G., & Pei, D. S. (2023). Plastic biodegradation by in vitro environmental microorganisms and in vivo gut microorganisms of insects. *Frontiers in Microbiology, 13*. Frontiers Media SA. https://doi.org/10.3389/fmicb.2022.1001750

Zeenat, Elahi, A., Bukhari, D. A., Shamim, S., & Rehman, A. (2021). Plastics degradation by microbes: A sustainable approach. *Journal of King Saud University—Science, 33*(6), 101538. https://doi.org/10.1016/J.JKSUS.2021.101538

Zhang, J., Gao, D., Li, Q., Zhao, Y., Li, L., Lin, H., Bi, Q., & Zhao, Y. (2020). Biodegradation of polyethylene microplastic particles by the fungus Aspergillus flavus from the guts of wax moth Galleria mellonella. *Science of the Total Environment, 704*, 135931. https://doi.org/10.1016/J.SCITOTENV.2019.135931

[illegible] A. [illegible] decomposition of [illegible]

[illegible] The formulation of [illegible] the pre[illegible] [illegible] https://doi.org/10.1016/[illegible]

Phycoremediation as a Green Solution for the Remediation of Metal Toxicity in Wastewater with Circular Bioeconomy Integration

Jayaprakash Nayak, Alok Prasad Das, and Sunil Kumar Behera

Abstract The escalating levels of metal contamination in aquatic systems, driven by industrialization and urbanization, pose significant risks to our ecosystem. Conventional physicochemical methods for metal remediation and wastewater treatment have been proven effective, though they have high operational costs, low selectivity, and are associated with the emission of secondary pollutants. This chapter cites the emerging role of algae as sustainable and eco-friendly biocatalysts for the remediation of heavy metals in wastewater. It explores the underlying mechanisms of metal removal, including biosorption, bioaccumulation, and biotransformation performed by the algae in the remediation of critical metals such as Pb^{2+}, Cd^{2+}, Cu^{2+}, Co^{2+}, Cr(VI), Ni^{2+}, Zn^{2+}, Mn^{2+}, As, and U. Furthermore, this chapter focused on the integration of innovative approaches like photosynthetic microbial fuel cells (PMFCs) and value recovery of metal in remediation of metal-contaminated water. This chapter emphasizes the transformative role of algae in environmental detoxification and resource recovery.

Keywords Phycoremediation · Algae · Heavy metals · Biosorption · Wastewater treatment · Circular economy

1 Introduction

The increasing pressures of urbanization and industrialization, combined with inadequate wastewater treatment techniques and rapid population growth, have led to severe water scarcity issues worldwide (Mishra & Das, 2023; Mishra et al., 2024). The past few decades have seen a concerning rise in environmental pollution due to heavy

J. Nayak · S. K. Behera (✉)
Department of Biotechnology, Berhampur University, Berhampur, Odisha, India
e-mail: skbehera2020@gmail.com

A. P. Das
Department of Life Sciences, Rama Devi Women's University, Bhubaneswar, Odisha, India

A. P. Das (ed.), *Bioconversion of Environmental Waste to Value Added Products*, Sustainable Environmental Waste Management Strategies,
https://doi.org/10.1007/978-3-032-14215-3_18

metals (Priyadarsini & Das, 2024, 2025), microplastics (Mishra et al., 2021; Sahu & Das, 2024), xenobiotic compounds, and organosulfur compounds in water and soil. Heavy metals generate long-term adverse impacts on plants, animals, and humans due to their strong toxicity, bioaccumulation, and non-biodegradability (Hu et al., 2020); Wang et al., 2022) (Mishra et al., 2022). The environment has been contaminated by heavy metals primarily sourced from industries such as mining, petrochemical refining, electroplating, dyeing, and battery manufacturing (Das & Ghosh, 2018; Das et al., 2023). To mitigate metal pollution, it is crucial to remediate heavy metal-contaminated wastewater effluents before their discharge. Various physicochemical approaches, including chemical precipitation, ion exchange, membrane filtration, and electrochemical treatment, are currently used in metal remediation (Priya et al., 2022). However, these techniques have disadvantages, such as high investment and operational costs and very low removal efficiency (Das & Ghosh, 2022; Ghosh & Das, 2017). Alternatively, microbial-catalyzed green techniques have been developed for treating mineral waste by using microbial leaching and flotation processes (Behera & Mulaba-Bafubiandi, 2015, 2017; Behera et al., 2022). Also, the microorganisms have evolved as a tool for the remediation and degradation of organosulfur compounds in coal and petrochemicals (Rout et al., 2022, 2024; Rout & Behera, 2024). In addition to this, various microorganisms have been reported for their ability to remediate plastic polymers (Behera, 2024). In the present context, the bioremediation technology, especially phycoremediation, has gained increased attention recently due to its environmental friendliness and cost efficiency. Algae growing in diverse habitats can be utilized as a noble natural agent for the treatment of wastewater bodies contaminated with various toxic metals. Conventional physical and chemical remediation technologies for heavy metals are costly, exhibit limited efficacy in large water bodies, and often have adverse effects on aquatic ecosystems (Gandhi et al., 2022). Phytoremediation (algal remediation) offers an emerging, eco-friendly, sustainable, and cost-effective approach, utilizing metal-accumulating terrestrial and aquatic plants, microalgae, and microalgae to remove toxic metals from contaminated soils and water sources (Plöhn et al., 2021).

2 Heavy Metal Sources and Their Effect

Metal pollution refers to the contamination of the environment by different types of metals. The metals can accumulate in soil, water, and living organisms due to anthropogenic activities such as mining, industrial effluent discharge, and improper waste disposal from agrochemicals. Several metals, such as arsenic, cadmium, chromium, copper, cobalt, mercury, nickel, lead, zinc, and manganese, have been reported for their toxicity and hazardous nature (Das et al. 2015a, 2015b, 2015c; Mishra et al., 2009). Arsenic is one of the toxic metalloids that exhibits both metallic and non-metallic characteristics. It is widely used across various industrial and production processes of fungicides, pesticides, and herbicides (Das & Mishra, 2008; Paul et al., 2023). Arsenic is recognized to be a hazardous as well as a carcinogenic element despite being extensively used in various fields, impacting human health and environmental safety. Arsenic reduces photosynthetic efficiency in organisms and impairs the integrity of the thylakoid membrane (Upadhyay et al., 2022). In microbes, it destroys cell organelles, disrupts proteins, DNA, and lipids (Upadhyay et al., 2016). Cadmium (Cd^{2+}), primarily used in agriculture, industry, and domestic applications, is one of the most hazardous waterborne contaminants (Das & Bissoyi, 2011; Das & Mishra, 2010). It is extensively used in the manufacture of paints, alkaline batteries, copper alloys, fertilizers, and during zinc refining processes. These heavy metals can enter the food chain through environmental exposure and subsequently affect living organisms through bioaccumulation, causing carcinogenic, mutagenic, and teratogenic effects (Chandrashekharaiah et al., 2021). The respiratory system gets hampered, particularly through inhalation, through cadmium exposure, leading to pulmonary complications and other related symptoms. In plants and microorganisms, it causes growth impairment, alteration in morphology, and disruption in cellular metabolism and photosynthetic activity, causing ultimate cell death. Furthermore, chromium is released to the environment due to the weathering of chromite ores and minerals. However, significant anthropogenic contributions arise from sewage sludge, electroplating operations, and tannery effluents, and improper disposal of chromium-containing wastes facilitates the leaching of toxic compounds into surface and groundwater systems. Chromium is used in various means such as pigments, dyes, tanning agents, and other industrial chemicals, carcinogens (Devi et al., 2023).

The oxidative state of chromium, Cr (VI), possesses high mobility, solubility, toxicity, and carcinogenic effects. It causes various clinical conditions like coughing, chronic bronchitis, wheezing, ulceration, hemorrhage, and pulmonary edema. It also harms the cell when it enters the cell membrane and manipulates genetic material (Leong & Chang, 2020). The heavy metals like copper (Cu) and cobalt (Co) present in the aquatic environment are of primary concern due to their toxic effect, and they are widely used in various industrial processes such as metal extraction, refining, casting, and alloying (Prabhakar et al., 2019; Raj & Das, 2023). Both copper and cobalt are important in trace amounts, but they become toxic when they cross the permissible limits in humans. It also poses considerable threats to aquatic biota, disrupting ecological balance and bioaccumulating through the food chain. Therefore, the efficient treatment of industrial wastewater is crucial not only to mitigate environmental contamination but also to ensure adherence to environmental and public health regulatory standards (Das et al., 2024a).

Mercury is another heavy metal widely used in industries like mining, smelting, coal, electronics, pesticides, and chemicals, and it is released to the environment. Mercury is added to agricultural land through manure, sludge, fungicides, or fertilizers. Organic mercury is 200 times more lethal than inorganic mercury, as it is formed by the combination of mercury and carbon, resulting in the formation of compounds such as methylmercury, which is neurotoxic in nature. It causes nervous disorders, gene expression alterations, autoimmune diseases, depression, gingivitis, Minamata, irritation in eyes, mental disorders, reproductive disorders, antibiotic resistance, chest pain, and interstitial pneumonitis, which ultimately leads to death (Huang et al., 2019). In microbial systems, mercury induces protein denaturation, compromises the structural integrity of the cell membrane, and inhibits both enzymatic activity and transcriptional processes (Salam et al., 2019). In plants, mercury also affects the plant by altering its cell envelope and causing abnormal seed germination patterns. In the context of nickel metal, it is a toxic metal, and it is capable of exerting adverse effects on biological systems. Also, nickel is a carcinogenic agent and impairs both reproductive and pulmonary health (Deng et al., 2016). Another metal lead is used in various industries like coal, batteries, gasoline, pipelines, pigment, chemical fertilizers, pesticides, combustion of fossil fuels, industrial wastes, urban wastes, and mining. Lead is known for its toxic effects and greater health hazards, such as neurological, immunological, and reproductive disorders.

The heavy metal zinc enters the waterbodies from the industrial effluents of the electroplating, battery manufacturing, the production of paints, dyes, and rubber (Das & Ghosh, 2023a; Ghosh & Das, 2018). At higher concentrations, zinc affects the gill tissues, reproduction, and causes behavioral changes in aquatic organisms. In humans, lower concentrations of zinc are essential for growth and development, but at higher concentrations, it leads to adverse health effects such as nausea, vomiting, diarrhea, and, in chronic cases, disruption of immune and nervous system functions (Das & Ghosh, 2023b; Das et al., 2011; Ghosh et al., 2016). Excessive zinc discharge into the environment causes substantial ecological and health risks (Liu et al., 2023). Manganese concentration increases in wastewater bodies due to anthropogenic activity, like industrialization, battery manufacturing, including mining operations, metal processing industries, and is also used for agricultural purposes, fertilizers, and pesticides (Oladimeji et al., 2024). In addition, municipal wastewater systems, manganese enters through the process of corrosion of water distribution pipes and leaching from soil and rocks under certain redox conditions (Das et al., 2015b; Ghosh et al., 2023). Manganese toxicity can severely affect aquatic organisms, particularly fish and invertebrates, and damage their reproductive and nervous systems (Sánchez-González et al., 2015). Higher concentration of manganese enters the human body through drinking water, and it affects the nervous system and causes neurological disorders such as Parkinson's disease, especially in children (O'Neal & Zheng, 2015). Consequently, controlling manganese levels in wastewater is critical to protect environmental and public health.

3 Phycoremediation Mechanism

Some of the heavy metals are essential micronutrient elements for the growth and development of algae; these elements are molybdenum, copper, zinc, nickel, manganese, iron, cobalt, and boron. They play important roles in cellular metabolism, enzymatic functions, and overall physiological processes (Salama et al., 2019). Other heavy metals such as silver, gold, aluminum, mercury, titanium, cadmium, lead, and arsenic are mostly regarded as toxic to algae and do not contribute usefully to their physiological processes (Jais et al., 2017). Algae are recognized as highly effective for bioremediators because they catalyze a wide range of reactions for metal

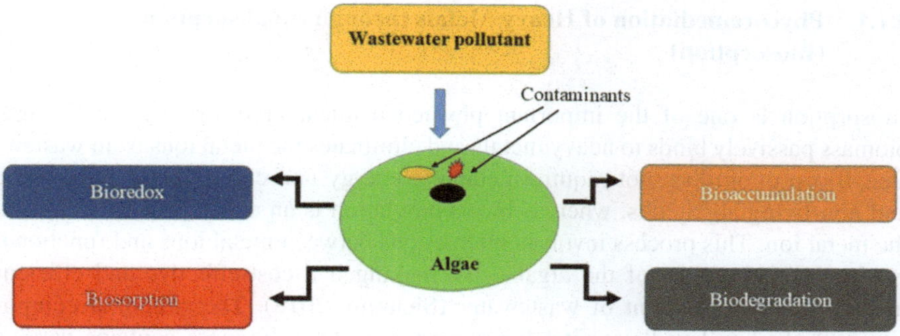

Fig. 1 Different strategies of metal remediation by algae

remediations such as biosorption, bioaccumulation, bio-reduction, and biodegradation (Fig. 1). Algae have been gaining an effective tool for remediation studies due to their remarkable tolerance to heavy metals, easy cultivation, high metal-binding affinity, large surface-to-volume ratio, environmental sustainability, and the ability to utilize both living and non-living biomass for remediation purposes, minimizing the need for additional nutrients or intensive effort inputs (Darda et al., 2019; Ramakrishnan et al., 2010). Moreover, algae engage various defense mechanisms against heavy metal stress, including gene regulation, metal chelation, and other adaptive strategies.

3.1 *Heavy Metal Tolerance and Detoxification by Algae*

Algae have intrinsic self-defense mechanisms that enable them to tolerate the toxic effects of heavy metals (Das et al., 2012, 2015a). These effects are gene regulation, complexation, ion exchange, chelation, the production of reducing agents, antioxidants, and the immobilization of heavy metals. Such processes mitigate the inhibitory effects of heavy metals on cellular biochemical processes, enzymatic activity, physiological and morphological development, protein structure, nucleic acids, and cell membrane integrity (Priyadarshini et al., 2019). These adaptive responses empower algal survivability in environments with higher concentrations of toxic heavy metals, such as contaminated wastewater, and contribute remarkably to removing these toxic substances (Das & Ghosh, 2023c; Das et al., 2012). Algal-defense strategies against these heavy metal stresses are typically classified into four major categories, such as biosorption, bioaccumulation and compartmentalization, biotransformation and detoxification.

3.1.1 Phycoremediation of Heavy Metals through Bioabsorption (Biosorption)

Biosorption is one of the important physicochemical processes by which algal biomass passively binds to heavy metals and eliminates the metal ions from wastewater. Biosorption does not require metabolic energy and can occur in both living and non-living algal cells, whereas bioaccumulation is an active process to uptake the metal ion. This process involves interactions between metal ions and functional groups on the surface of the algal cells, making it a cost-effective and efficient method for the treatment of wastewater (Shamim, 2018). The structural composition of algal cell walls is vital for biosorption. Algal biomass contains various biopolymers, including lipids, proteins, and polysaccharides, which are abundant in metal-binding functional groups. These groups are anionic in nature, for example, hydroxyl (−OH), carboxyl (−COOH), sulfate ($-SO_4^{2-}$), and phosphate ($-PO_4^{3-}$), and consequently, they attract the positively charged metal ions.

The process of biosorption mechanisms involves complexation, ion exchange, chelation, physical adsorption, and coordination. The positively charged metal ions bind to negatively charged functional groups with different bonding mechanisms, for example, ion exchange, covalent bonding, and van der Waals forces. Heavy metals like cadmium (Cd^{2+}), mercury (Hg^{2+}), lead (Pb^{2+}), arsenic (As^{3+}/As^{5+}), and chromium (Cr^{3+}/Cr^{6+}) form a stable interaction with these functional groups, and the functional group acts as a ligand with the heavy metal binding.

Depending on their taxonomic group, cell wall composition, and environmental conditions, different algal species show varying degrees of biosorption efficiency. For instance, brown algae exhibit high biosorption capacity due to the presence of alginates—a group of polysaccharides with strong affinity for divalent and trivalent metal ions such as *Fucus vesiculosus* is able to remove zinc with the help of carboxyl and hydroxyl groups present on the algal surface (Brinza et al., 2020), and *F. vesiculosus* is also able to remediate 143.2± 7.5 mg/g cadmium, 70.1 ± 1.9 mg/g nickel, 516.3 ± 12.5 mg/g lead, respectively (Moreira et al., 2019). Likewise, Khajavian et al. also observed that brown algae *Cystoseira* indicant was able to decrease 18.17 mg/g Nickel and 55.34 mg/g cadmium (Khajavian et al., 2019). Blue-green algae *Anabaena sphaerica* was demonstrated to remediate 111.1 mg/g cadmium and 121.95 mg/g lead with the help of amino, hydroxyl, and carboxyl functional groups present on the cell surface (Abdel-Aty et al., 2013).

Biosorption performance is influenced by different environmental parameters such as pH, temperature, metal concentration, and contact time. Optimum metal binding usually occurs at pH values between 4 and 6, where functional groups become ionized and can effectively interact with positively charged metal ions. Biosorption is also inhibited by the extreme pH levels, due to the competition from protons. The rate of heavy metal uptake increases with the increase in temperature due to the rise in kinetic energy. However, excessively high temperatures may lead to desorption or structural degradation of the biomass. In addition to increasing the surface area and pore volume, algal biomass can be pre-treated chemically (for instance, with NaOH,

$CaCl_2$, or formaldehyde), and the availability of functional groups helps improve its metal-binding efficiency. Biosorption process uses both live and dead algal biomass because this technique relies on surface biochemistry rather than metabolic activity. It is a cost-effective and environmentally friendly process for remediating the heavy metals. Further future research into the diversity of algal species and strategies for surface functionalization is expected to enhance the efficacy of biosorption in future bioremediation efforts.

3.1.2 Phycoremediation of Heavy Metals through Bio-oxidation

Bio-oxidation is an important process in algal bioremediation. In this process, the metal ions convert from one oxidation state to another through enzymatic or metabolic conversion. By this conversion, the heavy metal reduces its toxicity, increases solubility, or improves precipitation. This process is critical for detoxifying metals and transforming them into forms that are more manageable in ecological systems.

The various metabolic pathways facilitated by algae for the bio-oxidation process are the generation of reactive oxygen species (ROS), peroxidases, and oxidases that catalyze redox transformations. For example, in anoxic conditions, the highly toxic, more soluble divalent iron (Fe^{2+}) can be oxidized to ferric iron (Fe^{3+}), which successively precipitates as insoluble hydroxides or oxides, sustaining in its removal from the aqueous phase by algae (Rana & Prajapati, 2021). Metals like manganese (Mn^{2+}) and arsenic (As^{3+}) can be oxidized and converted into less toxic forms, such as Mn^{4+} and As^{5+}, through bio-oxidation processes facilitated by algae. Similarly, previous studies revealed that the oxidation of arsenite (As^{3+}) to arsenate (As^{5+}) is catalysed by Chlamydomonas reinhardtii, Chlorella vulgaris, and *Anabaena* sp. This redox transformation is vital because arsenate is less toxic, less mobile, and more easily precipitated or immobilized than arsenite ion (Jana et al., 2015; Yin et al., 2011). In this oxidation process, the important arsenite oxidase enzyme is secreted by the alga extracellularly or on its cell surface. Some important factors, like the pH and redox conditions of the surrounding environment, play a vital role in the mechanism of bio-oxidation in algal bioremediation, thereby influencing metal speciation.

In bacteria, the mechanisms of bio-oxidation are well-documented, whereas in algae, the study of bio-oxidation is just emerging. Some research evidence supports that algal species also possess the enzymatic machinery necessary for redox transformations of heavy metals.

In summary, algal bio-oxidation plays an important role in detoxification, and this process promotes metal precipitation, modifies bioavailability, and alters toxicity profiles, thereby improving the efficiency of algal bioremediation systems.

3.1.3 Phycoremediation of Heavy Metals through Biotransformation

Biotransformation is an important remediation process in which the highly toxic heavy metals are converted to less toxic forms and are successively removed from the environment (Das et al., 2024a, b; Pradhan et al., 2023). Algal biotransformation is a vibrant and metabolically active process that facilitates algae to detoxify and stabilize heavy metals through redox reactions, mechanisms like methylation, demethylation, complexation with biomolecules, or intracellular compartmentalization (Kanungo et al., 2024; Nag et al., 2023a). Biotransformation is likely different from biosorption or bioaccumulation processes because these two remediation processes are passive; biotransformation involves specific metabolic pathways and enzymatic reactions that change the chemical nature of the heavy metals (Danouche et al., 2021).

A key example of algal-mediated biotransformation is the redox cycling of chromium. In its hexavalent form (Cr^{6+}), chromium is highly toxic, carcinogenic, and mobile in aquatic systems. Algae such as *Chlorella vulgaris*, the ability to reduce Cr^{6+} to Cr^{3+}, a less toxic and more stable form that tends to precipitate and adsorb onto biomass or sediment particles, whereas *Galdieria sulphuraria, Selenastrum minutum*, and *C. fusca* produce the mercuric reductase enzyme to biotransform Hg^{2+} into elemental mercury and metacinnabar (Yan et al., 2022; Yin et al., 2011). It was reported that arsenate biotransformation occurs in *C. reinhardtii* in the presence of arsenate reductase. This reduction is mediated by intracellular reductase enzymes that utilize electron donors such as NADPH or glutathione. Similarly, the redox transformation of arsenic, particularly the oxidation of arsenate (As^{3+}) to arsenate (As^{5+}), has been observed in several algal species. Arsenite (As^{3+}) and arsenate (As^{5+}) are both toxic in nature, whereas arsenate is less mobile and more easily removed via precipitation or adsorption processes (Zhu et al., 2018)

In algae, the methylation and demethylation of mercury are an important biotransformation pathway. Methylmercury ($MeHg^{+}$) is a highly toxic and bioaccumulative form of mercury that can be produced or detoxified by algal activity. Some algae, mainly under anaerobic conditions, can methylate inorganic mercury (Hg^{2+}) using methyltransferase enzymes, producing $MeHg^{+}$ that may enter aquatic food webs (Li et al., 2022). Many algal species facilitate the demethylation of $MeHg^{+}$ to less toxic inorganic mercury under aerobic conditions, reducing ecological risks. The enzymatic mechanisms involved in this demethylation process include oxidative cleavage and the action of reactive oxygen species (ROS) generated during photosynthesis or stress responses. Biotransformation is also associated with metal complexation by intracellular biomolecules, such as metallothioneins, phytochelatins, and glutathione. These peptides and proteins are synthesized in response to metal stress and participate in binding and sequestration of metal ions, converting them into non-toxic complexes. For example, cadmium and zinc ions are often detoxified through their conjugation with phytochelatins, forming stable, non-reactive complexes that are subsequently compartmentalized in vacuoles or chloroplasts (Faizan et al., 2024).

Environmental factors such as light intensity, nutrient availability, pH, and other contaminants can influence the biotransformation capabilities of algae. In conclusion, biotransformation serves as a vital biological process that significantly strengthens the algae to reduce heavy metal toxicity.

3.1.4 Phycoremediation of Heavy Metals through Bioaccumulation

After first self-defense mechanism against heavy metals which prevents the entry of metal ions inside the cell, heavy metals manage to enter the cells if present in higher concentration in the surrounding. To fight back the metal ions' toxicity inside the algal cells, they show another mechanism of self-protection, termed as bioaccumulation or compartmentalization, which helps in the remediation of heavy metals. In bioaccumulation, metal ions are accumulated inside the microalgal cells. When these ions are accumulated inside the algal cell organelles (vacuoles or thylakoids), then this process is known as compartmentalization (Zohoorian et al., 2020). Bioaccumulation is a metabolically driven process by which living algal cells actively uptake and internalize heavy metals from their surrounding environment (Dey et al., 2024; Nag et al., 2023b). Unlike biosorption, which is a passive and surface-bound process, bioaccumulation involves the transport of metal ions across the cell membrane into the cytoplasm, often followed by sequestration in specific organelles or binding to intracellular molecules.

The bioaccumulation process is a two-step mechanism: initial rapid surface adsorption of metal ions onto the algal cell wall and then the slower, and energy-dependent internalization via active transport or facilitated diffusion (Gadd, 2009). Various membrane-bound transporters and channels help in mediating the uptake of specific metal ions, often utilizing existing ion transport systems that algae use for essential metals like calcium, magnesium, and iron. Unfortunately, due to structural similarities, non-essential and toxic metals such as cadmium (Cd^{2+}), lead (Pb^{2+}), and mercury (Hg^{2+}) can take advantage of these pathways and gain entry into the cell.

Metals are detoxified after internalization by becoming inert granules, complexing with chelating molecules such as phytochelatins and metallothioneins, or sequestering inside vacuoles. For example, copper (Cu^{2+}) and cadmium can be successfully bioaccumulated by Chlamydomonas reinhardtii and Scenedesmus obliquus. They achieve this by binding the metals intracellularly through thiol-rich peptides and then compartmentalizing them to lessen their harmful effects (Monteiro et al., 2012).

Numerous parameters, such as species-specific metabolic rates, growth phase, parameters like pH, temperature, and light intensity, and the presence of competing ions, affect the bioaccumulation potential of algae. Younger, actively growing algal cells typically exhibit better bioaccumulation efficiency because of their higher levels of transporter expression and metabolic activity. Again, stress-responsive genes associated with metal transport, chelation, and sequestration may be activated under specific stressors, which would accelerate the bioaccumulation process (Ali et al., 2013; Nowicka, 2022).

Strong bioaccumulative qualities have also been demonstrated by macroalgae like *Ulva lactuca* and *Gracilaria verrucosa,* especially in marine and estuarine environments where they can collect metals like copper, nickel, chromium, and zinc during extended exposure times (Geddie & Hall, 2019). Although bioaccumulation is an effective method of extracting metals from polluted streams, it also raises questions about bio magnification in the higher trophic levels that eat such metal-laden biomass.

Therefore, to prevent re-entry of metals into the food chain, proper post-treatment disposal of the metal-loaded algal biomass is essential.

In addition to this, bioaccumulation can be combined with other bioremediation processes such as biosorption and biotransformation to enhance overall metal removal efficiency. Genetic engineering of algae to overexpress metal transporter genes or stress-inducible chelators offers a better approach to further enhance this process for industrial-scale applications (Rai et al., 2019). Hence, bioaccumulation is an essential and physiologically active process in algae bioremediation that makes it possible for heavy metals to be permanently sequestered within algal cells. Its integration with other physiological and molecular techniques not only enhances metal removal efficiency but also opens new avenues for developing sustainable, algae-based remediation technologies. The global studies on the remediation of metals by using algae are presented in Table 1.

4 Phycoremediation and Microbial Fuel Cell

Microbial fuel cells (MFCs) represent an innovative and integrative bio-electrochemical system that synergistically combines bioelectricity generation with pollutant remediation (Fig. 2). Algae-based MFCs function dually as biosorbents for contaminant removal and as bioenergy resources. In particular, photosynthetic microbial fuel cells (PMFCs) exhibit substantial promise for wastewater treatment by facilitating concurrent electricity generation and metal ion sequestration via algal metabolic processes. Moreover, metal-enriched algal biochar as a cathodic electrocatalyst in PMFCs has been shown to improve oxygen reduction reaction (ORR) kinetics. This enhancement contributes to increased power densities and diminishes the dependence on expensive, noble metal-based catalysts (Sharma & Chhabra, 2021; Das et al., 2022). PMFC configurations have demonstrated high metal removal efficiencies, achieving elimination rates of up to 94% for Cu^{2+} and 88% for Co^{2+}. At an influent concentration of 50 mg L^{-1} Cu^{2+}, algal cultures exhibited a 1.2-fold increase in lipid content despite a slight decline in biomass productivity.Under heavy-metal stress, specifically in the presence of Cu^{2+} and Co^{2+}, microalgae not only accumulate these ions through biosorption and bioaccumulation but also exhibit enhanced lipid biosynthesis, thereby providing an additional route for biofuel production. The deployment of Cu–Co-enriched algal biochar as a cathode catalyst resulted in a maximum power density of 64.2 mW m^{-2} and a chemical oxygen demand (COD) removal efficiency of 84%, underscoring the viability of this approach for integrated energy generation and metal detoxification (Das et al., 2024b). Table 1 represents the application of algal strains' remediation of metals.

Table 1 Application of algal strains in the remediation of metals

Algae	Bioremediation	Reference
Scenedesmus acutus and *Chlorella pyrenoidosa*	The algal trains were used in the remediation of lead and cadmium. In this study, the algal strains were taken along with a consortium of bacterial strains of *Bacillus* and *Micrococcus*. The proto-cooperation between the consortia of bacteria and algae showed a higher rate of remediation lead and cadmium	Chandrashekharaiah et al., 2022
Chlorella pyrenoidosa	The algal strain *Chlorella pyrenoidosa,* in association with *Bacillus subtilis* bacterium and activated carbon, successfully remediated Cadmium from polluted water and the cadmium stress induced lipid accumulation in the algae. The accumulated lipid is a rich substrate for biofuel production	Huang et al., 2024
Freshwater microalgal mixed consortium	The consortium was used for the treatment of metal-contaminated wastewater. The microalgal-catalyzed wastewater treatment was performed in a photosynthetic microbial fuel cell for the simultaneous removal of heavy metals (copper and cobalt) and energy production	Das et al., 2024a, b
Chara vulgaris	Biomass of the green alga *Chara vulgaris* was used for the remediation of eutrophicated wastewater contaminated with copper metal. The algal biomass showed a higher absorption potential compared to activated carbon and proved to be a potential metal remediation agent	Mir and Rather, 2024
Cladophora sp.	A phycoremediation study was demonstrated for the remediation of chromium by using the algal strain *Cladophora* sp. The study revealed the chromium bioremediation potential of the algae by bioaccumulation of chromium in algal biomass	Ocinski et al., 2023
Chlorella vulgaris	Algal strain *Chlorella vulgaris* in association with a composite bacterial agent comprising bacterial strains of *Pseudomonas*, *Rheinheimera*, *Brevundimonas*, *Acinetobacter*, and *Delftia*, effectively used for the remediation of arsenic and mercury from a wastewater sample. The symbiotic association of algae and bacteria composites effectively absorbed the toxic metals from the wastewater	Zhao et al., 2024

(continued)

Table 1 (continued)

Algae	Bioremediation	Reference
Scenedesmus Sp	Arora et al. (2023) demonstrated that the use of microalgae species of *Scenedesmus* efficiently removes 75–89% of arsenite [As(III)] and arsenate [As(V)] from artificial seawater (ASW)	Arora et al., 2023
Neowestiellopsis persica	A blue-green algae *Neowestiellopsis persica* A1387 was effectively used for the absorption of nickel from a water sample. The presence of negatively charged exopolysaccharides (EPS) on the algal cell favored binding and sequestering positively charged metal ions during the bioremediation process	Nowruzi et al., 2024

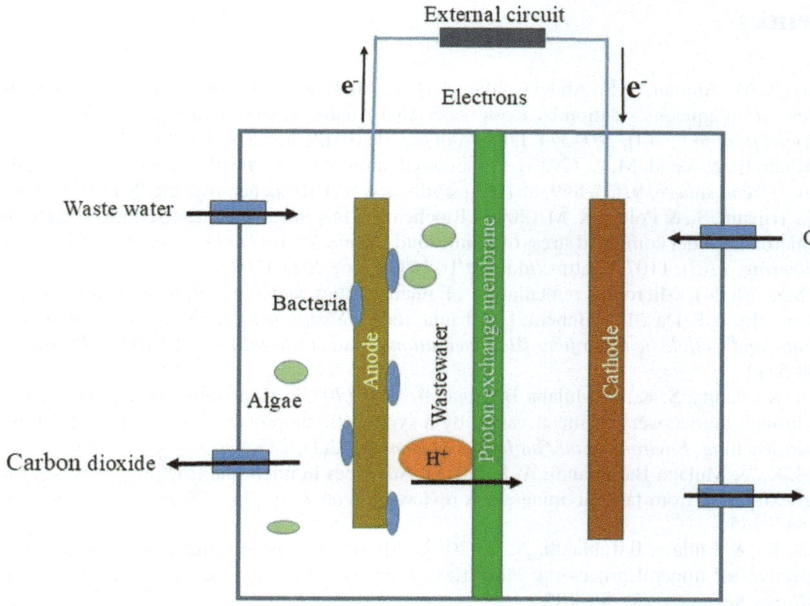

Fig. 2 Typical microbial fuel cell showing application of algae in waste remediation and energy production

5 Conclusions

Algae offer an eco-friendly and sustainable alternative to conventional processes for the remediation of heavy metals in wastewater. Their diverse metabolic activities performed by algae effectively remediate a wide range of toxic metals, including lead, cadmium, chromium, arsenic, uranium, and more. Integrating phycoremediation with emerging technologies such as photosynthetic microbial fuel cells and biomass valorization strategies enhances metal detoxification and opens pathways for energy production and metal value recovery. However, challenges remain in such processes for their large-scale application. The future of the phycoremediation technology depends on the research focus on the application of genetic engineering for improvement in algal strains, evaluation of process parameters for the development of optimized processes under real-world conditions, and development of circular models that integrate remediation with resource recovery. Overall, algal technologies represent a vital step toward green and circular solutions for the remediation of metal contamination in aquatic environments.

Acknowledgements The first author acknowledges the University Grants Commission, Government of India, for providing a Junior Research Fellowship to pursue PhD.

References

Abdel-Aty, A. M., Ammar, N. S., Abdel Ghafar, H. H., & Ali, R. K. (2013). Biosorption of cadmium and lead from aqueous solution by fresh water alga Anabaena sphaerica biomass. *Journal of Advanced Research, 4*(4), 367–374. https://doi.org/10.1016/j.jare.2012.07.004

Ali, H., Khan, E., & Sajad, M. A. (2013). Phytoremediation of heavy metals—concepts and applications. *Chemosphere, 91*(7), 869–881. https://doi.org/10.1016/j.chemosphere.2013.01.075

Arora, N., Tripathi, S., & Poluri, K. M. (2023). Biochemical insights into enhanced lipid production by a microalga under combined stress of salinity and arsenic. *Journal of Environmental Chemical Engineering, 11*(5), 110778. https://doi.org/10.1016/j.jece.2023.110778

Behera, S.K. (2024). Microbial remediation of microplastics and integration into the circular economy. In: A.P. Das, I.D. Behera, D. Bhanja, (eds.), *Sustainable Microbial Technology for Synthetic and Cellulosic Microfiber Bioremediation.* Cham. https://doi.org/10.1007/978-3-031-62600-5_11

Behera, S. K., Panda, S. K., & Mulaba-Bafubiandi, A.-F. (2022). Valorization of copper smelter slag through the recovery of metal values by a synergistic bioprocess system of bio-flotation and bio-leaching. *Environmental Quality Management, 32*(1), 233.

Behera, S. K., & Mulaba-Bafubiandi, A. F. (2015). Advances in microbial leaching processes for nickel extraction from lateritic minerals-A review. *Korean Journal of Chemical Engineering, 32*, 1447–1454.

Behera, S. K., & Mulaba-Bafubiandi, A. F. (2017). Microbes assisted mineral flotation a future prospective for mineral processing industries: A review. *Mineral Processing and Extractive Metallurgy Review, 38*(2), 96–105.

Brinza, L., Geraki, K., Cojocaru, C., Holdt, S. L., & Neamtu, M. (2020). Baltic Fucus vesiculosus as potential bio-sorbent for Zn removal: Mechanism insight. *Chemosphere, 238*, Article 124652. https://doi.org/10.1016/j.chemosphere.2019.124652

Chandrashekharaiah, P. S., Gupte, Y., Sarkar, P., Prasad, S., Sanyal, D., Dasgupta, S., & Banik. (2022). Algae-bacterial aquaculture can enhance heavy metals (Pb^{2+} and Cd^{2+}) remediation and water re-use efficiency of synthetic streams. *Resources, Conservation and Recycling, 180*, 106211. https://doi.org/10.1016/j.resconrec.2022.106211

Chandrashekharaiah, P. S., Sanyal, D., Dasgupta, S., & Banik, A. (2021). Cadmium biosorption and biomass production by two freshwater microalgae Scenedesmus acutus and Chlorella pyrenoidosa: An integrated approach. *Chemosphere, 269*, 128755.

Danouche, M., El Ghachtouli, N., & El Arroussi, H. (2021). Phycoremediation mechanisms of heavy metals using living green microalgae: Physicochemical and molecular approaches for enhancing selectivity and removal capacity. *Heliyon, 7*(7), Article e07609. https://doi.org/10.1016/j.heliyon.2021.e07609

Darda, S., Papalas, T., & Zabaniotou, A. (2019). Biofuels journey in Europe: Currently the way to low carbon economy sustainability is still a challenge. *Journal of Cleaner Production, 208*, 575–588. https://doi.org/10.1016/j.jclepro.2018.10.147

Das, A. P., & Bissoyi, A. (2011). Computational approach for comparative phylogenetic analysis of isolated chromium resistant strain Brevibacterium casei. *Journal of Engineering and Technology Research, 3*(3), 82–87. https://doi.org/10.5897/JETR.9000020

Das, A. P., & Ghosh, S. (2018). Bioleaching of Manganese from mining waste materials. *Materials Today: Proceedings, 5*(1), 2381–2390. https://doi.org/10.1016/J.MATPR.2017.11.459

Das, A., & Ghosh, S. (2022). Role of microorganisms in extenuation of mining and industrial wastes. *Geomicrobiology Journal, 39*, 173–175. https://doi.org/10.1080/01490451.2022.2038953

Das, A. P., & Mishra, S. (2008). Hexavalent Chromium (VI): Health hazards and environmental pollutant. *J Environ Res Dev, 2*(3), 386–392.

Das, A. P., Sukla, L. B., Pradhan, N., & Nayak, S. (2011). Manganese biomining: A review. *Bioresource Technology, 102*(16), 7381–7387. https://doi.org/10.1016/J.BIORTECH.2011.05.018

Das, A. P., Swain, S., Panda, S., Pradhan, N., & Sukla, L. B. (2012). Reductive acid leaching of low grade manganese ores. *Geomaterials, 2012*(04), 70–72. https://doi.org/10.4236/GM.2012.24011

Das, A. P., Ghosh, S., Mohanty, S., & Sukla, L. B. (2015c). Consequences of manganese compounds: A review. *Toxicological and Environmental Chemistry, 96*(7), 981–997. https://doi.org/10.1080/02772248.2015.1005428

Das, S., Das, S., & Ghangrekar, M. M. (2022). Efficacious bioremediation of heavy metals and radionuclides from wastewater employing aquatic macro- and microphytes. *Journal of Basic Microbiology, 62*(3–4), 260–278. https://doi.org/10.1002/jobm.202100372

Das, A. P., Mallavarapu, M., & Ghosh, S. (2023). Ecotoxicity of mining pollutants on the environment and their remediation. *Environmental Chemistry and Ecotoxicology, 5*, 165–167. https://doi.org/10.1016/J.ENCECO.2023.08.002

Das, S., Kumar, S., Kumar Mehta, A., & Ghangrekar, M. M. (2024b). Heavy metals removal by algae and usage of activated metal-enriched biomass as cathode catalyst for improving performance of photosynthetic microbial fuel cell. *Bioresource Technology, 406*, Article 131038. https://doi.org/10.1016/j.biortech.2024.131038

Das, A. P., & Ghosh, S. (2023a). Mechanism of manganese bioleaching. *Manganese Mining Microorganisms*, 87–98. https://doi.org/10.1016/B978-0-12-822147-1.00014-5

Das, A. P., & Ghosh, S. (2023b). Microorganisms in manganese biomining processes. *Manganese Mining Microorganisms*, 55–61. https://doi.org/10.1016/B978-0-12-822147-1.00013-3

Das, A. P., & Ghosh, S. (2023c). Role of metabolites and metabolomics for understanding of manganese bioleaching. *Manganese Mining Microorganisms*, 117–127. https://doi.org/10.1016/B978-0-12-822147-1.00011-X

Das, A., & Mishra, S. (2010). Biodegradation of the metallic carcinogen hexavalent chromium Cr(VI) by an indigenously isolated bacterial strain. *Journal of Carcinogenesis, 9*. https://doi.org/10.4103/1477-3163.63584

Das, A. P., Bal, B., & Mahapatra, P. S. (2015a). Chromogenic biosensors for pathogen detection. In P. Prokopovich (Ed.), *Biological and Pharmaceutical Applications of Nanomaterials* (1st ed.). CRC Press, Taylor & Francis.

Das, A. P., Ghosh, S., Mohanty, S., & Sukla, L. B. (2015b). Advances in Manganese Pollution and Its Bioremediation. In L. B. Sukla, N. Pradhan, S. Panda, & B. K. Mishra (Eds.), *Environmental microbe biotechnology* (1st ed., pp. 313–328). Cham. https://doi.org/10.1007/978-3-319-19018-1_16

Das, A., Dey, S., Bala, K., Gnanasekaran, R., & Das, A. P. (2024). Microbial bioremediation technology for sustainable treatment and management of synthetic microfiber waste. *Sustainable Microbial Technology for Synthetic and Cellulosic Microfiber Bioremediation*, 221–238. https://doi.org/10.1007/978-3-031-62600-5_12

Deng, J., Guo, H., Cui, H., Fang, J., Zuo, Z., Deng, J., Wang, X., & Zhao, L. (2016). Oxidative stress and inflammatory responses involved in dietary nickel chloride ($NiCl_2$)-induced pulmonary toxicity in broiler chickens. *Toxicology Research, 5*(5), 1421–1433. https://doi.org/10.1039/C6TX00197A

Devi, A., Verma, M., Saratale, G. D., Saratale, R. G., Ferreira, L. F. R., Mulla, S. I., & Bharagava, R. N. (2023). Microalgae: A green eco-friendly agents for bioremediation of tannery wastewater with simultaneous production of value-added products. *Chemosphere, 336*, Article 139192. https://doi.org/10.1016/j.chemosphere.2023.139192

Dey, S., Dash, A., & Das, A. P. (2024). Bioremediation as a sustainable and circular solution for mining waste management and bioenergy generation. *Sustainable Management of Mining Waste and Tailings: A Circular Economy Approach*, 107–119. https://doi.org/10.1201/9781003442455-6/BIOREMEDIATION-SUSTAINABLE-CIRCULAR-SOLUTION-MINING-WASTE-MANAGEMENT-BIOENERGY-GENERATION-SUDESHNA-DEY-AKANSHYA-DASH-ALOK-PRASAD-DAS

Faizan, M., Alam, P., Hussain, A., Karabulut, F., Tonny, S. H., Cheng, S. H., Yusuf, M., Adil, M. F., Sehar, S., Alomrani, S. O., Albalawi, T., & Hayat, S. (2024). Phytochelatins: Key regulator

against heavy metal toxicity in plants. *Plant Stress, 11*, Article 100355. https://doi.org/10.1016/j.stress.2024.100355
Gadd, G. M. (2009). Biosorption: Critical review of scientific rationale, environmental importance and significance for pollution treatment. *Journal of Chemical Technology and Biotechnology, 84*(1), 13–28. https://doi.org/10.1002/jctb.1999
Gandhi, T. P., Sampath, P. V., & Maliyekkal, S. M. (2022). A critical review of uranium contamination in groundwater: Treatment and sludge disposal. *Science of the Total Environment, 825*, Article 153947. https://doi.org/10.1016/j.scitotenv.2022.153947
Geddie, A. W., & Hall, S. G. (2019). An introduction to copper and zinc pollution in macroalgae: For use in remediation and nutritional applications. *Journal of Applied Phycology, 31*(1), 691–708. https://doi.org/10.1007/s10811-018-1580-5
Ghosh, S., & Das, A. P. (2017). Bioleaching of manganese from mining waste residues using Acinetobacter sp. *Geology, Ecology, and Landscapes, 1*(2), 77–83. https://doi.org/10.1080/24749508.2017.1332847
Ghosh, S., Mohanty, S., Nayak, S., Sukla, L. B., & Das, A. P. (2016). Molecular identification of indigenous manganese solubilising bacterial biodiversity from manganese mining deposits. *Journal of Basic Microbiology, 56*(3), 254–262. https://doi.org/10.1002/JOBM.201500477
Ghosh, S., & Das, A. P. (2018). Metagenomic insights into the microbial diversity in manganese-contaminated mine tailings and their role in biogeochemical cycling of manganese. *Scientific Reports 2018 8*(1), 1–12. https://doi.org/10.1038/s41598-018-26311-w
Ghosh, S., Tripathy, B., Dey, S., & Das, A. P. (2023). Comparative investigation of fungal and bacterial manganese biomining mechanisms. *Manganese Mining Microorganisms*, 99–115. https://doi.org/10.1016/b978-0-12-822147-1.00016-9
Hu, B., Shao, S., Ni, H., Fu, Z., Hu, L., Zhou, Y., Min, X., She, S., Chen, S., Huang, M., Zhou, L., Li, Y., & Shi, Z. (2020). Current status, spatial features, health risks, and potential driving factors of soil heavy metal pollution in China at province level. *Environmental Pollution, 266*, Article 114961. https://doi.org/10.1016/j.envpol.2020.114961
Huang, R., Huo, G., Song, S., Li, Y., Xia, L., & Gaillard, J.-F. (2019). Immobilization of mercury using high-phosphate culture-modified microalgae. *Environmental Pollution, 254*, Article 112966. https://doi.org/10.1016/j.envpol.2019.112966
Huang, J., Su, B., Fei, X., Che, J., Yao, T., Zhang, R., & Yi, S. (2024). Enhanced microalgal biomass and lipid production with simultaneous effective removal of Cd using algae bacteria-activated carbon consortium added with organic carbon source. *Chemosphere, 350*, 141088. https://doi.org/10.1016/j.chemosphere.2023.141088
Jais, N. M., Mohamed, R. M. S. R., Al-Gheethi, A. A., & Hashim, M. K. A. (2017). The dual roles of phycoremediation of wet market wastewater for nutrients and heavy metals removal and microalgae biomass production. *Clean Technologies and Environmental Policy, 19*(1), 37–52. https://doi.org/10.1007/s10098-016-1235-7
Jana, A., Bhattacharya, P., Swarnakar, S., Majumdar, S., & Ghosh, S. (2015). Anabaena sp. mediated bio-oxidation of Arsenite to arsenate in synthetic arsenic (III) solution: Process optimization by response surface methodology. *Chemosphere, 138*, 682–690. https://doi.org/10.1016/j.chemosphere.2015.07.055
Kanungo, J., Sahoo, T., Bal, M., & Behera, I. D. (2024). Performance of bioremediation strategy in waste lubricating oil pollutants: A review. *Geomicrobiology Journal, 41*(4), 360–373. https://doi.org/10.1080/01490451.2023.2245395
Khajavian, M., Wood, D. A., Hallajsani, A., & Majidian, N. (2019). Simultaneous biosorption of nickel and cadmium by the brown algae Cystoseria indica characterized by isotherm and kinetic models. *Applied Biological Chemistry, 62*(1), 69. https://doi.org/10.1186/s13765-019-0477-6
Leong, Y. K., & Chang, J.-S. (2020). Bioremediation of heavy metals using microalgae: Recent advances and mechanisms. *Bioresource Technology, 303*, Article 122886. https://doi.org/10.1016/j.biortech.2020.122886

Li, Y., Li, D., Song, B., & Li, Y. (2022). The potential of mercury methylation and demethylation by 15 species of marine microalgae. *Water Research, 215*, Article 118266. https://doi.org/10.1016/j.watres.2022.118266

Liu, X.-Y., Hong, Y., Liang, M., & Zhai, Q.-Y. (2023). Bioremediation of zinc and manganese in swine wastewater by living microalgae: Performance, mechanism, and algal biomass utilization. *Bioresource Technology, 385*, 129382. https://doi.org/10.1016/j.biortech.2023.129382

Mir, D. H., & Rather, M. A. (2024). Kinetic and thermodynamic investigations of copper (II) biosorption by green algae Chara vulgaris obtained from the waters of Dal Lake in Srinagar (India). *Journal of Water Process Engineering, 58*, Article 104850.

Mishra, S., Swain, S., Sahoo, M., Mishra, S., & Das, A. P. (2021). Microbial colonization and degradation of microplastics in aquatic ecosystem: A review. *Geomicrobiology Journal*. https://doi.org/10.1080/01490451.2021.1983670

Mishra, S., Dash, D., Al-Tawaha, A. R. M. S., & Das, A. P. (2022). A review on heavy metal ion adsorption on synthetic microfiber surface in aquatic environments. *Applied Biochemistry and Biotechnology, 194*(10), 4639–4654. https://doi.org/10.1007/S12010-022-04029-W/METRICS

Mishra, S., Dash, D., & Das, A. P. (2024). Isolation and molecular characterization of polyester-degrading bacterial strains from municipal wastewater and their application in microplastic degradation in contaminated wastewater. *Geomicrobiology Journal*. https://doi.org/10.1080/01490451.2024.2376101

Mishra, S., & Das, A. P. (2023). Microbial remediation of synthetic microfiber contaminated wastewater. In M. P. Shah (Ed.), *Microbial Technologies in Industrial Wastewater Treatment* (pp. 337–350). Springer, Singapore. https://doi.org/10.1007/978-981-99-2435-6_17

Mishra, S., Das, A. P., & Seragadam, P. (2009). Microbial remediation of hexavalent chromium from chromite contaminated mines of Sukinda Valley, Orissa (India). *J Environ Res Dev*, 1122–1127.

Monteiro, C. M., Castro, P. M. L., & Malcata, F. X. (2012). Metal uptake by microalgae: Underlying mechanisms and practical applications. *Biotechnology Progress, 28*(2), 299–311. https://doi.org/10.1002/btpr.1504

Moreira, V. R., Lebron, Y. A. R., LC, L., & Santos, L. V. S. (2019). Simultaneous biosorption of Cd(II), Ni(II) and Pb(II) onto a brown macroalgae Fucus vesiculosus: Mono- and multi-component isotherms, kinetics and thermodynamics. *Journal of Environmental Management, 251*, 109587. https://doi.org/10.1016/j.jenvman.2019.109587

Nag, M., Lahiri, D., Ghosh, S., Sarkar, T., Pati, S., Das, A. P., Ram, D. K., Bhattacharya, D., & Ray, R. R. (2023a). Application of microorganisms in biotransformation and bioremediation of environmental contaminant: A review. *Geomicrobiology Journal*. https://doi.org/10.1080/01490451.2023.2261443

Nag, M., Lahiri, D., Ghosh, S., Sarkar, T., Pati, S., Das, A. P., Ram, D. K., Bhattacharya, D., & Ray, R. R. (2023b). Application of microorganisms in biotransformation and bioremediation of environmental contaminant: A review. *Geomicrobiology Journal*, 1–18. https://doi.org/10.1080/01490451.2023.226144

Nowicka, B. (2022). Heavy metal–induced stress in eukaryotic algae—mechanisms of heavy metal toxicity and tolerance with particular emphasis on oxidative stress in exposed cells and the role of antioxidant response. *Environmental Science and Pollution Research, 29*(12), 16860–16911. https://doi.org/10.1007/s11356-021-18419-w

Nowruzi, B., Ghazi, S., Norouzi, R., & Norouzi, R. (2024). The impact of plasma-activated water on the process of nickel bioremediation by Neowestiellopsis persica A1387. *Ecotoxicology and Environmental Safety, 285*, 117101.

O'Neal, S. L., & Zheng, W. (2015). Manganese toxicity upon overexposure: A decade in review. *Current Environmental Health Reports, 2*(3), 315–328. https://doi.org/10.1007/s40572-015-0056-x

Ocinski, D., Jacukowicz-Sobala, I., Augustynowicz, J., Wołowski, K., Cantero, D. A., García-Serna, J., Pińkowska, H., & Przejczowski, R. (2023). Algae from Cr-containing infiltrate bioremediation for valorised chemical production—seasonal availability, composition, and screening studies on hydrothermal conversion. *Bioresource Technology, 389*, Article 129798. https://doi.org/10.1016/j.biortech.2023.129798

Oladimeji, T. E., Oyedemi, M., Emetere, M. E., Agboola, O., Adeoye, J. B., & Odunlami, O. A. (2024). Review on the impact of heavy metals from industrial wastewater effluent and removaltechnologies. *Heliyon, 10*(23), Article e40370. https://doi.org/10.1016/j.heliyon.2024.e40370

Paul, A., Dey, S., Ram, D. K., & Das, A. P. (2023). Hexavalent chromium pollution and its sustainable management through bioremediation. *Geomicrobiology Journal*, 1–11. https://doi.org/10.1080/01490451.2023.2218377

Plöhn, M., Escudero-Oñate, C., & Funk, C. (2021). Biosorption of Cd(II) by nordic microalgae: Tolerance, kinetics and equilibrium studies. *Algal Research, 59*, Article 102471. https://doi.org/10.1016/j.algal.2021.102471

Prabhakar, A., Mishra, S., & Das, A. P. (2019). Isolation and identification of lead (Pb) solubilizing bacteria from automobile waste and its potential for recovery of lead from end of life waste batteries. *Geomicrobiological Journal, 36*(10), 894–903. https://doi.org/10.1080/01490451.2019.1654044

Pradhan, G., Tripathy, B., Ram, D. K., Digal, A. K., & Das, A. P. (2023). Bauxite mining waste pollution and its sustainable management through bioremediation. *Geomicrobiology Journal*, 1–10. https://doi.org/10.1080/01490451.2023.2235353

Priya, A. K., Jalil, A. A., Vadivel, S., Dutta, K., Rajendran, S., Fujii, M., & Soto-Moscoso, M. (2022). Heavy metal remediation from wastewater using microalgae: Recent advances and future trends. *Chemosphere, 305*, 135375.

Priyadarshini, E., Priyadarshini, S. S., & Pradhan, N. (2019). Heavy metal resistance in algae and its application for metal nanoparticle synthesis. *Applied Microbiology and Biotechnology, 103*(8), 3297–3316. https://doi.org/10.1007/s00253-019-09685-3

Priyadarsini, S., & Das, A. P. (2024). Analytical and structural characterization of waste lithium-ion batteries for their effective recycling strategy. *Environmental Chemistry and Ecotoxicology*. https://doi.org/10.1016/j.enceco.2024.12.004

Priyadarsini, S., & Das, A. P. (2025). Lithium bioleaching: A review on microbial-assisted sustainable technology for lithium bio-circularity. *Journal of Water Process Engineering, 69*, Article 106744. https://doi.org/10.1016/j.jwpe.2024.106744

Rai, P. K., Lee, S. S., Zhang, M., Tsang, Y. F., & Kim, K.-H. (2019). Heavy metals in food crops: Health risks, fate, mechanisms, and management. *Environment International, 125*, 365–385. https://doi.org/10.1016/j.envint.2019.01.067

Raj, K., & Das, A. P. (2023). Lead pollution: Impact on environment and human health and approach for a sustainable solution. *Environmental Chemistry and Ecotoxicology, 5*, 79–85. https://doi.org/10.1016/J.ENCECO.2023.02.001

Ramakrishnan, B., Megharaj, M., Venkateswarlu, K., Naidu, R., & Sethunathan, N. (2010). The impacts of environmental pollutants on microalgae and cyanobacteria. *Critical Reviews in Environmental Science and Technology, 40*(8), 699–821. https://doi.org/10.1080/10643380802471068

Rana, M. S., & Prajapati, S. K. (2021). Resolving the dilemma of iron bioavailability to microalgae for commercial sustenance. *Algal Research, 59*, Article 102458.

Rout, P. G., & Behera, S. K. (2024). Evaluation of organosulfur compound degradation property of a noble bacterial strain of Proteus mirabilis. *Environmental Chemistry and Ecotoxicology, 6*, 223–228. https://doi.org/10.1016/j.enceco.2024.06.004

Rout, P. G., Mohanty, A. K., Pradhan, N., Biswal, S. K., & Behera, S. K. (2022). Study on the reaction mechanism of oxidative microbial desulfurization of organic sulfur-rich coal. *Geomicrobiology Journal, 39*(3–5), 210–218.

Rout, P. G., Singh, A., & Behera, S. K. (2024). Insights into the kinetics of the microbial degradation of dibenzothiophene (DBT) by Acidithiobacillus ferrooxidans. *Geomicrobiology Journal, 41*(4), 451–458.

Sahu, S., & Das, A. P. (2024). Microplastics degradation mechanism by aquatic Plastiomes: A review. *Environmental Science and Engineering, Part F2827*, 213–226. https://doi.org/10.1007/978-3-031-50840-0_12

Salam, L. B., Shomope, H., Ummi, Z., & Bukar, F. (2019). Mercury contamination imposes structural shift on the microbial community of an agricultural soil. *Bulletin of the National Research Centre, 43*(1), 163. https://doi.org/10.1186/s42269-019-0208-5

Salama, E.-S., Roh, H.-S., Dev, S., Khan, M. A., Abou-Shanab, R. A. I., Chang, S. W., & Jeon, B.-H. (2019). Algae as a green technology for heavy metals removal from various wastewater. *World Journal of Microbiology and Biotechnology, 35*(5), 75. https://doi.org/10.1007/s11274-019-2648-3

Sánchez-González, C., López-Chaves, C., Gómez-Aracena, J., Galindo, P., Aranda, P., & Llopis, J. (2015). Association of plasma manganese levels with chronic renal failure. *Journal of Trace Elements in Medicine and Biology, 31*, 78–84. https://doi.org/10.1016/j.jtemb.2015.04.001

Shamim, S. (2018). Biosorption of Heavy Metals. In J. Derco & B. Vrana (Eds.), *Biosorption*. InTech. https://doi.org/10.5772/intechopen.72099

Sharma, A., & Chhabra, M. (2021). Performance evaluation of a photosynthetic microbial fuel cell (PMFC) using Chlamydomonas reinhardtii at cathode. *Bioresource Technology, 338*, 25499. https://doi.org/10.1016/j.biortech.2021.125499

Upadhyay, A. K., Mandotra, S. K., Kumar, N., Singh, N. K., Singh, L., & Rai, U. N. (2016). Augmentation of arsenic enhances lipid yield and defense responses in alga Nannochloropsis sp. *Bioresource Technology, 221*, 430–437. https://doi.org/10.1016/j.biortech.2016.09.061

Upadhyay, A. K., Singh, L., Singh, R., Singh, D. P., & Saxena, G. (2022). Bioaccumulation and toxicity of As in the Alga Chlorococcum sp.: Prospects for as bioremediation. *Bulletin of Environmental Contamination and Toxicology, 108*(3), 500–506. https://doi.org/10.1007/s00128-020-02964-0

Wang, Z., Luo, P., Zha, X., Xu, C., Kang, S., Zhou, M., Nover, D., & Wang, Y. (2022). Overview assessment of risk evaluation and treatment technologies for heavy metal pollution of water and soil. *Journal of Cleaner Production, 379*, Article 134043. https://doi.org/10.1016/j.jclepro.2022.134043

Yan, C., Qu, Z., Wang, J., Cao, L., & Han, Q. (2022). Microalgal bioremediation of heavy metal pollution in water: Recent advances, challenges, and prospects. *Chemosphere, 286*, Article 131870. https://doi.org/10.1016/j.chemosphere.2021.131870

Yin, X., Wang, L., Duan, G., & Sun, G. (2011). Characterization of arsenate transformation and identification of arsenate reductase in a green alga Chlamydomonas reinhardtii. *Journal of Environmental Sciences, 23*(7), 1186–1193. https://doi.org/10.1016/S1001-0742(10)60492-5

Zhao, M., Zhang, X., Zhang, M., Guo, J., Zhang, J., Zheng, G., Feng, B., & Chen, Y. (2024). Bacterial and microalgal co-fixation for remediation of industrial wastewater contaminated with arsenic, mercury, and other pollutants. *Journal of Environmental Chemical Engineering, 12*(4), Article 113099. https://doi.org/10.1016/j.jece.2024.113099

Zhu, N., Zhang, J., Tang, J., Zhu, Y., & Wu, Y. (2018). Arsenic removal by periphytic biofilm and its application combined with biochar. *Bioresource Technology, 248*, 49–55. https://doi.org/10.1016/j.biortech.2017.07.026

Zohoorian, H., Ahmadzadeh, H., Molazadeh, M., Shourian, M., & Lyon, S. (2020). Microalgal bioremediation of heavy metals and dyes. In *Handbook of Algal Science, Technology and Medicine* (pp. 659–674). Elsevier. https://doi.org/10.1016/B978-0-12-818305-2.00041-3

The manufacturer's authorised representative in the EU is Springer Nature Customer Service Centre GmbH, Europaplatz 3, 69115 Heidelberg, Germany. If you have any concerns regarding our products, please contact ProductSafety@springernature.com

Printed and bound by CPI Group (UK) Ltd, Croydon, CR0 4YY
05/10/2026
02228057-0013